Words from the Lakeside

by

Howard Johnson

includes

Letters, Essays, Commentary, Short Stories, Memoirs, Quotes, and Poetry

Some of the Quotes, and Poetry are from other authors

**This book is available at special discounts for bulk purchases for sales
promotions, premiums, fund raising, or educational use. For details, contact:**

Special Sales - Senesis Word

PHONE: 904-687-1865   **FAX:** 904-825-0222   **CELL:** 574-265-3386

Website: www.Senesisword.com

Email: Senesisword@yahoo.com

WLSNU_txt13401AW

The cover:

A frozen Lake Tippecanoe in Northern Indiana greets a bright, very cold January morning in 2003. This photo was taken out our front window, and caught a small group of deer grazing on our lawn. The deer had come from the woods behind our house. They walked out into our yard where they grazed for nearly half an hour. When they were finished, they walked out onto the ice and crossed the lake to the other side, slipping frequently as legs spread wide and they appeared about to fall. Somehow, they managed to get across without falling, in spite of several near misses.

This is a quiet and peaceful time at the lake, a stark contrast to summer days when the lake is criss-crossed with noisy personal watercraft, skiers, and boats of many kinds.

 Here's another photo taken from the same spot on a warm, sunny fall day in October 2010. Notice the two stumps that are all that remains of the two trees at the left on the cover.

DEDICATION

This book is who I am, what I think, what I believe, what I imagine, what I dream of, why and whom I love . . . in short, it is a collection of bits and pieces of my life—of me. It is also a testimonial to all those beautiful human beings who, through love and some blunt trauma, helped me become the person I am. Therefore, I dedicate this book with great love and affection to all of my family, friends, and others, whose actions helped create, guide, inspire, stimulate, and mold my life into the person I am today. My passionate desire to please and never to displease them has guided me in positive directions throughout my lifetime. The family members, lovers, friends, mentors, and teachers who have left this earth are remembered fondly, appreciated greatly, and sorely missed. I am powerfully blessed to have known all of these incredible people, family, friends, colleagues, and acquaintances.

I have decided to list in this dedication as many of those people as I can without creating an entire new book. They are remembered in roughly chronological order. Many of these important people have roles in stories and essays in this book. I have been blessed with a close, loving relationship with members of my family, unique to each one. Those described as *special* were not loved any more than others. There was something different, maybe *magic*, about the two of us together. It defies definition. There were those, other than family, with which I felt a special connection as well. In this book, there is a description of a conversation with one of my grandparents that clearly illustrates my meaning. Those that are not listed are no less loved or appreciated. If I listed all of them, there would be no pages left for the burgeoning content already in place.

The first was of course, my mother, **Ethel Johnson**. A tiny woman, she was still a powerful and loving presence to all who knew her. Mom was a dedicated Christian with all of the best that can mean. She was a loving mother, in the best sense of that calling. She was also a shining example of a truly decent person to everyone she met for the entire 96 years she was alive on this planet. To my knowledge, every organization she joined in her life elected her president. She was loved by all family and friends who knew her.

My father, **Howard R. Johnson**, was a decent, honorable, Christian man. He was as terrific and faithful a father as a boy, then a young man could have. Our close, extended relationship continued when we were in business together for nearly twenty years. A stable, dependable man, he taught me that tears were a manly expression saying only brave and secure men cry openly. Many of the most joyous moments in my life were when I made him proud of me. I will never forget the countless happy hours we spent together, or the experiences we shared.

My grandparents, **Eva May** and **George Dickinson**, were the only grandparents I knew. My father's parents both died before I was old enough to remember. *Grandma Dick* was a strong and loving woman who taught me a great deal. *Granddad Dick* and I had a special, close relationship that was reinforced during numerous fishing trips. Granddad Dick was a master story teller and wove his magic on me frequently, when I was quite young. He taught me much about the realities of life, and how to deal with them.

Many a time when one of his stories was being woven on their porch I would hear Grandma Dick calling from elsewhere in the house, "George! You quit filling that boy's head with your nonsense."

Granddad would grow silent for a moment then resume with a much softer voice.

My sisters, one a virtual second mother, the other, my nemesis during my childhood, added their individual, loving touches to whom I am. Both, like our mother and father, were deeply Christian women, but quite different in their passions and how they practiced their religion. **Lois**, twelve years my senior was much like our father, even being born on his birthday. A strong willed yet gentle and loving force, she and I were extremely close. **Roberta** or **Bobbie**, six years older than I, had a temperament different from Lois's. We fought constantly when we were young, typical sibling battles, but often quite passionate. As adults we still battle occasionally over differences of opinions, but those differences have no effect on the strong bonds, of love and respect, we have for each other.

My two brothers-in-law, Lois's husband **Harold**, and Bobbie's **Robert** were as fine a set of brothers as a man could have. As different as my two sisters, they were a positive influence on others and especially on their *little brother* or *Bro*.

There were aunts, uncles, and cousins, who brought joy and companionship to early days. Though most are now passed away, they are all remembered fondly. Of those few remaining, several are still kept in contact if only with Christmas cards.

There are nephews and nieces and their families in almost countless numbers, the next generation now carrying the torch of family. They are also loved and treasured. There are many among the group with whom I have a special relationship. You know who you are and what I mean.

The *lovers* in my life have had a staggering emotional impact on the person I am. There are stories about most of them herein. I will not provide the details, some of which could be painful to many people involved. Needless to say, each of them was loved deeply and passionately. I still care deeply for each of them and know the love shared with each diminishes in no way the love for any of the others.

Dolores was my first love, professed at 17, who became my wife and the mother of five of my children. A dedicated and devoted mother, she sacrificed many times for our children. After many happy years with our large brood, our marriage fell apart, and we were divorced. With the well being of our children foremost in each of our minds, we kept our difficulties hidden as best we could. Neither of us ever said a bad or harmful thing about the other to any of our brood. Time softened our feelings and buried our differences. We enjoyed a friendly relationship until her death.

Caroline rescued me from the depths of depression and helped me regain my lost self esteem. Her love and compassion were the most powerful forces in turning my life from the angry, damaging path I had chosen. She presented me with a beautiful daughter in 1968. For several reasons and in spite of our great affection for each other, we parted company when Kristen was three. To my boundless joy, we have been reconciled and are now friends.

Iola came into my life a few years after I left Cleveland and moved to Chicago to try to put my life back in some kind of order. Once more I stumbled into a truly exceptional woman who helped me restore my devastated self respect. Iola has two delightful daughters who became, and remain to this day, as two of my own. After a number of years together, we drifted apart when I moved to Indiana.

Barbara, my wife and companion for the *golden* years, brought joy and her two delightful grown sons into our marriage. She also filled my life with love and spirited activity. When I began writing seriously, she became my editor and critic. She was positively brutal with a red pen. Her efforts contributed a great deal directly to this book.

After we were married, she became a Methodist pastor and led a small country congregation in a church "in the middle of three cornfields" as she always said.

A committed Christian, she took to the ministry with a vigor and determination that grew the small church considerably. With both of us far from any family, the congregation became our family, *warts and all* as she frequently remarked. I was so proud of her accomplishments in the pulpit and with the many members who loved her dearly and showed it. It was devastating to us both when she had to step down because of failing health. The outpouring of accolades and tears from the congregation on her last day in the pulpit was overwhelming. She left us at far too early an age and is now missed terribly, and will continue to be.

Daphne, who came into my life possibly through the efforts of my guardian angel (story in the book) is now my passion, my lover, and my dearest friend. It is our sincere hope we can enjoy many of the *golden years* together and then go peacefully. She brought her large family into my life as well as her circle of friends. With four daughters, two sons, and thirteen grandchildren, they are an impressive group. I feel a special bond with her children and their spouses who have each treated me with grace and warm affection. Those of her grandchildren I have had the opportunity to get to know have treated me in the same gracious manner. Each is now a vital part of my life. Her many friends have become my friends as well.

My children and grandchildren are a precious legacy of deep and everlasting love. Again, there is much about each of them in the book. I am so proud of what they have accomplished, and the individuals they have become.

Deborah, Debby, or **Deb**, is a delightful and energetic woman, the mother of two grown sons and grandmother of my first great-grandchild, Kelan. A dedicate career educator, she is a hard worker, leader, and friend to others. The winter I spent with her after losing Barb was a precious time of remembering, healing, and getting on with my life. It was a joyful, lifting experience at a time when I needed to be lifted.

Howard Michael, Mike, or **Noward** to his siblings, is the kind of son many men dream of having. He has three accomplished sons of his own. The oldest, Russ, and his wife presented me with my second great grand child, a girl named Jameson.

Roberta, Robbie, Rob is a full-time mother to three teenage girls. The quiet one of her siblings, she is a softly loving woman with deep emotions.

Diana, or **Dee Dee** is a vivacious bundle of energy and love. The ***Aunty Mame*** of our family, there is never a dull moment when she is around.

Melinda, or **Mindy**, is the delightful, loving mother to an active young son and a daughter who is a dynamo of loving energy. We have a particularly close and uniquely loving relationship.

Kristen, Caroline's daughter, came back into my life in the summer of 2009 after a long absence to my utter amazement, boundless joy, and incomparable loving delight. Mother of two adorable little girls I met for the first time a few months later, she, her husband Vince, and those two little girls have fulfilled my long-held fond dreams of reunion with buckets of tears of sheer joy.

To the various spouses and children of my children I am especially indebted and enamored. I could write pages about each of you and your spouses or *significant others*, but **Words from the Lakeside** would then take several volumes. Let it be known you are all treasures of my heart and enjoyable to be with. Sadly, our times shared together are far too few.

To the many friends I have enjoyed during each of the passages of my life I say, thanks for the memories. Though many of you from the early years have lost touch, I remember you fondly. I especially treasure the memories and renewed contact enjoyed at our fiftieth Heights High reunion in 1996 and the sixty-fourth in 2011. There are many friends from the forty-five years of membership in the Euclid Avenue Christian Church now in Cleveland Heights, Ohio. Then there are the new friends brought into my life by moves and relationships. My membership in the Leesburg United Methodist Church brought new friends. My marriage to Barbara and the church she served, Morris Chapel United Methodist Church, brought more new and dear friends. Many of those I mentioned are close and cherished to this day.

Since the latest passage of my life has taken me to St. Augustine, I have garnered many new and close friends. I have become a member of two singing groups here, Singers by the Sea, and the St Augustine Community Chorus. These and the Socrates discussion group I joined have each brought new friendships. I am actively involved in The Florida Writers Association as well as several critique groups of writers. I am a participant in a group of talented writers who meet and share memoirs and short stories each week at the Council on aging. I am pursuing my thespian activities in a drama group at the Council on Aging. Also, I give lectures on several subjects including energy and global warming. These are all important new parts of my life.

I close this dedication with a repeat of the true words with which I started.

This book is who I am, what I think, what I believe, what I imagine, what I dream of, why and whom I love . . . in short, it is a collection of bits and pieces of my life—of me. It is also a testimonial to all those beautiful human beings who, through love and some blunt trauma, helped me become the person I am. Therefore, I dedicate this book with great love and affection to all of my family, friends, and others, whose actions helped create, guide, inspire, stimulate, and mold my life into the person I am today. My passionate desire to please and never to displease them has guided me in positive directions throughout my lifetime. The family members, lovers, friends, mentors, and teachers who have left this earth are remembered fondly, appreciated greatly, and sorely missed. I am powerfully blessed to have known all of these incredible people, family, friends, colleagues, and acquaintances.

Contents

About 500 short quotes of about 300 authors are found scattered throughout the book. These include many of my own sayings, plus some 20 by unknown authors.

Index of known authors quoted - alphabetical by last name: Page

Contents

Contents

Contents

Contents

Contents

Contents

Contents

Contents

Section I - Miscellaneous Commentary, Letters, Essays, and Poetry.

Contents

Section II - Short Stories, Mostly SciFi

In Section II - Quotes within Short Stories

Contents

Section III - Memoirs

Contents

Contents

PHOTOS:

Contents

PREFACE

Over the years, I have collected, written, and saved many stories, quotes, comments, letters, and poems. These include facts, ideas, thoughts, hypotheses, or theories from my mind and soul. There are about 500 of these from at least 270 different individuals including myself. My purpose in collecting these quotes and in writing this book is to share these with others. I designed it to be a book one can pick up and read for a few minutes or for hours. Its content runs from single lines to multi-page stories, memoirs from my life. My own opinions on numerous subjects are sprinkled liberally throughout the book. Like every other human mind, I may be right or wrong. I try to think and also to write in a rational way, rather than emotional, especially about those subjects that require or could use serious, thoughtful effort. It is quite difficult to keep those emotions from breaking into even serious, rational discourse, but at least I make the effort. Things of the heart and soul, however, are tied much more to feelings and emotions. I hope the reader will feel my emotions as they have a powerful effect on this work in virtually every part of life where feelings participate.

Much of the first section of this book describes concepts that make sense to me and feelings I have personally experienced. I believe one's personal belief system will determine their social, religious, and political beliefs, their relationships with others, the kind of life they lead, and ultimately, the person they are at any given time. The first page has two short pieces about my personal belief systems. They describe the most significant of my guiding principles. The next page describes how I try to relate to my children, a most salient part of who I am.

Here are a few comments about my use of the word or term, "God."

"God" has been used by man for ages as the name or explanation for everything inexplicable—the unknown—the mysteries—the unfathomable. It is not a meaningless or empty term no matter what your belief system. Simply stated. "God" is a term that means and is interchangeable with "the order of things" or "natural law" or even "the inexplicable." Many anthropomorphize it to mean, "The man up in heaven." When fundamentalist atheists say, "there is no God," they are actually declaring their personal distaste for what the term implies religiously—the anthropomorphic father figure. As with many terms, individuals will attach their own meaning to "God." If you are one of those offended by prejudiced against the term for any reason, simply substitute, "the order of things" or any other term that fits when you come across the word, "God" in my writing. I'm sure there are other words that could be considered offensive to some readers for we all have our own abhorrence to different things, word included. Should you come across any of these, simply substitute a less offensive word or phrase. As some wise person once said, "Blessed are the flexible for they shall not be bent out of shape."

The following statements are a collection of my current basic beliefs relating to interactions with other individuals. It is provided so you can better understand the basis and origin of what I have to say. It explains how I see myself and how to understand my words.

I am a believer in myself and those individuals I trust.

I trust no politician, political operative, political activist, government official, celebrity, or media reporter or talking head I do not know personally, and very few of those I do.

I trust and admire the rational opinions and logical judgements of the so-called *common people*. Their wisdom is far greater than that for which they are given credit. I do not trust their opinions or judgements when based on emotions as they are too often persuaded by those described in the previous paragraph.

I see politics, religion, and culture as powerful belief systems often used by unscrupulous individuals to control others for their own purpose.

I am not a follower of or beholden to any ism, group belief system (religious, political, cultural, or other), political party, union, peer group, grant committee, dean or head of faculty, political or other boss, or corporate officer at any level. This is why I am free to express my own opinions without disrespect, concern for, or apologies to anyone or any group.

I consider myself a truly independent and quite liberal individual, a realist who knows what it means to conserve, an equal opportunity supporter or offender, although any offense taken is not intentional.

I am not ever in any way controlled, intimidated or cowed by any kind of political correctness. I believe it to be a creation of the many narcissist members of the entertainment world and in particular the TV news media. These self serving hypocrites use PC to coerce people into speaking and thinking the way they determine. It is merely one more system that elitist intellectuals use to try to control others, mostly the gullible, unthinking lemmings so many people, including especially Americans, have become.

I will not accept as a fact, any words, concepts or ideas that do not meet the tests of logic, reason and/or hard science as I understand them. My opinions and beliefs are subject to change when and if new information makes a change necessary. I see the inflexible, closed mind - the mind of the fundamentalist of any flavor, religious, political, social, cultural, or other - as an evil curse on the individual whose mind is closed for any reason.

It is clear to me that thousands of free and independent individuals and groups working in a favorable competitive environment, under a capitalist system with limited government in a democratic republic, are infinitely superior to a central decision making collectivist body or government of any kind. The bigger and more powerful the government, the less freedom individuals have to grow and improve their life and the lower will be the standard of living under that government. Examples of this reality abound now and throughout history for at least the last 3,000 years. Freedom works. Collectivism always leads to impoverishment, dependency, and ultimately some form of slavery..

PREFACE

I believe in treating every individual with respect and honesty. These both deserve respect and honesty in return. However, I see no reason to be bound to do the same when faced with disrespect or dishonesty, but I will expect respect and honesty first.

I try to deal with every person with consideration in all of these things.

—Howard Johnson - 2012

"There are many who find a good alibi far more attractive than an achievement. For an achievement does not settle anything permanently. We still have to prove our worth anew each day: we have to prove that we are as good today as we were yesterday. But when we have a valid alibi for not achieving anything we are fixed, so to speak, for life. Moreover, when we have an alibi for not writing a book, painting a picture and so on, we have an alibi for not writing the greatest book and not painting the greatest picture. Small wonder that the effort expended and the punishment endured in obtaining a good alibi often exceed the effort and grief requisite for the attainment of a most marked achievement."

—Eric Hoffer

My freely expressed opinions may or may not be in accord with the thinking of those who read my words. This especially includes my views on both of the no-nos of human verbal interaction, religion and politics. Because both areas can be so emotionally charged and can be quite devoid of rational thoughts, there is an opportunity to offend, bring to anger, and damage feelings. Those from many emotional persuasions will surely find themselves pricked by barbs from many directions.

I have much respect for the knowledge and wisdom found in the words of virtually every human being. I even include those deemed foolish and unwise by the multitudes, those whom elitists and intellectuals see as far beneath them in intellect, brainpower. This applies especially to those who populate flyover country. Genius or mentally challenged, corporate president or ditch digger, priest or sinner, person of any age, sex, culture, race, wealth or education—each of these and others have their own set of knowledge from which can be gleaned words of wisdom and truth if one listens.

I do not judge the worth of a person by any of these criteria. To do so is among the greatest faults of those who shut off all sources of knowledge and understanding that could be gained from those with whom they do not see eye to eye. It extends to even the lowliest among us. This fault is usually exhibited by political or religious elitists who refuse to be involved in communications of any kind that does not agree with or conform to their personal belief system. As a result, their inbred concepts shut out more and more good, even profound knowledge because it does not fall within the limitations of their beliefs, or confirm them. This is why *political correctness* is the political equivalent of *fundamentalist beliefs* in the broad field of religion including atheism. All of these are belief systems driven by emotion, and not necessarily based in reality. Simply stated, one man's belief is another man's anathema.

In 1969, I gave a talk on personal communication at the American Dental Trade Association annual meeting in Chicago. The following comment is from that talk. I was using one of my own strong beliefs to illustrate the often hidden but possibly immense value of listening to what even the lowliest among us has to say.

"My measure of a man or woman is not how much they agree with me, but rather, how logical and persuasive are their arguments when they disagree. I also consider what kind of emotions play in these arguments. Do they lash out in anger with words of resentment and condemnation, or do they listen and make rational judgements?"

—Howard Johnson, from a talk on communication in 1969

Especially in the areas of human thoughts and ideas, I much prefer to choose my own belief systems based on knowledge, experience, and logical thought processes, rather than adopt those of others. This does not mean I shun the wisdom or counsel of others. It means I accept such only after checking it through my own understanding of how the universe works. That may seem crazy to some. I address the following saying to them:

Those who dance are thought insane by those who can't hear the music.

—Angela Monet

Hopefully, you will hear and enjoy some of the music of my heart, soul, and imagination which has been liberally poured into these pages. There is one other particular quote that I find describes quite beautifully how I have tried to approach life, at least for the last fifty years. It has been attributed to a number of people including Alfred D. Souza whose name appears as the author on a cup I have had for some time. Some research I conducted attributes it to Mark Twain who preceded Souza by a hundred years. The cup displays the last four lines of the words that follow:

Work like you don't need the money.

Dance as though no one is watching you.

Love as though you have never been hurt before.

Sing as though no one can hear you.

Live as though heaven is on earth.

Mark Twain (Samuel Clemmons)

Why I Write

I am a story teller mostly, both fiction and memoirs: fabricated and remembered. I have six finished books published and several more that will be published this year. At least I am hopeful they will be finished this year. I have five more writing projects in stages from half finished to just started. There are many more in the idea stage. My writing dreams are far too big for me to accomplish in one lifetime. That alone should help keep me young at heart and always thirsting for another day, even at my advanced age. Some time back I told everyone on my email list that I discovered I was a writer, but I didn't say why. Then I read the following words of Samuel Taylor Coleridge:

"Poetry has been to me its own exceeding great reward; it has given me the habit of wishing to discover the good and beautiful in all that meets and surrounds me."

His words prompted thoughts reminding me how impossible it is for me to write all I have to say that I would like to write. Each story, thought, idea, or memory that I put into words brings forth from the depths of my mind and imagination, more stories, more thoughts, more ideas, and more memories. I am deliciously excited by writing these things. I have difficulty deserting my writing to take the time to do much else. This passion moves me so strongly that at night in bed, I often stay awake, planning how best to word this story, thought or idea.

I was an avid reader for many years devouring all kinds of literature. Once I started writing at age seventy, my reading time gave way to mostly writing time. It has been this way ever since. To me, writing is so much more rewarding. Certainly I would like my words to be read, but my main pleasure lies in the writing. I would write even if I knew no one would ever read my words.

So think about writing. If there is a story, memory, or idea in you, give it the wings of the written word. Who knows how many others you may touch.

—Howard Johnson, 2011

Section I

Miscellaneous Commentary, Letters, Essays, and Poetry, Mostly my Own Writing

Throughout the original edition of this book were white spaces at the end of chapters, stories, and other writing. I have moved many quotes and small comments written by myself and others from the original first section into these spaces. These are from my private store of quotes, comments, and poems I thought significant. Authors of quotes are acknowledged if known.

I believe there are no more fitting words with which to begin this work than those of Saint Francis of Assisi. They have been a guiding light for many decent lives and a beacon of peace and love for centuries.

> Lord, make me an instrument of your peace,
> Where there is hatred, let me sow love;
> where there is injury, pardon;
> where there is doubt, faith;
> where there is despair, hope;
> where there is darkness, light;
> where there is sadness, joy;
>
> O Divine Master, grant that I may not so much
> seek to be consoled as to console;
> to be understood as to understand;
> to be loved as to love.
> For it is in giving that we receive;
> it is in pardoning that we are pardoned;
> and it is in dying that we are born to eternal life.
>
> *—Saint Francis of Assisi*

Truth and Belief

When truth and belief come to conflict,

it is better to change one's belief to fit the truth

than to change the truth to fit one's belief.

Beliefs are the creations of men

while **Truths are the creations of God!**

—*Howard Johnson, July 7, 1986*

To all my dearly beloved children,

Your kind of father? I think only maybe. I will always try to be the best I can be for you while remaining my own kind of man. As a father, my kind of man will always try to realize his children are not his possessions, but are growing, separate human beings with their own lives to lead. He is, therefore, responsible for doing the best job he can to teach his children how to cope with the world. He does not have the right to impose his own will on them, but must protect them from danger. He must not be a pal, a dictator, a friend, a slave, or a slave-master to his children. Yet as occasions and situations dictate, he must be each of these and still more.

His relationship must be multidirectional and fluid in all respects. As the child grows, he must constantly adjust to the proper degree of control for both the child's education and protection. He must have the strength to let his charges be hurt so they learn some cautions are in order. He must carefully protect and gauge the amount of hurt to be allowed to both the child's age and constitution.

Likewise, in life's decisions he must grant more and more autonomy as the child gains the experience to handle it. He must maintain a benevolent dictatorship until his charges are on their own. Democracy is suitable for a nation or group of adult equals, but it is a disaster in a family of growing children. He must also recognize it is best to loosen the reigns too early than too late since this teaches the child responsibility for his or her actions. Above all, he must know love is not possession, but sharing.

A wise man was asked how to hold love, to which he replied, "Like a small bird in the hand. Hold it too tightly and it dies; hold it too loosely and it flies away."

I know not how you view your father now, but when you are a full person at whatever age, invite me into your life as you would a friend. If it comes to pass in a comfortable and loving fashion, I will have been the father I intended to be.

—*Howard Johnson, 1965*

To my dearly beloved grandchildren

Your kind of grandfather? Well, maybe! I wrote the previous message to your parents. Now it's your turn. Being a grandfather is a different experience and challenge. There is no choice, little direct responsibility, some commitment and yet, still more mixed blessings. Also, there is far more good than bad. For me, one of the hardest and most necessary things to do is to keep my mouth shut when I feel like spouting volumes, at least during the years before your maturity and independence. I hereby give notice, once you've left the nest and become fully adult, I no longer feel constrained and will freely share opinions about most everything. I urge you to pore through this book, *Words from the Lakeside,* or my other similar book, *Memoirs from the Lakeside,* which both have this letter. There is far more of whom I am in these books than I could include in any letter.

Note especially those words which urge you to be independent, self-reliant, your own person, and to make your own way in life. Don't be a second edition of anyone; be who you are. The comfortable nest, once abandoned, can never be regained. Make your own nest where and when you choose. The silver umbilical cord must be discarded, or you and your parents will never share an adult relationship. I take great pleasure in my relationship with your parents. It is one independent human with another. I would only hope you will some day enjoy the same kind of unfettered relationship with your parents and, of course, with your grandfather.

My maternal grandfather, Granddad Dick (for Dickinson) was a marvelous companion and teacher for me when I was small. We spent many hours together, often fishing as described in one of the stories. One incident when I was quite young, six or seven probably, had a lasting and positive effect on my life. He and grandma were at our house for dinner, and mom was serving stewed turnips. When they were passed my way, I turned up my nose announcing, "I hate stewed turnips."

Granddad turned to me and said, "Howard, you should never say you hate anything. Say, 'I love stewed turnips,' and you'll be surprised how good they taste. It works Howard. Try it."

That won't work, I thought to myself, but since I held my granddad in such respect, and even awe, I decided to try it. I then bravely announced, "I love stewed turnips" while smiling at Granddad and at the same time helping myself to a large spoonful of turnips.

I could hardly believe it. They tasted delicious! I looked at Granddad and announced incredulously, "I do love stewed turnips."

"You see?" Granddads said smiling knowingly. "It does work, like I said it would."

To this day I love stewed turnips and a whole lot of other things I tried the same trick on. I have always believed my grandfather's dinner table lesson is the reason why I like so many foods to this day. There is almost nothing I am served I don't eat with relish—if it is well prepared.

This carried over onto all parts of my life, enforced by my natural tendency to go against the crowd—to resist peer pressure. Many boys repeat the mantra, *I hate school,* and then feel bound to prove it. I went against strong peer pressure saying repeatedly, "I love school."

Well, guess what? I always loved school and learning. There is no doubt in my mind I have used this principle to good effect on many other situations in my life. It is a powerful motivational force. I suggest you try it. No, don't try it, DO IT!

There are several memoirs in this book about Granddad Dick. You may gain some valuable insight from these stories from a man who taught me many worthwhile lessons about life. He was your great-great-granddad so you carry some of his genes.

To the youngest and the next generation, even though you don't know me or maybe never even met me, I will live on in the pages of this book as long as there are those who read it. Hopefully someone in the family will give you a copy. It's the legacy I am leaving for you. My hope is it will live on long after I'm gone.

—Howard Johnson 2009

Letter to a Friend

Each of us must live within our own skins. No matter how much we want things for another, no how much we love and wish and cry and hope and pray, no matter how good our intent or noble our wishes, we are but bystanders to the pain, joy, sorrow, happiness, and love that another experiences—unless that other person lets us into their lives. So it is with children, parents, friends, and lovers. *Love them and let them be,* someone once said about children. So be it with others, even lovers. That decision lies fully within them as within us.

Sometimes it is extremely difficult to let go of that fierce loyalty one feels for one so dear we have lost. Sometimes I hurt terribly—tears flow, my heart aches, my mind reels! Barbara will always be who she was and still is to me. Yet I am moving on. I consider myself so fortunate to find out this new thing about myself—to love deeply and passionately once more and have it never touch negatively or diminish in any way what Barb and I had for the time we spent together. How wonderful that Daphne understands, and we can share moments of grief—cry together and then laugh together. I am indeed a fortunate human having met such women. God has been so kind to me in so many ways.

—Howard Johnson, November 2006

❧   ❧   ❧

Every heart sings a song, incomplete, until another heart whispers back. Those who wish to sing always find a song. At the touch of a lover, everyone becomes a poet.

—Plato

Sound Rainbow

Out of the nothing whiteness they explode. Purple-red sounds and fresh green smells—brittle, crackling things, bright crashing pain. The yellow sea rolls quietly against itself, and two green suns light endlessly a bright blue desert.

Somewhere, a quiet, fur-soft sound begins a trip. Small creatures watch and run, stop and listen, in and out, to and fro.

The fur-soft sound becomes an orange din and searches out the creatures for its own concentric pleasures.

Larger creatures now move and see and listen and slither warmly, silently outward.

The orange din invades the separation of the blue desert, and the yellow sea and the creatures follow for their own pleasures.

The orange din grows into a scarlet scream of terror and realization. The creatures flee and stumble and die. The scarlet scream stands alone between gentle rolls of yellow sea and the fierce blue desert sand. And the two green suns make pleasing smells. The scarlet scream deepens to the purple roar of a thousand purple cataracts of a thousand purple rivers rushing ever onward toward unknown whiteness.

The yellow sea stretches forever, guiding and beckoning the purple roar along its helix path both outward and inward.

Enormous creatures now roll slowly onward between the yellow sea and the blue desert and the purple roar.

The creatures' pleasures are infinite. But the whiteness fear is the bitterest and the sweetest of them all. The creatures lurch and throb; the purple roar turns grayish brown and the whiteness fear engulfs completely.

The yellow sea and the blue desert merge. A green fragrance is born. The green fragrance reaches the two green suns.

The creatures pulse and are gone.

All become one—

Silence returns,

Whiteness,

0.

—*Howard Johnson, 1974*

Dream Thoughts

The mind drifts in and out of sleep on a Sunday morning.

It's early morning, those hours before my normal waking time. I'm in that delicious hazy state of drifting in and out of sleep moment by moment. During longer periods of being awake, I read about quantum physics and new thoughts about application of quantum effects on a cosmological scale. I read a paragraph or so, then sleep catches me for a few moments again, and the book slips from my grasp. Hazy thoughts and visions flow through my dream sleep of singularities, quantum fuzz, expanding universes, and celestial objects mixing with odd shapes and flashes of colored light—sensations of moving, falling, flying, and many indescribable ones. I see what I can't see, feel what I can't feel, and know what I can't know. The potent grasp of concepts of a universe, created by God and far too vast to see, fascinates me.

With powerful telescopes, astronomers can now see objects in every direction moving away from us at more than half the speed of light. Visions of moving to one most-distant part of the visual universe and looking back across that universe waft through my dream thoughts. I cannot see those known objects farthest from my new position since they are now receding faster than the speed of light relative to where I am. The light from these known objects will never reach me. I am moving away from them faster than the light they emit is traveling. As I am moving to my new location, those objects will *wink out*" when our relative departure reaches light speed.

Are we in a huge universe, forever limited to seeing only a tiny spherical part of an unimaginably vast space? This limit of and by the speed of light shuts us off from knowledge beyond that limiting sphere. In much the same way, time shuts us off from knowledge of the universe before the big bang, if that theory is correct. Does the force of gravity move at light speed, or is it instantaneous? We know that gravity *warps* light into a nearly parabolic path in much the same way as it *warps* the movement of objects. Perhaps light moves through the universe in a wavy, irregular path, accelerated and decelerated by the mass of objects it passes. Could it possibly move in a circular path if the universe is finite?

I imagine a roughly spherical shape for the universe. At the surface of this shape, all mass would be to one side of any point on the surface. The center of mass of this universe would be roughly at the geometric center. Could not light be warped by this center of mass so it could not escape from the universe? Could not the true speed of light be a factor of its distance from this center of mass? Maybe our measurement of light speed is a function of our own distance from this center of mass. Perhaps light passing through this center of mass is moving much faster than when it passes us. Likewise, light reaching the limit of the universe would slow and *fall* back in the same way one celestial body orbits another, controlled by the force of gravity. The limiting surface of the universe would act, in effect, as a kind of mirror, returning light and holding it within the gravitational grasp of the universe.

If this were the case, perhaps those faraway galaxies we see with our telescopes are viewed with light that has coursed many times through and around the universe. We may even be seeing our own galaxy millions or billions of years ago, possibly seeing it at numerous positions in the sky. How does one measure the effect or determine if such is the case? Our senses tell us that light moves in a perfectly straight line. We know this to be untrue as there are numerous objects we see in several different positions in space at the same time. The light has taken different paths to reach our observation, warped by what is called a gravitational lens. Could not this multiplication of apparent objects be occurring on a grand scale in the universe as a whole? Could not that galaxy we see far beyond Polaris actually be in the southern sky with its light taking an enormous circle through the universe before it reaches us? Our eyes and senses would straighten that irregular, roughly circular path, placing that galaxy in the northern sky rather than in its true position.

Our thoughts, even our dream thoughts, planted in a fertile field of the minds of men can grow new ideas and concepts which challenge the accepted. All great ideas flow into being in this manner. What prompts these dream thoughts? Is it God's hand at work or mere chance? Whether these thoughts from my mind are nonsense or a giant step forward remains to be seen. There are few mental seeds among the tons of chaff, few mental diamonds among acres of stones. These I share have been gleaned from many years of manipulations of dream thoughts prompted by input from many sources during my life. The future will decide whether they have substance or are mere feathers in the wind. Perhaps God's handiwork is showing. Perhaps this is another view of his creation. Perhaps the big bang was how he created our universe. We may never know, but the search for answers, the search for truth has to be man's greatest godly striving.

—Howard Johnson, August 29, 1999

❖   ❖   ❖

Moral indignation is jealousy with a halo.

—H. G. Wells (1866-1946)

Glory is fleeting, but obscurity is forever.

—Napoleon Bonaparte (1769-1821)

The fundamental cause of trouble in the world is that the stupid are cocksure while the intelligent are full of doubt.

—Bertrand Russell (1872-1970)

There must always remain something that is antagonistic to good.

—Plato

Flaglar 1037

My friend, Flaglar 1037, an alien astrophysicist whom I met in February of 1934 through a timeslip he created, says we humans have it all wrong. The speed of light is not constant, but is a function of the total mass of the universe and the inverse of the square of the distance of the point of measurement to the effective center of mass of the universe. Because the mass in the universe is irregularly distributed, the speed varies somewhat randomly as the measurement is taken near massive objects like black holes, for instance. For this reason, light curves through space under the influence of gravity in a similar way as mass moves in curved paths. Like planets under the gravitational control of a star, or a star under gravitational control of a galaxy, light is under gravitational control of the center of mass of the universe.

When energy or matter (energy slowed and thus trapped in closed three-dimensional loops) reaches the edge or surface of the universe, all mass is in one direction. The light would then *orbit* or *fall* back toward the center of mass in exactly the same way any amount of matter would behave. This results in a universe shaped like a slowly pulsing irregular ball. The irregularities in the microwave background radiation could define the surface of such a universe, but only at one point in time.

Since we on Earth are in a relatively fixed position in the universe relative to its center of mass, light speed measurements always give the same results. For this reason, it will take millions of years for the speed of light to change enough so man can detect the change. We can, however, see the effect of changes in light speed every time we view light from a distant object passing close to a massive object or through a gravitational lens.

This means that light coming to us from a distant galaxy, which we assume came in a straight line, may instead have coursed several times or even hundreds of times through the universe. That galaxy appearing to be billions of light years away, could, in fact, be our own galaxy—its light coming back to us by a circuitous, curved-space-routed, irregular path. The red shift we see in distant objects is simply light that has been affected by countless time dilations and **lost momentum (energy)** through gravitational waves over billions of years of travel.

It follows that given enough time, every photon of light and every particle of matter in the universe will eventually pass through the event horizon of some black hole. When the last photon and the last bit of matter have disappeared, and the universe consists only of the remaining black holes, what then?

Flaglar also thinks it will be a long time before man will figure this out. He doubts our species will last that long.

—Howard Johnson, May 15, 1995

Comet Hale-Bopp

First visible in the early morning sky about 25° above the horizon in the northeast, comet Hale-Bopp may well be the most spectacular comet display since the appearance of Halley's comet in 1910. It is now at its highest point in the evening sky. Hale-Bopp's nucleus is at least three times the size of Halley's and will remain in view to the naked eye until after the first of May. The comet sank to the horizon in the morning sky, disappearing shortly after April 1. During the same period, it rose in the northwestern evening sky, reaching its highest point on April 1. It will remain visible in the evening until after May 1 (1997).

A comet is a fluffy ball of ice and rock, a sort of *dirty snowball*. As the comet nears the sun and comes inside the orbit of Jupiter, radiation from the sun causes the surface to evaporate, creating streams of dust and gas that make up the comet's tails. The curved tail is glowing dust, and the straight tail is ionized gas. The tails may vary in brightness, direction, and color as differences in the solar wind's magnetic field buffet the nucleus. Changes can be dramatic even in periods as short as half an hour.

Comets are visitors from two clouds of objects held in the sun's gravity out beyond the orbit of Neptune. Called the *Kuiper Belt* and the *Oort cloud,* the objects in these areas are mostly ice and range from tiny to bodies like Pluto and recently discovered Eris in the Kuiper belt. There may be even bigger bodies in the Oort cloud. a cloud of debris held in the sun's gravity out beyond the Kuiper Belt. Each object follows its own orbit around the sun. Those objects whose orbit brings them inside the orbit of Jupiter react to the sun's radiation and may become visible. When they are discovered, comets are named for their discoverers. Hale-Bopp is one of these, albeit a spectacular one. Comet orbits are sometimes altered by passing near a planet or other large body. Comet Shoemaker-Levy was one of these. It passed close to Jupiter, whose gravity altered its orbit so much that it broke up into a *train* of some thirty large objects which crashed into Jupiter with spectacular results. Like most comets, Shoemaker-Levy was not visible to the unaided eye but was tracked by astronomers with telescopes. Highly visible comets appear only once every one or two decades, so be sure to view this spectacular display of a solar system visitor whenever possible.

—Howard Johnson, April 2, 1997

Victory goes to the player who makes the next-to-last mistake.

—Chessmaster Savielly Grigorievitch Tartakower (1887-1956)

Don't be so humble - you are not that great.

—Golda Meir (1898-1978) to a visiting diplomat

A Parent's Prayer

Dear Lord make me a good parent! Teach me to be as courteous to my children as I would have them be to me. When I make decisions contrary to their will, let me be reasonable not arrogant, conciliatory not condescending. Forbid that I should ridicule their mistakes or humiliate them when they displease or confuse me. By my own acts of omission and evasion, let me not teach them to mislead and lie. Make me mindful of the value of direction by example and of the hypocrisy of preaching what is not supported by practice.

When they were young, I was mindful that I could not expect from them the judgment of adults. Now that they are adults, let me be mindful that I should not expect from them the judgment of children. Let me also understand that their adult judgments may differ from mine and keep a careful tongue about those differences.

Let me be to my children a pillar of strength rather than a leaning post. Insofar as it is practical, let me help them to point their ambitions high. Let me also applaud their concerns and encourage their self-fulfillment. Let me not rob them of their opportunities for rewards or deprive them of their responsibility to make decisions and accept the consequences.

Do not allow me to become so distracted by externals that I miss the significance of their inner concerns. Teach me to evaluate, but not exaggerate the shock and impact of hair and holler, beads and barbs, scene and obscene, fiction and friction. Let me share their concern for the world that was bequeathed to them—a world of anxiety and confusion, quite different in many ways from the world of my own youth.

Let me be receptive to ideas and responsive to ideals. Help me show through my own life that I value principles more than expediency, courage more than conformity, individual action with responsibility more than dependency, human values more than monetary. Let me also demonstrate that when I *do my own thing,* I am mindful of its effect on others, and that when my children *do their own thing* I do not sit in judgment. Let me enjoy with love those things we have and do together. Let me not disparage or condemn those things we do not do together for any reason.

Help me love them and leave them be. Make me a good parent for these times.

—Anonymous quote paraphrased by Howard Johnson, December 2000

His ignorance is encyclopedic

—Abba Eban (1915-2002)

Homework, Sports, and Other Education Problems

I'm sure there are parents who drive their children to be overachievers in both academics and sports. I'm equally sure there are even more American parents that take only a passing note of how their children are doing academically, usually to gripe at them at grade time. With the increasing emphasis on sports (prestige and entertainment) over academics, it follows that academics will suffer. This is particularly so in some school systems which glorify sports.

I recently read what to me was an extremely biased article on schools that dealt with homework. The thrust of the article and what most people will get from it is that our poor little darlings are overwhelmed with schoolwork and that something ought to be done about it. The article indicated that since 1981, homework has increased from seventeen minutes to nearly half an hour each school night. That totals two hours and sixteen minutes per week! I realize there is a wide variation between schools, but come on! How many hours a week do those same children watch the idiocy on TV? Or how about sports? Sports practice and games certainly take up far more time than any homework.

Read carefully, it points out how harmful, misguided parental demands of their children can be when filtered through our antiquated school systems. Ignoring the emotional thrust of the article and the two inserts, the obvious conclusion is that good parental interest and instruction at an early age make a tremendous difference in academic success later.

Most inequality in education is caused by parents too uneducated or uncaring to teach their children. The title of the insert alone, *Where It's an Unaffordable Luxury*, defines the content and will give many parents and children an excuse to fail to do what will help them most. A better title would be *An Absolute Necessity Being Ignored*. The poor and uneducated are truly the ones most in need of this luxury. There are solutions, but few are willing to pay the price.

Until people who are disadvantaged for any reason realize that it is up to them to correct the situation for themselves, they will continue to propagate the forces that keep them disadvantaged in the first place. No matter how much money or effort others (families, schools, government programs) spend, overcoming the causes of their disadvantaged status remains their own responsibility. Contrast the cultural attitude of many poor recent immigrant families with so many poor American families of many ethnic backgrounds. Many poor immigrant families pull themselves out of poverty in a single generation while so many poor Americans don't. The big difference? Attitude, expectations, and involvement of the family in the education of the young and the high value placed on education, even when they don't know the language! Historically, poor immigrants from Europe and Asia have succeeded in America from their own hard work, self-education, and determination in spite of massive organized discrimination. Our poor cultural communities of all ethnic types will succeed as they follow those examples, or fail as they concentrate on discrimination and government handouts. As long as they concentrate on excuses and supposed discrimination as the reasons for their sorry state, they will remain there.

Success in life is ensured by teaching children to overcome obstacles and thus preparing them to succeed. That includes teaching them to prepare for life as best they can instead of wallowing in self-pity and being entertained. Sure it's easier for some than for others! The playing field is not even and never will be. There will always be Albert Einsteins and Joe Dumbos albeit many more of the latter than the former. In all people, genetic differences will abound. There will be more basketball failures than Michael Jordans. Placing a high value on a good education (from birth onward) will be more important for the less-than-average than for the average and more important for the average than for the gifted. Education starts with the family. If they don't provide a leg up, life will be an uphill struggle!

—Howard Johnson, January 22, 1999

A Christmas Gift

I have tried hard to think of a Christmas gift I could send to all of you. This would be an impossibility in itself. Instead I send a wish to you, a special Christmas tree wish for this year. A tree that is tall and straight with boughs outstretched to shelter you from life's woes this coming year. Trimmed with love and surrounded with faith, the Christmas tree I wish for you has a bright, shining star at the top. This star will shine through the darkness around you and bring hope to the world as it did many years ago in Bethlehem. May its radiance in the days ahead flood your heart.

Under this tree, there are presents for each of you. A package filled with extraordinary memories of special people, of special times, and of other Christmas seasons from the past. While some memories may call for you to weep for times and people no longer with you, take joy in the memories of happy times and happy people. Give thanks for these happy memories and for faith in the building of new memories for the future.

Another package is filled with the peace of understanding and hope. Share that with everyone you meet and especially those dear to you. To settle for less is to make mockery of the words ***Peace on earth, to men of good will.*** Indeed, peace is the foundation upon which the tree I wish for you stands.

Included in my wish for you is that this tree will be part of all the good times that will be yours this season, helping to make this Christmas the merriest and most rewarding ever.

—Howard Johnson, December 1980

In theory, there is no difference between theory and practice. But in practice, there is.

—Yogi Berra

Letter to My Grandson on Graduation from High School - June 5, 1996

Dear Grandson,

You are passing a memorable milestone in your life. It is one of many already passed and other yet to reach. Your mother asked if I would write some *words of wisdom* for you on this occasion.

When I asked Barbara to proofread what I had written, she replied, "He's only seventeen! You should only write a short note!"

I replied, "What I have to say is too important to limit. I am certain that he will have no difficulty with it."

At seventeen, you are much wiser than most adults give you credit and much more foolish than you think you are. Time and experience usually adjust these opinions. I have a great deal to say, and on many other subjects, but will only write for you a small part of what's in my mind and heart at this time. I have two large books and many short articles in my head which are partially on paper. What I am giving you is a small part of several of these, not intended as quick reading. There is a lot which needs to be chewed thoroughly and digested to be fully understood.

When I say what I've written is important, I mean that it is important to me to say it. Whether you read it, to understand it, to believe it, to use it or not is entirely up to you. My purpose has been accomplished in the writing. It may be unimportant to you—that doesn't matter. It matters only that the effort has been made to say these things and share them with you. My father, your great-grandfather, shared many *words of wisdom* with me throughout the forty-five years we were together. He shared both the spoken and the written word. Sometimes I listened, sometimes I didn't. Though we often differed in opinion, sometimes strongly, we never lost respect for the other's opinion. I was lucky to have had such a father. Few do.

I often wish he could have known you, for he would have been so proud of you, as am I. I wish we could share more time, but life is busy for us both. Maybe in the future we can snatch a day together. I shared some precious time with my own grandfather, my mother's father, when I was twelve and then fifteen. We got to know each other during long fishing trips and while working together at the lake. The tree that now goes through the roof at the old cottage was planted by my grandfather and me when I was fifteen. He had a different view of life from my father and shared stories of his youth. He told me of traveling with a medicine show and then of running his own medicine show with an ex-slave as a sidekick. He told me many stories, some of which I'm sure were not completely true, but they were delightful tales.

Congratulations on your graduation! Well wishes for your future health, happiness, and prosperity! Prayers for the path you walk that the joys are much more than the sorrows!

With a great deal of love,

Granddad

—Howard Johnson, June 5, 1996

The Most Beautiful Will ever Written

I first heard *The Most Beautiful Will ever Written* in November of 1962 in Chicago at the American Dental Trade Association annual meeting in the Palmer House Hotel. Don't hold me to the date as it was a long time ago. The way it was presented was, a homeless man died on Maxwell Street, the *Bum's Row* of Chicago at the time. When he was taken to the morgue, they discovered this will in a paper in one of his pockets. The speaker had obviously taken poetic license as I learned later when I discovered the true origin of the will.

Williston Fish wrote **A Last Will** in 1897, It was published first in Harpers Weekly in 1898. Shortly afterwards it began to appear in a sporadic way in the newspapers.

From author Williston Fish:

"Whenever a newspaper did not have at hand what it really wanted, which was a piece entitled 'Reunion of Brothers Separated for Fifty Years' or 'Marriage Customs Among the Natives of the Fricassee Islands' it would run in this piece of mine. In return for the free use of the piece, the paper, not to be outdone in liberality, would generally correct and change it, and fix it up, often in the most beautiful manner; so that I am forced to believe that nearly every paper has on its staff a professor of literature and belles-lettres, always ready to red-ink the essays of the beginner and give them the seeming of masterpieces and gradually to unfold to the novice all the marvels of the full college curriculum. This simple work of mine has been constantly undergoing change and improvement. Sometimes the head has been cut off; sometimes a beautiful wooden foot has been spliced on. When a certain press at Cambridge reprinted it [Cambridge is undoubtedly the home of acute belles-lettres] it used a copy in which the common word dandelions was skillfully changed to flowers, daisies was changed to blossoms, and creeks, which is only a farmer boy word, was changed to brooks. When I said that I gave 'to boys all streams and ponds where one may skate,' this Cambridge printer added, 'when grim winter comes.' Some writers can boast that their works have been translated into all foreign languages, but when I look pathetically about for some little boast, I can only say that this one of my pieces has been translated into all the idiot tongues of English"

The name, Charles Lounsbury, of the devisor in the will, is a name in my family of three generations ago back in York State where the real owner of it was a big, strong, all-around good kind of a man. I had an uncle, a lawyer, in Cleveland named after him, Charles Lounsbury Fish, who was a most burly and affectionate giant himself and who took delight in keeping the original Charles Lounsbury's memory green. He used to tell us of his feats of strength: that he would lift a barrel by the chines and. drink from the bung-hole, and that in the old York State summer days he used to swing his mighty cradle, undoubtedly a *turkey-wing,* and cut a swath like a boulevard through incredible acres of yellow grain. His brain, my uncle always added, was equal to his brawn, and he had a way of winning friends and admirers as easy and comprehensive as taking a census. So I took the name of Charles Lounsbury to add strength and good will to my story."

WILLISTON FISH

Charles Lounsbury was stronger and cleverer, no doubt, than other men, and in many broad lines of business he had grown rich, until his wealth exceeded exaggeration. One morning, in his office, he directed a request to his confidential lawyer to come to him in the afternoon as he intended to have his will drawn. A will is a solemn matter, even with men whose life is given up to business, and who are by habit mindful of the future. After giving this direction he took up no other matter, but sat at his desk alone and in silence.

It was a day when summer was first new. The pale leaves upon the trees were starting forth upon the yet unbending branches. The grass in the parks had a freshness in its green like the freshness of the blue in the sky and of the yellow of the sun, a freshness to make one wish that life might renew its youth. The clear breezes from the south wantoned about, and then were still, as if loath to go finally away. Half idly, half thoughtfully, the rich man wrote upon the white paper before him, beginning what he wrote with capital letters, such as he had not made since, as a boy in school, he had taken pride in his skill with the pen:

A LAST WILL

I, CHARLES LOUNSBURY, being of sound and disposing mind and memory [he lingered on the word memory], do now make and publish this my last will and testament, in order, as justly as I may, to distribute my interests in the world among succeeding men.

And first, that part of my interests which is known among men and recognized in the sheep-bound volumes of the law as my property, being inconsiderable and of none account, I make no account of in this my will.

My right to live, it being but a life estate, is not at my disposal, but, these things excepted, all else in the world I now proceed to devise and bequeath.

Item: And first, I give to good fathers and mothers, but in trust for their children, nevertheless, all good little words of praise and all quaint pet names, and I charge said parents to use them justly, but generously, as the needs of their children shall require.

Item: I leave to children exclusively, but only for the life of their childhood, all and every the dandelions of the fields and the daisies thereof, with the right to play among them freely, according to the custom of children, warning them at the same time against the thistles. And I devise to children the yellow shores of creeks and the golden sands beneath the waters thereof, with the dragon-flies that skim the surface of said waters, and the odors of the willows that dip into said waters, and the white clouds that float high over the giant trees.

And I leave to children the long, long days to be merry in, in a thousand ways, and the night and the Moon and the train of the Milky Way to wonder at, but subject, nevertheless, to the rights hereinafter given to lovers; and I give to each child the right to choose a star that shall be his, and I direct that the child's father shall tell him the name of it, in order that the child shall always remember the name of that star after he has learned and forgotten astronomy.

Item: I devise to boys jointly all the useful idle fields and commons where ball may be played, and all snow-clad hills where one may coast, and all streams and ponds where one may skate, to have and to hold the same for the period of their boyhood. And all meadows, with the clover blooms and butterflies thereof; and all woods, with their appurtenances of squirrels and whirring birds and echoes and strange noises; and all distant places which may be visited, together with the adventures there found, I do give to said boys to be theirs. And I give to said boys each his own place at the fireside at night, with all pictures that may be seen in the burning wood or coal, to enjoy without let or hindrance and without any incumbrance of cares.

Item: To lovers I devise their imaginary world, with whatever they may need, as the stars of the sky, the red, red roses by the wall, the snow of the hawthorn, the sweet strains of music, or aught else they may desire to figure to each other the lasting-ness and beauty of their love.

Item: To young men jointly, being joined in a brave, mad crowd, I devise and bequeath all boisterous, inspiring sports of rivalry. I give to them the disdain of weakness and undaunted confidence in their own strength. Though they are rude and rough, I leave to them alone the power of making lasting friendships and of possessing companions, and to them exclusively I give all merry songs and brave choruses to sing, with smooth voices to troll them forth.

Item: And to those who are no longer children, or youths, or lovers, I leave memory, and I leave to them the volumes of the poems of Burns and Shakespeare, and of other poets, if there are others, to the end that they may live the old days over again freely and fully, without tithe or diminution; and to those who are no longer children, or youths, or lovers, I leave, too, the knowledge of what a rare, rare world it is.

Significant Quotes of Talbot Mundy
from: *Tros of Samothrace* and *The Purple Pirate*
Most are from the log of Lord Captain Tros of Samothrace.

✳   ✳   ✳

It never was my view that women are the worse for audacity. To be mothers of sons worth weaning, they should have the manly virtues in addition to the qualities that charm and tempt.

Manners? They are like a cloak that either illustrates its wearer's self-respect or masks his vileness, popinjays his vices, or reveals his taste. I have observed that decent manners are invariable befitting the occasion, blunt and direct when causes are at issue, civil to the verge of gentleness where nothing but another's momentary comfort is at stake. Too-smooth manners in the face of issues is a sign of fear or treachery or weakness or of all three.

I know of no justification for the wars that men wage on one another. On the other hand, I know no reason and perceive no wisdom in the floods and famines, pestilence and earthquakes, fire and hurricane, which priests say the gods devise.

If a friend in friendship errs, it is vile to retaliate. Regret is stupid. Recrimination is a waste of time and breath. There is nothing to be done but to redeem the error. Friendship is not measurable by an error—no—no matter how great or how disastrous.

A man may be a murderer and faithful. Many are. A man may be a courtier and faithful. Some are. But the courtier-murderer, disarmed and faced with the alternative of cold steel in his throat, will babble all he knows to avoid the kind of death he has meted out to others. But first, disarm him. Armed, he believes himself an honorable man. Disarmed, he knows he has no honor.

They who laugh at a commander's failure usually lack ability or will to understand the nature of his problem. Detail, detail, detail—each dependent on another's or a hundred others' loyalty, devotion, skill, intelligence, obedience, and health. One sick man, fretting faithfully to do his stint, unknowingly unknown, may wreck a well-imagined strategy before its details unfold. I have heard self-styled critics speak . . . aye, and I have read the books of some historians who write, as if a warship can put to sea without a thousand cares first well attended. And if a ship, what of a fleet? What of an army? It is a pity, for their foes' sake, that some critics are not taken at their own evaluation and entrusted with command.

When I ask myself, as I think all thoughtful men inevitably do, have I done my duty? Have I acted manly? I perceive it is impossible for oneself to answer. That is something that only other men can do until the gods' day comes to issue judgment—aye, and beware of flattery! Man's speech is seldom sheeted close to truth's wind. But their deeds are eloquent.

So that when ignorant dogs of bawdy seamen, who I have shepherded and thrashed and loved and led, behave like loyal comrades behind my back, then I take comfort. My men shall judge me. Gods, if gods there be, may judge me by the good foul-weather friends who have stood by.

It was the sea, with its roaring rage and smiling treachery, that taught me sometimes to appear to yield. Many a time I have luffed and let an enemy believe me to be beaten. I have avoided battle. I have run. But I have never struck my flag. Storm lover though I have ever been, and conqueror of storms though I have had to be . . . aye, though I pray for a storm if I must meet an enemy at sea . . . I see no wisdom in opposing storm and enemy. Rather, I use the one to help me defeat the other. And if it seems advisable, I run from both to await my moment.

Certain philosophers, some priests, and many women have accused me of loving war. I hate it. I despise it as an arbiter of quarrels. Would that my intelligence and vigor might be put to a more creative use. But I have seen that they, whose speech is most contemptuous of warriors, are also they whose blunders, acrimony, ignorance, and malice aggravate the quarrels that produce war. To avoid war, for the sake of friendship—aye, to prevent a quarrel—I am willing to risk all that I have and to forego my own ambition. But I will yield to no tyrant. And when I find myself at war, I choose to win.

Two heads are better than one, and three than two, but when a plan is reached, let there be one commander. One only. Let the others obey. I would rather obey a man, whose talent for command I thought inferior to mine, than make the unwise effort to attempt to share authority.

The one test of a commander's competence is battle, no other. There is no denying a defeat. No argument annuls a victory.

The incomparable depth of stupidity is that of the commander who does that which his enemy expects because tradition justifies it. The only time when traditional strategy and tactics are fit to employ is when the enemy expects something else and therefore mistakes old methods for a ruse.

It is unwise to expect a clever opportunist to obey, if given opportunity to disobedience to serve himself. Your aims, your plans—aye, and your dangers also, should he know them—would be the natural means by which he would secretly seek to advance himself, inevitably to your cost and perhaps your ruin. There is one wise way, and only one, to make use of such men. Study their natural cunning, as the hunter studies animals, in order to be able to predict their probable behavior when free to follow their inclination.

Money to pay for provisions is more important to a ship's commander than the wind. He can wait for a fair wind. He can hunt a lee in stormy weather. But unless he can pay for supplies, there is no alternative but piracy, disguised or open. And whoever thinks that pirates avoid paying for their depredations is either very ignorant or void of common sense.

It is not the unpredictables that govern issues. It is the steady, unwavering day-by-day persistent exercise of judgment, always hewing nearer to the line of wisdom. Far though it may be from wisdom, yet that effort rarefies its maker's thought until he fits himself for swift and right decision in emergencies that baffle them who envision only purpose and let wisdom wait, as if it were not, or as if it were a poet's word for something unattainable or unknown.

Half of human history was made by drunkards in their cups and written down by slaves of one imposter or another in the hope of table leavings.

I was born and taught upon the threshold of the holy mystery, and all my days I have been faithful to the duty laid upon me to pursue peace . . . aye, and to forego my own advantage if thereby peace might come. But I have found no peace on earth or any honorable way of avoiding war.

✳   ✳   ✳

Notable Quotes from the NIV Bible

Those who discount the Bible because of religion, or because it is a religious work, deny themselves learning a great deal of ageless knowledge collected and preserved for thousands of years. This knowledge, mostly about human behavior and about interacting profitably and successfully with others, is real, practical, and valuable. A great many life rules and lessons explained and demonstrated, apply to almost any human situation. Here are a few excerpts from the Bible that apply across the scope of human actions.

—Howard Johnson

The proverbs of Solomon, son of David, king of Israel: For attaining wisdom and discipline; for understanding words of insight; for acquiring a disciplined and prudent life, doing what is right and just and fair; for giving prudence to the simple, knowledge and discretion to the young—let the wise listen and add to their learning, and let the discerning get guidance—for understanding proverbs and parables, the sayings and riddles of the wise. The fear of the LORD is the beginning of knowledge, but fools despise wisdom and discipline. Listen, my son, to your father's instruction and do not forsake your mother's teaching. They will be a garland to grace your head and a chain to adorn your neck.

—Proverbs 1:1–9

Moreover, no man knows when his hour will come: as fish are caught in a cruel net, or birds are taken in a snare, so men are trapped by evil times that fall unexpectedly upon them.

—Ecclesiastes 9:12

Blessed is the man who finds wisdom, the man who gains understanding, for she is more profitable than silver and yields better returns than gold. She is more precious than rubies; nothing you desire can compare with her. Long life is in her right hand; in her left hand are riches and honor. Her ways are pleasant ways, and all her paths are peace. She is a tree of life to those who embrace her; those who lay hold of her will be blessed. By wisdom the LORD laid the earth's foundations, by understanding He set the heavens in place; by His knowledge the deeps were divided, and the clouds let drop the dew. My son, preserve sound judgment and discernment, do not let them out of your sight; they will be life for you, an ornament to grace your neck.

—Proverbs 3:13–22

Wisdom calls aloud in the street, she raises her voice in the public squares; at the head of the noisy streets she cries out, in the gateways of the city she makes her speech: "How long will you simple ones love your simple ways? How long will mockers delight in mockery and fools hate knowledge? If you had responded to my rebuke, I would have poured out my heart to you and made my thoughts known to you. But since you rejected me when I called and no one gave heed when I stretched out my hand, since you ignored all my advice and would not accept my rebuke, I in turn will laugh at your disaster; I will mock when calamity overtakes you—when calamity overtakes you like a storm, when disaster sweeps over you like a whirlwind, when distress and trouble overwhelm you.

"Then they will call to me but I will not answer; they will look for me but will not find me. Since they hated knowledge and did not choose to fear the LORD, since they would not accept my advice and spurned my rebuke, they will eat the fruit of their ways and be filled with the fruit of their schemes. For the waywardness of the simple will kill them, and the complacency of fools will destroy them, but whoever listens to me will live in safety and be at ease, without fear of harm."

My son, if you accept my words and store up my commands within you, turning your ear to wisdom and applying your heart to understanding, and if you call out for insight and cry aloud for understanding, and if you look for it as for silver and search for it as for hidden treasure, then you will understand the fear of the LORD and find the knowledge of God.

—Proverbs 1:20–33 & 2:1–5

I also saw under the sun this example of wisdom that greatly impressed me: There was once a small city with only a few people in it. And a powerful king came against it, surrounded it and built huge siege works against it. Now there lived in that city a man poor but wise, and he saved the city by his wisdom. But nobody remembered that poor man. So I said, "Wisdom is better than strength." But the poor man's wisdom is despised, and his words are no longer heeded.

—Ecclesiastes 9:13–16

The quiet words of the wise are more to be heeded than the shouts of a ruler of fools. Wisdom is better than weapons of war, but one sinner destroys much good.

—Ecclesiastes 9:17–18

Enjoy life with your wife, whom you love, all the days of this meaningless life that God has given you under the sun—all your meaningless days. For this is your lot in life and in your toilsome labor under the sun. Whatever your hand finds to do, do it with all your might, for in the grave, where you are going, there is neither working nor planning nor knowledge nor wisdom.

—Ecclesiastes 9:9–10

(Read also the King James, below)

Notable Quotes from the KJV Bible

Live joyfully with the wife whom thou lovest all the days of the life of thy vanity, which he hath given thee under the sun, all the days of thy vanity: for that is thy portion in this life, and in thy labour which thou takest under the sun. Whatsoever thy hand findeth to do, do it with thy might; for there is no work, nor device, nor knowledge, nor wisdom, in the grave, whither thou goest.

(Read also the NIV above) —*Ecclesiastes 9:9–10, KJV*

A continual dropping in a very rainy day and a contentious woman are alike. Whosoever hideth her hideth the wind, and the ointment of his right hand, which betrayeth itself.

—*Proverbs 27:15–16, KJV*

A foolish son is the calamity of his father: and the contentions of a wife are a continual dropping. House and riches are the inheritance of fathers: and a prudent wife is from the LORD.

—*Proverbs 19:13–14, KJV*

Truly the light is sweet, and a pleasant thing it is for the eyes to behold the sun: But if a man live many years, and rejoice in them all; yet let him remember the days of darkness; for they shall be many. All that cometh is vanity.

—*Ecclesiastes 11:7–8, KJV*

It is better to dwell in a corner of the housetop, than with a brawling woman in a wide house.

—*Proverbs 21:9, KJV*

Rejoice, O young man, in thy youth, and let thy heart cheer thee in the days of thy youth, and walk in the ways of thine heart, and in the sight of thine eyes: but know thou, that for all these things God will bring thee into judgment. Therefore remove sorrow from thy heart, and put away evil from thy flesh: for childhood and youth are vanity

—*Ecclesiastes 11: 9–10, KJV*

Who can find a virtuous woman? for her price is far above rubies. The heart of her husband doth safely trust in her, so that he shall have no need of spoil. She will do him good and not evil all the days of her life. She seeketh wool, and flax, and worketh willingly with her hands. She is like the merchants' ships; she bringeth her food from afar. She riseth also while it is yet night, and giveth meat to her household, and a portion to her maidens. She considereth a field, and buyeth it: with the fruit of her hands she planteth a vineyard. She girdeth her loins with strength, and strengtheneth her arms. She perceiveth that her merchandise is good: her candle goeth not out by night. She layeth her hands to the spindle, and her hands hold the distaff. She stretcheth out her hand to the poor; yea, she reacheth forth her hands to the needy. She is not afraid of the snow for her household: for all her household are clothed with scarlet. She maketh herself coverings of tapestry; her clothing is silk and purple. Her husband is known in the gates, when he sitteth among the elders of the land. She maketh fine linen, and selleth it, and delivereth girdles unto the merchant. Strength and honour are her clothing, and she shall rejoice in time to come. She openeth her mouth with wisdom, and in her tongue is the law of kindness. She looketh well to the ways of her household, and eateth not the bread of idleness. Her children arise up, and call her blessed; her husband also, and he praiseth her. Many daughters have done virtuously, but thou excellest them all. Favour is deceitful, and beauty is vain: but a woman that feareth the LORD, she shall be praised. Give her of the fruit of her hands, and let her own works praise her in the gates.

—Proverbs 31:10–31, KJV

Make a joyful noise unto the LORD, all ye lands. Serve the LORD with gladness: come before his presence with singing. Know ye that the LORD he is God: it is he that hath made us, and not we ourselves; we are his people, and the sheep of his pasture. Enter into his gates with thanksgiving, and into his courts with praise: be thankful unto him, and bless his name. For the LORD is good; his mercy is everlasting, and his truth endureth to all generations

—Psalm 100, KJV

❧   ❧   ❧

Søren Aabye Kierkegaard (5 May 1813 – 11 November 1855) was a Danish Christian philosopher and theologian, considered to be a founder of Existentialist thought and Absurdist traditions. **Here are a few quotes of his words:**

The highest and most beautiful things in life are not to be heard about, nor read about, nor seen but, if one will, are to be lived.

What is a poet? An unhappy person who conceals profound anguish in his heart but whose lips are so formed that as sighs and cries pass over them they sound like beautiful music.

Boredom is the root of all evil - the despairing refusal to be oneself.

A man who as a physical being is always turned toward the outside, thinking that his happiness lies outside him, finally turns inward and discovers that the source is within him.

How absurd men are! They never use the liberties they have, they demand those they do not have. They have freedom of thought, they demand freedom of speech.

Since boredom advances and boredom is the root of all evil, no wonder, then, that the world goes backwards, that evil spreads. This can be traced back to the very beginning of the world. The gods were bored; therefore they created human beings.

Anxiety is the dizziness of freedom.

Faith is the highest passion in a human being. Many in every generation may not come that far, but none comes further.

Once you label me you negate me.

Here are a few suggestions for a more pleasant and enjoyable life.

Recognize that life is never fair. Most of it is a random process.

Never look back. They may be gaining on you.

Once you've fallen in love, never stop loving that person.

Dream all the wild dreams you can dream, but plan and decide rationally.

Always show everyone you care about, even your pet, that you love them.

Use emotions for love. Use logic for decisions.

Savor and cultivate pleasant emotions. Minimize and weed out unpleasant ones.

Laugh every day, even if you don't feel like there is anything to laugh about.

Take frequent walks—through the woods if possible.

Get rid of everything you don't need or that doesn't hold treasured memories.

Use your last cent to buy a flower for someone.

Each and every day, share a treasured memory with someone you care about.

Tell everyone just how special your special someone is, and do it at every chance you get..

Consider that every statement you make to anyone is a commitment to make it true.

Never lose that child-like wonder at the world. Keep it alive and active.

Rid yourself of everything that is not useful, beautiful, joyful, or filled with wonder.

Forgive everyone of everything and especially discard grudges.

If you have a dog, give it all the love you can and as much freedom as possible.

Say something pleasant to everyone you meet every day.

—Howard Johnson, 2011

Every normal man must be tempted at times to spit upon his hands, hoist the black flag, and begin slitting throats.

—Henry Louis Mencken (1880-1956)

Now, now my good man, this is no time for making enemies.
—Voltaire (1694-1778) on his deathbed in response to a priest asking that he renounce Satan.

He would make a lovely corpse.

—Charles Dickens (1812-1870)

I've just learned about his illness. Let's hope it's nothing trivial.

—Irvin S. Cobb

I worship the quicksand he walks in.

—Art Buchwald

Wagner's music is better than it sounds.

—Mark Twain (1835-1910)

A poem is never finished, only abandoned.

—Paul Valery (1871-1945)

God gave men both a penis and a brain, but unfortunately not enough blood supply to run both at the same time.

—Robin Williams, commenting on the Clinton/Lewinsky affair

If you were plowing a field, which would you rather use? Two strong oxen or 1024 chickens?
—Seymour Cray (1925-1996), father of supercomputing

#3 pencils and quadrille pads.
—Seymoure Cray (1925-1996) when asked what CAD tools he used to design the Cray I supercomputer; he also recommended using the back side of pages so that the grid lines were not dominant.

Interesting - I use a Mac to help me design the next Cray.
—Seymoure Cray (1925-1996) when he was told that Apple Inc. had recently bought a Cray supercomputer to help them design the next Mac.

Fill the unforgiving minute with sixty seconds worth of distance run.
—Rudyard Kipling (1865-1936)

What Is Love and What Is it about

Love is a word used to describe and define an almost infinite number of sets of emotions related to a sense of strong affection and attachment. The word love can refer to an endless variety of types of feelings, states of mind, feelings of the heart, and attitudes, ranging from generic pleasure —"I love that suit," to intense, personal devotion and attraction —"I love my spouse." This wide and varied range of feelings, uses, and meanings make *love* impossible to define out of context. The intense variations and complex nature of the reactions and feelings associated with *love* make it wide ranging and extremely difficult to communicate or understand. In many instances, it is such a broad and nebulous thing it defies description and even comparison to other emotional states.

In most instances, it comes from the deepest instinctual, almost primordial parts of our being. Because of this, it can bring individuals to do incredibly beautiful as well as terribly evil things. The dedication and self sacrifice of love is frequently equaled on the dark side by cruel and evil acts. This powerful force of central psychological importance is one of the most common themes in the creative arts. Our literature is filled with love themes from the earliest writings, both fictional and true. The quotes in the section to *that* follows span the width and depth of love, mostly romantic love between a man and a woman.

—Howard Johnson 2009

Greek words for love:

* Eros (ἔρως érōs) is passionate love, with sensual desire and longing. The Modern Greek word *erotas* means *(romantic) love*. However, eros does not have to be sexual in nature. Eros can be interpreted as a love for someone whom you love more than the philial love of friendship. It can also apply to dating relationships as well as marriage. Plato refined his own definition. Although eros is initially felt for a person, with contemplation it becomes an appreciation of the beauty within that person, or even becomes appreciation of beauty itself. It should be noted Plato does not talk of physical attraction as a necessary part of love, hence the use of the word platonic to mean, *without physical attraction.* Plato also said eros helps the soul recall knowledge of beauty and contributes to an understanding of spiritual truth. Lovers and philosophers are all inspired to seek truth by eros. The most famous ancient work on the subject of eros is Plato's Symposium, which is a discussion among the students of Socrates on the nature of eros.

* Philia (φιλία philía), which means friendship in modern Greek, a dispassionate virtuous love, was a concept developed by Aristotle. It includes loyalty to friends, family, and community and requires virtue, equality and familiarity. In ancient texts, philia denoted a general type of love, used for love between family, between friends, a desire or enjoyment of an activity, as well as between lovers. This is the only other word for *love* used in the ancient text of the New Testament besides agape, but even then it is used substantially less frequently.

* Agapē (ἀγάπη agápe) means *love* in modern day Greek, such as in the term s'agapo (Σ'αγαπώ), which means *I love you*. In Ancient Greek it often refers to a general affection rather than the attraction suggested by *eros*; agape is used in ancient texts to denote feelings for a good meal, for one's children, and for a spouse. It can be described as the feeling of being content or holding one in high regard. The verb appears in the New Testament describing, amongst other things, the relationship between Jesus and the beloved disciple. In biblical literature, its meaning and usage is illustrated by self-sacrificing, giving love to all--both friend and enemy.

It is used in Matthew 22:39, "Love your neighbour as yourself," and in John 15:12, "This is my commandment, that you love one another as I have loved you," and in 1 John 4:8, "God is love."

However, the word *agape* is not always used in the New Testament in a positive sense. II Timothy 4:10 uses the word in a negative sense. The Apostle Paul writes," For Demas hath forsaken me, having loved (agapo) this present world...." Thus the word *agape* is not always used of a divine love or the love of God. Christian commentators have expanded the original Greek definition to encompass a total commitment or self-sacrificial love for the thing loved. Because of its frequency of use in the New Testament, Christian writers have developed a significant amount of theology based solely on the interpretation of this word.

* Storge (στοργή storgē) means *affection* in modern Greek; it is natural affection, like that felt by parents for offspring. Rarely used in ancient work and then almost exclusively as a descriptor of relationships within the family.

Poems of and about Love

It is not, Celia, within our power to say how long our love will last;
It may be within this hour may lose those joys we now do taste.
The blessed that immortal be, from change in love are only free.
Then since we mortal lovers are, ask not how long our love will last,
But while it does, let us take care each minute be with pleasure passed:
Were it not madness to deny to live because we're sure to die?

—George Etheridge

Memory is the power to gather roses in winter, snowflakes in July, and to taste a loving kiss from long ago.

—Howard Johnson, 1970

A First Impression - Cool, devastating eyes across the room. Drawing . . . attracting. Deep, poised eyes, searching . . . reaching. Lean, taut body moving catlike, possessing space, not using it. Finely chiseled face, controlled smile—exotic, mysterious, promise of depth.

—Howard Johnson, 1982

To say that one will perish without love does not mean that everyone without adequate love dies. Many do, for without love the will to live is often impaired to such an extent that a person's resistance is critically lowered and death follows. But most of the time, lack of love makes people depressed, anxious, and without zest for life. They remain lonely and unhappy, without friends or work they care for, their life a barren treadmill, stripped of all creative action and joy.

—Smiley Blanton

O! let me have thee whole,—all—all—be mine!
That shape, the fairness, that sweet minor zest
Of love, your kiss,—those hands, those eyes divine,
That warm, white, lucent, million-pleasured breast,—

—John Keats

Love me not for comely grace, for my pleasing eye or face;
Nor for my outward part, no, nor for a constant heart:
For these may fail or turn to ill, so thou and I shall sever.
Keep, therefore, a true woman's eye,
And love me still, but know not why;
So hast thou the same reason still
To dote on me ever.

—George Etheridge, 1635–1691

The human heart has hidden treasures, In secret kept, in silence sealed; The thoughts, the hopes, the dreams, the pleasures, Whose charms were broken if revealed.

– Charlotte Bronte

Memory is the treasury and guardian of all things

—Cicero, 80 BC

Oh, the comfort, the inexpressible comfort of feeling safe with a person; having neither to weigh thoughts nor measure words, but to pour them all out, just as they are, chaff and grain together, knowing that a faithful hand will take and sift them, keep what is worth keeping, and then, with the breath of kindness, blow the rest away.

—George Elliott

It's wonderful to have someone who really understands . . .
Someone who gives the tenderness your heart sorely demands . . .
Someone to tell your troubles to when evening lights are low . . .
Who with a smile can drive away the dreary clouds of woe . . .
Understanding is a treasure gold can never buy . . .
For it has a magic power to lift the spirit high . . .
Those who proceed without it lose out at every turn . . .
Like souls adrift upon life's sea they will ever yearn . . .
So if you have someone, love them with all your heart . . .
For understanding people are few and far apart!

—Smiley Blanton

LOVE - IMPRESSIONS

Love is a passion of the heart, the soul, and the body. When new, it quickens the heartbeat, excites the soul, and stimulates one physically and mentally. It brings about intense desire to be with the one loved, great joy when together, and intense longing when apart. There is no antidote without mental anguish, no relief without deep sense of loss, no cessation without intense pain. Even when not returned in kind, it causes one to smile constantly and weep tears of pure joy until it fades away. It is a powerful and motivating force for good when well directed.

As the years pass, its burning intensity is gradually replaced by deep feelings of comfort and warmth of the soul and heart. It gets better and better, only ending in the final, tragic parting that brings such terrible pain. For these reasons, love is not the domain of cowards. It takes genuine bravery to risk it. Such is love and the price we must someday pay for loving even until that final moment—pain. Fear of that pain cannot stop me from loving deeply and sincerely—even at an advanced age.

—Howard Johnson, 2006 at age 78

Remembering Easter Sunday, 1945 - 67 years ago today.

Iawakened early to what was to be a busy day. It was still dark, but the birds were announcing the day was about to begin. The first thing on my schedule was the big, downtown sunrise service at the Cleveland Public Auditorium. The Cleveland Heights High a capella choir was to sing, and I was in the second tenor section. The service was to begin at 6:30 and we were to be there no later than 6:00. I was granted the use of the family car as my parents were to be taken to our church services by some church members who were also neighbors. That meant I would have to leave home by about five in order to have time to pick up a couple of other choir members, get downtown, park, and walk to the auditorium by 6:00.

I would have picked up my steady girl, Dolores, who also sang in the choir, but her parents were attending the service and she was to go with them. The drive downtown was brightened by the clear, blue sky of a warm and gorgeous spring day. Daffodils and narcissi were blooming everywhere. Even a few early tulips were showing off their colors. One spectacular passage, Cedar Hill, was down a small gorge through the Euclid Escarpment. It was ablaze with bright yellow forsythia clinging to the sides of the gorge. The air was filled with the fresh fragrance of spring. I don't remember, but I know the birds were singing their hearts out as we drove down town.

We parked the car, walked to the auditorium, and to our dressing room in good time. After we donned our choir robes, I had the chance to talk to my sweetie. We made arrangements to meet with her and her parents when the performance was over. Soon Strick, our choir director, George F. Strickling, lined us up for a warm up before our stage entrance. I don't remember much about the concert, or even any of the songs we sang. As a teen, deeply in love, I was probably too busy trying to catch Dee's eye while we were singing. From my position in the back row on the extreme left of the semicircular arrangement of the choir, I was in her line-of-sight from where she stood in the first row on the extreme right.

I was to meet Dee and her folks after we finished and changed back into our Easter finery. We were to meet outside the side entrance. We actually ended up meeting in the hallway on our way out. One look at her in her Easter outfit and I was overwhelmed. Her bright yellow dress was set off by a spectacular, dark blue, wide-brimmed, straw hat. She was positively the most beautiful girl I had ever seen, and she was my girl. I couldn't get over looking at her. To use some of today's vernacular, she was drop-dead gorgeous. To make it even better, we were about to walk up Euclid Avenue in Cleveland's Easter Parade, me, walking with the most beautiful girl in the world. I was walking on air, proud as a young man could possibly be.

The aura of her in that gorgeous yellow dress and that spectacular hat, spun a magic spell that held me all day long. I cannot remember another thing about the middle of that day other than watching her. After her folks headed for home, Dee joined me while I took my two passengers home. Soon after we dropped the second one off, Dee carefully removed her hat as I pulled to the curb. Soon we were wrapped in an embrace and a lingering kiss. We just couldn't wait until we could park in front of her house. I remember later events clearly, sitting with her in front of 2471 Saybrook Road, sharing tender love words, and kiss after kiss until she had to go inside. Tomorrow was a school day and her curfew was 10:00.

As I drove home down Meadowbrook Boulevard, visions of the days events flashed through my mind. I was totally and deliriously in the grasp of young love, and loving every minute of it. All I could think was, how could I possibly be so lucky?

—Howard Johnson - Easter Sunday, 2012

I have not failed. I've just found 10,000 ways that won't work.

—Thomas Alva Edison (1847-1931)

Political correctness is tyranny with manners.

—Charlton Heston (1924-2008)

You can avoid reality, but you cannot avoid the consequences of avoiding reality.

—Ayn Rand (1905-1982)

Sex and religion are closer to each other than either might prefer.

—Saint Thomas More (1478-1535)

Religion and politics are closer to each other than either might prefer.

—Howard Johnson, 2008

How wrong it is for a woman to expect the man to build the world she wants, rather than to create it herself.

—Anais Nin (1903-1977)

You may not be interested in war, but war is interested in you.

—Leon Trotsky (1879-1940)

Writers and Women
A warning for women about writers, male writers.

Beware of starting a relationship with a writer. They are human and have all the foibles, the charms, the weaknesses, the strengths, the nastiness, the kindnesses, and the frequent inconsistencies exhibited by the male of the species. That being said, there are things writers share with all artists, and a few peculiarities attributable to writers alone. All of them have a number of these sometimes pleasing and sometimes distressing traits. Many have all of them. I repeat, **all of them have a number of these sometimes pleasing and sometimes distressing traits. Many have all of them.**

Writers are Artists and thus unconventional people. They do not often follow the norms of polite society and this often puts them at odds with "normal" persons. If they follow rules they are often of their own making or at least manipulating, and do not necessarily follow society's rule book. The very few who gain fame and fortune are especially afflicted with these attributes.

Writers are frequently unkempt and unconcerned about their personal appearance, their place of residence or their workplace. Since they work alone, they do not have to worry about interference of any kind from other people. As a result they often lack social skills and can be opinionated, rude, and even uncivil.

They treat money and finances in the same manner and rarely have enough to live on. Those who inherited money or have a sponsor who provides for their needs, are the fortunate ones. When a windfall comes for any reason including getting successfully published, they give freely to friends and family. They are apt to give to groupies or other leeches who disappear as quickly as the money. Frequently, portions of their life are spent in poverty and they sometimes die penniless and alone.

Writers are emotional. Most are driven by emotion more than by rational thought. Most people share this trait, but it is greatly amplified in writers. They are known for volatile, often unpredictable actions that can become very damaging, even self destructive. Their emotions frequently override their logic, particularly in situations where the powerful emotions of love, hate, revenge, remorse, and self pity are aroused. They frequently get themselves into situations where there are no possible favorable solutions. Those who do learn to leash in their emotions can do so up to a point. Once that point is passed, emotions can burst forth with tremendous force, drowning all rational responses.

Their lives are often erratic and in turmoil. Many are subject to most acts of debauchery and in some extreme cases, can be found in a drunken stupor in an alley or gutter. For this and other reasons, writers often die quite young. Quite a few die of drug overdoses, or by their own devices. A few die at the hands of the husband or lover of one of their conquests. They are not a good risk for life insurance or investment.

Though they often woo and marry women they love deeply, they are prone to cheat, given the opportunity. Writers have almost a magical charm. Why women start a romance with a known bounder and expect to change him is beyond understanding. Tragic stories of these foolish adventures abound in literature and folklore.

Writers have all of the vices of other artists plus a few peculiar to the species. They are trained and practiced in the art of using words to paint pictures and scenarios of imaginary people and things. They are completely unbounded by facts and are not even bound by natural laws. Science fiction writers may adhere to most of the rules of science, but fantasy writers in particular are limited only by their considerable imaginations. When their work turns out badly, what do they do? They either discard it or rewrite it. There is a common belief and saying among writers that is touted in books and lectures, "rewriting is the key to success."

That would pose no problem if it were not carried over into the writer's life experiences. What happens in real life is that the writer jumps in and acts in a similar manner to his writing. It's, "If it doesn't work out right, rewrite it!" But life cannot be "rewritten." Things once said cannot be "unsaid." The vase once broken cannot be returned to its original condition. Real life happenings cannot be "rewritten."

The situations in his stories can become quite factual and the characters in his creations become as real people to the writer. In his mind he can change things at will, hate characters or love them. He can also use his talent as a wordsmith to manipulate their lives, kill them if he so chooses, with no real consequences. He can even resurrect the ones he killed if he wants. Writers can come to view real live people the same way. They can become extremely disturbed when people don't follow their "script." There are many other problems associated with this imposition of creative writing into real life. There is one horrible true story about a writer who became so enamored with one of his heroines that he committed suicide after writing of her sudden death. That's quite obviously carrying "method writing" a bit too far.

Just think about writers as normal people whose emotions are much stronger and more able to create actions than most and who can shift into emotional hyperdrive in an instant. The writer's emotional fish story would not turn the small fish into a bigger one, but into a virtual monster.

On the positive side, writers can be quite wonderful friends or companions. They are quite sensitive and try not to be hurtful to those they care for. Emotions directed to positive action can be wonderful resulting in deep, abiding love, and unbelievable acts of kindness. They can become quite altruistic.

They are also quite sensitive and can be hurt deeply and quickly by unintentional acts. Those without the negative traits can make affectionate and loyal friends, lovers, and husbands. Ah, but finding those with all positive traits borders on the impossible. So if you take one into your life as a friend or lover, be prepared to accept a few warts . . . maybe lots of warts.

—Howard Johnson, December, 2009

Perception, ah yes, perception, it is what drives our decisions, controls our emotions of love, anger, joy, disappointment, friendship, hatred, virtually everything we think or react to. Perception overrules facts, logic, and reality. Whether from love, avarice, or foolishness, and no matter how removed perception is from truth, it still rules us and determines our life decisions. We do not live in a real world, but live totally in a world created by and subject to our perceptions.

—Howard Johnson, 1960

When one person suffers from a delusion it is called insanity; when many people suffer from a delusion it is called religion.

—Robert Pirsig (1948-)

When one person suffers from a delusion it is called insanity; when many people suffer from a delusion it is called politics.

—Howard Johnson, 1967

I can write better than anybody who can write faster, and I can write faster than anybody who can write better.

—A. J. Liebling (1904-1963)

Happiness and moral duty are inseparably connected.

—George Washington

People demand freedom of speech to make up for the freedom of thought which they avoid.
—Soren Aabye Kierkegaard (1813-1855)

Give me chastity and continence, but not yet.

—*Saint Augustine (354-430)*

Not everything that can be counted counts, and not everything that counts can be counted.

—*Albert Einstein (1879-1955)*

Hold yourself responsible for a higher standard than anybody expects of you. Never excuse yourself. Never pity yourself. Be a hard taskmaster to yourself and be lenient with everybody else.

—*Henry Ward Beecher*

We are all atheists about most of the gods humanity has ever believed in. Some of us just go one god further.

—*Richard Dawkins (1941-)*

The artist is nothing without the gift, but the gift is nothing without work.

—*Emile Zola (1840-1902)*

This book fills a much-needed gap.

—*Moses Hadas (1900-1966) in a review*

The full use of your powers along lines of excellence.

—*definition of happiness by John F. Kennedy (1917-1963)*

Only two things are infinite, the universe and human stupidity, and I'm not sure about the former.

—*Albert Einstein (1879-1955)*

A lie gets halfway around the world before the truth has a chance to get its pants on.

—*Sir Winston Churchill (1874-1965)*

A slender acquaintance with the world must convince every man that actions, not words, are the true criterion of the attachment of friends.

—*George Washington*

I do not feel obliged to believe that the same God who has endowed us with sense, reason, and intellect has intended us to forgoe there use.

—*Galileo Galklei*

Relativity and a Completely New Concept of Our Universe
A new and different way of thinking about our universe

In 1905, Albert Einstein published his first dissertation on his new Theory of Relativity. This theory was hardly noticed when he first published it. Among its new concepts, it proposed the path of light is bent or curved by gravity, in particular the powerful force of gravity exerted by large objects like the sun, black holes, or groups of galaxies.

On July 6, 1918, the following comment by Sir Arthur Eddington was published in *Scientific American Supplement*, "The position we have now reached is known as the principle of relativity. In so far as it is a physical theory, it seems to be amply confirmed by numerous experiments (except in regard to gravitation)."

The theory put forth in the 1916 paper lacked experimental proof. Several astronomers, including Arthur Stanley Eddington, in charge of Cambridge Observatory, used a solar eclipse of May 29, 1919, as an opportunity to test one prediction: light rays from a star would be bent as they passed close by the gravitational field of the sun. When the prediction appeared to be proven accurate, Einstein was hailed by the science community and achieved almost an apotheosis in the public mind. The following is an excerpt from the *Scientific American Supplement* of December 6, 1919:

The results of the total solar eclipse of May 29 last were reported at a meeting of the Royal Astronomical Society, held on November 6. These results were most satisfactory. The star images are well defined, and the resulting shift at the limb is 1.98". That indicates a probable error of 0.12". This result agrees closely with Einstein's prediction of 1.75". It was acknowledged at the meeting that this agreement went far to establish his theory as an objective reality.

I am an amateur cosmologist and claim no fame or special expertise to back up my own theory, a concept developed over many years of thought on the subject. For this reason, I am presenting my theory as a thought-provoking variation on the accepted theory of how the universe operates. It is a combination and simplification of the many theories and ideas proposed over the years concerning the speed of light, the expansion of the universe (as conceived from information on red shifts of distant galaxies), the relation between gravity and the speed of light, and the perceptions we seem to take for granted as true and factual. It is important to recognize perception is reality to most people. The perception of the absolute linearity of light to the mammalian eye is the basis for our sense of space and time whether it is true or not. Humans and some primates are the only ones who can understand the spacial displacement of light passed through lenses and reflected by mirrors.

Add to this our own extremely complex movements. Consider we are flying on a jet between Chicago and Sau Paulo. (1) We are moving in an arc (roughly) near the surface of the earth. (2) We are moving in a circle around the earth's center of rotation. (3) This center of rotation is moving in an elliptical path around the sun. (4) The same center of rotation is moving in another elliptical path (roughly) determined by the rotation of the galaxy. (5) The center of rotation of the galaxy is moving in another path determined by the galaxy's gravitational attraction to the rest of the universe diminishing roughly as the inverse ~~cube~~ *square* of the distance between the center of mass of the galaxy and all of the other dispersed mass of the universe. We still haven't taken into account the gravitational effect of the moon, the Sun, the rest of the planets and asteroids, the Oort cloud, the rotating arms of the galaxy, or our position relative to the rest of the mass of the entire universe, but then who's counting?

I base the reasoning for my theory on the following:

The trajectory of light is bent (the speed changes) when light passes near a large object (the sun, a star, or a black hole). Perhaps it never goes in a straight line, but is constantly wandering, bent by gravity as it passes by or through all kinds of collections of matter.

On Earth, we are in a specific location in the universe. This means our movement, relative to the universe, is infinitesimal, even over long (to us) periods of time.

The universe in its entirety has an enormous mass and thus an extremely large gravitational force. There is no question but that this force affects the path and the speed of light within the universe. This is true even though the dispersion of known mass (mostly in galaxies) in the universe is lumpy and irregular.

We perceive light as traveling in a straight-line path no matter how circuitous is the actual path. Star photos taken during the May 29, 1919, eclipse proved this. The true position of those stars appearing close to the sun was different from the apparent, observed position during the eclipse. Gravitational lenses in space sometimes cause multiple images of the same objects to appear in several positions, all different from the true position. No matter what the true path light takes to reach a human observer is, the observer perceives the light to be coming directly to him in a straight line from its source. This would be true, even if the actual path were randomly distorted and irregular, spiral, helix, circle, or arc. The red-shifted galaxy we see so clearly could even be our own in the distant past where the light has circled back and intersected our current path.

Would it not be true that at a position at or near the outermost reaches of the universe, the pull or force of gravity of the entire mass of the universe would be entirely in one direction, toward the center of mass of the universe? Would this force of gravity *bend* light back into the universe, keeping it from escaping? This would, in effect, provide an event horizon for the universe, much like the event horizon of a black hole, where light cannot escape, but would be turned back into the universe. Were this the case, the visual effect on an observer within the universe would be the same as what we presently see. Our universe could even be considered similar to a ***black hole*** from which energy or matter can never escape.

If the above facts are true, all light or radiation we are aware of is traveling in a path not straight but curved because of the gravitational effect of mass—the *curvature* of the universe. At the precise point in space-time where we are now, the speed of light is relative to the masses of all objects in the universe. It is relative to the mass of planets, stars, galaxies, and the entire universe. The force of gravity (G) exerted by a mass at any given location is proportional to the mass of the object and inversely proportional to the ~~cube~~ *square* of the distance between the center of mass of the object and the location of the observer. The net G force on us is a function of our position in the universe. More correctly, the ***vector sum*** of the G forces of all objects with mass in the universe. If there is a way of calculating the relationship of the mass of the universe, our position relative to such mass and its dispersion, and our measurement of the observed speed of light, it is beyond my knowledge.

If this theory is true, the universe is a finite object with a specific mass/energy value, a specific size, and a specific set of physical laws. It may or may not be expanding as the *red shift* of distant galaxies could be an effect of changes related to the position of ourselves as observers relative to the overall forces of gravity of the universe. Such being the case, there could be many other universes of varying sizes distributed throughout space-time in a fashion similar to galaxies within our universe. Energy/mass from those universes could never reach our universe, so we would never know of their existence.

I see confirmation of this overall concept in the cosmic microwave background radiation initially discovered in 1964 by Arno Penzias and Robert Wilson. They had this serendipitous finding when they were doing radio astronomy with a giant horn developed by Bell Labs for experiments with communication satellites.

Penzias recalled, "No matter where we looked, day or night, winter or summer, this background of radiation appeared everywhere in the sky. It was not tied to our galaxy, or any other known source of radio waves."

This information was confirmed and advanced by NASA's Cosmic Background Explorer or COBE satellite in 1992. Regardless of its source, this radiation was coming at us from every part of the sky in nearly the same strength. If it was generated inside the universe at any time, it meant the radiation had turned around from its outward path and was now heading toward us. This is precisely what it would do if my theory were correct.

Two narrative descriptions—more personal reflections—of my realization or creation of this theory are included in this book. These two, titled *Dream Thoughts* and *Flaglar 1037*, represent unbounded thinking, perhaps unscientific or even whimsical. I leave it to those theoretical physicists or cosmologists who know more about these things than I to apply known mathematical and physical tests to prove or disprove my theory. That is in the event they would even consider it. This would, however, provide a framework for the detailing of the relationship between the force of gravity and the speed of light. It might even lead to a better understanding and explanation of such things as the apparent accelerating red shift of faraway galaxies, dark matter, dark energy, and other theoretical creations aimed at answering cosmological questions. It certainly meshes with most of current knowledge on the subject.

More information and an update for 2008:

I seriously wonder if the astronomers and cosmologists don't have their understanding of the expansion rate of the universe backward. If, as they say, the farther a galaxy is from us, the faster it is moving away, then those galaxies farthest away from us are actually accelerating. They decided this based on measurements of the red shift of light coming from distant galaxies. The results indicate these galaxies are moving faster away than geometric expansion would indicate. Therefore, they are accelerating away from us because, as they say, the expansion rate of the universe is **accelerating**. If this is so, our universe will expand faster and faster until it disperses entirely. When the recession rate reaches the speed of light, no light from any star or galaxy will reach any other star or galaxy.

But wait a moment. The light we are measuring from those galaxies indicates they were moving away faster a billion or more years ago, much farther back in time than physically closer galaxies. If this is so, then the rate of expansion of the universe may actually be **slowing** or **decelerating.** This would also indicate the expansion will eventually slow to a stop, and the universe will start **contracting**. As some theorized a while back, this would eventually lead to a *big crunch*, an occurrence wherein matter would become pure energy and then *bounce* back as another big bang. The universe would be reborn and start over again in much the same manner as the *bounce* of matter collapse of a giant star which results in a supernova explosion. I personally believe that to be a rational theory. Proper application of this theory could counter the necessity to construct forces like *dark matter* and *dark energy* that are now so popular with theoretical cosmologists.

Not long after I wrote these last paragraphs I received the October 2008 issue of *Scientific American*. Lo and behold the title of the cover story was, ***Forget the Big Bang: Now it's the Big Bounce.*** It went on to describe such a scenario as I described in the earlier paragraphs and in my stories and essays written years ago. The quote, "Quantum gravity theory predicts the universe will never die," confirms my past reasoning. It then goes on to explain this new theory, similar to my own. My theory was written down first in 1992 and then expanded in 2001. It is gratifying to hear my theory at least partially confirmed, and my doubted ideas exonerated. Still, no scientist would even consider these theories coming from an unlettered amateur.

Who knows? It could become valid one day. Stranger things have happened. Oh well, at least it gave me a good feeling.

Sailing with the Lake Tippecanoe Sailing Club

It's January 2003, and I am sitting in our dining room. A sun-brightened panorama of snow-covered ice on Lake Tippecanoe fills my view beyond the bird feeder at the window. I watch as juncos, nuthatches, house finches, sparrows, chickadees, titmice, cardinals, doves, and gold finches in winter garb flit between nearby trees and the feeders. The scene is a far cry from summer sailing, but memories of many sailing experiences on Tippy fly through my mind. When Jim Hayes first asked me to write about the sailing club, I started with a descriptive account of the club's history. Then I decided to take another, quite different tack. I would tell a story about sailing on Tippy and about the sailing club from a sailor's point of view.

My first sailing experience on Tippy was probably in 1935 when I had my first ride in a sailboat. It belonged to the family of my friend Paul Harruff on Walker's Park. Sailing all the way down past Silver Point and back was an exciting and amazing experience for a seven-year-old. The boat was immense to a small boy. It was moored to a buoy out away from shore in front of the Harruff's cottage. During the off-season, they hauled it out of the water on a large dolly running on steel rails. The rails led from the storage driveway, down to the beach and far out into the lake to where the water was at least five feet deep. I can't remember much about the boat which was probably sold or otherwise disappeared the year after my one ride. While the steel rails on the beach and yard disappeared many years ago, those submerged remained. I remember seeing them on the bottom many times while passing over in a boat. For all I know, they may still be there, lost and long forgotten.

My first actual sailing experience was in 1945 and 1946 in a Snipe owned by my friend Ed Emrick. We would sail up and down the lake picking up girls who wanted a sailboat ride. I learned how to sail all on my own because Ed preferred to sit on the deck and talk to the girls. Ah, but that's another story. My first sailboat was a molded plywood tub of a boat made by Dunphy. It had oarlocks and seats and converted into a rowboat by removal of the mast, sail, and tiller. The transom was even designed for an outboard. In it, I took my son and his entire Cub Scout pack sailing. This boat may have been slow and unwieldy, but it held many kids.

About this time, I learned about the sailboat races on Tippy. Since the Dunphy would not be suitable for racing, I acquired a used red-and-white cat-rigged boat. Soon I was racing with other small boats as a member of the sailing club. I also discovered my new boat was far too slow and unwieldy to compete. The next year, I traded it for a brand-new, bright yellow, faster and more maneuverable Barracuda. The Barracuda was a cat-rigged boat with a tall mast. I watched with undeniable envy as Harry Bishop, John Bundy, Herb Gawthrop, Jack Thompson, Jim Murray, George Buckingham, Howard Webb, Ken Zinzer, and others competed in their Lightnings—the boat to race on Tippy. It was thrilling to see a dozen or more Lightnings racing downwind, their colorful spinnakers ballooning in the following breeze. Alas, my new Barracuda was little more competitive than my previous sailboat. My enthusiasm was not dampened by my poor performance. I was definitely hooked on sailboat racing.

Determined to be a winner, I studied a book on racing and looked for a new boat that would be competitive. I found what I had been looking for, the Flying Fish, built by the maker of the Sunfish. With a planing hull, a single sail, and room for two in the cockpit, it looked like the winner it was to be. I raced the Flying Fish for several seasons, learning the thrill of my first trophy in 1969. During our Labor Day regatta, I took my oldest daughter, Deb, with me as my crew. I soon learned the extra weight was holding us back. I couldn't catch Mick Case in his Glastron. Finally, Deb abandoned ship to lighten it. An excellent swimmer and with her life jacket on, she would be safe until one of the pickup boats got to her. Now I was gaining on Mick. I knew I could catch him before the finish line, but catching Mick and passing him are two different things. One rule of sailing is that when attempting to pass, a following boat must maneuver around a leading boat. The leading boat has the right to turn up or downwind to block the boat trying to pass. The following boat must draw even with the leader to where both masts align. This is a point called *mast abeam*. The skipper of the passing boat hails the other, calling out, *mast abeam* at which point the boat being passed can no longer block. I learned several new rules of racing that day. Mick kept me at bay and won the race. I was disqualified for two infractions.

Under the able guidance of Jim Murray, the Lake Tippy Sailing Club grew with two general classes of boats racing together against boats in the same class. The *open* class included larger boats like Lightnings, Y-Flyers, E-scows, and others. Most of these boats carried both a *jib* (front sail) and a *main*, a configuration called a *sloop* rig. The other class was the *small boat* class of one-man boats with a single sail. Porpoises, Super Porpoises, Sunfish, and several others used a *lateen rig* with a single triangular sail spread between a boom at the bottom and a top spar attached at the front, all hoisted on a relatively short mast. My Flying Fish was a *cat rig* with a single sail hoisted to the top of a tall mast and with a boom at the bottom. For years, the Lightning was by far the most popular large boat while the Super Porpoise was the most popular small boat on the lake and in the races.

Small Boat Regatta on Labor Day 1973

Lightning Regatta on Labor Day 1973

After several seasons with the Flying Fish, I noticed a new and completely different sailboat on the lake. This fast-moving Hobie 16 catamaran (two slender hulls with a trampoline to hold the crew) both fascinated and excited me. One windy day in July in 1971 Bruce Pierce took me out on the Tippecanoe Boat Company's Hobie. One thrilling ride and I was definitely hooked on the new boat. During one outing, we *pitch poled* twice, once throwing me through the air and far up the mast as it slammed into the water. Pitch poling is an unusual and exhilarating experience, unlike any other way of *flipping* or turning over a sailboat. Although it can happen to other sailboats, fast-moving catamarans are most likely to do so. It usually occurs during *heavy air* (relatively high winds, fifteen knots or more). The downwind hull gets buried under the water to the point where it catches in the water and virtually trips the boat, sending the prows deep into the water and the stern high into the air. The skipper and crew, usually as far back on the stern as possible, are literally catapulted through the air as the mast slams down into the water. It's a thrilling and thoroughly startling experience as it can happen so quickly.

Sometime later, I was thrilled when he asked me to sail it whenever I liked. He wanted the new boat out on the lake, so people would see it in action and thus promote sales. During the same summer, Bruce sailed the boat in a few races. Then he asked me if I would race it for him as he was far too busy to with the marina to spend Sundays racing. This started my love affair with Hobie racing lasting more than forty active years and continuing to this day. The following season, I purchased the blue and white Hobie 16 number 1128 I had been racing for the Tippy Boat Company. I raced my Hobie for 24 summers until 1995 when a sudden windstorm destroyed it. After winning frequently in my Flying Fish, I began experiencing frequent last places in the Hobie. A different handling sailboat, the Hobie proved difficult to master for a single hull sailor. With no other Hobie racers on the lake, I had to learn by trial and error.

Pointing or sailing close to the wind is a technique of sailing as far into the wind as possible. The key is finding the balance of direction where the wind fills the tightly drawn in sails and the boat moves fairly quickly. *Point* too close into the wind and the sails will luff (flutter in the wind), and the boat will lose forward speed. Then it will stop moving as it faces directly into the wind. Sailors colorfully refer to this cessation of forward motion as *being in irons*, an ancient sailing term. Far too often, those first summers, I found myself in irons, frustrated, and watching other boats glide swiftly and cleanly past. I sometimes even backed up.

During my second summer of Hobie racing, I learned to release the jib while *coming about* (changing directions to *tack* or zigzag upwind) caused the boat to stop turning and face directly into the wind and to be dead in the water or *in irons*. By holding the jib in place until the wind caught it and pulled the boat to the opposite tack, I found the solution to the problem. Forward motion was not lost, and I saved a lots of time by using this maneuver. Within a few years, there were more Hobies on the lake than any other sailboat, maybe more than all other sailboats combined. Mick *Killer* Case started racing a Hobie. John Emrick appeared in his purple Hobie and immediately became the leading winner. About this time, the club began to use the international Portsmouth handicap system. In this system, the race committee adjusts the time of each type of boat according to its performance ability. This leveled the playing field somewhat allowing all types of sailboats to compete with a chance to win, even beating much-faster boats. This system favors the best sailors rather than the fastest boats. As the Hobies became more popular, there were enough to have a third and separate class for Hobie 16s alone. This was in addition to the large and small boat classes. Besides the Hobie 16 sailors already mentioned, there were many others over the years. These competitors include Ken and Carolyn Davidson, Jim and Verlyn Hearn, Jim and Susan Hayes, Bill Jared, George Buckingham, Daren Baier, J. B. Van Meter, and many others no longer active or in the club. For several years, we had a small fleet of Hobie 14s raced by Ron Brown, Phil Jung, Bill Allen, and several others.

Rod Keesling in the Flying Fish, Bill Christine in his Snipe, and HoJo in his Hobie 16 rounding a mark.

While competing, I learned the sailing techniques peculiar to racing on Tippy with its high shores and jutting points. In almost any wind, there are blind spots where those high shores change the wind direction or block it completely. Silver Point, a high bluff jutting out into the lake, creates wind shadows and eddies that must be dealt with, often as far from shore as the middle of the lake. Strong winds from the southwest create rotating horizontal cells that roll well out into the lake. One picture in particular demonstrates the unbelievable effects of one of these invisible cells. In it, two Lightings—one skippered by Harry Bishop, one by Jack Thompson—are running dead parallel. Each is on a different tack, heeled over in opposite directions away from each other and headed directly toward the photographer.

With more racing, I learned to *read* the surface of the water, which reveals wind direction and speed to those who learn the technique. This understanding provides a considerable advantage as changes in wind direction and force can often be anticipated. The sails can then be trimmed before the change reaches the sailor. This is particularly important when the wind is light and variable, a common condition on Tippy. Learning and practicing these techniques often makes the difference between first and last in a race.

Over the years, we tried several new types of races. Most lasted but once. We had watermelon races, moonlight races, and several others. One race, the Commodore's Cup, suggested by Al and Norma Hayes in 1986, has become a regular event in mid-July. This race is one heat, the full length of the lake from in front of the dance hall to the flats at the east end and back. I had the pleasure and thrill of winning the first two Commodore's Cup races in my Hobie 16. This occurred during my most successful seasons of racing.

One of the values of being a member of the sailing club was the rehash of the race day with other sailors. We learned so much from each other and especially from those competitors with years of experience. A truly friendly group in spite of being highly competitive, each willingly shared secrets of racing techniques with other members. I learned most of what I know about racing during those meetings or on the water, watching other competitors. I formed numerous friendships among Tippy sailing club members I treasure to this day.

Each season, we held an opening get-together at a member's place, usually in June. For years we had another party during the Fourth of July races. We don't hold the races and the party anymore because of competing activities. Immediately after the Labor Day regatta, we hold our annual meeting, awards dinner and election of officers.

Today, we have two classes racing: monohulls and multihulls or catamarans. Each class uses the Portsmouth handicap system, so all boats compete on an equalized basis. Even inexperienced skippers soon become competitive with the help of our seasoned members. If you have an interest in sailing and good fellowship, come join us. Contact a member or show up at the committee boat—the pontoon flying a colorful windsock at the starting line of the race most Sunday afternoons.

For more information about the sailing club including a schedule of the current season, look up our Web site at clubs.kconline.com/clubs/LTSC.

—Howard "HoJo" Johnson, 1998

About the Author (updated in January 2011)

Howard Johnson's first summer on Tippy was 1929 when he was one. His family rented a cottage on Walker's Park. After several summers renting, the Johnson's purchased a lot on Walkers Park and built a cottage there. In 1958, Howard bought the cottage from his dad, also a Howard Johnson, when the senior Johnson built a year-round home on Willow Bend, around the point from the cottage. Howard moved into his mother's house in 1982, several years after his father passed away, to take care of his mother during the summers. In 1985, he became a full-time resident in the house his parents built. In his life, he missed spending summers at Tippy only in 1928 when he was a baby and then in 1952 and 1953 when he lived in California. He now spends winters in St. Augustine, Florida, where he is active in several writers' groups and sings in a local chorus, Singers by the Sea, in Jacksonville Beach.

A Purdue engineer, businessman, and most recently a writer, Howard published his first novel, Blue Shift, in 2002. His novel received a rave review in the Times Union newspaper. Currently, he has several other books in the publishing process, including *Energy, convenient Solutions,* scheduled for release in the summer of 2010. This is a nonfiction book. It describes the many possible and practical ways to solve our energy crisis. Another book, *Words from the Lakeside,* is scheduled for publishing late in 2010, or early in 2011. An anthology of primarily his own writing, it also contains some quotes from others. The author describes it as a several decades long *labor of love.* It contains sayings, essays, poems, letters, and short stories, many from his life. He has several other novels in progress.

He also gives talks on several subjects about science, religion, and the environment. One presentation, *Science and Religion: A Reconciliation,* has been constantly updated since its first presentation in 1975. He gives this talk to churches and other groups. Two popular new lectures are about energy and global warming. The one about energy is based on his book, *Energy, Convenient Solutions.* The one about global warming deals with the realities of global warming theories and the current political and media frenzy on the subject titled, *Global Warming: Fact, Fiction, or Somewhere in Between?* This talk is described as "a realistic and objective look at the human effects on climate."

Tragically, he lost his wife, Barbara, in 2005 after she fought a losing battle with Post Polio Syndrome. Barbara was a Methodist pastor serving Morris Chapel UMC in Pierceton from 1995 through 2000 when she had to step down because of her illness. He says of Barbara, "She was not only my wife, but my best friend, my proofreader, my editor, and a major contributor and supporter of my writing. Incidentally, she was absolutely brutal with a red pen." There is more about Barbara in several memoirs near the end of this book.

It makes no sense, but to the easily led masses, ***Getting even*** seems preferable to the status quo even when those who do so know they are certain to be rewarded with severe loss, pain, or even death. They become one of Eric Hoffer's *True Believers.*

—*from, Energy, Convenient Solutions II by Howard Johnson, 2012*

Time goes, you say? ... ... Ah no! Alas, Time stays, we go.

— *Henry Austin Dobson*

The judgements of men are formed not from facts as they are, but as they wish them to be. They root through tons of good wheat to find three pieces of chaff if the chaff lends weight to their beliefs and argument. It is not that they want others to know the truth, but to have those others believe as they do. Beyond this, they do not care. The conceit of man ordinarily forms his criterion of truth.

—Unknown

> He who knows not and knows not that he knows not is a fool.
>> Shun him!
>
> He who knows not and knows that he knows not is a child.
>> Teach him!
>
> He who knows, and knows not that he knows is blind.
>> Lead him!
>
> He who knows and knows that he knows is wise.
>> Follow him!

—Many versions and sources, Persian saying, Sanscrit, Confucius

Power tends to corrupt, and absolute power corrupts absolutely. Great men are almost always bad men.

—John Emerich Edward Dalberg Acton

Any form of government is good that actually governs and not offers opportunity to rogues to buy and sell preferment. Let a ruler rule, and let the ruled obey. But woe betide a ruler who is faithless to the lonely task of ruling firmly, justly, decorously, wisely, and to sum the terrifying total . . . well.

I trust or mistrust, having found no middle course worth following. But the charlatan zone between these courses is a wilderness wherein another's treachery by no means can be held to justify my own bad faith. A man must stand or fall, judge and be judged, by his own faith, always.

To an honest man, though I may veil or dissemble my thoughts, I will never leave in doubt the main question: am I for him or against him? Honesty deserves honesty. But I have yet to be persuaded that a lying scoundrel has a claim on me, that I should feel in duty, bound to guide his guessing.

—Lord Tyros of Samothrace in **The Purple Pirate** *by Talbot Munday*

The Political Challenges Now Facing the People of America

Quoted directly from, Energy, Convenient Solutions II, by Howard Johnson

John Stossel, consumer advocate and 20/20 reporter, explains much of what we hear—and what the media say—are myths. In his book, *Myths, Lies, and Downright Stupidity subtitled, Get Out the Shovel Why Everything You Know Is Wrong*, Stossel points out how politicians and activists use anecdotal evidence to prove the truth of concepts factually untrue and often downright harmful. This is one of the nearly two-hundred common myths debunked in his book. It is in the chapter about business.

MYTH: Government must make rules to protect us from business.

TRUTH: Competition protects us if government gets out of the way.

It took me a long time to learn that regulations can't protect consumers better than open competition. After all, I worked in newsrooms where *consumer victimization* was a religion and government its messiah. But after fifteen years of watching government regulators make problems worse, I came to understand that we didn't need a battalion of bureaucrats and parasitic lawyers policing business. The competition in the market does that by itself. Word gets out. Angry customers complain to their family and friends; consumer reporters like me blow the whistle on inferior products and shoddy services. Companies with bad reputations lose customers. In a free society, cheaters don't thrive. (At least in business.)

Once I learned more about economics, I saw how foolish I had been. Government uses force to achieve its ends. If you choose not to do what government dictates, men with guns can put you in jail. And clever lawyers will then remove gobs of money from you to get you out. Business, by contrast, cannot use force, no matter how big or powerful they are. So all business transactions are voluntary—no trade is made unless both parties think they benefit. In 1776, economist Adam Smith brilliantly realized that the businessman's self-centered motivation gets strangers to cooperate in producing a multitude of good things: "He intends only his own gain, and he is in this, as in many other cases, led by an invisible hand to promote an end which was no part of his intention".

Few of us appreciate the power of that invisible hand. I don't give my pencil a second thought, and yet I could spend years trying to produce one without turning out anything as good as the worst pencil available."

He goes on quoting an essay, *I, Pencil*, by Leonard Read of the Foundation for Economic Education. The essay describes the people equipment and organizations actively involved in gathering the components of the pencil from all over the world. This includes many items: the machines and workers who combine those components, the machines and workers who make those machines, the trucks and ships that carry the raw materiels and finished product around the globe, and the systems that distribute the pencils to the end users. It's an interesting and enlightening bit of prose. I highly recommend reading Stossel's book. It will provide an instructive background for how to accomplish much of what the information contained in these pages describes.

Sadly, politicians will have a great deal to say about what we do and how we address the growing problems involving energy. These problems include many that have major economic effects now and will in the future determine the health of our economy. These politicians have little understanding of energy problems and solutions, or the invention, development and manufacture of fuels and energy systems. As a result, they are poor judges of what might and might not work. The reason for the political emphasis in this section is to make clear the problems facing any new technology and business entity that must run the political gauntlet.

Politics is a totally emotional game with virtually no rational component. Many politicians are without any practical ideas or effort not supportive of their ideology. They are so totally engrossed in ideology they accept nothing outside of the purpose of their closed minds. They are far more likely to propose and vote for those things that promote their agenda or benefit their backers and constituents rather than practical or creative proposals that address the problems. Sadly, all but a few politicians deplore solutions to problems. They would much rather use problems to further their agenda. The last thing these politicians want is a solution to the problem since a solution removes the usefulness of the problem. Is it any wonder they will argue with, fight with, and denounce any solution proposed by any opponent, regardless of the efficacy of the solution proposed. Entrepreneurs and their investors will be the ones who will solve these problems if government and politicians will get out of the way and stay out..

Unfortunately, politics and politicians inject themselves into virtually every problem with the support of their party members and, if you are on the left, the constant adoration of the media including the entire entertainment world. These politicians will have a great deal to say about where we go with energy, which systems we *approve* and fund, and which we discard. This is unfortunate because many important decisions will be made by these people who haven't a clue as to the real actions, values, or cautions any energy system requires. Their decisions will almost certainly be decided by those who provide them—the *governmentalists* (my apolitical term with obvious meaning)—with the most power and/or money in response to their almost pure emotional appeal. Politics has so invaded many of our institutions (education, unions, entertainment, the media, and many professions) and is so controlling, little can be accomplished without the money to buy political clout.

No one can accuse any politician of providing an honest, forthright, nonemotional proposal on anything. In fact, even to question some of their pronouncements—to ask for answers to rational questions and concerns—is to invite public ridicule at the hands of politicians and celebrities including those in the mainstream media. The public's growing worship and adoration of celebrity dooms us to succumb to the charms of charlatans who promise the moon while lining their pockets and empowering themselves. The ancient worship of *royalty* is not dead. It is merely substituting a new set of *royals* so enamored of media celebrity worshipers.

It is my considered opinion and experience most Americans can rise above these unreasoning passions if given the opportunity to open their minds, to think and reason. I am a sincere believer in the basic goodness of most people, Americans in particular. I have seen a great deal of evidence in the outpouring of compassion accompanied by free giving of physical aid during times of catastrophe virtually anywhere in the world. This has often been twisted by those who envy and thus despise America. Recently, after a serious natural disaster, one news service reported the US government had given far less in disaster relief funds than the governments of a number of smaller nations. The truth (never mentioned) was that the private, non government gifts from Americans dwarfed the private giving of the entire rest of the world. When added to the government's gifts, the total was more than the next three or four nations total giving combined. That's one example of how a hostile news service can twist the news and skirt the truth to serve their agenda.

Free, independent Americans are the ones who will find interest in this book. They are the ones who will understand and appreciate it. These are the open-minded thinkers and doers, workers and organizers, creators and builders, who made this country the greatest, most free and independent nation on Earth by their energy and hard work. I urge those who belong to this shrinking majority of real Americans to read with an open and objective mind. I'm sure many will find statements that run counter to long held personal beliefs, no matter where their political loyalties lie. When this happens, please read the entire section as objectively as possible and consider the logic of the arguments presented. The understanding gained will be surprising.

Many new and revolutionary materials, systems, and combinations of these are now available. Application of these innovations could help solve our energy crisis in a few years, a decade at most. We must investigate these and bring the best through development and into production and use. This will only happen to those that gain attention and favor in the eyes of the influential and then the public. No matter how excellent, the best ideas and proposals will fall by the wayside if they don't gain this favor Unfortunately, politicians, the media and entertainment people, who are mostly ignorant of the complexities and nuances of energy problems and systems, could have a great deal to do with the choices made that will impact our immediate future considerably and our long range future drastically. These ideologues will simply not support proposals that do not further their agenda no matter how practical or advantageous they are. Entrepreneurs will have an important part, but only if their efforts earn rewards and they are not shackled by emotionally driven government regulations. The next section, [starting on page 137 in the original *Energy, Convenient Solutions* book], deals with these realities in our nation as I see them.

It is my fondest hope and wish that, within our nation, we can find cooperation and respect between groups that now see each other as enemies, or sources of political power. At their best, these are fellow Americans trying to do something positive by building for a mutually advantageous future. The groups I speak of include the following: academia and their excellent researchers, private enterprises from individuals to the largest corporations, governments and government agencies, local and national, and all of the entertainment world including the media, Hollywood, New York, and the world of sports. I realize this is a big order, particularly since the political wars have ratcheted up to fever pitch and emotions run high. The human energy and wealth consumed by these growing conflicts incited by angry rhetoric, are enormous and terribly wasteful. Emotions and resentment for past real or imagined injustices are powerful and deeply held. Nevertheless, peaceful cooperation and a little understanding and tolerance along with some give and take can work wonders.

Doing things exactly the wrong way: The growing rancor of political campaigning is one example of the power of inciting hatred to sway voters. The preponderance of personal attacks over substantive proposals shows that it is far easier to lash out at a political opponent than to build one's own stature and make serious proposals. Negative campaigning is easy and especially effective in the age of the *sound bite*. Like war, it is easy. All you have to do in war is break things and kill people. Any idiot can do that with little training. In political campaigns, it is pure emotion that drives voters. Serious proposals, even solutions for all-important problems, rarely get the media play and public attention that hate-filled rhetoric aimed against those same proposals receives.

For similar reasons, it takes far less skill or organization to demolish a home or even the World Trade Center than it does to conceive, design, and build the same thing. Cooperation and creative building are far more demanding, require careful consideration, intelligence, dedication, creative effort and hard work. They are infinitely rewarding and promise a bright and successful future for all, not just the few. This is what we sorely need right now. Unfortunately, conflict is easy. Hate rhetoric and negative campaigning have become so pervasive because their use sways the unthinking masses, mostly the uneducated and poor, those easiest to influence. This is particularly true of the use of class and race hatred. Look at the poor economic results of anger and hatred wherever it surfaces all over the world. Hate is the tool of choice for despots because it is easy. Also, there is a strong sense of human pleasure at seeing those we deem better off than ourselves, damaged, destroyed, or made miserable. It makes no sense, but to the easily led masses, ***Getting even*** seems preferable to the status quo even when those who do so know they are certain to be rewarded with severe loss, pain, or even death. They become *True Believers* as so well described by Eric Hoffer.

Often attributed to Lincoln in error are these words penned by William J. H. Boetcker, in 1916.

You cannot strengthen the weak by weakening the strong.

You cannot help small men by tearing down big men.

You cannot help the poor by destroying the rich.

You cannot lift the wage earner by pulling down the wage payer.

You cannot keep out of trouble by spending more than your income.

You cannot further the brotherhood of man by inciting class hatreds.

You cannot establish security on borrowed money.

You cannot build character and courage by taking away a man's initiative and independence.

You cannot help men permanently by doing for them what they could and should do for themselves.

Yet are those not precisely the short lived, instant gratifications politicians and media personalities regularly wield against those they oppose for any reason?

A prediction of where we seem to be headed may have come from far back in history, when the 13 colonies were still part of England. The following quote is often attributed to a Scottish Historian, Alexander Tytler or Tyler. The true origin of the quote is obscure and might have originated in the early 20th century from an unknown politician or writer. Nevertheless, this does not detract from its accuracy.

One version of this quote on why democracies always fail is:

A Democracy cannot exist as a permanent form of government. It can only last until the citizens discover they can vote themselves largesse out of the public treasury. After that, the majority always votes for the candidate promising the most benefits from the public treasury with the result that the Democracy always collapses over a loose fiscal policy, to be followed by a dictatorship, and then a monarchy.

A version of the second part of the misquote, often attributed to Arnold Toynbee is:

The release of initiative and enterprise made possible by self-government ultimately generates disintegrating forces from within. Again and again, after freedom brings opportunity and some degree of plenty, the competent become selfish, luxury-loving and complacent; the incompetent and unfortunate grow envious and covetous, and all three groups turn aside from the hard road of freedom to worship the golden calf of economic security. The historical cycle seems to be:

from bondage to spiritual faith; from spiritual faith to courage; from courage to liberty; from liberty to abundance; from abundance to selfishness; from selfishness to apathy; from apathy to dependency, and from dependency back to bondage once more.

But the person who appears to be the actual author of the second part is Henning Webb Prentis, Jr., President of the Armstrong Cork Company. In a speech entitled *Industrial Management in a Republic*, delivered in the grand ballroom of the Waldorf Astoria at New York during the 250th meeting of the National Conference Board on March 18, 1943, and recorded on page 22 of *Industrial Management in a Republic*, Prentis had this to say:

Paradoxically enough, the release of initiative and enterprise made possible by popular self-government ultimately generates disintegrating forces from within. Again and again after freedom has brought opportunity and some degree of plenty, the competent become selfish, luxury-loving and complacent, the incompetent and the unfortunate grow envious and covetous, and all three groups turn aside from the hard road of freedom to worship the Golden Calf of economic security. The historical cycle seems to be: From bondage to spiritual faith; from spiritual faith to courage; from courage to liberty; from liberty to abundance; from abundance to selfishness; from selfishness to apathy; from apathy to dependency; and from dependency back to bondage once more.

At the stage between apathy and dependency, men always turn in fear to economic and political panaceas. New conditions, it is claimed, require new remedies. Under such circumstances, the competent citizen is certainly not a fool if he insists upon using the compass of history when forced to sail uncharted seas. Usually, so-called new remedies are not new at all. Compulsory planned economy, for example, was tried by the Chinese some three millenniums ago, and by the Romans in the early centuries of the Christian era. It was applied in Germany, Italy and Russia long before the present war broke out. Yet it is being seriously advocated today as a solution of our economic problems in the United States. Its proponents confidently assert that government can successfully plan and control all major business activity in the nation, and still not interfere with our political freedom and our hard-won civil and religious liberties. The lessons of history all point in exactly the reverse direction.

These are the real malignancies we must overcome if we are to solve the rapidly growing problems facing not just the US, but the entire world.

It's about time members of groups that constantly denigrate and condemn others out of class or economic envy began to honor and respect the achievements and rewards of those others. I see it as vital we recognize the realities of our situation and the real reasons we are where we are. There has been enough of this debilitating blame game and all of its political distortions and emotional, hate-filled activities. We are engaging in terrible inner political warfare while our enemies stand on the sidelines urging on the various sides and gleefully watching our self destruction.

This is the background of where we are, why we are there, and what we must do to go forward. It has everything to do with what we must overcome in order to counter this growing menace before it destroys us.

Vote early and vote often.

—*Al Capone (1899-1947)*

There are some who are too proud to yield until compelled by force. They are not to be blamed. It is their privilege; I also, if I think my cause is just, maintain it to the last breath. But let them not blame me when I accept the challenge. I will yield anything for friendship's sake, except a principle that I believe is right.

As to whether there are gods or not, I am ignorant. I have never set eyes on a god or seen or heard anything, anywhere, that seems to me to justify the belief in gods or to suggest that, if gods there be, their doings justify respect. But I have been observant all my days. Whoever believes there is no such force as destiny directing us and our occasions would waste breath seeking to unconvince me. I have been in the grip of destiny, have seen its shape, have felt the weight of its hand. I know.

—*Lord Tros of Samothrace in* **The Purple Pirate** *by Talbot Munday*

If thou of fortune be bereft,

And in thy store there be but left,

Two loaves, sell one, and with the dole,

Buy Hyacinths to feed thy soul

—*Muslih-uddin Sadi*

When plunder becomes a way of life for a group of men living together in society, they create for themselves in the course of time a legal system that authorizes it and a moral code that justifies it.

—*Frederic Bastiat*

Enigma

We place the pieces in the puzzle randomly,

Fitting each together with the one before it.

One doesn't fit. It is taken out,

Turned around. Replaced,

Only to find that it doesn't fit again.

Can the pieces be altered?

Or the puzzle changed?

Or is the only solution in

Putting the pieces into a different maze?

—Deb Archer to her father, Howard Johnson, 1972

Epilog to Enigma

The puzzle is nearly complete. The picture almost whole.

 Only a few random spaces remain.

Too many pieces are left over and none of them fit

 And we keep finding more pieces

And more pieces and still more pieces!

 Another puzzle? Another picture?

     More pieces, more puzzles, more pictures!

The puzzles that were wholes

 Become pieces, small random pieces

     That seem to fit still greater puzzles.

We find more puzzles that are pieces

 And few fit . . . and the enigma starts over . . .

     Full cycle . . . at another level . . . ?

—Reply to "Enigma" sent to Deb Archer by her father, Howard Johnson

Reflections On and About Columbus Day

Ah, Columbus Day—when Europeans celebrate their discovery of this land. To some of my distant relatives, this is a day of sadness and mourning. A day recognizing a painful turning point in the land where America's original inhabitants lived for hundreds of centuries before the invasion by Europeans. For me, this is a day of mixed emotions. I thank God my Native American blood made peace with my European American blood a long time ago. However, I still get upset when reminded how my European forebears treated my Native American ancestors, particularly the repeatedly violated treaties, but all that is behind us. It's history, and we move forward.

It is time for the forgiving of old injustices and evil acts by long-forgotten ancestors. I firmly object to the recurring demand for reparations for the descendants of slaves. If anyone deserves reparations, it is certainly Native Americans more than and before any other group. I am not in favor of those kinds of reparations either. The time of hatred and recriminations is long gone. It is a time for healing.

Yes, Columbus Day reminds every Native American of repeated raw deals at the hands of Europeans for most of the last five hundred years. Yes, there are frequent grumblings and protest marches on Columbus Day, but they receive little press coverage. In spite of some places where discrimination still exists, most Native Americans now walk proudly as fully accepted members of American society and are proud to be called Americans. They are increasingly returning to their tribal roots for reawakening of their cultural heritage and language. Even more important, the public are now honoring them and their culture as they are increasing numbers of other ethnic, racial, and religious groups. This is a kindly human movement for the most part with expressions in dance, literature, festivals, foods, and song. America's multinational and multi cultural people are more and more honoring their differences, respecting their variety, and enjoying their fellow human beings of all kinds.

Sadly, there are those who never forget and still hold hate in their hearts. Not only do they hold on to hate, but they pass it onto their children and others they can influence. Violent grudges held for centuries and passed down through generations are what make the hells in places like The Middle East, Bosnia, Rwanda, Israel, Afghanistan, and sometimes even New York, Washington, and Pennsylvania. Racial, ethnic, religious, cultural, social, language, political, income, and many other differences can be used to foment hatred, fear, and mob action. This is particularly true among the young or uneducated. Use of these protracted hatreds and twisted religious and political beliefs by unscrupulous leaders to enslave followers is everywhere, including the United States. From small-scale operations like Jim Jones and his cult to David Koresh and the Branch Davidians to some militant blacks and their followers to bin Laden, Al-Qaeda, and all the angry fundamentalists Muslim groups, they have a similar pattern. Adolph Hitler and his Nazis are one monstrous example from the past. Only the scales of death and destruction are different.

The personal question becomes obvious. How do you fight against hatred without using hatred and becoming the very thing you despise? That is a knotty question with few ready or obvious answers. The only real hope is to rally the entire world to help eradicate all manner of hate mongering and the terrorism it fosters wherever it exists. Many members of the media and in political life would do well to curb their own hate speech. They by themselves are not so apt to do direct damage, but there are many among us who become so inflamed by such talk they take actions that are terrorism on whatever scale. The recent, numerous incidents involving Middle Eastern–looking people are examples.

Currently, we are reeling from the results of a major diabolical attack by a group who have been indoctrinated since youth with an unreasoning hatred of our way of life. Make no mistake, groups of evil men are using this calculated and pernicious hatred to gather support for the complete destruction of individual freedoms of all people. They want to impose a false and evilly convoluted version of Islamic law on the entire world by any means possible. Among other things, this law places women in a state worse than slavery. Under their law, these men have the power of life and death over women without question. Women must be completely hidden by clothing and veils when outside. In extreme cases, homes with women must have their windows painted over with black paint or other nonremovable opaque covering so there is no chance of seeing them from outside.

In Afghanistan, the Taliban committed horrible atrocities against women in the name of Islam. Among the worst examples, a group of men dragged a woman out of her car and stoned her to death because she accidentally exposed her arm as she drove. The sports stadium in Kabul was used solely for public executions, usually of women. Many women are killed by their husbands or a relative by having their throat slit with the popular jambiya: a short curved dagger carried by many men.

The following is a quote from a man who spent much of his life in the Middle East and Central Asia. The italics are my words. "These inhuman monsters *by our view* are from a culture that places little value on human life. In the middle east, if you show concern for human life, they conclude that you're a patsy and act accordingly. One example: the Iran hostage affair. They understand murder as we understand humanitarian acts. I don't mean for us to actually commit the murder, but it does deter those people who think you will *commit murder on them*. Then you don't have to do it. Murder is reliable, feasible, and affordable, so the preparations *for war we are now making* send the kind of message that those people understand."

This culture is an extreme example of the cruel subjugation of women by men. It is so radically different from western culture as to be beyond our understanding. It is very much like the cruel male treatment of females found in many troops of baboons. In fact, it probably has the same origins of males so insecure and unable to compete with other males they must take their anger and frustration out on the much weaker females. They do this both individually and in groups because of their inadequacy. It's an example of macho male activity in the extreme.

Recently, my wife, Barbara, and I watched in horror at the TV views of Afghan men beating and executing women in a sports stadium. These so-called Islamic fundamentalists are no more followers of Islam than the average barnyard pig. They are inhuman, satanic monsters—cruel, extreme misogynists who use women as objects for their frustrated hatred and anger. They have been indoctrinated since youth in a satanically twisted version of Islam by teachers whose convoluted, pent-up hate is an expression of their own inadequacy and weakness. The actions of these subhuman creatures display mob mentality of the worst kind. They are in stark contrast to the Muslims who brought forth the light of education, mathematics, astronomy, architecture, and art during the depths of the European Dark Ages. Those great men of knowledge would probably be stoned to death by these slaves of satanic masters if they were around today.

In his book, *The True Believer*, Eric Hoffer describes men who think so little of themselves they can only gain self-esteem by abandoning *self* to a *cause*. These *true believers*, as he calls them, will do anything, including committing suicide, for their cause. Following their *leaders* who enslave them to serve the leader's own and often undefined purpose, these are not men of free will, but true slaves of those who manipulate them. Such are the enemies free men always face.

Man has a natural instinct for enslavement. All movements, large and small, utilize this *pack animal* instinct as tools of opportunistic leaders to control masses of people. Humanitarian civilization tends to counter this instinct while mobs, movements, charismatic leaders, and fundamentalists of many kinds`` tend to nurture and expand it.

The real power in mobs, movements, fundamentalism, and other uses of instincts to control lies in a simple, irrefutable fact—it is infinitely easier to damage or destroy to change things than to build or create. A few men used only the most rudimentary skills to bring down the World Trade Center in a few hours. Contrast this with the immense time and effort required to design and build those same structures. In the same vein, it is far easier to make angry criticisms of ideas that differ from your own than to listen to those ideas and then make calculated judgments. Closed minds can be true agents of evil.

A simpler illustration, which many have experienced firsthand, is the frequent reaction of small children to sand castles, even those created with hours of careful work. With glee and a real sense of power, a small child will rush through and demolish the creation. It is the rush of power—the instinct for destruction—that creates such childish joy. On any scale, it provides those who feel relatively powerless a form of power over those whom they fear or to whom they feel inadequate for any reason. Vandalism, terrorism, murder, rape—all real crimes—are examples of the destructive efforts of those who feel weak or inadequate in some way directed at those toward whom they feel weakness or inadequacy. Mob action is the lowest form of human expression, but therein lies its power. It is the easiest way for an individual to abandon decency with anonymity and *get back* at real or imagined sources of power.

It will be infinitely more difficult for us to hold our dignity, our respect for all life, our love of freedom, reason, and humanity while engaged in this battle. A battle that is indeed for survival against an enemy that holds an opposite view of almost everything and demands our annihilation in the name of blind subjugation of self to a religion without reason or rationality. Whatever our course, let us pray we do not become like those satanic leaders or their blind followers. Above all, let us take care not to condemn all of Islam and thus fall into the trap these evil men are trying to spring. Islam is not the enemy. The true enemy is ignorance, prejudice, anger, fear, and genuinely evil men who are *true believers* in a twisted fundamentalist Islamic *cause*.

I believe it was Thumper who said, "If you can't say somethin' nice, don't say nothin' at all!"

—*Howard Johnson, November 2001*

Too much planning is the commonest cause of defeat. The mediocre strategist conceives a plan and, like a pregnant woman, thinks the offspring of his belly, and his mood shall set a heel on destiny. A true commander's plans are changeable, adaptable, reversible, sudden, frequently surprising even to himself. They are the means that his genius seizes, to employ his full strength, at a well-considered moment, to a foreseen, unflinched from, and undeviating purpose.

—Lord Tros of Samothrace in **The Purple Pirate** by Talbot Munday

O suns and skies and clouds of June,
And flowers of June together,
Ye cannot rival for one hour
October's bright blue weather;

When loud the bumblebee makes haste,
Belated, thriftless vagrant,
And goldenrod is dying fast,
And lanes with grapes are fragrant;

When gentians roll their fingers tight
To save them for the morning,
And chestnuts fall from satin burrs
Without a sound of warning;

When on the ground red apples lie
In piles like jewels shining,
And redder still on old stone walls
Are leaves of woodbine twining;

When all the lovely wayside things
Their white-winged seeds are sowing,
And in the fields still green and fair,
Late aftermaths are growing;

When springs run low, and on the brooks,
In idle golden freighting,
Bright leaves sink noiseless in the hush
Of woods, for winter waiting;

When comrades seek sweet country haunts,
By twos and twos together,
And count like misers, hour by hour,
October's bright blue weather.

O sun and skies and flowers of June,
Count all your boasts together,
Love loveth best of all the year
October's bright blue weather.

This was one of my dad's favorite poems. He repeated the first four lines to me many times, especially in October at the lake. I always had it in my mind that it was written by James Whitcomb Riley, but researching it to put in my book I discovered the actual author. She wasn't even a Hoosier. Helen Hunt Jackson was born in Amherst Massachusetts and lived much of her life in Colorado Springs and southern California. A fiery and prolific writer, Jackson engaged in heated exchanges with federal officials over the injustices committed against American Indians. Several of her books reflected her efforts supporting Indian causes. Her novel **Ramona** was a very popular book about the troubles of an Indian family. More details of her life and writing can be found on the Internet at: **http://en.wikipedia.org/wiki/Helen _Hunt_Jackson**

—Helen Hunt Jackson (1830-1885)

Section II
Short Stories
Mostly Sci Fi

The Loop

*S*haar slowly became more and more aware of herself. *"What's happening?"* She thought as a wave of unease flowed through her mind just as she realized she had arms and legs. Her mind was so sluggish, like trying to run in a dense gravity field.

Shaar tried to move, but couldn't quite remember how to make a limb respond, or why she should. This whole experience was starting to feel familiar, which was comforting. *"It'll be all right,"* she thought. *"I'll figure this out in a moment or two."* If only she could remember where she was, or who she was. Then it all came back with a flash, and she screamed.

As the scream died in her throat, and her mind climbed back into sanity, Shaar once more evaluated her circumstances and options. She had lived this same déjà vu so many times. Fear-filled thoughts of insanity again flashed through her mind and were gone. Furiously, she fought for control and immediate action. Her hands scrambled for the computer console as plans and actions found order and demands in her mind.

The time loop reconstituted her body and ship to exactly what and where it was when she began the test. Her memory alone continued in linear time, each rerun starting where the last one had completed. No matter how many physical records she made during a loop, they were all gone when the next one started. Computer memory, logbook, notepads, camera images, voice recordings, even computer programs, all returned to the precise condition they were in when she first reached the point of no escape. The only thing that did not return to the starting point was her memory. Each loop lasted two hours, thirteen minutes, and twelve seconds—the exact time it first took her ship to go from the point of no escape to the event horizon of the black hole.

By this time, she knew the drill. She would be mentally alert until about thirty seconds before the end. During those thirty seconds, her senses would grow duller and her mind would *fuzz* out until she lost all mental faculties. She became a consciousness with no input, no memory, and no senses—a mental black hole. The reverse of the process at the start of the next loop brought on a massive surge of incredible fear as her senses and memory returned. Each time, her hands whitened as the surge of fear closed her grip on the console a bit harder. Immortality in an unending cycle of a bit more than two hours at a time promised a maddening future. She often thought of suicide, but feared the outcome when she would be reconstituted in the next cycle.

It was not the same experience each time, just the same point of restarting. She tried countless strategies to break out of the loop using the main jump drives in every conceivable configuration. Frustration gripped her a bit more at each failure. It was frustrating to realize that no matter how much power she used, the fuel charge was always back to 89 percent when a new loop began. She wondered if she was cycling in universal time, and if each new start was the same as the last. If so, how could her memory be linear? Her mind crawled with questions of how and why she could remember actions she took in twenty, fifty, or several hundred previous loops.

She tried sleeping once, but it had been an emotional disaster. It was both maddening and frustrating wondering why her memory continued for all of the 274-time circles since the first. Shaar finally decided she didn't need sleep, being in effect rejuvenated every few hours.

She thought about Kiaho and their daughter Minia'i and cried. For Shaar, volunteering for this dangerous mission was her response to the emotional pain of their tragic deaths. The engineers and physicists who designed this entire black hole research project gave her a fifty-fifty chance of survival. Theoretically, she was to *slingshot* around the black hole right at the point of no escape, and then enter a return trajectory to make it back. Things do not always go as planned. In the few minutes the calculated path of the ship took to *slingshot* around the black hole while avoiding the event horizon, disaster took control. Unexpected forces overwhelmed everything she did to maintain the calculated trajectory. Instead, she spiraled from the point of no return and into the event horizon in an almost infinite number of circumnavigations, each one a bit faster than the last. The hopelessness of her efforts in this gravity maelstrom ate at her mental control as she spiraled into what she knew would be her doom. The cycle first repeated, and started a growing mix of wonder, incredulity, frustration, fear, and a thousand other emotional blasts which, by this time, ricocheted through her brain, creating stabs of pain at each impact.

At the current moment she fought to control her mind. She had to train herself in setting up the computer to try ways to break the loop. Each time, she managed to be a bit faster, to get a bit farther. Infinitesimal hope grew and overpowered the demons of failure that dogged her as she drove her mind faster and faster. Maybe this next time she would succeed. Hope was all she had along with a generous dose of determination and grit. Holding all this information in her memory and planning for the next cycle was all she could do. She was learning and gaining, but the damning fear of impossibility clawed at her vitals.

She was now working at a frenetic pace, knowing the end of the current loop would soon engulf her in the unknown. She was memorizing what she was doing so her effort would go faster and farther next time. Still, fear stalked her every step, no matter how she tried to empty it from her mind. The gnawing fear of continuing in this loop forever was a real terror hiding just below the surface, ready to engulf her. She would prefer death, but that might prove impossible.

Thoughts of what she might find if ever she broke out of the loop also plagued her. Would her world still exist? Maybe she would come out in a distant time and place. The death that type of scenario ensured would be preferable to living forever in an infinite time trap.

She was making gradual headway with her programming and training. If she could time a strong blast from the main control thrusters close to the start of the loop, maybe it would move the ship far enough out from the event horizon that the main jump drive would work into the originally planned trajectory. Unfortunately, she would have to complete the entire sequence extremely close to the beginning of a loop. This meant coming out of that state of bare consciousness quick enough to enter the program from memory and execute it within the first minute or so. Each time she missed, she spent the two-plus hours driving herself, training her mind and body to enter the program quickly and without mistakes. She was practicing entering the program when a slight fuzziness heralded the end of the current loop. Shaar wasn't ready for that yet. "Damn!" she cursed as she faded into nothingness and became a bare consciousness once more.

Shaar slowly became more and more aware of herself. "What's happening?" She thought as a wave of unease flowed through her mind just as she realized she had arms and legs. Her mind was so sluggish, like trying to run in a dense gravity field.

Shaar tried to move, but couldn't quite remember how to make a limb respond, or why she should. This whole experience was starting to feel familiar which was comforting. "It'll be all right," she thought. "I'll figure this out in a moment or two." If only she could remember where she was, or who she was. Then it all came back with a flash and she screamed.

The really dangerous people are not those who believe in violence as a means to every end or they who believe in treachery as a means to most ends. Those can be overcome by resistance and by alertness. The truly deadly menace is the intelligent man or woman whose central vision, has been indoctrinated in the accuracy and supremacy of their belief system, their view of reality. They then impose their views as controls on others. This is particularly true of many elitist intellectuals, particularly when they gain political power and control. So often they become misdirected and confused until suspicion becomes their guiding principle and pure power their only end.

—*Howard Johnson, 1992*

Images of Pain

A Satanic burst of flame - Screaming, burning flesh - Bright tinkling shards of glass - Another monstrous flash of fire - Black smoke billowing - Heart-rending phone calls - Humanity in the stairwells - Electronic pictures burned into brains - A rumbling, crushing, obliterating collapse - Terrible showers of stone, steel, glass, dust, and flesh - Lives painfully obliterated as millions watch in horror and disbelief - Booming clouds of smoke and dust, then dooming silence.

Heroic thousands in vain efforts - Photos of lost loved ones - Withering hope - Veils of tears - Anguish a billionfold, but a few scream with joy - Faces of horrible pain of loss - Electronic images of child faces of evil - I cry, you cry, millions cry, God cries. Satan laughs!

—Howard Johnson, September 11, 2001

One midnight, deep in starlight still,
I dreamed that I received this bill:
 (-------- in account with life:)
Five thousand breathless dawns, all new;
Five thousand flowers, fresh with dew;
Five thousand sunsets, wrapped in gold;
One million snowflakes, served ice-cold;
Five quiet friends; a baby's love;
One white-mad sea, with clouds above;
One hundred music-haunted dreams
Of moon-drenched roads and hurrying streams;
Of prophesying winds, and trees;
Of silent stars and browsing bees;
One June night in a fragrant wood;
One heart that loved and understood.
I wondered when I waked at day,
How . . . how in God's name

   . . . I could pay!

—Courtland Sayers

The Great One

And the priest said to me, "Pray to the Great One, and he will guide and protect you. In our homeland, he alone withstood the terrible time of destruction." So I prayed to the Great One, to the all-powerful visage of the one standing in the sacred cave temple, and my prayers were answered. Indeed, I walked through battles unscathed. No harm fell upon me or mine. Triumphantly did I walk over my enemies who fell before me.

And the priest said further, "Follow the way of the Great One, and all things will be yours. The land will do your bidding, and all the creatures thereof will be servants unto you."

So I did follow the way of the Great One as ordained by the great and powerful words of the priests. My land did become fruitful, and the great and the small creatures of the land paid homage to me and did my bidding. My days were full of joyful toil, and my nights did ring with laughter and rest with quiet peace.

One day I asked of the priests, "Why giveth the Great One these things of joy and pleasure to me?"

And the priests answered, "Do not question. Only pray, obey, and follow the way of the Great One, and you will be always happy."

But sadly, my heart was troubled. Question after question bred still more questions, and I became obsessed with finding the answers. I sought among the people, asking of them the answers to my questions.

Even those who had been my friends turned from me, saying, "Do not ask these questions! We are not to know!"

And it came to pass that I made the long pilgrimage to the sacred cave of the Great One to ask the questions. Had not he protected me in battle and destroyed my enemies? Had he not done as the priests and the people said and given me more than I asked from the land and the creatures? Surely, the Great One would answer the simple quest of a loyal follower. I entered the sacred cave and walked among the worshipers. As I stood before the Great One in his temple, my knees shook. He stood so tall and powerful in his brilliant reds and quiet blues. His four green eyes shone from his orange forehead, high as the tallest trees in the forest. His upper two powerful arms crossed on his broad chest while his lower arms held the sacred symbols. Cradled in his crossed legs was the pot of plenty with its golden brown cover, gleaming in the dim light. *Surely the all-powerful Great One would be pleased to answer my simple request*, I thought.

Then in quiet, careful tones, I began asking all the questions of my mind and heart and soul. I waited resolutely for the answers. Being patient and obedient, I knelt and meditated while the Great One pondered my questions. After some time, the silence began to weigh heavily on me, and I again asked the troubling questions. Once more, I knelt and patiently waited the answers.

The darkness came and then the light and yet again the darkness, yet still no answering words came from the Great One. I asked and waited again and again with less and less patience each time. *Perhaps the Great One is asleep*, I thought, though his green eyes shone in the dim light of the temple in the cave. I crept up to the altar and thrust my staff to strike the Great One gently on the knee to awaken him. Several times did I tap his knee, each time more vigorously than before.

Growing bold with eagerness and anxiety, I vaulted onto the altar and shouted to the Great One, "Awake! Awake! Your loyal servant seeks of you some answers to his quest!"

The silence bore down on me like an enveloping black cloud of fear, stirring me to more desperate and violent action. I screamed again and again and beat mightily upon the Great One with my staff in growing madness. The people fled from the temple in terror as a thunderous rumble filled the high-walled room and a blinding cloud of reddish dust engulfed me. I realized I had seen my staff pass into the Great One's body at my last savage blow. A large crack shot upward in blackness to his face in the instant before I was blinded by the red-brown cloud. As quickly as the rumble began, it ceased. Had the terrible sound been the Great One's answer? I stood in fear and trembling as the reddish dust settled slowly and quietly to the floor of the temple, my eyes locked closed in fright. Would the Great One destroy me for this insolence?

When the silence bore too heavily on my patience, I opened my eyes. A new terror seized and froze my heart; the Great One was gone! My eyes opened wide to survey the frightening scene. In the growing light, I could see the altar clearly, but there was nothing behind it. Looking down at my feet, I saw potsherds strewn about the altar and the floor of the temple. Behind the altar where the Great One had been was a great pile of broken pottery of many colors. I stood transfixed. There in the midst the potsherds, glowing green and unmistakable, was an eye of the Great One. Overcoming my fear, I jumped to the floor behind the altar and picked it up. It was common brown pottery covered with bright green glaze. There on the floor, leaning against the back of the altar, was a large brass plaque with words engraved in the ancient and long-forgotten tongue. I had once seen similar words before, but could not understand their meaning:

The largest piece of fired pottery ever created. This piece designed and fabricated by the East Liverpool Pottery Combine, East Liverpool, Ohio. Made for the Great Lakes Exposition, August 23, 1935.

Charles "C" Miller

Dr. Charles Botkin dropped his lean frame into the window seat of the 757 as he headed home for a visit. He looked forward to seeing the farm and his family again. As the plane flew east from Los Angeles toward Chicago, the loquacious Charlie became engaged in animated conversation with the young woman in the next seat. The usual exchange of destinations, reasons for traveling, and a little idle chitchat, led to Charlie started telling her about the uncle he was named after.

"I've run across a difficult technical problem that I hope my uncle can help solve. That's the main reason for my trip."

"Are you a student? I believe you said you were at Cal Tech."

The youthful-looking Charlie chuckled. "Actually, I'm sort of a grad student," he replied. Charlie found strangers treated him more openly as a student than as a professor and required much less explaining. "I have a wonderful uncle who has a unique way of looking at things and may be able to help."

Crazy Charlie, as his friends called him, was sort of a maverick genius. He graduated from Purdue at seventeen with degrees in physics and math. He then went to Cal Tech for his doctorate and has been there ever since. A genuine wild card, an unconventional thinker and personality, and a capable rock musician, his hair and dress make you think of him as anything but a serious scientist. One of the pioneers in the application of quantum physics to cosmology, he was a world-renown leader in his field by the time he reached twenty-seven.

"And you're going clear back to Indiana to see him? He must be someone special."

"My mother named me for him, and he is definitely special. To all but a handful of people, my uncle Charlie is an eccentric old man with a checkered past, a tendency to tell tall tales, and an unlikely source of unusual knowledge. We all call him C. I was at least seven before I knew his real name. He's my mother's older brother, considered a real maverick by the entire family, except my mom and me. My mom told me many tales about his life before I knew him.

"As a young man, C married a glamorous young woman who left him a year later to pursue a career as a movie actress. Mildly successful, she had small roles in many pictures. *Uncle C* as I called him as a youngster, never tried marriage again although he had several relationships that eventually went sour. He had one long relationship with a lady named Carla. She must have been something special because whenever C talked about her he would get misty eyed.

"I remember once asking him why he hadn't married her if he cared so much. He looked off trance-like, and said quietly. 'I was young, foolish, had a severe case of low self esteem. I didn't realize what an incredible person she was or what a beautiful relationship we had. Letting her go and ending that relationship was one of the stupidest things I ever did. When I came to my senses and realized the gross mistake I had made, it was too late.'

"He was light-hearted, thoughtful, or studious. He seemed never to take life too seriously and was always upbeat. Nothing ever seemed to get to him except thoughts about that lady. These were the only times I ever saw my uncle in a melancholy mood. Everyone in the family suspected he never got over Carla. After that, he embarked on a nomadic life of searching. He did so many offbeat things in his life he's hard to describe in a few words. He made a great deal of money on one business venture, only to lose it all on another, ill-fated one. He worked hard to pay his debts and managed to build himself a small fortune of some two hundred thousand dollars. He gave a portion of this to my folks so they could acquire some additional land adjoining their farm. In his late forties, he embarked on a six-year trip through the Far East which exhausted his remaining fortune. When he found himself stranded and out of money in the Philippines, my folks scraped up enough money for his ticket home. With no place to live, he stayed for several months on our farm. I had just started school when he came to live with us."

"My, he sounds like a wild one."

"He is unconventional, but not wild. He's a wanderer, a thinker, and a builder. He built a neat little cabin next to a pond in a wooded area on the farm. The woods and the pond had remained undisturbed since loggers cut all the useable timber during the late eighteen hundreds. It was 1979, and I was fascinated and delighted to be helping C build his cabin. During the next few years, I spent many wonderful times listening to stories of Uncle C's adventures around the world. One of C's passions was his lifelong collection of books, which overflowed the shelves on two walls of the main room of the cabin. There were texts on a wide variety of subjects from astronomy to zoology, several sets of encyclopedias, and quite a bit of fiction, from the classics to Jules Verne and several modern authors. Stacks of *National Geographic* and several scientific magazines filled the lower shelves on one wall of the room. All were stored at the farm until he built his cabin. I often read from those books and magazines."

"Sounds like a wonderful experience. He must care for you a great deal."

"And I for him. That's for sure," Charlie said, a bit misty-eyed. A momentary pause of wistful remembering and Charlie continued. "C has a great workshop too. It is the largest of the three rooms in the cabin. He has tools fitting many trades. When I was little, he devoted one corner of his workbench to the assembly of one of the new computers that were beginning to hit the electronic hobbyist market. This new gadget fascinated me and I worked on it with C every chance I had. By my ninth birthday, I was a bona fide computer whiz."

"Is that what you're studying, computers? I'm fascinated by them, but only use one at work. I haven't a clue how they work. It's all magic to me. Are you going back for help on some computer problem?"

"No, I use computers, but my work is with basic physics."

"Wow! That's another complete mystery to me."

"A lot of it is still a mystery, even to the experts, but we're learning."

"Is your uncle an expert in physics?"

"Sort of. He's hard to describe. He's done about everything and been all over the world. He spent a year as a preacher for a nearby evangelical, nondenominational Christian church. This led to a short career as a DJ and talk show host on a local radio station. His offbeat views from all over the spectrum earned the animosity of a great many listeners, so they fired him. This was done in spite of the fact that his audience grew to huge proportions for the local area and in a short time. His short-lived, local-celebrity status was a mixed blessing. It earned him a reputation as *that crazy old coot who lives in a cabin in the woods*. His meager odd-job income barely covered his living expenses, including fuel for the old pickup he drove. I remember many family discussions about that when I would come home from school for a visit."

"When you were in college?"

"No, I was in a military school, a grade school, when this was going on."

"Were you in the military?"

"No, it was a private school for boys which taught military discipline. I'm not much on that."

"I can tell that by your clothes. I'm guessing the military part didn't stick."

"I was grateful for the wonderful education I received there. I endured the military discipline. Truthfully, what I learned about discipline helped my scientific education. There's a great deal of discipline required in physics."

"Please go on about your uncle. He sounds like a fascinating person."

"That he is. After I graduated from high school and before leaving for college, C sat me down and shared with me his concern about the unusual knowledge he possessed. There were many things he knew about, but had no idea how he gained the knowledge. Often, when I asked him how he knew so much about so many things, he would answer, 'Just my insatiable curiosity, I guess.' This time it was different. He told me he was often frightened and upset when he would explain some strange or unusual phenomenon and realize he had no way of knowing how he knew about it. I always thought he had picked up all that information reading and from his travels. He explained that was part of it, but why was it he 'knew' certain scientific discoveries or theories before anyone made the discovery or published the theory? That's why I'm going to see him. I always thought he was being modest until one time when I made a new discovery during my research only to realize my uncle C had explained the same thing to me accurately, many years before. That realization startled me."

"I can see why. Are you sure it wasn't one of those *déjà vu* experiences? They can be scary."

"I know what you mean, but no. It couldn't have been. I had written documentation from years before that I found and compared with my research. They were an exact match. It was a complete mystery to me."

"Well, I hope he can help you out with your new problem."

The look on her face and her body language told Charlie she was through talking, so he settled back and let his mind wander back to when C told him an amazing tale. He had relived that special time almost verbatim, trying to make sense of what C said. It was before he first headed out for college. He was visiting Uncle C at his cabin and helping him repair a TV set. Once more his mind slipped back to that time, and he relived the experience as he asked C the question that started it all.

"You were going to tell me something secret before I went off to college? Well, I'll be leaving in a few days, so how about it? Will you share your big secret now?"

Uncle C leaned back on his stool and looked Charlie in the eyes. "I did promise that, didn't I? I tell you what, let's finish this TV and go sit on the sofa. This story needs a private place with no distractions because I want you to think hard about what I will tell you."

They finished and tested the TV in about fifteen minutes and went into the main room. On the way, C opened the fridge for a pair of cool drinks to accompany the tale. When he sat down on the sofa, C leaned back and looked off into space for a few minutes while Charlie sat and watched him in rapt expectation.

"I'm going to tell you about something that happened to me a long time ago, something I never told your mother or anyone else still alive for that matter. This is for your ears only, and I hope it will remain between us. You are the one person on this Earth whom I know will hear this story without prejudice. After you hear it, you can ask anything you want. I won't have many answers for you, but it may explain how so many of our little talks have gone where they have."

Charlie remembered looking at his uncle in wonder and amazement. They had spoken of so many things. He couldn't imagine what new marvel was about to be revealed. Enthralled, almost enchanted, he waited, eager and expectant.

C began, "I was walking the few blocks home from grade school for lunch one cold, crisp, blue-skied January day in 1935. As I passed the Buhers' house, about halfway home, I happened to look up and notice a shiny object through the naked branches of a wintering tree. At first I though it to be a new kind of balloon caught in the branches, but as I took a few more steps, I realized it was above the tree and almost overhead. It was far from the winter sun hanging low in the southern sky. I was fascinated, for this was an exciting new wonder for an inquisitive second-grader still in the intoxicating time of life when the days are full of new discoveries. I leaned against the concrete wall separating the Buhers' yard from the alley, right where Mrs. Buher would place food for wandering beggars during the Great Depression.

"The object was round and the same apparent size as the full moon in the sky. It was shiny, and looked like a mirror, yet I saw no reflection. Transfixed, I watched it for a long time, for a small boy—maybe a minute or two. Then it began to move toward the west. It accelerated and disappeared over the western horizon a few moments after it started moving. As soon as it was gone, I ran home to tell my mother, and find out from her what I had seen. This was at a time when people ran outside to look when an airplane flew over, and my favorite was the new Douglas DC-3. This object was not like any airplane I had ever seen before.

"When I arrived home for lunch, my mother was angry with me, and she was rarely angry. 'Where have you been? I've been worried sick!' Her words astonished me. I didn't get a chance to ask her about the marvelous shiny object. She chided me, 'You'll go without lunch young man. Now hurry back, or you'll be late for school.' I didn't understand how I could be late. I had a full hour at least for lunch, and school was five minutes away. I stopped to watch the strange object for just a few minutes. What had happened to the missing time?

"I remember nothing of the rest of the day until I came home and my mother again questioned me about why I was so late. When I asked her about the object I had seen, she thought it was another of the 'stories' I used to invent to liven up my life and amaze others. I had discovered the price you pay and the pain of what happens when you become a 'story' teller, and people learn not to believe you. Every single person who heard my story laughed at me and ridiculed my tale except one, my grandfather. A storyteller himself, he listened to my tale and wondered with me what the object was and what happened to the missing hour.

"I was so humiliated at every attempt made to find out about the object that I gave up. Since then I told no one else about my experience. My grandfather and I discussed it a number of times over a number of years, often when we were fishing together on the lake. I was around fifteen the last time we talked about it. It was our little secret.

"When I was in junior high school, I began to experience a strange phenomenon which has continued unabated to this day. In Mr. Armstrong's science class, I discovered that I understood a great deal about things I had not read about or had not been explained to me by anyone. For me, this started a fascination for things of science that would last my entire life. I also found that I knew the answers to many questions I should not have known. My classmate and buddy, Fred Hunziker, who sat next to me, was amazed at my knowledge. Even Mr. Armstrong was flabbergasted to the point where he quit letting me answer questions in class. Another classmate, a bright girl, told me she thought I knew more about science than our teacher. They began calling me 'the brain' and not always in a complimentary fashion.

"Your aunt Matty, who was six years older and ahead of me in school, brought home her chemistry book and let me have it. In my mind, I can still see the diagrams in that book of atoms with electrons in circular orbits about a solid, compact nucleus. I knew those diagrams were wrong, but of course, my sister thought I was nuts when I asked her about it. I was crushed, but what could I say? She was a high school senior and far wiser than I about everything. That book piqued my interest in chemistry, which then led to my selection of chemistry for my college studies. It was many years later when I read a similar description of the indefinite *cloud* structure of electrons about a tighter cloud of protons and neutrons, the nucleus, that I had tried to explain to my sister. It seemed this was the latest concept of atomic structure, developed many years later I tried to explain the very same concept to my sister as a boy of twelve.

"There are many other concepts of our physical world that I *know* without any idea from where the knowledge came. I keep searching and reading to gain confirmation of many of these things. For example, I 'know,' or at least can conceptualize, an understanding of our universe that has yet to be discovered or explained by anyone. I think the universe has a roughly spherical shape with an irregular, changing surface. At the surface of this shape, all mass would be to one side of any point on the surface. The center of mass of the universe would be near its physical center. Light, warped by this center of mass, does not escape from the universe. The true speed of light is a factor of its distance from this center of mass. Our measurement of light speed is a function of our own distance from this center of mass. Light passing near or through this center of mass is moving much faster than when it passes us. Likewise, light, on reaching the limit of the universe where all of the mass is to one direction or side, slows and finally *falls* back in the same way one celestial body orbits another, controlled by the force of gravity. The gravitationally limiting surface of the universe acts to return light and hold it within the gravitational grasp of the universe.

"The first time I heard a flying saucer story, I thought of my childhood experience. It was so similar to the descriptions of many of these tales. The ridicule heaped on those who saw a UFO caused me to rethink my experience and continue not to talk about it to anyone. The first time I heard an *abduction* story, I thought about that missing hour so long ago and of the things I *know* that I have no reason to know.

"I have come to no conclusions, nor do I make any claims other than those I have described. The mystery to me is now greater than ever, and I am sure that I will not have an answer in my lifetime. I search and ask in every way I know how, yet the mystery continues to deepen. If there are others with similar experiences I would like to meet them, yet I hesitate to admit what I experienced. I am still a bit apprehensive about any ridicule that might come and destroy my own knowledge. I tell you this now because you know and understand me. Besides, I am nearing the end of my life and have so much less to fear than when I was younger. You can decide for yourself if the story is for real or the ravings of a crackpot. A discovery that confirmed the view of the universe I described would change the acceptance of my story now, wouldn't it?"

"Would it ever!" Charlie replied, then asked, "What do you think it was?" Charlie was amazed, grasping the implications of the story, but never doubting a single word his uncle spoke.

"I gave up speculating on that years ago," C replied. "I only know what I told you. No more than that, but no less either. Many years later, when the first UFO stories began appearing, I thought I might find an answer, but soon realized all I would do would be to make a fool of myself if I came forward with my story. It's clear to me that whatever or whoever they are, they have a purpose to their activities. I've wondered for years what that purpose might be without ever coming up with anything logical. I can't determine if it bodes good or bad for humanity. It's a real mystery. Why would they imbue a small child with advanced knowledge? I'm convinced that is what happened to me so long ago. There must have been others who received similar treatment. Though I've searched my whole life, I've never found another person who shared my experience. I've met and spoken with people who reported sightings and abductions, but none were anything like mine. In fact, I find myself doubting the truth of their stories like everyone else."

"The truth is like the proverbial needle in the haystack, isn't it?"

"That's almost an understatement. Over the years, I've had many incidents where the announcement of a discovery was something I already *knew*. Did I actually *know* it, or was that a trick of the mind, a déjà vu experience? I've pondered that question many times. Once, while talking with a group of engineers about a particular metallurgical problem, I posed a solution to them. It was a solution that I didn't have to think about. I just 'knew' it. Several months later, the specific problem was solved by the method I proposed. One of the engineers from that group contacted me and asked how I had come up with the exact solution. He knew I was not a metallurgist and wondered how I knew that particular answer to a problem no one else could solve. I was at a complete loss to explain it. Had my lack of knowledge in the field let me think outside the limits imposed by an expert understanding? Was my solution one of those serendipitous 'aha' experiences we all have on occasion, or had I *known* the answer? I don't have a clue as to how, and would not claim to understand where the answer came from.

"There are two concepts that I feel certain came into in my mind by an extraordinary process. The understanding of the true nature of the particles in the atom is one. The general makeup of the universe with the gravitational effect on light and other electromagnetic waves or particles, is the other. The first was postulated by particle physicists many years after I knew and described it. The second seems to be a theory in my mind alone. No theory I have read about is remotely similar. I don't know if it's correct, and I have no idea how to prove it."

Charlie determined he would set that as one of his goals, to prove or disprove C's theory about the universe. It would prove to be a daunting task.

Charlie came back to reality as the woman next to him shook his arm.

"Wake up! Raise your seat back and fasten your safety belt."

"What? Oh yes! Thanks! I guess I went to sleep."

"You certainly did. You've been sound asleep for at least an hour. Never moved a muscle."

It took him a few minutes to reorient from the where-am-I, what-time-is-it daze. By the time the plane arrived at the gate he was wide awake. He walked through O'Hare Terminal toward his next flight still thinking about his Uncle C. One more quick jaunt in the small

commuter airplane to South Bend and his folks would be there to pick him up. He hoped C would be with them.

As he walked down the steps from the small plane that brought him from Chicago to South Bend, he noticed his folks were not waiting there. That was unusual. He hoped nothing was wrong. He walked to the baggage carousel, waited for, and picked up his luggage. Still, no one appeared even as he walked outside. His folks were never late, so visions of accidents or other calamities stalked his mind. After waiting about ten minutes, he turned to go inside and call home using his cell phone. As he started to go back in, his mother's car came around the corner and pulled up to the curb. She was the only one in the car.

"Where's everyone?" he asked as he placed his bags on the rear seat. When he sat in the front seat, his mother stared straight ahead and said not a thing. "What's wrong, Mom?" he asked again, sensing something was wrong.

"It's your uncle C." she said through tears. "He's disappeared. We drove out to his cabin to see if he wanted to come with us to pick you up, and he was nowhere to be found. His door was standing open, and last night's supper was still sitting on the table, uneaten. Ralph was sitting on the porch, whining, and you know he always takes that little dog with him wherever he goes. I'm afraid something terrible has happened."

"I'm sure he'll turn up. Maybe he walked somewhere."

"No! He just disappeared. His pickup was still there, and with Ralph on the porch, we knew something strange had happened. Your dad stayed there to search for him while I came to pick you up. Maybe he'll have found him by the time we get back."

On the way home, Edith drove much faster than usual. "Slow down, Mom! I know you're in a hurry to get back, but let's not add an accident to today's problems."

"I'm sorry! I didn't realize I was driving so fast. I'm worried to death about C."

"I know. I'm concerned myself. Is it usual for him to go off without telling you?"

"Never! Even when we're away, he'll leave a note on the door about where he's going and when he'll be back. He's really good about that."

"I'm sure it's something simple. Maybe a friend of his came over and picked him up."

"Not a chance. Anyone driving would have to go right by on the driveway. We'd have heard them no matter when they went by. Besides that, how do you explain the uneaten meal on the table? C would never have left the place like that under ordinary circumstances. He eats at about dark this time of year, so you know he must have left last evening between six-thirty and seven."

"What were you doing about that time? Were you home?"

"We were eating dinner about then ourselves. There was one strange thing, the lightning flash. We saw a bright, sustained flash of lightning. You know how sometimes, when it's dark out, a lightning flash lights up the whole sky for almost a second and you can see everything, even in the pitch-black?"

"Yes, I know what you mean. You must have had a thunderstorm last evening."

"That's it. We didn't! After that flash, your dad and I went outside to close the windows on the car and saw the sky was crystal clear. That's when we heard Ralph barking. Also, there was no thunder. We decided it was a bright flash of lightning so far in the distance we couldn't hear the thunder. We thought Ralph was barking at some strange dog or critter. You know how he carries on when something's around the cabin. What else could we think?"

"It sounds strange to me. As soon as I get home, I'm going to check the neighbors to see if anyone else saw that flash," Charlie commented. A dark cloud of silence hung in the car during the rest of the trip home. Charlie was hoping C would be there waiting when they arrived.

As they crested the rise in the road north of the farm, they could see two sheriff cars parked in the driveway. Ray was standing in the backyard with two deputies, and they were all looking back toward the woods where C's cabin was hidden. The two deputies were friends of the Botkins and longtime members of the local sheriff's department. As soon as his mother parked the car, she and Charlie stepped out and walked to where the men were standing.

"What have you found?" Charlie asked as soon as he joined them.

"Nothing," his dad replied in obvious distress. "I called Pete and John here to see if they could find anything I might have missed."

"We looked over everything and couldn't find anything suspicious," Pete said. "Wherever he went, C didn't leave a trace. We don't make missing person reports until the individual has been missing for at least twenty-four hours, but this seemed unusual. We hightailed it out here as soon as Ray called. John and I combed the area for more than an hour and found nothing, nothing at all."

"We called for some dogs to come out to track him in case he walked off somewhere and can't make it back," John added. "They'll soon be here. Pete is going to stay and work with the dogs. I've got to get back right away, but I leave you in good hands. If he's here, those dogs will find him."

After John drove away, Charlie took his bags up to his room. When he came down, his Mom, Dad, and Pete were sitting in the living room and talking about C's possible whereabouts. It wasn't long until a truck drove up in the driveway. It was the dogs and their handler, Tara Bailey. Tara was a well-known dog trainer and breeder who had two hounds that were often used by law enforcement all over the northern part of the state. Ken Bailey, her husband, was a veterinarian. Together they had an animal hospital, training center, and boarding kennel in the next county about thirty miles away. All four of them went outside to greet her.

After talking to Tara for a few minutes, Pete said, "She asks that none of you go with us to the cabin. The less people around, the better the dogs work. We'll drive the truck back to the cabin and release the dogs there. Is there any article of clothing you can remember him wearing recently? We'll need something to give them his scent."

Edith thought for a moment and then said, "He puts his dirty clothes in a hamper in his bathroom closet. He hasn't brought them up to wash for a while, so there should be plenty in that hamper."

"That's perfect. I'm sure the hounds can get a good scent from those clothes," Pete answered.

"Where did you leave Ralph?" Edith asked her husband.

"I think we left him inside the cabin," Ray answered then added, "You'd best leave his little dog inside. He'll be friendly enough to you folks, but he wouldn't take kindly to a couple of hounds poking around his property."

"We'll see to it he's okay and kept out of the way," Pete remarked as he stepped into the truck with Tara.

—— The Dogs Get Buffaloed ——

"How about some details, Pete? When John called, all he said was someone disappeared and asked me to get the hounds here as soon as I could," Tara said as she guided the truck down the bumpy lane toward the cabin.

"The missing man is Charles Miller. Everyone calls him C around here. He's an unusual, somewhat eccentric old man about seventy-five, the brother of the woman who lives here. I'll say one thing about him. The man can build or repair about anything. You'll see what I mean when we get to his cabin. He built it all by himself about sixteen years ago. It's a neat little place, perfect for a man living alone."

"Sounds like an interesting man," Tara commented. "Do we turn here?" She asked as they reached the end of the lane where it crossed a tiny, dry streambed.

"Just follow the stream right into the woods. You can see where he's driven his truck over the years. It's a bit bumpy but high enough above the streambed to be out of the water in the spring and early summer when the stream runs full of water."

After they entered the woods, the trail made an abrupt right turn and led about fifty feet to a cleared area by the cabin. As they stepped out of the truck, they heard Ralph inside, barking furiously. He knew there were strangers outside and was giving them what for. As they walked up to the door, Tara asked, "He's not a biter, is he?"

"He was quite friendly when we were out here an hour or so ago. I'm sure he'll be okay."

Tara crouched down as she entered and extended her hand along the floor, palm up in a friendly gesture. The little dog stopped barking, came over, and sniffed her hand, his tail wagging furiously. He did not like to be left alone and was genuinely happy to see them."That's a good boy," Tara said as she gave him some soft pats. She was rewarded with several doggie kisses. Ralph was a typical affectionate pooch. With the little guy now at ease, she retrieved several shirts from the hamper to use for the scent. "Why don't you stay inside and keep Ralph company while I work my dogs? If he can be kept from barking, he won't distract the hounds. That will make my job easier."

"Okay! Give me a call if you find anything. Use your two-way. I see you have one. Set it to our standard frequency."

Tara closed the door behind her and headed for the rear of the truck. She opened the small door and released the two hounds who bounded around exuberantly for a few moments, glad to be out of their confinement. Soon they were back at her side, knowing full well what their job was. They sniffed the shirt Tara held out, then sat down to announce they were ready. Her hounds were well trained. She didn't need to run them on leashes as they knew not to outrun their handler. They waited for her at times so she could catch up. She directed them to the porch where she wanted to start the search. Hand signals were her method of directing the dogs who worked in relative silence.

As soon as she gave them the signal to begin, they headed off the porch, trailed about fifty feet to the center of the clearing where they began circling. Several times, they started off on a track only to stop and return after going twenty or thirty feet. It was obvious these were old, cold tracks. After returning to the porch several times and looking for other tracks, the dogs returned to the center of the clearing and sat down. There was no ambiguity to their message. The track ended right there in the center of the clearing. It was the only fresh track the dogs could find. She had them try several more times with the same result. It indicated several things. There was no way of determining which direction the short track was laid down. It could have been from the house to the clearing or the reverse. It showed that C had either walked from a vehicle parked in the clearing to the cabin or walked the other way, from the cabin to a vehicle. There was no other possibility. Tara examined the ground for tire tracks or other markings around where the track ended. Other than what looked like a single faint set of footprints in the soft ground, she found nothing. She took a marker flag from the truck and stuck it in the earth to mark the spot. Knowing they would stay nearby, she let the dogs roam as she headed for the cabin to tell Pete what she found.

"Did you forget something?" Pete asked as she reentered the cabin so soon.

"No, but the dogs did find the end of a short trail in the middle of the clearing," Tara replied.

"What do we do now?" Pete asked.

"I marked the spot where the track ends. I let the dogs roam to see if they could find something else. When we drive back to the house, I'll let them search the way back and around the house. Maybe they'll find something there. I think we should take this little guy with us. He'll not be happy if we leave him alone; I can tell." She called Ralph over to her. When she picked him up, he gave a low growl to let her know he didn't like it. He would endure the indignity without complaint after she reassured him with soft words and a gentle touch and headed for the door. Tara was a master at handling dogs.

As soon as she walked outside, the two hounds romped over to investigate this little pooch their master held. A few words and a hand command and the two hounds sat still while Tara took Ralph and placed him in an empty kennel in the back of the truck. Signaling them to begin tracking again, Tara got in the truck with Pete, and they headed back, following the dogs as they searched back and forth across the stream side driveway and then the lane. When they reached the house, the dogs crisscrossed the entire yard, pausing several times by C's pickup where there was an obvious, but weak scent. It had been several days since C had driven the pickup and he had last walked to the house on Friday, so all the tracks were old and weak. The dogs followed tracks up to the house and the truck, but it was apparent these too were old tracks. Tara retrieved Ralph from his cage in the truck and carried him into the house. He never made a sound, resting peacefully in her arms.

"I didn't think this little guy should be left alone in the cabin. Is it okay to bring him in?" She asked Edith as she stood in the doorway.

"Certainly," Edith answered with a smile as she opened the door for them. "C brought him here often. He knows his way around. We have a bed for him and food and water bowls. He can stay with us 'til we find C. Incidentally, what did your dogs find?"

As they walked into the living room, Tara relayed what happened with the dogs and how she marked the end of the scent trail in the clearing. "That's in case anyone else wants to look for signs of what happened. I couldn't see any indication of tire tracks or anything other than that single set of faint footprints. There were tire tracks near the cabin where cars parked and from there to the drive, but nowhere else in the clearing. If your brother walked anywhere other than that one scent trail, the dogs would have found it. There's been no rain to wash the scent away, and the one scent track we did find was fresh and definite. He had to have walked that track one direction or the other within the last twenty-four hours. The dogs told me that in no uncertain terms."

Ray looked puzzled. "That's strange. It looks almost like he disappeared into thin air, right at that spot. Since that is impossible, we must try to come up with something else, something that makes sense."

"I have no idea what that could be," Pete commented. "If it wasn't a wheeled vehicle, it would have to have been a helicopter, but no chopper could have set down in that clearing. It's far too small."

Charlie replied, "The only way a chopper could have picked him up would be on the end of a cable lift dropped down from above the trees."

"We would have heard any chopper hovering over the trees at that time of night," Ray commented.

"I don't know about that," Pete said. "The military have birds that can hover almost silently. They make some noise when they fly fast, but at slow speeds, and while hovering, they are silent. You wouldn't have heard them from here. The cabin is nearly half a mile away, isn't it?"

"At least," Charlie answered. "But what would the military want with C? He's never been in the service as far as I know. He traveled all over the Pacific Rim for a number of years, and he's done some contract work with the navy out there, but that was twenty years ago. I can't imagine what the military would want with him. Anyway, they could have driven up in a car and gotten him."

"A good point," Pete replied. "It's a real mystery. It still looks like the only way he could have been taken away. There was no sign of a scuffle anywhere. I looked for those signs, particularly where the scent trail and footprints ended. Those footprints were hard to see. Tara noticed them while watching the dogs sniffing at several of them. They were slight indentations spaced as normal walking prints would be. We couldn't be sure which direction they headed. Then we found one complete print in some softer ground toward the porch. It showed they were headed away from the cabin. We looked around but found no other prints within ten or twelve feet of the end of the marker Tara placed."

"What about your own footprints?" Ray asked. "Couldn't they have covered or obliterated other footprints?"

"A good point," Pete replied. "One of the most important aspects about inspecting a crime scene is how to do so without destroying evidence. Extensive training and practice on how to walk so as to avoid areas we believe may contain clues is part of our forensic discipline. I can assure you there were no other footprints around that marker. We can't consider that a crime scene, but we did treat it in the same careful manner."

"My dogs were the only ones who left prints in the area, and the ground is quite hard, so I doubt they left much of a mark," Tara commented. "It would take a person weighing more than a hundred pounds to leave any mark on that ground. I'm a hundred and forty, and I barely left a print where I walked."

"So that leaves us without a clue," Charlie said. "Your dogs found significant information, but it merely added to the mystery. C couldn't have just evaporated into thin air, but that's what it looks like so far. Where do we go from here?"

"Well, I have to go write a report," Pete said resignedly. "I don't look forward to that, seeing as we have no conclusions with any merit to write down. I hate to leave things so unresolved, but I think we all need to get on with things. I'll do some checking in a couple of areas including the nearest military bases, but I am at a loss as to how to proceed from here. So far we have generated a lot of questions and no answers. I'm sorry folks, but that's the reality of it all. If you think of anything else, please call me."

Tara stood up. "I'm sorry we didn't find more, but I must gather up my wandering hounds and head for the barn. I hope you find him and soon. I'd like to know the answer, so please call me when you learn anything new. I'll leave my card for you."

After Pete and Tara left, Charlie and his folks sat in the living room and talked about the last twenty-four hours. The last time either of his folks saw C was on Friday evening when he joined them for dinner. He walked up to the house two or three times a week and stopped in whenever he went off in his pickup. Charlie thought the meal left on C's table could have been from Saturday until he asked his dad.

"No, it couldn't have been from Saturday or the meat would have smelled. In this warm weather, it would be okay for about eighteen hours. Another twenty-four, and it would have smelled bad," his dad explained. "That was the first thing I checked. He fixed his dinner Sunday evening and never ate it. There is no doubt about that."

"Did he seem worried or preoccupied Friday during dinner?" Charlie asked. "Was anything bothering him, anything at all?"

"No, nothing. He was his usual happy self. He talked about your coming visit to which he was so looking forward. You know he thinks you are the greatest scientist in the world, don't you?" his mother told him.

"And I think he's about the greatest uncle and friend a young man could have," Charlie said with worry in his voice. "I hope and pray nothing's happened to him. Run through Sunday evening again. Try to remember the slightest thing out of the ordinary that happened no matter how insignificant."

They talked about the strange lightning flash and Ralph barking, but that was all they could remember that was unusual in any way. Then his mother said, "We missed our favorite Sunday evening TV show. The satellite signal must have been messed up. Remember, Ray, you tried adjusting the dish? No matter where you searched, there was no signal. You shut it off to silence that awful hiss, so we could eat our meal in peace. Then we saw that unusual lightning flash with no thunder. Half hour later we tried the TV again, and everything was okay."

"That's significant," Charlie said. "Was there anything else strange that evening? Think! It's important. Dad, did you see anything at all unusual when you went outside?"

"I noticed it was dead quiet. The only sound I heard was Ralph barking. If I hadn't gone outside, I wouldn't have heard him. I didn't see anything unusual, at least nothing I remember. After that I went in to eat, and everything seemed normal. We watched TV for a while and went to bed. We discovered C was missing when we stopped out to see if he wanted to go with us to meet your plane. You know the rest."

"I've got to get hold of Matty and tell her about C," Edith said. "She'll be worried sick. I think I'll ask her to come over. I'd like to tell her in person what we know. Her whole family will take the news hard. What will we do when this hits the news? I hate to think of how some of the local news people will treat it. I hope they don't come out and want to go over C's cabin."

When his mother mentioned C's cabin, Charlie remembered the new portable PC his uncle showed him during their last visit. He wondered if C had entered in it anything that might throw some light on what happened. He explained to his folks he was going to get the computer from C's cabin and see if he could find anything that might help.

Charlie grabbed a flashlight as he walked out to the cabin. As he headed for the door, he called Ralph to join him for the trek. Ralph bounded eagerly out the door and trotted along in front of Charlie as he made his way to the cabin. As they entered the clearing, Ralph put his nose to the ground, found the end of the old scent trail, sat down next to the yellow marker Tara had placed in the ground, and began to howl. Charlie flashed the light all around the clearing and nearby woods but saw nothing. He called Ralph who stopped howling, but stayed where he sat. "Okay, little buddy, you stay there. I'm going inside," he said to the dog as he turned and headed into the cabin.

Once inside, he went straight to the workshop to get the PC. When he turned on the lights, he noticed it was not on the desk where C had shown it to him. The printer and scanner were there with their cables among the organized clutter of the workshop, but the PC was gone. Maybe C had taken it with him. He continued searching back in the main room. The PC was nowhere to be seen. He next went into the tiny bedroom where he opened the closet that covered an entire wall. There were many things in the closet, but no computer. As he started to leave the bedroom, he noticed C's briefcase on the floor behind the open door. Next to the briefcase was a shipping box all sealed and ready for shipment. The box was about twice the size of the briefcase and had a label with Charlie Botkin written in C's printing. He picked up the box and examined it. There was nothing else written on the box, no address, just his name. He placed it on the bed, took out the pocketknife C had given him many years before, and cut the box open.

It contained the computer and a note in C's handwriting dated Sunday, September 23, 2001. He sat down on the bed and read the note.

Dear Charlie,

It is early Sunday morning as I write this note. Since you are reading it, I am either gone or dead as I plan to destroy it if I'm still okay when you arrive. Some strange things have happened the last few days. I've been having dreams about the past like I never had before. They've been almost like TV shows where I am reliving that experience when I was seven and saw that strange object. I've had that same dream at least four times in the last few days. I never dreamed about that before in my entire life. I wonder why now. During the same time I felt weird at times. I know I told you about those spells I used to have once or twice a year where bright, twinkling purple and orange lines like C-shaped battlements would interfere with my sight. Well, in recent weeks, they have been coming more and more often. When they do, I feel dizzy and lightheaded, and my heart seems to beat erratically for a few minutes. Then things go back to normal. A couple of times I thought I might be having a heart attack but then it would go away, and I would feel fine again. I have an appointment with Doc Markley for a physical next week, but until then I plan on taking it easy. Ralph must be noticing something as well. He follows me around and stays right by me when we're outside. That's unusual as he usually heads off for a jaunt in the woods when we go out. This morning, I had one of those *spells*. When I sat down on the couch, Ralph sat down on the floor facing me and let out a howl. He has never done that before! He must sense something. I have no idea what is happening, but it is strange.

I've logged those spells once I noticed they seemed to be coming with increasing frequency and intensity. The results startled me as the time between spells has been decreasing regularly by about one-third each cycle. Projecting this forward, the time between spells will disappear at about seven this evening. I have no idea what this means, but there must be some significance. I'm not in fear, but I have a strange, sad feeling about this. Each time a spell comes, I have an overpowering feeling of sadness which makes no sense at all. I look at Ralph and almost burst into tears as he looks back at me. Maybe I have an unusual hormone imbalance. Hopefully it will pass, and things will return to normal.

I want you to take the computer. Consider it a gift. Take the briefcase too. Tell your mother and dad and your aunt Matty I love them and thank them for being such a wonderful part of my life. Share with them whatever of this note you feel is appropriate. I love you very much. You have been the son I never had. I am so proud of you, what you've done, and of the man you have become. Keep that spirit of adventure and hunger for knowledge alive throughout your life. I hope you will live a full one.

In the briefcase and also in the *Charlie* directory on the computer, you will find a collection of sayings, poetry, essays, letters, and miscellaneous quotes and writings I have saved over the years. These are ideas, happenings, experiences, and concepts that have been of great value to me and tell about the man I tried to be. I hate to give advice, as you know, so consider this collection of words as a sharing for you to use as you see fit. I finish this note with a quote from Alfred Adler that I applied liberally to my life. It may explain a great deal. "There is only one danger I find in life. One may take too many precautions."

The note was signed, "With love and respect, C."

Charlie finished reading the note and sat there on the bed, tears streaming down his face. He felt certain he would never see his uncle again. He sat on C's bed for a long time, staring into space and recalling memories of times with his uncle. He missed C terribly, and now there was a big empty place in his heart. This would hurt for a long time. It was the first loss of someone close Charlie had ever experienced, and he knew he wasn't handling it very well. Sharing the note with his folks and Matty would be a painful necessity. Convincing them C was not coming back would be difficult. The troubling thing was wondering what had happened? Not knowing that answer would be maddening, but he believed they would never know.

It was past ten when Charlie heard the door to the cabin open. It was his father. "Are you okay, son?" Ray asked. "We were beginning to worry when you didn't come back. When Ralph showed up at our door, we thought maybe something else happened, so I came to find out."

"I found this note from C," Charlie replied, holding the paper in his hand. "I don't think we'll ever find him. He's gone, and not under his own power. Right now I feel tired. Let's get back to the house. It's late, so let's hold off examining and talking about the note until tomorrow."

Charlie left the computer and briefcase where they sat and headed out the door. He put his arm around his dad, and the two of them headed home. For once his dad didn't stiffen at his touch, but responded by placing his arm around Charlie. It was a powerful, loving message that warmed Charlie's heart.

When they reached the house, his mom opened the door, saying, "Is everything all right? We were worried."

"Yes, everything's okay," Ray answered as they went inside. Neither of them mentioned the note to Edith. They would tell her about that tomorrow.

When you do the common things in life in an uncommon way, you will command the attention of the world.

—George Washington Carver (1864-1943)

If a man does his best, what else is there?

—General George S. Patton (1885-1945

In any so-called *equal* society, there will always be those who are more *equal* than others.

*—the words of George Orwell in **Animal Farm** as adapted by Howard Johnson, 2001*

I do not feel obliged to believe that the same God who has endowed us with sense, reason, and intellect has intended us to forgo their use.

—Galileo Galilei (1564-1642)

I found solace in nursing a pervasive sense of grievance and animosity against my mother's race. There was something about her that made me wary, a little too sure of herself, maybe and white.

—From Dreams of my Father - Barack Hussein Obama

The real, the sweetest taste of victory comes when you win in your adversary's battlefield, fought with his weapons and his set of rules at a time of his choosing, when losing would cost you no loss of stature. Even more so when you are your own adversary!

—Howard Johnson, 1972

Honest work bears a lovely face, for it is the father of pleasure and the mother of good fortune. It is the keystone of prosperity and the sire of fame. And best of all, work is relief from sorrow and the handmaiden of happiness.

—Unknown

Indeed . . . Man does not live by bread alone,

but without bread, man does not live at all!

—Howard Johnson, 1983

Socialism has a record of failure so blatant that only an intellectual could ignore or evade it

—Thomas Sowell.

Love is friendship set on fire.

—Jeremy Taylor

The Mask

Alexis carefully slid the package into her leather purse, slung its strap over her shoulder, and casually walked around the corner into the heart of the space port and up to the customs desk. The fat albino behind the desk looked up from his monitor to stare at her. She noticed his right hand sliding slowly out of sight and down by his side.

"Ah, Ms. Stereo, back so soon?"

From behind her, a familiar voice shouted, "Alexis! Don't!"

Alexis spun around, "Schad! Where have you been? I've been worried something had happened to abort our trip. Were you afraid I was about to retrieve my stuff from customs and cancel our flight?"

"Something like that. I... I... took the wrong shuttle and had a frantic time getting back as soon as I did. My communicator quit working, so I couldn't contact you."

Ignoring Schad's lame comment, Alexis turned back and glared at the albino as she spit out, "Touch that alarm button, and you'll have more trouble than you can imagine."

The albino raised both hands, palms forward. "Why did you think I was going to hit the alarm, Ms. Snotty?"

"Enough of the swazzo crap you fat slob! I'm not blind. I gave you enough cash to cover any contingency, and you took it. It's all recorded right here." Alexis snarled, patting her AV. "One wrong move on your part, and you'll spend a long time rotting in the Ranko penal colony."

The albino's pink eyes morphed from arrogance to radiated fear. A bit of drool ran down his chin from a loose, shivering lower lip. "I'm sorry. I'll do as you told me. It's just that I didn't expect you back so soon. You told me you were leaving on the two-ten, and it's past three."

"So you figured something went wrong and started to turn me in and keep the cash. Definitely not a good thought!"

Whirling back to face Schad, Alexis narrowed her eyes. Something was not right, and she knew it. She noticed two obvious security cops moving toward them, trying to appear casual. "How in hell could you have taken the wrong shuttle?" She snapped, trying to decide, run, fight, or wait for a better opportunity. Reality hit her like an LK blast. "You bastard!" she shouted at Schad. "You pulled a switch and turned me in for the reward."

As one of the cops pulled his Galbo blaster, she dove at Schad. In a single motion she pulled him in front of her, grabbed the LK from her leg, and rolled into firing position. The Galbo cut Schad in two as the stutter from her tiny LK knocked down the cop who had fired and leveled the other one before he could raise his weapon.

Back on her feet and running at the end of her move, Alexis headed for the neutral zone at the end of the terminal. Galbos didn't work within the force field there, but her LK did. She would be safe for the moment. It would take the two cops about an hour to recover, and by that time, she would be long gone. Fortunately, running in the Ranko space port was quite common and drew little attention. The bloody mess that had been Schad and the two unconscious cops drew all of the attention. People hardly noticed a small woman in a dark-blue Cirec suit running from the mayhem. Many others nearby did the same when the excitement hit. The tiny LK, unnoticed in her palm, didn't signal any threat.

As she approached the neutral zone, Alexis slowed down to a trot and chanced a look back. No one was looking in her direction as a police shuttle streaked toward the gathering crowd, ordering people out of the way with a blaring PA. Numerous guards on foot were hustling toward the scene and moving gawkers out of the way.

So far so good, she muttered under her breath. *Now, if I can get aboard one of the outbound shuttles before that damned albino spills his guts.*

She scanned outbound flights and chose the one headed for Stentor 7. It was about two hundred yards from the neutral zone. She was sure she could make it before all hell broke loose in the main port. If her fake ID chip cleared the security scanner, she would be home free. The ancient Telurian mask contraband in her purse was another matter. It was the reason for the whole setup with Schad. Worth several hundred million on the open artifact market, it was her ticket to freedom she was to have split with Schad after he smuggled it through security, his specialty.

How in hell am I going to get this past security? Alexis wondered as she neared the gate. A smile crossed her face as a brilliant plan popped into her mind. She stopped for a few moments in one of the Icom booths, made an adjustment to her purse and threw the now-empty mask packaging into the trash vac.

As she entered the security scanner, the operator, an attractive young blonde, commented, "What an unusual purse. Where did you get it?" as she opened the purse and examined the contents.

Alexis tried to appear casual as the inspector closed the purse and put it on the *passed* counter. "It was a gift from a dear friend," she commented sweetly. "I have no idea where he got it, but I do like it." Then she had another idea. "Incidentally, would you be a dear? His wife is a vindictive bitch and may have found out about it. If anyone comes looking for me, tell them I'm on the flight to Aldebaran Three, over there. Her brother is a security cop, and I don't need the hassle."

"I know what you mean," she answered with a wink. "Angry wives can be a real bummer."

As soon as her ID chip cleared, Alexis picked up the purse and headed for the flight at a rapid walk. *So far so good* she thought. Once the boarding shuttle cleared the terminal, she would be home free.

Taking a window seat, she watched the security gate she had just cleared. Two security cops ran up to the blonde, pointing and talking rapidly. "Good girl!" she said out loud as the blonde pointed to the other boarding shuttle and the cops took off at a run. The shuttle doors spun shut just before they got there. The blonde looked at her shuttle and gave a thumbs-up. *Sometimes you need a little impromptu luck to complete a good plan,* Alexis thought as she leaned back in her seat for the short hop to the IS craft.

As the shuttle lifted off, she began dreaming of happy times on Stentor 7 with the fortune the mask would bring on the open market. She wouldn't have to share it with Schad as planned. The Telurian mask stared up at her and seemed almost to smile in spite of the Zepok fasteners that held it seamlessly and flush on the front of her purse.

In human intercourse the tragedy begins, not when there is a misunderstanding about words, but when silence is not understood.

—*Henry David Thoreau*

An untruth that conveys a true meaning from one person to another is, in fact, a truth—a truth that conveys a false meaning is in fact an untruth. Truth or untruth is not in the medium, only in the message!

—*Howard Johnson, 1968*

There is only one danger I find in life. One may take too many precautions.

—*Alfred Adler*

> Slow me down, Lord!
> Ease the pounding of my heart by the quieting of my mind.
> Steady my hurried pace with a vision of the eternal reach of time.
> Give me, amidst the confusion of the day,
>   the calmness of the everlasting hills.
> Break the tension of my nerves with the soothing music of the singing
>   streams that live in my memory.
> Help me to know the magical restoring power of sleep.
> Teach me the art of taking minute vacations of slowing down;
> To look at a flower; To chat with an old friend or make a new one;
> To pat a stray dog; To watch a spider build a web;
> To smile at a child; To read from a good book.
> Remind me each day that the race is not always to the swift;
> That there is more to life than increasing speed.
> Let me look upward into the towering oak and know that it grows
> great and strong because it grew slowly and well.

—*Orin L. Crain*

Heavens! How many obstacles there are between a resolution and its fulfillment! How much compromising to be done with unessential issues to preserve the main thing whole and worthy! Each new obstacle to be surmounted in its turn, its smashed entanglements converted into means toward the main end! And the main end never to be overlooked, forgotten, substituted, changed, abandoned, nor once dishonored by a coward doubt! The worst hour is the eve of the final effort, when the goal that seemed so near, seems passing out of reach, and all the work done hitherto that seemed so wise, appears ill done and ill-conceived, and all the unpredictable, imponderable dangers suddenly invade the mind like specters. Then a man needs courage. Aye, he needs the courage to believe his vision all along, from the first until now, was clear, and all his efforts well aimed to a good conclusion.

—*Talbot Mundy in* Tros of Samothrace

The most fundamental fact about the ideas of the political left is that they do not work. Therefore we should not be surprised to find the left concentrated in institutions where ideas do not have to work in order to survive.

—*Thomas Sowell*

The Switch

Timothy O'Brien never expected to find himself in such circumstances. Whoever would? He paused on the narrow path on the side of the grey granite mountain, the wind pressing on his backside as if it wished to push him to his death on the ragged rocks a kilometer below. A sound caught his attention over the rush of wind.

A sound heightened my fear responses and made me shudder. "How could they have found me?"

The sound was unmistakable. Vordanay thrusters have a unique noise profile and the only units that use these old but effective thrusters are Old Earth military police RG vehicles. hard to handle in Earth gravity, RG vehicles would be fast and quite maneuverable in the light gravity of Stentor seven where I had hidden for seven years.

I looked for a place to hide, but saw none. In a panic I started running down the path as fast as I could. The light gravity stretched my running steps to ten meters, but I had to plan a landing place for every huge step. One misstep and I would be off the path with a kilometer of air between me and the rocks at the base of the huge cliff. My mind raced trying to find an answer, but none was forthcoming. Then I came to a slight curvature to the right on the almost flat and vertical granite face of the mountain. I shortened my paces to stay on the path.

On my third giant step, I misjudged and missed the path by a meter. I was hurtling away from the cliff at a slight angle and beginning to drop. In the low gravity of Stentor Seven I would still be falling fast enough to be killed after falling a kilometer. For a moment I wondered if terminal velocity would be low enough to let me land without a major injury. Knowing that, I spread my arms and loosened my shirt to slow my descent.

The sound of the Vordanays grew louder. "They've spotted me." I muttered to myself as I tried to look in the direction of the sound. "Shit! Even if I survive the fall, they'll have me." They were coming from my blind spot above and to my rear. Then something flat and heavy hit me with considerable force and everything went black and silent.

As dazed consciousness returned, I realized I was out of the wind and in the vehicle with the Vordanay thrusters. Then I opened my eyes and looked into the face of my bride of six weeks, Enid. She had her finger to her lips, so I complied by lying still and quiet. Enid pulled a Gleary laser pistol, leveled it my head and said, "He's coming to. I've got him covered," to the pilot. Looking at me she snarled, "Lie still and don't move."

I was devastated! Had my sweet lady sold me out in order to collect the 400,000-credit reward for my capture? Something was not right. This was impossible. Then I noticed her little finger was waving in front of the pistol. The safety switch, right where her finger was pointing, was on. With the safety on, the Gleary was as harmless as a toy.

The pilot said, "Lady, I have no idea what you have against this guy, but I do appreciate your help. We'd never have caught him without your tip. He's in for some rough treatment when we get him back to Earth. I'd rather have blasted him when we found him and only had to deal with the body. Now he'll get prisoner treatment, and that means a lot more work for me."

"I have my reasons for wanting him alive. You can't know why." Enid said, winking at me. "I'm sure not going to tell you, Jack."

"I can say one thing," Jack remarked, smiling. "I'm glad you're on my side. I'd hate to have you working against me."

I had no idea what was going on, but that safety and Enids's wink were reassuring.

Jack glanced over his shoulder. "Watch him closely. He's the only one ever to escape our holding center. I have to report his capture and send him on his way to Earth within the next five days, or those charges we have against him will expire. He'll no longer be a fugitive."

"Incidentally, how are you going to get us past security here at the terminal? He destroyed his ID chip and mine's specific. Tim's popular with the locals and until I get cleared of some legal crap, I'm not."

"You leave that up to me. I've got connections." Enid said, smiling knowingly.

"Well, you got me in, so I suppose you can get me out OK. We're approaching the terminal."

"Remember what I told you." Enid ordered. "Set her down outside the confinement zone so you two can change clothes. When we walk into the zone, security will think it's Tim covering you, the renegade Earth agent, and let all of us through. Once through security we'll be in International territory and your jurisdiction will take precedence. Then you can process him, we can split the reward, you can retire here as you planned, and I'll have my revenge. Tim will be shipped back to Earth, and after that, who cares."

"Lady, I wouldn't want you working against me. You have a devious mind."

"If he only knew." Enid whispered to me as the set down sequence began.

As soon as we landed, Enid ordered, "Come back and cuff him to the hand rail. Then you'll be able to change clothes without danger. I'll sit back here and cover you." Then she moved to a seat in the rear of the vehicle. It took about ten minutes for us to change clothes, everything including underwear, which I thought a bit much.

Jack turned around and went to the front to retrieve his own pistol before getting out. As he did, I noticed Enid released the safety and cranked the control on her pistol up to max. As soon as Jack turned around, one blast from her Gleary caught him center chest. He was dead before hitting the floor. In an instant, Enid handed me her pistol, took a knife out of her pocket and began slicing Jack's chest open.

"What the hell?" I shouted. "He's already dead!"

With a flourish Enid held up a small, bloody ID chip and tossed it to me. "This is the key to the success of our mission. You are a free man with a new identity no one can crack. Stick it in your pocket. That will cover you for the five days until those charges expire. After that you're free and clear of those trumped-up charges."

She took the pistol out of my hand, cut the power to half, and seared the open wound on Jack's chest to close it. Then she hit his face with enough wide laser spray to make it unrecognizable. She handed me the pistol as sirens of approaching security announced a tense visit shortly. In a single motion Enid grabbed Jack's pistol, set the control on low, and blasted my left shoulder enough to burn Jack's uniform and sear my skin a bit.

"Damn it, Enid, that hurt!"

"Sorry about that," she muttered as she replaced the pistol in his hand.

As soon as she stepped back, the door flew open and two uniformed officers of Stentor Seven security entered, pistols drawn. "Don't anyone move. Touch the trigger on that pistol and you'll be fried," one of them said to me. I dropped it on the floor.

"He's a security agent from Earth." Enid panted, feigning fright. "His prisoner managed to grab one of his pistols and burned him. Jack had no choice but to terminate him. Scan him with your reader and you'll get his ID. That one doesn't have an ID, but check him anyway."

"Both of you stay still while we check this out. You know you're on sovereign Stentor soil, and your special privileges don't amount to anything." He said looking at me.

He activated his scanner and read the ID chip in my breast pocket. "Jack Evans, EAPD 17685 it says. It also says you're retired. What's that all about?"

"He's a bit dazed after flat-face there popped him with a laser pistol. This was his last assignment. He planned it to get him here to Stentor Seven to end his stint where he planned to retire. I'm his fiancé and can answer most of your questions."

The other officer looked at me and sneered, "If you're the Jack Evans we know about, you're in some trouble here."

Enid snapped back, "Not anymore. That's the guy who caused all the trouble and he's dead. Now, can't you get Jack some medical attention?"

Getting into the act and thinking about the bloody ID chip in my pocket I protested, "I don't need a medic. It's just a slight burn. He shot before turning up the power and grazed me. I didn't miss him."

"You sure didn't" the second officer muttered. "His mother wouldn't recognize him now. How are we going to be sure of whom he was?"

"The word of a field agent of the Earth Allied Police Department should be good enough for that. Besides, Jack will have to ship the body back to Earth on the next flight to complete his mission." Enid remarked. "That will save you a whole lot of paperwork. All you have to do is move us inside the zone and out of your jurisdiction. We can do the rest."

"You're right about that. It would save us a lot of grief."

"I'll reassign this old RG vehicle to you." I said with a sudden inspiration. "I don't need it anymore, and it's gotta be worth something."

The two officers looked at each other, nodded their heads, and in unison said, "Deal!"

✳   ✳   ✳

By nightfall of Stentor's thirty-hour day, Enid and I were sitting at the table in her apartment. Enid explained the entire plan and how she had arranged so much.

"My position at Nebson Security Research gave me not only the information, but enabled me to pull this off. You did know it was Jack Evans that offed your brother and faked those charges against you, didn't you?"

"I was quite certain he was the one."

"That's why he wanted you so badly, that and the reward the EAPD paid us for terminating one Timothy O'Brien. Poetic justice don't you think?"

"I have one comment."

"Oh? What's that?"

"To quote one Jack Evans, 'I'm glad you're on my side. I'd hate to have you working against me.'"

Enid grinned.

No Bomb Needed

Abdu Rahman answered the door to his small rented Chicago house. It was FedEx with a package from Germany. He grinned as he signed the papers for the lanky, hawk-nosed deliveryman. The package contents were the essentials for his project and the only item he couldn't purchase locally. Shaking with excitement, he took the package into the bathroom and placed it on the counter next to the sink. Thoughts of the next steps in his project and the deadline four weeks away raced through his mind. *At last, I can do my part,* he thought as he headed for the kitchen.

He opened the refrigerator and removed several of the five-pound packages of hamburger which filled the inside. *How proud my father will be,* he thought as he picked up the large, heavy-gauge plastic trash bag from the kitchen counter, tucked it under his arm and headed back for the bathroom. Placing the bag and the packages of hamburger on the counter, he directed his attention to the bathtub. After checking the silicon sealant he used on the drain, he turned on the hot water. When the tub was about half full, he unfolded the trash bag into the tub and ran several gallons of water into it. Next, he opened one of the packages of meat and emptied it into the trash bag with the water. The pungent smell of the warmed meat assailed his nose. Ten days in the frig and it was beginning to spoil. *Excellent,* he thought as he picked up a second package and dumped it into the bag.

A few trips to the kitchen and all one hundred pounds of meat were in the bag, supported by the water in the tub. He secured the top of the bag to the wall above the tub with duct tape to keep it free and away from the slurry of meat inside. The next step would be tricky. Abdu opened the FedEx package and removed the plastic bottle from the bubble wrap. He glanced at the short piece of garden hose he had placed on the floor by the tub, checking to make sure it would be within reach. Then he carried the bottle filled with a thick greyish brown liquid to the tub. To calm his jumpy nerves, he paused, holding the bottle above the bag opening. *Now for the tricky part,* he thought as he unscrewed the bottle cap and removed the seal. He lowered the bottle into the bag, inverted it, and dumped the contents into the meat slurry. When the bottle was empty, he dropped it into the bag.

Abdu turned to the sink and washed his hands with disinfectant soap. He breathed a sigh of relief as he dried his hands to prepare for the next step. Picking up the seven-foot section of garden hose that was cut from the one he had purchased, he thrust the cut end slightly inside the bag opening and held it there as he wrapped the bag tightly around the hose. Then he used duct tape to seal the bag to the hose so the only opening would be the other end of the hose. Raising the open end of the hose near the vent above the bathtub, he fastened the hose to the wall with duct tape in many places. The vent fan would remove the noxious fumes soon to be generated in the bag as the meat decomposed. He then turned on the new vent fan he had installed in the ceiling. There would be four weeks of waiting as the evil brew cooked and ripened.

During this period, Abdu did a bit more shopping at the local plumbing supply store. He purchased two six-foot pieces of metal-covered, high-pressure hose with standard female hose connections on each end and two tap-in saddle-valves with hose connections. These saddle valves were the kind which cut their own entry hole into the pipe with a sharp hardened point on the tip of the valve stem. He also bought a small high-pressure pump, the kind that powers the water jets used to strip paint. Abdu smiled as he paid the bill. "Only $486," he commented smugly to himself. "Add that to the twenty dollars at the hardware store and rent for three months, and it comes to about $2,000. Who says weapons have to cost billions?" He was quite pleased.

Once back at the house, he went about connecting and testing the new equipment. He replaced the aerator on the bathroom sink faucet with a hose adapter and connected the high-pressure hose between the adapter and the pump on the floor. A short suction hose cut from the other end of the garden hose was also attached to the pump. The open end of the garden hose was placed in a bucket of water Abdu had placed on the floor. "Now to try it out," he said as he opened the cold-water faucet of the sink. The sound of the sudden flow of water into the high-pressure hose stopped as soon as the hose filled. The positive displacement pump acted like a closed valve and prevented the water from flowing through the suction hose.

"So far so good," he mumbled as he reached to plug the power cable of the pump into the wall socket. The pump whirred into life, then objected loudly and strenuously as the check valve in the meter stopped the back flow. He immediately pulled the plug to stop the pump and headed for the basement with the other metal hose and the two saddle valves. "Now to bypass the meter and its check valve," he murmured under his breath as he walked down the stairs.

Abdu turned on the naked light bulb and lay the hose and valves on the floor next to the water meter. *How simple,* he thought as he fastened the first valve upstream from the water meter. When it was securely in place, he twisted the valve handle, driving the point through the pipe wall. He watched as he cracked the valve open. A trickle of water brought a lusty "Allah be praised" from his lips. Then he installed and tested the downstream valve, connected the metal hose between the two valves and opened them. When he saw there were no leaks, he headed back to the bathroom.

"Now it should work," he remarked to himself as he once again plugged the power cable of the pump into the wall socket. The pump spun once more, groaning a bit noisily as its pressure fought the system pressure. Soon, the higher pressure of the pump drove the water from the bucket into the faucet and then into the water system. "Praises be to Allah! It works!" Abdu shouted as the bucket emptied. He pulled the plug, refilled the bucket, and tried it again. Several tests later, he shut off the faucet, disconnected the hoses, and moved pump and hoses to the end of the room. All he could do after that was wait.

As he sat watching the TV news for any signal that would cancel his project, he thought of how easy it had been to gain access to the details of the Chicago water system. It pleased him that it took only three months to find and rent a small house fed directly by the main water line near the treatment plant. He was particularly pleased since he was allowed six months by his cell leader at the cell meeting where the project was given to the members. At the meeting, all members had been forbidden any further contact with any member of the cell until the project was completed. For the rest of the project, Abdu was on his own.

As he continued watching, the news shifted to a panel of defense experts discussing the costs associated with the Star Wars missile defense system. When the spokesman stated, "Only about a hundred billion dollars to keep America safe," Abdu laughed out loud. "Stupid American infidels with your useless expensive toys," he snarled at the TV. "Wait until you learn what we have in store for you."

On the appointed day, Abdu awoke early. He was so excited he could not sleep. His hands trembled as he put on his clothes. It was Wednesday, November 21, 2001, the day before America's Thanksgiving Day. On this date, many others of his jihad cell would be doing exactly what Abdu was doing, hidden inside small unobtrusive homes in quiet neighborhoods. They were acting in large cities all over the United States. He hated having to wait until eleven o'clock. The time went excruciatingly slow. To help pass the time, he again checked everything: the bypass hose in the basement, then the rig in the bathroom, the pump to the faucet. Once more he stuck the suction hose into the water-filled bucket, and repeated the test done earlier. When everything worked, he went to the living room, sat in front of the TV to watch the news. He jumped from channel to channel, watching for an interview with the code words which would abort the project. "Allah! Don't let those words come," he said repeatedly while watching.

When the clock reached eleven and with no words to abort, he headed for the bathroom, placing a disinfectant-soaked face mask securely over his face. He opened the door. Moving swiftly, he took a knife, slit the plastic bag, and emptied the contents into the tub with the water. The putrid smell of the reddish brown foamy mass assailed his nostrils through the mask, almost driving him out of the room. He held his breath as he lifted the now-empty, dripping bag from the tub and placed it in the bucket on the floor. He secured the suction hose in the tub, turned on the pump, and watched as the pump groaned and began to pump the foul mess into the water system. Before shutting the door, he made certain everything was working properly. All that remained to be done was to check the pump periodically to be certain it continued to pump. With the suction side restricted, it took several hours to empty the tub. As soon as the tub was empty, Abdu shut off the pump, closed the bathroom door, took his belongings, and headed for the international terminal at O'Hare Airport. As he boarded the international flight that was the first leg of his trip home to Yemen, he smiled smugly to himself. He was certain that his efforts would help destroy America.

Epilog

The Next Few Days - The evil biological soup flowed through the house pipes, the street junction, and into the large main under the street which served a major section of Chicago. Mixing with the flow downstream from the treatment plant, the reddish brown color disappeared as it was diluted by the large amounts of water flowing in the main.

A mile or so downstream on the tainted main, Beth Sosa drew some water to mix formula for her two-month-old daughter, Maria. It was early Thanksgiving Day, and many of her family would be coming for dinner. Not far away, Alan Black drew water to make coffee for himself and his brother Carl who lived with him. Across the street, Adrian Melchior mixed concentrated orange juice with tap water for his family's breakfast. It would be months before the deadly prions in their blood would complete their relentless rage of doom.

May 2007 - It was May when the CDC first realized something was wrong. Reports of a strange neurological disorder of small children began coming in from all over the country. Slowly at first, but then in growing number, infants and small children were losing control of their limbs and those walking were starting to fall down. In July, ten-month-old Maria Sosa of Chicago died of the disease. By September, the death toll was rising, and the first older children and adults were beginning to show signs of the disease. In November, Beth Sosa, the two Black brothers, and Adrian Melchior died. They would be the first known victims of a mysterious neurologic disease which was soon claiming thousands of victims in large cities all over the nation. Then came a sobering announcement from the CDC.

The symptoms of this new disease are the same as for variant Creutzfeldt-Jakob disease (vCJD), a type of bovine spongiform encephalopathy (BSE). BSE is a transmissible, neurodegenerative, fatal brain disease of cattle. The disease has a long incubation period, but ultimately is fatal within weeks to months of its onset.

BSE, commonly called mad cow disease, first came to the attention of the scientific community in November 1986. At that time, the appearance in cattle of a newly recognized form of neurological disease appeared in the United Kingdom (UK).

BSE is associated with a transmissible agent. The agent affects the brain and spinal cord of cattle and lesions are characterized by sponge-like changes visible with an ordinary microscope. The agent is highly stable. It resists freezing, drying and heating at normal cooking temperatures, even those temperatures used for pasteurization and sterilization.

vCJD was first reported in March 1996 in the UK. In contrast to the classical forms of CJD, vCJD has affected younger patients, has a relatively longer duration of illness (median of 14 months as opposed to 4.5 months) and is strongly linked to exposure, through food, to BSE. Recent studies have confirmed that vCJD is distinct from sporadic and acquired CJD.

The larger the infected person, the slower the disease progresses. As of this date, there is no known cure once the agent has been ingested. The disease is always fatal.

2008 and 2009 - By mid-2008, the disease had decimated the United States and spread to Canada and Mexico. More than a hundred million were dead, and the disease showed no sign of slowing. Deaths soon overwhelmed disposal facilities, and city streets became littered with the dead and dying. All warm-blooded life was affected, which added more decaying flesh to the streets and fields everywhere as birds and small mammals succumbed. Then began a series of events unforeseen by the terrorists. It began appearing in isolated areas all over the globe, first in Europe then Africa, Asia, South America, and Australia. It seemed to be transmitted in bottled drinks, then foods, and finally by air. The year 2009 saw the disease in every corner of the globe, and no one was left to try to stop it. Decaying birds and mammals covered every continent. The sea was littered with bloated decaying cetaceans as porpoises and finally the whales began to succumb. Reptiles, sharks, bony fishes, and other cold-blooded life-forms did not contract the disease.

Final Note - On July 8, 2013, the last human died. Mammals larger than humans lasted a bit longer, but by 2015, the last whale died, and the Earth was free of all bird and mammalian life. All warm blooded creatures were extinct.

By doubting we are led to inquire;
    by inquiring we perceive the truth.

—*Peter Abelard*

Choose, and take the consequences. Choose to command, and learn the pain of the barbed treachery of envy. Choose to obey, and learn how soon obedience begets contempt. Choose the philosopher's life, and learn the famished waste of thought that, like a barren woman, lusts unpregnant. Choose . . . or become the victim of others' choosing.

—*Talbot Mundy in* Tros of Samothrace

> If a child lives with criticism He learns to condemn.
> If a child lives with hostility He learns to fight.
> If a child lives with ridicule He learns to be shy.
> If a child lives with shame He learns to feel guilty.
> If a child lives with tolerance He learns to be patient.
> If a child lives with praise He learns to appreciate.
> If a child lives with fairness He learns justice.
> If a child lives with security He learns to have faith.
> If a child lives with approval He learns to like himself.
> If a child lives with acceptance and friendship
>   He learns to find love in the world.

—*Dorothy Law Nolte*

Tradition means giving votes to the most obscure of all classes, our ancestors. It is the democracy of the dead.

—*G. K. Chesterton*

The best argument for Christianity is Christians: their joy, their certainty, their completeness.

But the strongest argument against Christianity is also Christians: when they are somber and joyless, when they are self-righteous and smug, when they are narrow and repressive, then Christianity dies a thousand deaths.

—*From a sermon by Pastor Barbara Johnson*
quoted from Sheldon Vanauken, April 1997

There only two possible results when the lion shall lie down with the lamb. The lamb will be eaten, or the lion will starve.

—*Howard Johnson, 2004*

A Doctor from Detroit

Dr. Francis Lane drank in the lovely view before him. Lush tropical plants crowded the edge of the pond both on land and in the water. Several Hawaiian geese, or nenes, glided across the surface, along with many ducks. White wading birds dotted the edge wherever the vegetation was sparse. The pond was fed by a small stream which flowed in at one end and out over a concrete spillway at the other. The stream continued into a large culvert under the main road and then meandered off on the other side. To the east of the pond and the bordering street was a small hillside where their home merged into the lush vegetation. It was early evening, and Dr. Lane sat quietly on his patio with his wife, Oona, enjoying an after-dinner drink and gazing down across the pond in front of their home in Hilo, Hawaii.

Though the house was rather large, it didn't stand out but merged into and became part of the hillside. Looking up the slope from the street, all that could be seen was the edge of the hipped roof and the upper part of some huge picture windows. It seemed almost a wilderness, undisturbed by human invasion. From the cul-de-sac in front of the house, the driveway curved to the right and upward, disappearing behind the sloping front yard as it turned toward the three-car garage beneath the house and hidden by the slope of the yard. There was a doorway to the right of the garage doors and a stone walkway that climbed lazily up a slope between the garage and the mound which hid it. Walls on both sides of the walkway were faced with flat, casually stacked lava slabs. Plants grew profusely, drooping down the irregular stone walls and nearly closing the view to the sky above. At the top of the walkway, there was a surprisingly large patio overlooking the yard, pond, and cul-de-sac. The patio and main entrance to the house could not be seen from the street or driveway below. Line of sight from the street made the house roof appear to come to the ground at the edge of the patio. The view west from the patio was breathtakingly beautiful.

The main entrance to the house consisted of two huge glass doors centered between two equally large glass windows. It was a wall of clear glass. When the doors slid apart, the house opened itself to the outdoors and became part of it.

Beyond the patio, hidden by a wall of vines growing on a broad trellis, was a lanai with a pool. Entrance to the pool was through a two-gated arch in the trellis which almost hid the pool from view. The poolside lanai was ideal for casual entertaining with several wrought-iron tables and chairs mixed in with beach loungers. Like the entrance patio, it could not be seen from the road below. A glass wall opened from the large recreation room to the pool area exactly like the patio wall. On the far wall of the room was a huge picture of a football running back, exploding through a group of would-be tacklers.

Beyond the recreation room, the house tumbled up a gentle slope, half a floor at a time for a total of four levels, including the garage. It was as spectacular inside as it was out with the clever, tasteful blending of art and decorations reflecting the four cultures of the owners with a local Hawaiian flavor.

As Dr. Lane continued looking silently to the west, Oona followed her husband's gaze toward the pond and the sun beginning to drop behind the mountain. She finally remarked, "It's such a beautiful scene. I never get tired of it. I can't imagine a lovelier place to live."

"Yes, it is lovely, spectacular in fact. A far cry from where I originally lived," Francis commented as he thought of how he came to be here at this point in his life.

✻   ✻   ✻

Francis Lane had not always been so fortunate. He grew up in the inner city of Detroit during the late fifties and early sixties. His father, an ethnic Chinese, had been a math teacher in Beijing, China, who escaped to the West from Czechoslovakia while on a cultural exchange there. He waited a long time in England and then Canada. Finally, he was admitted into the United States as Charles Chang. He came to live with relatives in Detroit and worked as a cook in a Chinese restaurant in the city. His mother, Louanne, an inner city African American, was a waitress in the restaurant. Less than a year after his birth, both parents died horribly when the restaurant was firebombed one evening shortly after closing.

Francis, named by his mother, was brought up by his maternal grandmother, Annabel Lane. She lived in a neat, clean little house which stood in contrast to the row of dilapidated old houses that populated the rest of the block. Annabel worked as a cleaning lady for an office building downtown to supplement the small pension she received as the widow of a soldier killed in World War II. There were five other family members living in Annabel's house: her brother and his wife, two daughters, and an elderly uncle. Annabel ruled the household with a firm, but gentle hand. Everyone was expected to do their part in keeping the house clean, neat, and in good repair. Theirs was the only yard on the block with grass and a neat flower garden in front. There was a productive vegetable garden in the back.

Under Annabel's firm guidance, Francis grew into an honest, capable young man and an excellent student at school. With a great deal of effort, she finally convinced his Chinese relatives to accept him and teach him something of their different culture. He spent each Saturday with the family of a cousin of his father who lived in a Chinese neighborhood not far from Annabel's home. The blending of the two cultures gave Francis a unique vantage point for growth and learning which fed his insatiable appetite for knowledge. It also created inner conflicts which led to growing problems as he matured. Knowing both cultures, yet not accepted by either outside of his two families, Francis felt isolated in spite of his happy, outgoing personality. A brilliant student in a difficult school, he was frequently involved in scuffles over name-calling. A week after his high school graduation, he was seriously injured in a brawl with several gang members. After a short hospital stay, he completed his recovery at home where Annabel nursed him back to health.

"Now, Francis, you'd best tell your grandma Annabel what this is all about," she asked repeatedly as she tended his injuries. "I know somethin' o' what goes on in the streets 'round here, and I don't want you dead. Who was it, and why'd they beat you?"

"Grandma, you don't want to know."

"I've lived in this house for almost fifty years and seen some bad things goin' on in the neighborhood. Grandma Annabel knows how to keep her yap shut when it needs to be . . . Tell me what this is all about and maybe I can help . . . I won't do anythin' without your okay . . . I can't help if'n I don't know what you're up against . . . Talk to me, chile . . . It was that Ahmed and his bunch, wasn't it? I hear the rumors. Why'd they beat you?"

Francis finally opened up. "They beat me because I refused to be a runner for their drugs, wanted me to hook up with some people in the Chinese community. They figured I could get to people they couldn't."

"Did you talk to any relations in Chinatown?"

"Nah! I don't want to get involved. Sooner or later, I'd get caught or killed."

"Won't they come after you again? Why don' you go to the police?"

"They're sure to kill me if they think I ever talked to the police. I've got to get out of Detroit as soon as possible. They'll kill me if I stay and refuse to run drugs for them. I have absolutely no choice about that."

"Where you gonna go, an' how you gonna live? What about that scholarship to college? Wouldn't you be safe away at school?"

"If I take that scholarship, they'll know where I am and come after me sooner or later. I can't take that chance. If I head to the West Coast and hide my tracks, they'll have a hard time finding me. Hopefully, they'll give up after a while."

"You crazy! I'm afraid you'll get into trouble what with no friends, no money, and all. How you gonna live?"

"Grandma, you taught me to always try to be a decent person. You did a good job. I won't do anything to get into trouble. You can count on it."

"Bless you, chile. You know how proud I am of you. I still don' want you to go, but where you thinkin' 'bout?"

"I thought I might try Chinatown in San Francisco. I'm obviously part Chinese and know enough of the language to pass. I'll make out okay. I'll keep in touch somehow."

"I still don' like it. You get healed up an' we'll talk some mo' 'bout Ahmed and his gang."

Several weeks after he recovered from the beating, Annabel found a long, loving goodbye note on her grandson's pillow. He left during the night with a plan that started a chain of events leading to many changes in his life.

Francis was both frightened and fiercely angry with the three who had beaten him. His passion for revenge and intended flight west were combined in his plan. The leader and local drug boss, Ahmed, had been a classmate of Francis. Ahmed made quite a bit of money in a short period of time and was building his organization with relentless force. Those who opposed him either disappeared or were found beaten to death. Francis was alive only because Ahmed believed he could still get him as a courier. The group had taken over an abandoned house several miles away from Annabel's. There were seven of them in the house who drove around in two fast cars kept in the backyard. Around the yard was a six-foot high chain-link fence with a locked gate. There were always at least two members on guard, one in an upper front window, the other in the rear.

Word among students at the high school was that the drugs were kept in the basement near the front of the house. Ahmed supposedly kept most of his cash in his favorite of the two cars, a high-performance black Camaro. He frequently bragged about his car, once telling Francis about its hidden ignition lock beneath the shifter cover. With this sketchy knowledge, Francis made his plan.

About four in the morning, dressed completely in black, he crept up to the fence in the back, lugging a suitcase filled with clothing and a satchel filled with tools. He took a large bolt cutter from the satchel and opened the fence with a hole big enough for him to move through. He then crawled through the hole and crept silently up to the Camaro. Remembering about the door wiring from his auto shop training in school, he slid under the car to a position beneath the front of the driver's door where he could reach the inside lighting wire harness. Using small diagonal cutters, he snipped the wire to the door switch, opened the door, reached inside, and felt for the ignition lock under the shifter cover. His heart stopped as he heard the loud click of the solenoid as it snapped into place. He knew the car was ready to roll. Next he felt for and found the starter button which replaced the key switch in the dash. The exact location of the switch would need to be known for later. He placed the suitcase on the back seat of the Camaro, left the door unlatched, and crept toward the fence.

The bolt cutter made quick work of the fence, creating a hole directly in front of the Camaro and big enough for it to drive through. He hoped it would give him room to run in, start the car, and get out with a minimum of problems. Next, he headed for the front of the house, pleased that the lookout had not seen or heard anything. As he approached the side of the porch, he heard snoring. One of the gang was sleeping on a couch right under the open window to the porch so he would have to be extremely quiet. Taking the gallon jug of kerosene he brought, he poured it across the edge of the wooden porch floor, hoping it would spread evenly across the porch. By the dim light of the one streetlight at the end of the block, he could see the reflection on the shiny surface of the kerosene as it ran to the middle of the porch and began to puddle there. *Perfect*, he thought to himself. *The porch sags in the middle.* Taking a large rag from the satchel, he spread it out on the porch in the kerosene. He then placed a dozen shotgun shells on the rag, spreading them evenly and facing the porch wall. After pouring the rest of the kerosene over the rag and shells, he closed the satchel, grasped it tightly in his left hand, and prepared for a rapid, silent escape to the rear. With his right hand, he took a lighter from his pocket and started to light the rag.

Before he could snap the lighter, a car spun around the corner and sped up the street toward the house. His heart pounded as he moved away from the porch and flattened himself against the ground. The lights shone brightly on the side of the house as the car approached. Thankfully, it sped on past the house into the night. The sentry on the second floor leaned out the window and cursed the car as it sped by. Francis could see him clearly by the light of the nearby streetlight. He hugged the ground, remaining immobile until the lookout withdrew from the window. He thought he would never leave. When he finally disappeared inside, Francis crept back to the edge of the porch and again prepared to light the rag. The porch roof hid the entire porch from the sentry, but as soon as the fire was lit, its reflections would be seen.

He lighted the corner of the rag, moved away to the rear to be out of the light, and paused for a moment to make sure the rag caught fire. Seeing the porch corner post glow orange from the light of the burning rag, he moved silently back toward the opening in the fence where he paused to try to catch a glimpse of the rear lookout. By then the nearby house was glowing in the light of the expanding flames. Next he heard the front lookout shout "fire!" By now Francis could see the rear lookout peering intently in his direction. As the light of the fire on nearby houses brightened the area, he remained motionless while wondering, *Have I been seen?*

The loud bang of the first shell to explode took the lookout out of the window toward the sound of the explosion. The lookout gone, Francis headed straight for the Camaro.

Showtime! It seemed to take forever to cover the thirty or so feet from the hole in the fence to the door of the Camaro. Several more of the shells went off before he reached the door. Throwing the satchel in the back, he jumped into the driver's seat and pushed the starter button. *Damn*! Nothing happened! He pressed it several times more. Absolute silence! By now he could hear shouts from the house. *Have I been seen?* he wondered once more as his mind screamed for action. *Try the ignition lock again.* He felt under the cover for the lock switch and snapped it off and back on. Still, the starter button produced silence. Maybe the ignition lock was sticking. Holding down the starter button, he flipped the switch off and on. Each time he did so, the starter jumped and fell silent. Then it dawned on him. *The damned lock was left on for a fast start, and I have been turning it off!* As soon as he flipped the switch once and pressed the starter button, the Camaro's engine exploded into life. He slammed it into gear, turned on the headlights, and vaulted toward the fence. As he zipped through the fence, the crack-crack of bullets hitting the Camaro's rear window sounded. *Why didn't the glass break?* he thought for an instant.

By the time he hit the end of the quarter mile alley, a pair of headlights sprung up behind him near the growing glow of the now-blazing house. They were in pursuit. He had planned his escape route well. At the end of the alley he took a left, then three blocks and a right, three more blocks, and a left would take him to Livernois. A right turn on Livernois and about a mile of main thoroughfare would take him past a police station. If they were still behind him as he went down Livernois, he would drive normally by the station, hoping his followers would not. If he could get the police in the chase, he should be able to get away. Just beyond the station was a short side street that curved to the left and ended at another where you could head back to Livernois or go in the opposite direction. There was a narrow, enclosed alleyway off that street which led to a doorway in the rear of a small manufacturing plant. With the windowless building above and on both sides of the alleyway, it seemed to be a hole in the building not large enough for a car. It was a footpath to the rear door and was wide enough for a car. In the daylight, it was dark and hidden. At night, it would be the perfect hiding place for a black car.

When he reached Livernois and headed for the station, his pursuers were a good half mile behind. There were more cars than he expected on the road. As he slowed to normal speed approaching the station, he could see them gaining rapidly. They were moving at high speed once on the straight, wide road.

It seemed to take forever to get to the station. As he passed, he could see several policemen outside on the station steps. In a sudden inspiration, he honked and waved as he passed. During the time he took to drive the single block to the side street, the other car covered the distance to the station. They were only a block away when he turned off and headed for the alley. At the speed they were traveling, they would not be able to negotiate the turn into the side street. That should give him time to dive into his hiding spot. As he approached the alley, he heard the loud screech of tires as brakes were applied in a panic stop. As he drove into the alley and up to the door, he turned off the lights, killed the ignition, and stepped out of the car into the pitch-blackness. The sound of tires spinning in reverse told him they had indeed missed the corner. More burning rubber as the car spun into the side street and accelerated past the alley. They missed his hiding place and continued. He could hear their engine as the car accelerated away on the next street. His plan worked. As he turned to examine the Camaro more closely, the wail of sirens filled the night air. Two police cruisers sped past the alley, chasing his pursuers. He smiled to himself at the astonishing success of his plans.

Taking a small flashlight out of his satchel, he stepped to the rear to inspect the trunk of the car. As he turned the light on the back of the car, he noticed several bullet holes in the trunk. There were several marks on the rear window where bullets must have hit and glanced off. *Bulletproof glass*, he thought to himself with a smile. Shining the light on one of the holes, he noticed something behind the hole. The bullet had penetrated the sheet metal of the rear only to be stopped by a heavy steel plate behind the sheet metal. It was an armored car! There was no place for a key to the trunk. It must be opened by a latch on the inside. He remembered how several boys had worked on hidden trunk latches in auto shop class at school. He spent fifteen fruitless minutes trying to find the latch.

The rear seat upholstery had four large upholstered buttons, two on each side. After fruitlessly examining the floor, roof, and sides both inside and out for a pull ring or lever, Francis began feeling around the buttons. The upper ones did not move as easily and freely as the others. He pulled and pushed one button and finally twisted it and heard a loud *click*. The trunk latch advertised its opening. Stepping to the back, he took hold of the trunk lid and tried to move it. Nothing happened. Maybe that wasn't the latch after all. Going back inside, he reached for the button to twist it again. It was then he noticed the seat back seemed loose. An easy pull and the left seat back came down, revealing a compartment more than a foot deep behind the seat. A repeat effort on the other side revealed a similar compartment. Both were empty.

Obviously there was a space several feet deep between the back of these two compartments and the rear of the trunk. He couldn't find a single seam in either one. How do you get to that space? The money had to be hidden there. As he closed the compartments, he noticed the rear seat was split to match the backs. It, too, had the large buttons on the front panel of each side. Twisting one brought the same *click* he heard from the seat back buttons. Lifting the seat revealed a thin black briefcase in a compartment just large enough to hold it. Under the other seat was another identical briefcase.

He gasped as he opened the first briefcase. It was packed with neat bundles of one-hundred-dollar bills—a lot of money. Opening the other revealed about half as many bills.

The rumors among students at school had been correct. Apparently, Ahmed couldn't keep things to himself and had to brag about them, he thought as he placed the cases back in their compartments and closed the seat. The next few moments he sat in the car, trying to decide his next course of action. He had planned to drive the car west to Chicago, and if he found the money, dump it there and buy another plain, older car to drive onto the West Coast. San Francisco was his goal. Although he was rather dark for Chinese, he did have strong Chinese features and planned to get lost in Chinatown. The problem was the bullet holes in the trunk. A high-performance, late-model Camaro with bullet holes in the trunk driven by a young black man would be an invitation for every policeman he passed. How could he hide the holes?

Suddenly inspired, he dove under the dash. With his flashlight, he spotted several wiring harnesses held together with shiny black electrical tape. It took him about ten minutes to peel off six short strips of tape and place them over the bullet holes. The trunk didn't look bad, rather like a few scratches on an otherwise smooth black surface. While he was deciding when to leave his hiding place, he turned on the ignition lock and then the radio to see if he could catch any news. It wasn't an ordinary radio. It was a police band radio set to the local police frequency. He was startled to hear reports about the end of a chase where five black teenagers were killed. Their car went through a filling station and struck a parked dump truck at high speed while trying to elude police. It happened about ten miles away. The driver of the car was identified as a local drug dealer named Ahmed.

What a waste, Francis thought. *Ahmed had been such a bright kid in grade school with me. Too bad he turned that good mind into the wrong direction.* Relieved of the fear of being chased across the country, Francis headed west, deciding to drive the Camaro all the way to San Francisco.

By the end of the week it took to make the trip, many things happened. A number of Detroit charities received anonymous cash donations of ten thousand dollars in boxes sent by first-class mail. The Chinese community received a donation of fifty thousand for their proposed youth center. Annabel's church received a fifty-thousand-dollar donation earmarked to be used for youth programs. Annabel was the treasurer of her church and would administer the donation. She had no idea about the source of the funds. A string of five-thousand and ten-thousand-dollar donations were received by youth charities in inner cities across the country.

There were many bank accounts opened in small town banks from west of Chicago to Reno, Nevada, by one Charles Chang. Francis explained he would be moving there soon and wanted to have a local account. Francis Lane, as Charles Chang, was using his dead father's social security number on all these bank accounts. He noticed he was accepted as Chinese without question when he gave his name as Chang. The forty or so bank accounts totaled in excess of one hundred thousand dollars. Annabel had taught him well about bank accounts and squirreling away money.

In Cheyenne, he stayed long enough to have the holes in the rear of the car fixed and the car painted metallic silver. He paid cash to have it fixed. He explained the holes saying some hunters were shooting at a billboard, not realizing his car was behind it. His explanation was greeted with, "We have those kinds of problems with crazy city hunters all the time." To avoid obvious questions, he asked that the holes merely be filled from the outside, smoothed and then painted. He didn't want anyone poking around and trying to open the trunk. When asked about the steel plate in the trunk, he explained he added it to provide two hundred pounds in the rear to balance the power of the engine for slippery streets during Michigan winters. That seemed to make sense to the body man.

When he reached San Francisco, there was about fifty-thousand dollars left in one of the briefcases. Once there, he found a small furnished apartment above an import store and moved in. He asked about a job at the import store when he noticed a posting on the back of the front door. The owner, who was his landlord, said he was holding the job for a relative coming from China. When he asked if he could fill the job temporarily until the relative arrived, he was put off for a few days. He tried again, asking to be able to work enough to earn his rent.

After a lengthy discussion in Chinese with his wife, he turned to Francis, saying, "You start job next Monday. I pay you five dolla' an hour. You pay me rent as usual, okay?"

Francis agreed with a broad smile. Apparently his persistence paid off. Francis decided to do something about the car soon. He wanted to find out what was in the space between the rear seat compartments and the steel plate at the back. Opening both compartments, he searched for a way to get behind them. While banging on the back of the compartment, he realized there was a steel plate there as well. There was more than two feet of space between the two steel plates. Apparently, access was only possible through the bottom of the trunk. While on the ground and looking at the trunk bottom, he found a tire well on one side and gas tank on the other. He had missed something.

Again he went inside and opened the rear compartments. Examining them, he discovered a one-quarter-inch threaded hole at the top outside edge of the rear panel, flush with the carpet covering the back of the compartment. Francis remembered the two wing-screws in the glove compartment. He reached over and removed them, screwing one into this threaded hole. When he tightened the wing-screw, it went in all the way and kept on turning. As he continued to turn the screw, he realized it was beginning to come out, bringing a three-eighths-inch round rod with it. Withdrawn, the rod was about twenty inches long. Checking the other side, he discovered the same threaded hole at the extreme left top. Again, screwing in the wing-screws brought out the three-eighths-inch rod. With both rods withdrawn, it was clear that the trunk lid was free. Stepping out of the car, he took hold of the edge of the trunk lid which moved a little. Then he remembered the other objects in the glove compartment, two heavy plastic handles. From their shape, it was obvious they were used to lift the trunk lid. Taking the two handles and placing them on either side of the trunk and engaging the thin graspers with the side edges of the trunk, he lifted the lid straight up, the only way it would move. As he raised the lid, he could see four more steel rods protruding down from the lid. He lifted the heavy trunk lid until the rods came free from their guide holes and placed it on the ground.

Peering down into the now-open trunk, he found two soft brown leather zippered bags. As he lifted the bags from the trunk, he realized one was empty. Opening the other one, he found several plastic bags with white powder. He had been transporting a large quantity of drugs! After a moment of panic, he wished he had dumped the car a long time ago. Leaving the two bags on the ground, he replaced the trunk lid and closed it with the locking rods. After the seats were back in place, he picked up the two brown bags and headed for his apartment where he flushed the contents of the plastic bags down the toilet. He then went to the Dumpster behind the restaurant next door where he dumped the brown bags and covered them with trash hoping no one had noticed.

The car seemed such a liability he decided to get rid of it as soon as possible. Simply abandoning it was not an option as it could be traced back to him through the body shop where it was painted. Francis had to find a way to obliterate the car with no trace. It had to be done with no help from another person. *Maybe it could be dumped in the ocean in a secluded spot where it would never be found.* After studying maps, he found several promising spots. Checking each of them out, he selected one not far from the Golden Gate Bridge. Roadway construction of some kind had made a narrow dirt road sloping down to the edge of a cliff which dropped straight into the water. Late that night, he removed the temporary wooden barrier and stood on the dirt road while the Camaro rolled over the cliff into oblivion. Taking a branch from a nearby brush pile, he obliterated the Camaro's tracks from the edge of the cliff to well beyond where he replaced the barrier. He then walked a mile or more to where he caught a bus to make his way back to Chinatown.

Francis Lane, now Charles Chang, would work hard to establish himself in Chinatown, away from the dangers of Detroit gangs. He called Annabel to let her know he was safe and would talk to her often in the coming years. He did not tell her about the car, the money, or what happened the night he left. She told him Ahmed and his friends died in an accident, but several others had asked about him. She told them she had no idea where he was but suggested he might have gone to New York. Annabel was no fool and would keep his whereabouts to herself. Francis was grateful he had such a grandmother.

It took him nearly a year to be accepted into the Chinese community. By his twentieth birthday, he was given a party by the family who owned the import store where he worked. Then a job opened up working in a local hospital. There he took training as a medical technician. After several years, he became an emergency room technician in the hospital. He stayed there, enjoying the work and the people for the next five years.

Francis thought hard about his future and finally decided to go back to school and try to get a college degree. He phoned Annabel to ask her help in getting his records for college entrance. She was thrilled when he said the University of Michigan would be his best bet. With her assistance, Francis enrolled at Ann Arbor the following fall under his own name. The decade since his departure gave him a good separation from the past. No one had asked about him for at least five years.

Francis finished his undergraduate work in three years, graduating with honors. Accepted into the U of M Medical School, he worked hard, and by his thirty-sixth birthday, he had begun his internship at Detroit City Hospital. During his education, the many Charles Chang accounts across the country had been transferred one-by-one to a bank in Ann Arbor and used to pay for his schooling. By the time his internship was completed, there were three accounts left totaling less than ten thousand dollars. The drug money had been put to good use. About this time, Annabel grew gravely ill.

She smiled proudly at Francis from her bed on her last day. "Dr. Francis Lane! What a glorious sound for a name. I am so proud of you!" she said softly.

Francis knew he was losing his greatest life asset. With Annabel gone and his Chinese cousins moved away from Detroit, he decided to make a major move. The one cousin with whom he kept contact over the years lived in Hilo, Hawaii. On a hunch, he called her to ask about coming out for a visit. Dee Chang was a jolly woman of about forty who never married. Yes, she would be delighted to see him. Francis sold his car and other possessions, arranged for Annabel's house to be titled to her church for use as a youth hostel, and left for Hilo with all his remaining things packed in two suitcases.

With no openings for new doctors available anywhere on the islands, Francis applied for an EMT position at the University Hospital Trauma Center in Hilo. Two years later, he was given the next opening as staff physician. During that time, he met and married Oona Lee, a nurse at the center. He would remain there for many years, earning a reputation as an excellent physician. He also gained the community's respect as an active supporter and promoter of youth programs aimed at troubled young people.

If you give a man a fish, he has food for a day.
If you teach him to fish, he has food from then on.

—Chinese Proverb

Somehow, we always get back to the basics. Right and wrong, good and evil, like beauty, are in the eye of the beholder (or doer). Their rules are not immutable. They are lifestyle—cultural, social, or religious creations. They depend entirely on one's own situation—whose side you are on, to what group you belong, or who eats whom. I am sure Genghis Khan, Hitler, and Saddam Hussein had and have quite different views from their victims of right and wrong.

Good and evil, right and wrong have very different meanings for a zebra than for a lion.

—Howard Johnson, May 8, 2001

If you can start the day without caffeine or pep pills,

If you can be cheerful, ignoring aches and pains,

If you can resist complaining and boring people with your troubles,

If you can eat the same food every day and be grateful for it,

If you can understand when loved ones are too busy to give you time,

If you can overlook when people take things out on you when, through
    no fault of yours, something goes wrong,

If you can take criticism and blame without resentment,

If you can face the world without lies and deceit,

If you can conquer tension without medical help,

If you can relax without liquor,

If you can sleep without the aid of drugs,

Then . . . You're actually a dog.

—E-mail from Brenda Shears to Nancy Grimm, October 30, 2001

Heroes and Oracles, Where Have They Gone?

Emerson remarked, "Each man is a *hero* and an *oracle* to somebody."

Noah Webster describes a *hero* as "a prominent or central personage taking an admirable part in any remarkable action or event; as the *hero* of a romance; hence, a person regarded as a model of noble qualities; as, Washington is more than a national *hero*."

He describes an *oracle* as "the medium by which God reveals hidden knowledge or makes known the Devine purpose; also, the place where the oracle is given."

Like it or not, we all have heroes and oracles who we use to shape our actions, character, and beliefs. They are the winds that bend the twig into the tree.

Pick carefully your heroes for as they are, so will you become!

—Howard Johnson, 1998

Time Trap

*A*lexis *carefully slid the package into her leather purse, slung its strap over her shoulder, and casually walked around the corner into the heart of the space port and up to the customs desk. The fat albino behind the desk looked up from his monitor to stare at her. She noticed his right hand sliding slowly out of sight and down by his side.*

"Ah, Ms. Stereo, back so soon?"

From behind her, a familiar voice shouted, "Alexis! Don't!"

Turning she saw Dr. Stanford out of breath from running to catch her. "What do you mean, don't?"

"Just don't!" Dr. Stanford stammered through gasping breaths as he almost stumbled up to her side.

"You said it yourself, I have to go back twenty-four hours to stop the release of that RESO virus or the whole planet will be dead, and . . . and as soon as possible."

"They were . . . wrong! The RESO virus is . . . harmless and didn't cause the deaths . . . another experimental virus the two women . . . had in their bodies was triggered by the RESO . . . The medical team found it and destroyed it . . . easily before there were any more deaths. . . . But the TCD is **not** harmless . . . There are some unusual side effects if it is used on a large scale," Dr. Stanford blurted out between gasps for air.

Alexis paled. "What's wrong with the time collapse driver anyway? I already triggered it before slipping the package into my purse. It's set to cycle twenty-four hours back at precisely eleven o'clock, a few moments from now, and we both know it can't be stopped."

The albino pulled up what he had been reaching for and leveled a Galbo blaster at Alexis. "I've been waiting for you to return with that gadget. I saw it in the security scan and knew I wanted it. It will be my ticket out of this hell hole."

"You jerk!" Alexis snarled. "The damned thing will be useless to you. The start mechanism requires my hand print before it will fire."

"Hand over the purse and we'll see about that. Those scabs in the Telurian black market think they can reverse engineer anything. They pay highly for any new or unusual device."

"You don't know what you're getting into." Alexis stalled and held her purse tightly.

"And I don't give a crap! Give it to me or I might pop the trigger on this Galbo and you know what it can do."

"Let him have it, Alexis, it doesn't work right anyway."

"What do you mean it doesn't work right?"

"Ms. Stereo, give me the damned purse NOW! Otherwise, you and the doctor here will be splattered all over the customs office and I'll take your purse out of your dead hand."

Reluctantly Alexis handed over the purse. The albino took it and disappeared through a door into the bowels of the space port.

"Okay, Doctor. What's wrong with that expensive little gadget?"

Dr. Stanford finally caught his breath and explained, "We had only tested it on small bursts of power—ten or twenty seconds' worth. We didn't notice any problems, and it only warped a small section of space. Gravity waves from that small disturbance leveled out, and no harm was done. We decided upping the power to warp enough space for a person would work the same"

Alexis looked puzzled. "So what's the big deal? We'll have a little blip, and everything will be okay."

"That's not exactly correct."

"So what does happen?"

"We have theorized that a large-enough gravitational disturbance could cause a cascading G-wave effect that would create a major rift in the space-time continuum. The entire universe could be pulled through that rift in an instant and revert back to whatever time it was set to."

"That little box could do the whole universe?" Alexis shuddered as she said it.

"Yes . . . theoretically."

✳   ✳   ✳

Then there was an instantaneous physical change, absent of sensory information.

✳   ✳   ✳

Alexis looked at the clock on her desk. It was precisely 11:00 a.m., and she needed to hurry to get to the meeting on time. The meeting at the Stanford Gravitation Field Experimental Station was to discuss results of the testing of the new portable TCD unit.

At about the same time in the university hospital research facility nearby, the RESO virus was about to be tested on human volunteers. The Replicating Exchanging Self-Organizing virus would search out all nonstandard DNA cells in a person's body and replace the faulty

DNA with the correctly sequenced DNA. It was a cure not only for cancer but for many other diseases caused by aberrations in cell DNA. The promise and possibilities were staggering in their depth and breadth. Two women with advanced cancers were selected from thousands of volunteers. Its creator, Dr. Chan Ling, estimated it would take thirteen hours for the virus to convert all nonstandard DNA. Once that was done, it would disappear when all nonstandard DNA, its food in effect, disappeared. No trace of the virus would survive. One volunteer was injected with one type of RESO virus. The other with a slightly different strain.

"Now all we have to do is wait," Dr. Chan murmured as the two women got up to return to their rooms in the hospital section.

Three hours later both women were dead, their bodies riddled with a rapidly developing cancer. Seemingly, the RESO virus had not destroyed itself but had invaded every living cell in or touching the women. Near panic gripped the staff when they realized the gravity of the situation. The entire wing of the hospital was sealed off.

Searching frantically for a viable solution, Dr. Chan remembered that his friend, Dr. James Stanford, had been working on a time-warping device using gravity waves. "Get me Dr. Stanford at the Stanford Gravitation Field Experimental Station," he shouted at one of his assistants as they rushed toward the office.

In less than half an hour, he had explained the situation to Dr. Stanford who promised to see if his equipment might be useful in providing a solution. Soon he was explaining how the new and powerful TCD could send someone back twenty-four hours. This might enable them to stop the release of the RESO virus.

Dr. Stanford was coaching one of his assistants, Alexis. "Set this digital readout to 24.00, indicate hours and then press your hand against the actuator. The nuclear distorter will start building up G-pressure, and in about ten minutes, it will fire a monstrous G-wave that should bang you back twenty-four hours instantaneously."

Alexis thought for a moment. As Dr. Stanford's assistant, she knew there were some special location and time constraints. "When and where do I do this?"

Dr. Stanford took out a map. "The best place is inside the space port . . . here, in this open area." He pointed as he spoke.

"I'll have to go through security. What do you suppose they'll do when they see this device?"

"Tell them it's a new medical device using nuclear power. Dr. Chan would confirm that should they get testy."

"I suppose I'd better get going."

"The sooner the better."

Half an hour later, she stood by the security scanner and waited as the TCD unit was put through without incident. "That was easy," she muttered to herself as she picked up the TCD unit, hand printed the actuator, and slipped it back into its protective package.

Alexis carefully slid the package into her leather purse, slung its strap over her shoulder, and casually walked around the corner into the heart of the space port and up to the customs desk. The fat albino behind the desk looked up from his monitor to stare at her. She noticed his right hand sliding slowly out of sight and down by his side.

"Ah, Ms. Stereo, back so soon?"

From behind her, a familiar voice shouted, "Alexis! Don't!"

❖   ❖   ❖

On Idealism

To be idealistic and naive is to lay oneself open to charlatans and con men of all types. To discard naivete only stops the most obvious of these, and at what price?

To discard idealism in order to keep oneself safe from frauds and connivers creates total constipation of the soul, a deplorable condition.

I believe it is best to be idealistic and naive, but secure wise counsel before making life-changing decisions. Ah, but securing wise counsel, there's the rub. A quote from Talbot Mundy: "When a number of men, for a number of different reasons, counsel me to turn aside from danger, I have usually found it wise to recognize the danger, but do the opposite of what they urge. Although they likely know it not, their counsel is directed either by their own necessity or by their love of comfort, good repute and profit."

If you remain true to your idealism, you will surely be duped on occasion. But if losses are kept small while ideals and naivete are held, you will be able to smile at life and have numerous dear friends. Or perhaps, you search for something else?

—Howard Johnson, May 25, 2001

To be a friend a man should not attempt to reform or reprimand, but should strive only to make others happy if he can.

—Wilfred A. Peterson

Toys Are U.S.

March 11, 2002

Corporal Lance Mugambi sat studying the color display screen in front of him. His hands were on the keyboard of a Gunslinger remote controller. He and about a hundred other newly trained weapon operators, dubbed gunslingers from the weapon they used, were aboard a refitted AWACS plane high above the mountains of Afghanistan. He was deeply engrossed in the scene on the display. In one window was a video taken by a tiny Sky Eye TV camera mounted on a small Super Chopper deployed from a much-modified Tomahawk cruise missile far below them. In another window, a photo map of the mountainous terrain was displayed. A bright yellow dot moved over the map, indicating the exact position of the tiny helicopter as it was taking the video being shown.

"Damn! There it is," shouted the corporal as an Al-Qaeda training camp appeared on the display. It was in a canyon among the mountains. "Look at those guys running for cover. They must think it's an attack," he reported to his group of four as they watched the scene unfolding on their screens.

"Now, they're firing," another gunslinger remarked as bright flashes appeared on their screens. "They waste lots of ammo shooting at tiny things they can barely see. I wonder what they think is happening."

"There's one with a Stinger missile," another remarked. "Is he going to fire it at our choppers? This should be interesting."

"Wasting a $20,000 missile to attack a $300 drone helicopter is great for our side," the corporal commented. "Shoot, damn you, shoot."

They all cheered as a bright flash and smoke trail erupted when the Stinger burst into action. "Anyone hit?" Someone asked. There was no answer. A moment later, another one was fired from the camp.

"They got mine," another called out when his screen went black. "No . . . wait . . . it might be okay." The scene on his screen reappeared, tumbled wildly before it slowed, and then stabilized. "It still looks like I'm going down. The ground's coming up too fast." Indeed the ground rushed up to meet the camera and the screen went blank. He switched to an alternate camera on another Super Chopper in his assigned group of four. One of the four green active camera indicator lights on his screen was red.

"Anyone else get a red light?" the corporal asked. Silence indicated a negative response. "Great! So far the score is our $300 to their $40,000. At this rate, we'll bankrupt them in a hurry. Let's get on with the mission. We need at least three head counts to plan tonight's attack. Also, use your cursors to pinpoint any possible military target. Don't forget objects like buildings, vehicles, gun emplacements, caves, deep ravines, or any place they can hide. We don't want to miss anything. Don't worry about major roads or bridges. We've already got a handle on them, and they should be destroyed by smart bombs within an hour or so. First we isolate them in their camps, then we eliminate them."

Zalmi looked up at the sound of the Tomahawk missile as it blasted by the camp and then disappeared over the hill to the north. No one had time to raise a gun toward the missile. "Wakil! Wakil!" Zalmi shouted to his partner in the lookout station. "The Americans! They have come calling, and we must repel them."

Wakil cranked the ancient siren to arouse the camp which immediately burst into action. Soldiers with weapons poured out of buildings and took positions among the rocks. The fifty or so trainees were rushed from the conditioning field into a deep cave in the rocky hill to the east of the camp. Some were given weapons and posted in the rocks outside the cave entrance. Unnoticed were sixteen tiny helicopters dropped from the Tomahawk far down the valley to the south of the camp.

"I don't like the silence," Zalmi said to his partner. "Maybe that was a single missile headed north. If it was aimed at us, it missed."

"The stupid Americans. They'll blow up some mountainside. That's all there is north of us where it headed."

They both laughed. Zalmi cupped his hand to his ear. "Listen! Do you hear that? It sounds like a swarm of bees far in the distance."

Wakil frowned. "I don't hear a thing . . . Wait! Now I hear it. It does sound like bees. I wonder what it could be."

After about five minutes, the sound level jumped as the flight of tiny helicopters topped the rise a mile or so to the south and descended toward the camp. Zalmi grabbed an ancient pair of binoculars and pointed them toward the growing, high-pitched din.

"What is it?" Wakil asked. "What do you see?"

"Nothing at all. Clear sky and no dust on the ground."

The tiny low-flying helicopters were impossible to see at this distance, even with binoculars. Their sound, enhanced deliberately to create confusion, grew to an intense, almost-painful level as they neared the camp.

"You must be blind! Give me those so I can see our enemy." Wakil grabbed the binoculars from Zalmi as the noisy choppers passed over their outpost a few hundred feet above their heads. The size of a small bird, the tiny choppers were still nearly impossible to see. This was especially so when most of them reached their observation area with the sun directly behind them.

Zalmi watched as wild gunfire erupted from the many automatic weapons in the hands of his fellow Afghans. A few took aim, most simply fired at random into the air. Like Zalmi, many of these hardened fighters were confused and disorganized. "In the name of Allah, what is this?" he shouted to Wakil, a look of panic on his face.

Wakil was angrily emptying his AK-47 into the air and couldn't hear. The unmistakable flash and roar of a handheld Stinger missile rent the air as it blasted off from a few feet away. Zalmi followed its corkscrew smoke-defined path as it disappeared high in the sky. A second missile was fired only to explode almost immediately a few hundred feet in the air. Zalmi watched a tiny black dot descend from the area of the explosion. It fell slowly at first and then picked up speed as it headed for the rocks about thirty feet from his post. It struck the ground and exploded with much less force than a hand grenade. The self-destruct mechanism in the fragile tiny chopper blew it to thousands of tiny bits when it hit the ground.

Zalmi climbed down from the observation tower and rushed over to examine the site of the explosion. He picked up a tiny piece of green plastic and showed it to Wakil who followed him to help in the inspection. "What could it have been?" he said, turning the tiny flat object over in his palm. "This tiny piece of plastic is all I can find. Look for yourself."

Wakil examined the ground and came up with two tiny metal screws, the type used to fasten to plastic. "This is all I've found so far. What can it be?"

While the unearthly din roared above them, Zalmi left to check with Colonel Mustapha, the camp commander. Wakil climbed back up the ladder to man the observation post. Zalmi went in person as once more the ancient telephone connection failed to work. As he arrived at the headquarters building, Colonel Mustapha ordered a cease-fire and sent a boy to check the camp for casualties.

"What have you to report?" the colonel asked of Zalmi.

"Here's all we could find from the site, sir," he said as he handed the small piece of plastic and the two screws to the colonel. "There was a small explosion when the object hit. It was the size of a blackbird and fell quite slowly, so it must have been small and light. Do you have any idea what this is all about?"

At this point a young boy ran up impatient with news for the colonel. "Spit it out, child. What did you discover?"

"There are no casualties, sir," he answered. "No one was killed. No one was injured except a few who fell on the rocks while rushing to their posts. Sir, what is that terrible noise?"

"We'll tell everyone about it when we have all the details. Go, spread the word to look for any unusual objects on the ground and bring them here to the headquarters building. Go!" Turning back to Zalmi, he said, "Continue your report."

"Wakil and I examined the spot where the object exploded and found nothing but the three objects I gave you. That was not a bomb. It was much less powerful than a hand grenade. I doubt it would cause injury unless it exploded in contact with a person. We were less than thirty feet from the explosion and nothing reached our post."

Colonel Mustapha looked puzzled. "That is strange. Return to your post and try to find more pieces."

Zalmi was returning to his post when he noticed the sound seemed to be dying down. Reaching the observation post, he found Wakil searching the sky with the binoculars.

"They're definitely moving away," Wakil said as he put down the glasses. "I can no longer see the tiny black dots. The source of the sound seems to be moving north toward the mountains."

They watched their northern view for some time after the buzzing faded away. There were a number of tiny bright flashes high in the mountains. They heard no sounds from the distant explosions.

"Let's get them out of there," Corporal Mugambi ordered. "Head them for that high mountain peak and destroy them so no one can get their hands on them. We've got thousands more waiting to be used."

"I almost hate to destroy them that way," an eighteen-year-old gunslinger said as he headed his four Super Choppers north into the mountains.

At twenty thousand feet above raged peaks, the remaining fifteen tiny Super Choppers were deliberately blown into thousands of indistinguishable pieces the size of sand grains which rained down over a broad area of the deserted mountains.

It had been many hours since the tiny Super Choppers surveyed the terrorist camps, the sun had set and it was quite dark. Wakil and Zalmi were watching from their posts in the windswept silence. Two large explosions created a flash from several miles to the south, and the camp went dark as the explosions cut the power. The flashes and resulting smoke were clearly visible from their post. Twenty seconds later the unmistakable rumble of the distant explosions reached the camp. Zalmi reached for the battered phone to report to the colonel as the generator kicked in and the emergency lighting came up.

"Praise to Allah!" Zalmi uttered as the phone worked. When Sergeant Mahkmud, the colonel's aid answered, Zalmi reported, "Two explosions about two miles south of us, sir. We saw the flashes, and lots of flames and smoke. They appear to be at the camp entrance in the valley below us."

"I'll check with the guard post," the sergeant replied.

A few minutes later the phone buzzed. It was the sergeant. "We've lost contact with the guard post. The phone is dead. I'm sending a weapon carrier and men to see what's happened. I'm also sending all men to their defense positions. Keep your eyes open for anything unusual. This may be the Americans."

They watched the lights of the weapon carrier wind its way down toward the guard post until it went out of sight. About twenty minutes later, it returned. Zalmi could see by the outside lights that there were only four men on board as the carrier returned to the headquarters building. He waited tense and expectant until the phone rang.

"They've destroyed the bridge and a large portion of the roadway above the bridge. The guard post has disappeared with the roadway collapse. All direct communications lines were severed as well as the power lines. Our only means of communication is now by radio, and we're having problems with that. Many radio stations have become silent. We expect an air attack, so the lookouts will be doubled at the next changeover."

Far out of sight and missile range of those below, the modified AWACS plane was at fifty thousand feet for the entire operation. As it turned toward home base, the mission leader, Captain John Mook, rose to speak to the entire group. "That was a job well done. If phases two and three go as well, our terrorist enemies will lose about half of their members. The twenty battle groups of between four and five men on this aircraft used the Super Choppers for phase one of project WellCo. This one aircraft holds all airborne members of phase one surveillance. The entire project depended on the success of our mission, and you all performed well. As you know, members of the entire WellCo project have been branded WellCo Warriors. At this point in time, there is talk your existence will be kept secret permanently. Though your exploits remain unknown and unexplained, the results of your efforts will be a major historical event." At this point, an aide handed Captain Mook several papers.

"Here's a report on the last part of phase one. Laser-guided and other smart bombs have destroyed most of their communications network, at least the major installations, and isolated most of the training camps. The cruise missiles that delivered our choppers then struck at hardened targets. Damage assessment will come in about two hours from pictures taken by Eye Spy aircraft dropped with the smart bombs and from the cruise missiles. So far, it's better than we hoped."

With that, a cheer went up from the men.

"There's more," Mook said. "Here are the results of your efforts. Of the twenty possible camps we surveyed, four were small seemingly peaceful villages, three had been training camps in the past, but were abandoned sometime before our mission. We have detailed pictures and plots of the other thirteen, including personnel counts and weapon assessments. As the result of your efforts, phase two will start at about midnight local time. Fortunately, the weather is staying clear."

As the men began to celebrate, Captain Mook held up his hand for silence.

"Hold on for a minute while I give you the numbers and the response from our enemies. You can celebrate as long as you want, just don't damage the aircraft. Now for the numbers. Eight modified Tomahawk cruise missiles deployed 324 Super Choppers at twenty locations over literally thousands of square miles of rugged terrain. Seven of the choppers failed to respond and are assumed to have been damaged in the drop and self-destroyed. During the mission, five of the choppers were damaged by ground fire and six destroyed themselves, we assume from hitting an object like a hill or building. The enemy fired a total of forty-one Stinger missiles at them, destroying only eighteen. Apparently one Stinger took out two choppers and twenty-four missed. As far as we can tell, all Super Choppers were then destroyed by their fail-safe self-destruct mechanism after completing their mission. In our operation, the enemy spent about $820,000 in Stinger missiles to destroy $5,400 in our hardware that would have self-destructed later anyway. The total cost of the hardware expended in our mission was only about $96,000. Who says a high-tech war has to be expensive?

"Now, the news. These are the first reports from the Taliban in Kabul, which, incidentally, has not yet been attacked, at least not by us. The report read, 'Armed forces of the evil satanic Americans attacked the peaceful citizens of Afghanistan with bombs and troops. Raining death and destruction on Kabul and other cities, the Americans have killed and maimed thousands of innocent civilians including women and children. Their soldiers attacked many of our peaceful youth camps killing mostly civilians. Brave Afghan soldiers repulsed the Americans, inflicting massive losses. Cowardly American blood flowed deep in the streets at these camps as our fighters defended Islam. All attacks were repulsed, and not a single American soldier is left alive on our soil.' At least that last statement is true," he added with a smile. "Okay, men, we can relax and wait for the second wave."

✱   ✱   ✱

Zalmi was awakened by the sound of a series of small explosions quite different from gunfire. Vaulting from his cot fully clothed, he grabbed his AK-47 and headed for his post. He was running toward the observation tower when a huge explosion came from the cave. The whole camp lighted by the billowing flames as fuel stored in the cave burned, amplifying the explosion. Frozen in place for the moment, he saw the headquarters building disintegrated in a white blast that knocked him flat on the ground. As he crawled to retrieve his weapon, he saw his barracks go the way of the headquarters. During the time it took him to crawl to the ditch beneath his post, all the rest of the buildings in the camp exploded. As he dropped into the ditch, a ball of fire erupted above his head. He huddled deep, covering his head with his AK-47 as debris from the exploding observation tower rained down all around him.

"Praise Allah!" he remarked as the rain of debris stopped and he found himself unhurt.

The silence after the final explosion was interrupted at first by moans and cries of the wounded, then by shouted commands, "Go to the rocks! Go to the rocks!"

Zalmi vaulted out of the ditch and headed for the rocky hillside east of the camp where the rocky terrain would provide excellent cover. At the top of the ditch, he stumbled over another soldier. It was his friend, Wakil, who had not been so lucky. A large timber from the tower had impaled his chest. Zalmi hesitated for a moment; then realizing Wakil was beyond help, he ran for the rocks. As he ran, he saw many others heading the same way, black figures outlined in smoky silhouettes against the yellow flames. Some ran alone; some were helping wounded comrades. He finally dropped into the shadow of a large rock and leaned against it while regaining his breath. After a few minutes, he concentrated on staying hidden while listening for commands. All he could hear was the crackling of the flames and the moans and cries of the wounded. Then he heard a high-pitched buzz from high above. It grew louder and louder until finally a voice from the sky spoke in his own tongue, Pushto.

"Lay down your weapons and come to the center of the camp. Anyone who does not obey will be killed." The voice was answered with several bursts of AK-47 fire from the rocks to his south. This was followed by a number of streaks of fire high overhead. The streaks went straight to where the AK-47 fire originated and were followed by several small explosions.

A few moments' silence, and the voice spoke again, "Resistance is futile. Throw down your weapons, and gather in the center of the camp. Those who do not will be killed. Bring your wounded comrades with you, and they will be cared for."

Another round of gunfire from the rocks was followed by the same small streaks in the sky and then the small explosions. The same scenario repeated three more times until there was no more AK-47 fire. Everyone stayed where they were in the rocks.

Another five minutes of silence and the voice spoke again, "All who remain hidden in the rocks will be killed. We will start killing in ten minutes if you do not throw down your weapons and move to the center of the camp."

Zalmi was quite unnerved by this voice in the sky. His thoughts raced wildly about. *Should I follow their order? Will I be killed if I don't? Will others shoot me if I do? How can they know where I am?* For the moment, he chose to stay put and wait.

The voice again boomed out, "We can see all of you clearly. We have the means to search out each of you and kill you if you hold a weapon. Each one of your soldiers that fired on us was killed. If you do not throw down your weapon and head for the camp as directed, we will start killing in five minutes. Any soldier raising a weapon against those who choose to surrender will be killed before he can fire."

Suddenly Zalmi noticed two soldiers walking toward the center of the camp. Almost as soon as they appeared, several small explosions burst in the rocks.

Again came the voice, "The three who raised their weapons to fire at those surrendering were killed. In three minutes, all who hold weapons will be killed. We can and will do what we say."

Hearing this, Zalmi threw his AK-47 out from his hiding place. Many more soldiers walked toward the center of the camp, empty arms held high above their heads. Zalmi soon followed. Despite the Islamic fundamentalist promise of instant paradise, Zalmi wanted to live.

✱   ✱   ✱

Far above the camp, the mission command plane circled. Twenty gunslingers worked their keyboards to control eighty attack Super Choppers. Each little helicopter carried eight RATTLER antipersonnel rockets the size of a pocket pen as well as a tiny Sky Eye TV camera. The camp below was displayed in twenty sections, each section controlled by a single gunslinger keyboard. Mission commander, Captain Wesley Charron, spoke in Pushto over the speaker in a Super Chopper hovering several hundred feet over the targeted camp. As he spoke, the gunslingers watched their screens as bright green images of men moved in their fields of view. When an image raised and fired his weapon, the gunslinger clicked on the image, sending a RAP rocket on its body-heat detecting path to terminate the enemy soldier with a small shaped charge.

"I count eighty-three without weapons gathered in the middle of the camp," Captain Charron announced. "I don't think any more are coming in, so you know the routine. Don't target any man unless you're certain he's holding a weapon. Okay, fire!"

As the group waited in the middle of the camp, a number of small explosions among the rocks were followed by an eerie calm. The sounds from the still-burning buildings, the steady buzz from the sky, and the low rumble of voices of the soldiers were all that could be heard.

Suddenly the voice boomed out, "Thank you for following our directions. Now please face east and sit on the ground. A rescue mission is on its way to carry you out and treat your wounded. You will not be harmed if you follow our orders."

In about half an hour, the sound of fast-moving helicopters came from the south. Within a few minutes, two attack helicopters disgorged twenty armed troops who secured the camp. Soon after, two more large troop carrier helicopters set down. A contingent of five men approached the prisoners and stopped about thirty feet away, while armed guards flanked them on both sides.

Once again, the voice boomed from the hovering chopper. "Form a single line and approach our delegation one at a time. The first to be processed will be provided with stretchers to bring in the wounded. You will not be harmed if you cooperate and do as you are told. Any sign of resistance will be met with lethal force."

The careful processing of the eighty or so prisoners through metal detectors and onto the troop carriers took three hours. When told to give up all knives and metal objects before going through the detector, many argued they were valuable family heirlooms. In spite of promises they would be returned later, a few scuffles broke out. They were subdued and the resistors placed in handcuffs and leg irons. One zealot broke for the interrogators with a raised knife. He was cut down before getting within ten feet of his intended victims. After the excitement cooled down, a ground search of the area found eleven wounded men who were loaded on stretchers and placed aboard the helicopter with the medical team.

For the last time, Captain Charron's voice boomed from above the camp. Speaking in English, he said, "The mission is complete. Let's head for home."

In a secret meeting room at Andrews Air Force Base, five men waited at a table. As Andy Wells approached to take his seat, the men all stood up and applauded. This was the same group he met in this same room at the beginning of the project on September 13, 2001. He still knew only one of the men, his old friend and partner in the project, Dusty Adams. The other four included two men in uniform and two civilian representatives of the president. As always, no one wore a name tag, decorations, emblems of rank, or service branch identifications. When finally the applause quieted, one of the uniformed men spoke, "Welcome, Andrew. Please be seated."

One of the two civilians arose to speak. "The president and his cabinet are positively ecstatic. Accordingly, I bring congratulations to you all. The results of our first attack on world terrorism are better than we had hoped. All objectives were met. Other than those who may have been in the camps, there was not one civilian casualty. The only casualties we suffered were three minor injuries caused by falls as our men exited their aircraft in the dark after the mission. The mission was so successful, the president has ordered a hold on the second attack until the effects on the terrorist organizations and their supporters can be assessed. Because of our success at destroying their communications infrastructure, the news from Afghanistan is slow coming in, so we have adopted a wait-and-see attitude for the present."

Turning to Andy, he said, "Before we begin, here is a letter from the president of the United States, thanking you for your help. I believe you will be quite pleased with what it says."

Everyone waited silently as Andy read the letter.

Dear Mr. Wells:

Thanks to your unique talent and the unprecedented success you have helped us achieve, we have an outstanding initial victory over the terrorists and their supporters. When your original project was cancelled eight years ago, you left government service to start your own company. Who would have guessed that in a few months, your reinstated project could produce unique weapons invaluable in this new kind of war. Your country has a friend of mine to thank for telling me about your cancelled project. That was a most fortunate bit of information from a trusted friend. That you took your weapons technology and converted it into peaceful civilian use building an extremely successful toy company in a competitive business gives testimony to your talents.

Unfortunately, your specific efforts and the many useful products you provided will go unheralded for security reasons. Rest assured, however, the nation will know the results of your efforts and you will be accordingly rewarded. I counted on you, your country counted on you, the entire free world counted on you; you let no one down. We all thank you from the bottom of our hearts.

Who would have guessed that the thousands of young people who grew up playing electronic games using Welco Toy Company gunslinger keyboards would have been so effective in actual combat? They used those skills to control the military versions of America's most popular toys in one of the nation's most successful military operations.

The President of the United States.

The Hike

Timothy O'Brien never expected to find himself in such circumstances. Whoever would? He paused on the narrow path on the side of the grey granite mountain, the wind pressing on his backside as if it wished to push him to his death on the ragged rocks two thousand feet below. A sound caught his attention over the rush of wind.

At first he thought he knew what it was, a small animal moving through the bushes on the ledge below where he stood, but as it grew louder, he realized it was coming from much farther above. It soon sounded like many irregular hoof beats, the sound a herd of horses or buffalo would make running in a stampede. He leaned harder against the rock away from the sheer drop on the other side of the path as the sound grew louder and louder. He was horrified when he realized the cause of the sound. It was an avalanche of rock plummeting down the mountainside above him.

Tim realized how exposed he was on the narrow ledge with no protection above. Thinking quickly in that time-slowing pace of near panic, Tim remembered passing a deep indent in the path with a substantial protective overhang. He hurried back toward the safe haven as quickly as he dared on the narrow path. *If only I can make it in time.* Small rocks, dislodged by the vibrations from the approaching mass of plummeting rock, began pelting him as he ran. *God! How far back was that spot?* he thought as he rounded bend after bend without seeing the safe haven.

Finally! There it was on the opposite side of a deep ravine that cut back into the mountain. He knew it was much too far to reach in time. Then he noticed a narrow break in the rocks beneath the path about fifteen feet ahead. By then the small rocks dislodged by the earthshaking deluge far above were falling more often and getting bigger. One hit his arm and drew blood, then another. He ran to the edge above the break, looking for a way to get himself into the crack. It would be difficult. One slip and he would drop to the jagged rocks below. A huge boulder bounced on the path a few feet from where he clung, and he knew the main part of the avalanche would soon be smashing into him.

Grabbing a small rock handhold on the edge of the path, he rolled over the edge and into the crack. He was dangling in space inside the crack, supported only by the handhold. He searched the sides of the crack for another handhold or foot support. The sides of the crack were smooth. Then he realized the crack narrowed both downward and inward. If he could swing far enough in, he would be able to jam himself into the crack and hold himself there by pressing against the sides. After swinging several times to gain momentum, he prayed and let go as he swung inward. As he let go, a huge piece of rock smashed the ledge he had just been holding, smashing it and a piece of the path into small pieces, and showering the area with small jagged projectiles. He was deep inside the crack, but slipping downward. After dropping at least twenty feet, he was able to stem his fall by jamming his legs against one wall and his back the other. About ten feet inside the crack and under the path he hung on tightly as the massive avalanche crashed onto the path where he stood moments before.

The mountain shook, dust filled the air, and the sound was deafening as the main part of the avalanche thundered by a few feet from his refuge. Then he couldn't breathe because of dust whirled at him by the air blast from the avalanche. He used his free hand to pull his shirt out and over his face as a filter to keep the dust out of his lungs and eyes. Breathing was difficult, but possible, and the shirt kept the choking dust out. *How long would the thundering continue?* he wondered. Then he felt himself slipping and pressed harder with his legs, jamming himself tighter in the crack.

He wondered, *will the thundering never stop?* Then everything began quieting down. The thundering was rapidly moving down the mountain, and the dust was settling. Tim eased his shirt off his face and surveyed the scene before him. There was still a pall of fine dust in the air and an occasional small rock plunged off the shelf above him, but at least he was still alive. Looking up about thirty feet, he noticed the crack was now open at the top. The avalanche had obliterated the path that capped the crack. He wondered how much of the path still traversed the mountainside and if he could use it to get down the mountain.

His legs were beginning to ache from pressing so hard against the sides of the crack. He would have to move into the narrower section of the crack and jam himself in so he could rest his legs or they would eventually give out. It took him about fifteen minutes to move far enough into the crack so he could wedge his hips and rest his legs. As he relaxed a bit, he began examining his body. Blood oozed from several cuts on his arms and legs, and when he wiped his forehead, his hand came away quite bloody. Fortunately he was not losing much blood, just a bloody mess of small cuts and abrasions.

As he rested, Tim considered his predicament and options. *At least Alicia wasn't with me, so she was safe. By now she would be frantic and asking for help. Surely the avalanche was noisy enough and near enough to alert people back at the lodge. Alicia would be driving everyone to find me. Well, I might as well start finding a way out. Any help is surely hours away.*

With that thought, Tim decided he should work his way up the crack, keeping to the narrowest part so he could rest periodically. After about an hour, he was only five feet below the open top of the crack. Unfortunately the crack narrowed at that point in such a way that his only way up was out beneath an overhang with a long drop below. One slip and he would fall at least a hundred feet. He decided to rest for a while to prepare for the exertion he was about to have to make.

Fortunately his hiking boots were strong, so he decided to use them to grip the narrow portion of the crack as he worked his way out almost upside down. He jammed one boot into the crack and swung the other around and jammed it into the crack about three feet ahead of the other. Four maneuvers like that and he should be able to reach the top with his hands and pull himself up. It would be difficult, but doable. Each maneuver took him farther away from the wall hanging above a huge drop. If he missed—well—he couldn't. After the fourth maneuver, his feet were jammed into the crack a foot from the top. He would rest again for the final move where he would fold his body forward and reach for the top with his right hand. Once he had a hold, he could release his farthest foot and pull himself onto the path. It had taken him three hours to get this far, and he wasn't taking any more chances than he had to.

After resting for about fifteen minutes, Tim started making his move, bending forward as far as he could and reaching for the edge. It was a few inches out of his reach. He repositioned his feet to get a bit closer. Once more he bent up and reached for the edge. A third try provided a precarious hold as his fingers finally caught the edge. He worked his hand around the edge until he found a decent hold and began to release his left foot for the final move. Once it was released, there was no way to put it back. He would be holding on by one foot jammed into the crack and a precarious hold on rather smooth rock with his right hand. He couldn't move far enough into the narrow part of the crack to jam his torso without releasing his other foot. That would leave him dangling with but one tenuous handhold as he swung his other hand up for another hold. If there wasn't one, he was done for and he couldn't see where he had to reach.

Finally Tim took a deep breath, released his other foothold and reached over the edge of the crack with his left hand searching with his fingers for any projection or small opening he could catch with one finger. With his elbows over the edge of the crack, he held for an instant, but knew he must find a hold or he would slip off the edge. Frantically his fingers probed the smooth surface searching in vain for anything to hold. His right hand began losing its grip and started to slide toward the edge. He was losing it. "Damn" he said out loud. In a few seconds, he would be waiting for that crushing pain as his body hit the rocks below. He thought of Alicia. "Damn I hate losing!"

As his hand slipped slowly and agonizingly toward the edge, it seemed as if time was slowing. Was he imagining things, or had he heard a voice from above him? His left arm slipped off, and he hung for a moment with one hand. As his right hand began to lose its grip and ever so slowly slip toward the edge, he knew it was all over. Astonished when he didn't fall, Tim felt something snap around his wrist before his hand could slip over the side. He looked up to see two faces above him. Jack had grabbed Tim's hand before it disappeared. The rope in his hands was secured to others on the path.

"Hang on there, ole buddy," Jack said as he reached down with his other hand and grabbed Tim's left. They pulled him up over the edge to safety. As soon as he was safely on the ledge Alicia was sobbing and holding him like she would never let go.

✹   ✹   ✹

Back at the lodge, showered, cleaned, and patched up, Tim sat with Alicia and his friends and ran through the day's adventure. Alicia wouldn't let him out of her arms.

Jack reported, "This gal of yours was a whirlwind when we heard the avalanche. It was less than a mile from the lodge, and when we heard it, she shouted 'Tim!' and got us going right away. We were headed up the mountain before the avalanche stopped, I swear."

Rory added, "She ran up that mountain like a mountain goat. We had a hard time keeping up with her."

"Don't you ever go anywhere like that without me?" Alicia ordered, her dark eyes flashing.

"I don't plan to." Tim smiled to think what an unbelievably lucky guy he was.

♣   ♣   ♣

The problem with socialism is that eventually you run out of other people's money.
—*Margaret Thatcher*

Light appears in the darkness . . . a filmy, tenuous, reaching thing that invades the being that is the searcher. This delicious, tantalizing something, like the aroma of fresh food to a starving creature, draws the searcher inexorably onward. First, an awareness like the quick surge of the stag's head to an unfamiliar scent as he chooses between fear-driven flight and the search for knowing.
Light draws, destroying the nothing darkness, gradually tapping the entire concentration of the searcher until a delay of the blink of the eye becomes an agonizing pain of loss. Drawn as the ancient seamen to the sirens, the searcher bends his entire being toward the light. He must know . . . is it merely a firefly or perhaps a star?
—*Howard Johnson, 1970*

The greatest test of courage on the earth is to bear defeat without losing heart.
—*Robert G. Ingersoll, 1833–1899*

The Gold Feather

I deliberately studied each passenger who came through security and into the waiting area for the shuttle. It was from force of habit. The years working on various highly sensitive projects taught and trained me to look carefully for anything unusual. On my first vacation in years, I was headed back to my favorite place in this quadrant of the galaxy. Officially "on vacation" I remained, as always, an active though relaxed member of the Eegis project.

My relaxation ceased and my mind sprung to attention when a tall, and beautiful redhead strode catlike into the waiting area and flowed into a seat. From the feline way she moved I was sure she was a Scentar, a rare, homo variant. I'd heard about this advanced human subspecies, but had never seen one. Her simple dress clung to her like a second skin, moving enough to show it was not attached. It was an unusual color, a deep red with amber overtones, almost Titian.

As she sat, our eyes met and locked for a moment. A sudden, intense feeling of pleasure ran through my body as I imagined her moving sinuously against me. It was more emotion than thought and caught me off guard. I am **never** caught off guard. The thought, *something is not right*, sent a chill through me for an instant and was gone.

When they called my group to board, she stood and walked toward the gate right in front of me. She looked so slender, almost fragile, as she moved fluidly up the steps and into the shuttle. I have never seen anyone whose body moved so smoothly. She almost seemed to have extra joints in her limbs.

This is one lucky day. I thought to myself as she slithered into her assigned seat next to mine. She turned and looked directly into my eyes.

"I'm Leura Clauson. Who are you please?"

Her directness and the musical sound of her voice surprised me more than her exotic appearance. "Uh Draxel, Draxel Syl—call me Drax." I was uncomfortable and ill at ease—certain that my voice betrayed my discomfort.

"Have you been to Stentor Seven before? This is my first visit to the Vegan star system."

"Been there several times," I struggled to say. Her breath held the faint aroma of warm milk. She wore a perfume that hung on the edge of awareness. It was there, but as soon as I thought about it, the scent was gone. I was in uncomfortable territory without a secure mental foot hold. "I'm going on my first vacation in years and this is my favorite place to visit. Are you on vacation?"

"No, I'm a botanist on a research project. I plan to study plants growing in the low gravity and artificially controlled atmosphere."

The lilt of her speech was enthralling. It wasn't an accent, just different and quite musical. "A scientist! I'm impressed!" I smiled as I spoke thinking that was a huge understatement. "How long will you stay, on your project, I mean?"

"At least one stellar year. My grant may be renewed for an additional year. It's my first major assignment. . . . What was that little smile about?"

"A little private joke—on me." Her perception was amazing.

"A secret?"

"No, just a laugh at myself." Her directness, too, was a surprise.

"Tell me."

Now I was getting irritated. "Let's say it's something I'd rather not tell someone I've just met."

Disregarding my irritation, she switched the subject. "What's your profession?"

"I'm a gravity propulsion engineer. Do design work on the propulsion systems on craft like this one we're on."

"That must be interesting. Gravity propulsion is a highly complex technology, is it not? I know it takes a great deal of education. Tell me about it, please."

"You want those boring details?"

"Absolutely! And where did you get your education?"

"I took advanced gravity propulsion at the AGP center on Earth."

"And how long did that take?"

"On top of a basic engineering degree, state licensing requirements include two more years of advanced schooling with lots of math and physics. Then we have a year of training on the equipment, two more of working in the field and finally, passage of an examination before the state grants a license."

"That's five years. Botanists have it much easier."

"I don't know. Biochemistry is an intricate and demanding science involving complex living systems. That must demand a great deal of effort."

"It is also fascinating and rewarding."

"I'll bet it is."

"So you are vacationing here?"

"Yep! This vacation is long overdue and Stentor Seven's my favorite place to visit."

"Tell me about it. I've seen the digirecords, but those are quite bland. No beauty or poetry. You said you've been there?"

"Yes, and it is beautiful, so spectacularly beautiful it must be experienced."

The shuttle's engine hum increased and it rose from the pad to start the four-hour trip. The motion was quite noticeable, but would disappear as soon as we cleared the atmosphere and the main drive kicked in.

"How did it come to be? The records were sketchy about the planet's origins; they mention it was artificially created with no explanation. What does that mean?"

I was becoming more comfortable. Maybe it was because I was on familiar territory. "It was once a small, sterile planet a bit smaller than Mars and about two thirds its mass. It lies the right distance from the red dwarf star, Stentor, for a life supporting environment. Focused gravity beams were used to tow huge ice planetesimals in from the Stentor Oort cloud. They melted and became the oceans and created the atmosphere, mostly carbon dioxide. Special vegetation was introduced to consume the Carbon dioxide and add oxygen to the atmosphere, but you should know all about that, don't you?"

"Yes, I studied, the conversion of primordial atmospheres. All botanists study that early in their schooling since it has been used to modify many planets."

"Then you should also know about the biota from earth-like environments and that it took almost six-hundred years for the growth of these plants on the land and plankton in the seas to bring the atmosphere to its present mixture. It's much like earth's. Am I right?"

"The introduction of the biota, yes, but the six hundred years it took? I don't remember being taught about that."

"That's because the exact time for the change varied from place to place. Temperatures, pressures and everything else were adjusted for human habitation and the biota thrived. Since then, many larger life forms were introduced and flourished. The combination of optimal rotation rate and distance from Stentor, along with lots of work over the years gave us a semitropical paradise covering the entire surface."

"It sounds wonderful."

"Because of the low gravity, plants grow to immense size and spectacular proportions. That, I trust will be the focus of your research project."

"You are correct. Please tell me more."

"Better yet, I can show you. When we come in to land, you'll see what is possible in this light gravity. Watch for mountains that rise seventy thousand feet with sheer cliffs and unbelievable waterfalls. You should be able to see a lot of unusual geology and geography. It's quite spectacular from above."

"Point those things out to me, will you?"

"Gladly! Once on the ground be sure to take in how water behaves. It's quite different from Earth. The muted sounds of the slow waterfalls and of the unusual rivers are like a chorus of musical mumbles. Waves on the oceans can grow huge, yet they seem to roll in slow motion. The surf amazes everyone with spectacular thirty foot breakers tumbling slowly and gently onto the sand."

"I can see why it's such a popular vacation spot. How about the weather?"

"The weather is marvelous, sunny and warm with fractal-like white clouds moving across hazy, pale blue sky. In order to have adequate surface pressure, the atmosphere is kept many times deeper than on your home planet. Because of this, no stars can be seen at night and the central star, Stentor appears bright red. Clouds can rise as high as a hundred miles and the winds always drift by gently."

"That's amazing, quite different from Earth."

"Then there's the rain, the unbelievable warm rain. Because of the low gravity, raindrops fall slowly, congealing into large blobs which grow to near tennis ball size before they blow apart by the air as they fall through it. The soft pelting of large blobs of warm water feels great."

"I heard about the rain. I can hardly wait to experience it. I want to run through it—without clothes."

I would like to see that, ran through my mind, but I didn't mention it. Her next comment drew vivid mental pictures in my mind.

"If the chance comes up, could we run through the rain together? I'd like that."

It was said so innocently, so matter-of-factly, she caught me speechless. I paused to calm my imagination and struggle for composure. "Uh—yeah—sure. That sounds like a great idea."

"It sounds like true paradise. I hope I can spend my leisure time enjoying a few of the things you describe. Would you show me around some while you're on vacation? I don't mean to interfere with your plans, but I know no one else here."

I was beginning to believe my good fortune might overwhelm me. "Why, yes! I would enjoy it. I have no specific plans, none at all."

"Wonderful. I won't have much to do for the first few weeks so I want to look around a lot. I'm certain to find many new things to experience. It all sounds so exciting," she said as the main drive took over and the hum and vibrations ceased. We soon cleared the atmosphere and were on our way.

Over the next hour I relaxed as we spoke about families and friends. She drew pleasant experiences out of my memory and shared her experiences as a child and about growing up. There was an unusual quality to her stories. They were softly emotional. Incredibly, I could almost feel her joys and pains as she described them.

A long pause in our conversation and I realized she had fallen asleep. Her head against my shoulder brought on pleasant sensations, as did her snuggling down against me several times during the flight. I examined her closely. Her hair was extremely fine with individual hairs growing unusually close together. It was the same dark red as her dress with no hint of a color change near the roots. If it was dyed, it was an absolutely perfect job. She turned a bit and put her hand ever so gently on my right arm. Her pale amber skin was baby-soft and unflawed. When I touched her hand, it felt like satin, almost frictionless. I knew she was far too perfect for a normal human. Scentar were reported to have unusual emotional abilities. She certainly seemed to possess those.

I noticed a Gold pin high on her dress, the only adornment she wore of any kind. It was a feather, about an inch long and quite fragile. It looked like a real feather, but tiny and clearly gold. When it moved, it displayed faintly the many colors of the spectrum. One moment it seemed to be gold, another to flash color, and another to catch and reflect any light source like a diffraction grating. Colors flashed so vibrantly it seemed almost alive.

A slight bump was followed by vibrations and the hum of the landing drive. Leura sat upright without the slightest hint she had been asleep. "We must be arriving."

"Check out the scene below. Like I described earlier, it's spectacular."

She leaned toward the window. "It is amazing. The mountains—everything you said—they're so different."

When she sat back from the window, I looked at her. "You slept the last hour of our trip almost without moving. I wish I could do that."

"Concentrate on pleasant thoughts and close your eyes. You'll go right to sleep."

I smiled at her easy answer, still concentrating on the lovely gold feather pin. "What's that pin your wearing? It changes, sparkles with colors that seem to vibrate."

"A gift. My mother gave it to me when I completed my studies. It's the only jewelry I ever wear. It's supposed to signify fidelity."

"That's one I never heard before."

"Actually it's a special kind of fidelity. Fidelity to a common, usually treasured experience with someone you love. My mother loved me, and I her. It's about the wonderful life we spent together before I left home. Specifically, it's commemorating our last day together. That experience will never happen again."

"That tears me up, it's sad, but beautiful." I felt undeniably and intensely morose for a moment as she spoke. That nagging wariness of unknown origin again troubled me.

"Yes, I gave her a similar pin. It's a family custom. We both knew we would never see each other again."

I'm sure my shock showed. "Why not?"

Her voice had changed almost painfully. "It's a bit complicated. We knew our paths would never cross again."

The sorrow within me became almost overpowering. "How can you be so sure?"

Leura had the tiniest hint of melancholy for an instant. "Please, I'd rather not talk about it anymore."

I experienced a sudden intense change to terrible anxiety. It was almost overwhelming. Then it was gone and I felt fine. "What was that all about?" I said out loud in reaction.

"What was what all about?" her clear, silky voice had returned.

"Sorry. I had a strange feeling for an instant and it startled me."

Once more Leura shifted mental gears without hesitation. "Would you be able to help me to my hotel? This is all so new to me and I'm a bit nervous about going there alone."

With my luggage scheduled to be delivered, I was free to go where I wished. "Certainly!"

"You're sure it won't be an inconvenience?"

"Positively. I'd love to see you to your hotel." Once again I could hardly believe my good fortune. By this time I was beginning to grow accustomed to her soft, musical speech.

As we approached the hotel I remarked, "Buildings like this hotel are constructed in ways unimaginable on planets with normal gravity. Giant overhangs, huge spans, delightfully fragile overhead structures with plazas, walkways and open spaces."

"Yes, it is quite extraordinary," she said as the air car dropped us at level 196 of the hotel. It landed smoothly on the cantilevered plaza. Leura picked up the one small bag she carried and danced across the plaza right to the edge. She was a little girl spinning with excitement from one side of the outside walkway to the other as I led her to her room.

"I've never been up this high in the hotel. How'd you manage such a room? I thought the upper floors were reserved for foreign dignitaries?"

"And foreign botanists," she quipped as she flipped her hair and, with a flourish, hand-printed the key pad. The door slid soundlessly into the wall and then closed silently behind us after we walked inside.

I was dumbfounded. The room was decorated in shades of the exact same colors as Leura's dress and hair. "This can't be accidental. How'd you get your room decorated to match—you?"

Her look and demeanor changed and she laughed in that sensuous, lyrical way, no longer the little girl. Her voice also changed its timbre and now sounded almost like a flute or muted violin, terribly emotional.

"I plan on being here for at least a year so they let me have my choice of decoration. Do you like it?"

"It takes some getting used to, but it is beautiful." Once more I smiled as an intense feeling of warmth and pleasure flowed through my entire body. "Wow!" came out of my mouth as an involuntary expression.

Leura stepped lightly to the entertainment console and turned on music I had never experienced. In its unusual tones and mixed rhythms I sensed more than heard the plaintiff cry of a loon, the rustle of pine trees in the wind, the crashing of waves on a rocky shore and even the sounds of passion. It bordered on being visual and was pleasant to hear. Leura smiled as she switched the glass outside wall from clear to one way. We could see the beauty of Stentor Seven stretched out before us, but no one outside could see in.

Once more I became aware of her delicate perfume, on the edge of my senses as she walked over and looked straight into my eyes. The warm milk-like fragrance of her breath was erotically stimulating—emotionally intoxicating. She reached up and gently placed her wrists on my shoulders. Her hands hung loosely, barely touching my back. I hated the shirt that lay between her hands and my skin.

"Now, Mr. Syl, I want us to dance together. Would you like that?"

Completely out of my element and on the edge of losing any hint of control, I replied lamely, "Yes, I would."

I was totally beyond rational control. She slipped her slender fingers around my neck, took my hand and moved to the music. I looked directly into her eyes. They were a dark blue with a hint of red to the black of her huge pupils.

"Pull the little ring at the back of my collar," her soft voice commanded.

With a slight pull her dress changed from the dark red-amber to an iridescent blue-green. She began moving rhythmically against me to the hypnotic beat and sound of the strange music. The sensation penetrated my whole body which flushed with warmth.

"Now, dear Drax, I want to show you my appreciation for what you are going to do for me."

She pulled me gently into the bed where cool satin sheets caressed my skin. I could hardly tell the difference between those sheets, her skin and her satiny dress. Something akin to fear, surged through my being. I was perceiving everything with intensely heightened senses and enjoying every delicious moment.

"Lie on your stomach. I want to give you a massage," she urged.

Ecstatic, I complied. Her long, slender fingers were working up and down my spine, around my shoulder blades and neck and finally down the back of my legs. Never have I felt so good, so totally aware, not in my entire life. Just when my body had turned to jelly, she stopped the massage and began dragging her fingers lightly over my bare arms. I felt her lips moving up and down the back of my neck. The stimulation to my skin was ecstatic. She stopped and lay down on her stomach beside me.

"My turn."

I was overcome with passion and amazement. "What do you want me to do?"

"Do to me what I did to you. Don't you think that's fair?"

I remembered a line from the distant past and uttered it under my breath, "Resistance is futile."

I began in the middle of her back. The fabric of her dress seemed like a second skin. Unbelievably soft and satiny, it moved smoothly to my touch. She had no taut muscles. After

I massaged her for a while she rolled over on her back and looked up at me, those dark eyes boring into my essence.

"Tickle me please. Slide your fingertips slowly and gently over my skin. Barely touch me. Just like I did you. You liked that didn't you?"

"I prayed you'd never stop."

"Do it until I can't stand it anymore. Then we can weep together."

"Weep? What do you mean, weep?"

"Weep for joy. Ultimate joy."

"I'm game. Joy sounds wonderful right now."

"You're being fantastic. Then, when both of us are overwhelmed with joy—then we will weep."

I felt as if I would explode. Every touch of my fingertips on her silky body drove me to new heights of ecstatic pressure. It seemed like hours later Leura rose, slid over beside me and began brushing my hands and arms with her fingers as I continued touching her. When I could stand it no longer, I stopped moving my hands.

She sensed the change and rolled ever so slowly onto her back pulling me down with her.

Those dark blue eyes continued to bore into my soul while her soft voice hummed quietly, "Weep my love. Weep for all time,"—her voice trailed off into silence.

My mind and senses virtually exploded, one long explosion of complete and delicious abandon. I lost my sense of gravity and seemed to float in the midst of continuing soundless reverberations. I had never before felt such intense pleasure. The center of my being separated from my head and floated through my body. Intense feelings ricocheted between joy and melancholy, then pleasure and despondency, never remaining for long in any single state.

Leura's near whisper floated through my head. "Thanks dear Drax. Thanks for life and love." I opened my eyes and looked at her for an instant and was surprised to see narrow streams of tears running from the corners of her eyes. Once more, I drifted in complete, all-engulfing, feeling-filled silence.

Things changed—suddenly and drastically. Normal gravity had returned. When I reached for her, all my grasping hands found was a slightly damp, rumpled cotton sheet. *What the . . . I* thought as I opened my eyes to the shock of a bright, sunlit window in a beige room. I was alone and in a different bed in a different hotel. Outside, the sun was rising over the unmistakable skyline of Cleveland Ohio. "My God!" I said out loud incredulously. "I never . . . almost forgot who I was," came stumbling out of my mouth.

A flash of realization made me check my watch. I saw there was barely enough time to get to my breakfast meeting with Arlo Trippy, the engineer who was my NASA contact. He was working with me on their part of the Eegis project. I dressed, grabbed my suit coat and headed for the dining room. Arlo was waiting as I walked in.

"Right on time. I like people who are punctual."

"I almost wasn't. You wouldn't believe the wild dream I had last night or rather this morning. At least, I think it was a dream. It seemed so unbelievably—alive."

"Sometimes dreams can seem unbelievably real."

"This one sure was." I shook my head. Still, bewildered. "Well, let's get down to far out physics. That's reality."

"Certainly." Arlo paused and gazed intently at my coat lapel. "What's that pin you have on? You weren't wearing it yesterday."

I glanced at my lapel. Firmly attached was a tiny gold feather.

I will stand with the Muslims should the political winds shift in an ugly direction.
—*from: Audacity of Hope - Barack Hussein Obama*

Thought for the day: In today's topsy-turvy world, it is **not** okay to call a person a slut or whoremonger, but it **is** okay to be one!

—*Howard Johnson, 1999*

> To bear up under loss;
> To fight the bitterness of defeat and the weakness of grief;
> To be victor over anger;
> To smile when tears are close;
> To resist disease and evil men and base instincts;
> To hate hate and love love;
> To go on when it would seem good to die;
> To look up with unquenchable faith in something ever more to be.
> That is what any man can do, and be great.

—*Zane Grey*

Lyriel's Decision

A threat and a menace are not the same thing. A threat is the mere possibility of danger or something without danger that may have the appearance of danger for a time. It can usually be dealt with or avoided by clever counteraction. A menace, on the other hand, is a real danger that must be dealt with and is as certain as the rising of the sun. Though empty threat it may seem, beware the true menace that hides under the cunning mask of a threat.

Lyriel, leader and the oldest of the five in the control group at Far Station 322, was alone in her quarters and deep in troubled thought. The New Life Project she headed had gone terribly wrong. As usual, she would have to make the final decision to try one last time to redirect the project or terminate it and start over. The other four were divided, two on a side as usual. The project that seemed so promising now posed a terrible threat. It had burgeoned into a serious menace to her sector and maybe the entire galaxy.

A Muerr, Lyriel was by far the largest of the group and the farthest from her home planet. Despite her large size, Lyriel did not dominate the group, but worked diligently, almost gently, at leadership. The other four included two Fallons: Farcos, and the only other female in the group, Shremon: Stagus, a Thrack: and Llalimeno, a Torbun. The Fallons usually took positions opposing each other, rarely agreeing on anything and arguing constantly. Stagus was loud, aggressive, and stubborn while Llalimeno was almost the opposite: quiet, thoughtful, and open-minded, but not easily swayed once he took a position. Now the four were deeply entrenched, two to a position, leaving Lyriel to resolve the situation.

Finally deciding on an action, Lyriel stood up and popped her communicator. "Everyone to the observation deck. Let's have one more try at a solution we can all live with." She belted her tunic, pushed back her long reddish hair, and headed for the meeting. The door to the lift swished open to reveal Farcos and Shremon standing opposite each other, staring in cold silence. Their black eyes and grey faces were tightly framed by their straight black hair. As she stepped silently onto the lift, Lyriel smiled as she thought of these two having sex and wondered how they did it. They looked like clones with no obvious differences showing through the skin-tight coverings that could hardly be called clothes. Their ID patches were the only method of telling them apart. Even their voices were identical.

"After you," Lyriel said softly when the door of the lift opened, extending her hand to indicate the door. Usually talkative, the Fallons stalked off the lift, side-by-side in cold steely silence. As she followed them down the hall, Lyriel thought to herself, *This is going to be a difficult meeting.*

They took their seats in front of ULDI's display screen and waited for the others. Almost a sixth member of the group, the computer was an interactive, free-thinking entity called by the acronym for its unlimited logic database interface. Lyriel knew there was no point in trying conversation, so she set up project simulations on her input console. Before it swished open, the muffled sound of a loud angry voice penetrated the door. Stagus was about to enter.

"What the hell are we doing now?" The small but burly Thrack bellowed as he entered. "I say terminate the damned project so we can get started again with a clean slate. We've wasted far too much time in endless bickering already. I see no reason to waste any more time." Stagus had a loud, booming voice despite his small body. He plopped his compact frame firmly in his seat, crossing his arms defiantly. His long scraggly white hair hung over the back of the seat, completing the caricature of the stubborn, bellicose diehard—an immovable object.

"Now, Stagus, try to see another viewpoint," Llalimeno said quietly, but firmly. The tall slender Torbun constantly tried to persuade his companion to be more open, usually without success. He almost slithered into his seat, his body flexing as he adjusted to the seat's contour. The pale blue skin of his hairless head was in contrast to the ample hair of the other four. His pale blue head turned brilliant blue on those rare occasions when his emotions got hold of him. As usual, it was quite pale, indicating he was in control.

Lyriel gazed up at the information on the display. "Let's go over this once more to see if we can find a way to avoid termination."

"Damned waste of time!" Stagus muttered.

Lyriel stood up, turned, and towered threateningly over the smaller Thrack. She was quite angry. "I've had my fill of your closed-minded, non helpful attitude and comments. If you've nothing constructive to offer, keep silent!" she demanded. "And don't answer!" Though physical violence was not her way, the threat was there, and at three times Stagus' size, the threat worked. The rest of the group registered surprise and submission at this unusual, for Lyriel, display. She returned to her console and the business at hand.

"Let me review the situation. As I do, listen thoughtfully for anything we may have missed in our previous efforts. Try an open-minded approach. I know this is difficult for some of you, but do it anyway. I would like your acknowledgment that you will at least try to put aside any strong feelings and consider things objectively. Rather than answer by voice, please stand to show your agreement to this simple request."

To Lyriel's surprise, Farcos stood up immediately, followed by Llalimeno. Shremon, surprised at Farcos' quick move, glared steadily at him for a moment before finally rising. It was several tense, suspense-filled moments later when Stagus got out of his seat.

Lyriel smiled when Stagus rose. "Thank you all, now let's get at it. Please resume sitting, and I will review the situation. When we found this planet, we discovered it had been seeded with life by natural processes. The life was quite normal in every respect, save one. The evolution of new life-forms was quite rapid, about a thousand times faster than on any other known planet. We knew it was only a matter of time before intelligent life evolved, so we watched and waited. After several species with dawning intelligence were wiped out by rapidly evolving microscopic forms, we decided to try intervention and the New Life Project was created."

"We know all that. I think we would have been better off if we let things take their natural course," Farcos commented in his shrill voice. "Our interference created the problem in the first place. We should terminate it before it gets beyond our control."

"I agree, do it now," Stagus added in an uncommonly subdued, almost-reticent comment as he looked directly at Lyriel. Her threat had dampened his usual fervor.

Lyriel was livid, but gave a controlled response. "None of that. You both agreed to keep an open mind, and I intend to hold you to that agreement."

"Let's hear the rest of the review. We might have missed something. It might trigger an idea in someone so we can solve the problem," thoughtful Llalimeno commented. Shremon remained silent, staring blankly at the display.

"Please don't interrupt unless you have something constructive to say," Lyriel said firmly then resumed the review, following the outline on the display in front of them. "The next time a reasonably intelligent form appeared, we created this project and helped evolution by tinkering with the DNA of several of the most intelligent creatures. The end result was a bipedal creature, quite similar to us physically. A bipedal omnivore and tool user that stumbled along for a while before it learned the skills of an effective predator and exploded over most of the planet. Tool use, language, clothing, and growing intelligence rapidly changed the creature into a highly organized pack animal living in family groups that steadily grew in size and power. When written language and then the tools of science appeared, we moved our station from its orbit around the planet and tethered it on the edge of their moon to hide it, yet permit our direct observations. I arrived at Far Station 322 shortly after it was moved to the moon. With me came two more scout craft, and the station complement was increased from three to five. Llalimeno and I are the only two remaining from that first contingent of five. We increased our surveillance with the new scout craft and stopped all direct contact. About this same time, we began calling them the Leutra from the extinct inhabitants of that planet. The Leutra were beginning to look at the sky with better and better telescopes, and we knew it would not be long before we must move the station again."

"That's about the time we two Fallons came aboard relieving those horrid Kleps," Shremon commented. She was clearly repulsed by the hairy Kleps with their ugly eating habits. The Fallons were almost antiseptic with their personal habits, even eating in private. They avoided any gathering where food or drink was available. "Our first scouting mission analyzed the Leutra's great preoccupation with reproduction. That was when the problem we are facing now was first predicted."

"Correct!" Lyriel said, regaining control of the review. "Their knowledge and technology grew at an unbelievable rate. We believe that their short lives and rapid development were one effect of the high rate of evolution. It took a long time for us to realize everything about life on this planet was moving at an unbelievable rate of change. They had terrible wars, plagues, and were set upon by rapidly evolving microscopic life-forms, but still their numbers steadily increased. When they began conquering the diseases that were holding their numbers to a slow rate of growth, the rate exploded unbelievably. That caught us off guard, and we instigated a number of control factors, primarily new diseases they had trouble fighting and were specific to the Leutra. It wasn't difficult considering the rapid normal rate of evolution of the planet's tiniest life-forms. We tweaked the DNA of a few bacteria and viruses, thinking we could use that to keep a balance of populations of the many interesting life-forms that had evolved. We had slowed the explosive growth of their population, but only for a short period. With amazing speed and efficiency, they found cures and preventive measures for these diseases while engaging in several major wars. They developed machines that flew, created fission and fusion weapons, and had the beginnings of space travel. About the time they began flying rockets, we moved the station away from their moon and placed it in its present position, matching the planet's own orbit and hiding behind its star, 180 degrees away."

"It's their damned preoccupation with procreation that is the problem," Stagus boomed, finally unable to hold back. "As intelligent as they have become, and with the knowledge they have, why can't they control their population growth? They have the technology to do so. They've already far outstripped all other large animal species in numbers, and the creature mix has become terribly lopsided. They recently reached the limit of growth of their food supply and now starve by the millions. Yet, they remain so passionately involved in reproduction they can't see, or won't consider, it will ultimately lead to their destruction, and to the destruction of all other large animals as well. Any Thrack with half a brain could see that and insist that such madness be stopped. The Leutra have evolved into a high-enough intelligence level to be considered for contact and possible membership in the union, but they are still controlled far too much by instincts. Loosed on the union as they are now with their present knowledge, and particularly with their unusually rapid life processes, they would reproduce themselves into numeric control of the entire galaxy. I can't see us letting that happen. It would be genocide for all other intelligent species."

"Stagus is right," Shremon began. "Historically, look at those few planets with dominant intelligent species whose uncontrolled population growth created similar problems. Animal life on every single one I have studied eventually lost diversity, leaving only a few species which eventually went extinct. There was one planet where the intelligent species, a pure predator, eventually became its only remaining food supply. Horror stories of the resulting cannibalism are legion. Thankfully, they didn't achieve hyper–space travel or gain membership in the union before going extinct."

Lyriel was pleased the group was loosening their entrenched minds and beginning to open up a bit. "Good points, but are there any new ideas of what we can do to turn around the direction the Leutra are headed? They will surely learn the means of hyper–space travel and Trias Teleportation in the near future. We cannot let them expand beyond their planet as they are. In their hands, TT technology would create a huge menace. We need a means to change them, and soon. The alternative is to annihilate the species and start over. That would be a terrible setback to the project, not to mention the elimination of a promising species."

Llalimeno unfolded his sinuous form from his seat and stood up. Torbuns use so much body language with their long lithe bodies, they rarely spoke while seated. "Unfortunately, legal limitations on our actions prevent us from using some methods directly. Couldn't we develop a means to get them to take action themselves? What they need is a major cultural, emotional shift. Are their decisions so instinctively controlled, so emotionally charged, so illogical that we cannot find a method to cause them to overcome their passionate, illogical preoccupation with reproduction?"

"It seems to me, their tremendous drive is what caused their rapid development in the first place," Farcos brought up. "If we can't find a way to get them to greatly reduce or eliminate that reproductive drive without reducing it in other areas, the whole reason for the project will be lost. So far, we haven't a clue as to how that can be accomplished legally. We pushed the legal envelope long ago when our first DNA tweak resulted in the Leutra. I see no reason we can't push the legal envelope again. Can't we bypass a few of those bureaucratic limitations to get the results we seek?"

In a rare moment of harmony with her fellow, Shremon agreed, "A little legal latitude and we might find a workable answer. For such an advanced intelligence, they are still almost totally controlled by instincts. This is unusual. Most advanced, intelligent species learn to think logically rather than emotionally. The Leutra think they are so far above the other species on their planet and that they alone have overcome instincts, replacing emotion with rationality for decision making. How wrong they are."

"They are less rational than some of the species that live in their oceans," Lyriel remarked as she again took control of the discussion. "We must be wary of ignoring our legal limitations. If we find a method that 'pushes the legal envelope' as you've said, we had best examine it thoroughly and have immutable evidence of its effectiveness. With that caveat, I'm open to new and innovative ideas."

Stagus stood up smiling. "Let's merely inform them by using their visual communication system that unless they limit their population, they will be destroyed. Llalimeno could be our spokesperson. One look at that bald blue head would scare them into compliance."

Most of us laughed. Llalimeno, his head a brilliant blue, stood up and looked at Stagus. "I doubt my visage could instill one-quarter the fear as would one look at your ugly pink face and scraggly white hair."

After several moments of accusations, shouts, and curses, Lyriel stood and raised her hands in a command of silence. "I believe we have just demonstrated we, too, are subject to instinctive, emotional behavior at times. Let's not condemn a species that exhibits similar lack of control in a different behavioral arena. Now, I am going to demand effort. Stagus and Shremon will work together as will Farcos and Llalimeno. Go wherever you wish to work. I'll give you forty hours to come up with one or more new concepts for possible development into a solution. Push the envelope if necessary, but thoughtfully. Whatever we do, we had better have unimpeachable reasons behind it. We will be judged by results, not effort. I will work on one of my own, so when we reassemble here there should be at least three new concepts. It's 11.0710 now, so I will expect you all back here at 12.1310. That gives you the full forty hours to leave, do as you wish about developing a solution, and return with a full report. Now, get to it."

Lyriel left no room for questions or argument. With some shuffling about and a few grumbles, the two teams arranged for their places to work and left the room. Lyriel remained on the observation deck, her favorite place on the station. For a few moments she sat in silence, looking at the huge display screen. "ULDI! Show me the current full view of Leutra."

A view of the planet from the observation camera in another part of the planet's orbit filled the display. She marveled at its beauty. A bit larger than her home, Muerr, it was bright blue with patches of tan, green, and white. Muerr had more than twice the land area and about a third of the oceans. There was much less greenery on Muerr since most of the planet was desiccated deserts. Located closer to the center of the galaxy and at the trailing edge of the next arm, Her home planet orbited a star almost identical to this one and at about the same distance away. The two planets were near twins with similar atmospheres, temperatures, and life-forms. Life on Muerr was limited to the oceans, the islands in the oceans, and the edges of the three continents where rain fell. Unlike Leutra, there were vast areas of Muerr in the continental interiors where no rain ever fell and no life existed. She knew the Leutra well from data gathered by their manned research vessels and unmanned scout craft over many years. Once the Leutra developed electronic communication, including video transmission, the bulk of their knowledge was obtained from the Leutra themselves.

She continued musing about the focus of the project. Forbidden from having direct contact of any kind once the Leutra developed written records, they became observers only. Persistent reminders of earlier contacts were part of the lore of many Leutra cultures, coming from oral records repeated through many generations. Lyriel was amused every time a Leutran reference was made to the grossly changed, but unmistakable information about those contacts. No harm had been done. Sightings of their research and scout craft was another matter. Though there were few actual sightings, the Leutra expanded those with creative imagination and outright fabrication into a discredited phenomenon. The research craft, used to examine and experiment with Leutra individuals directly, created the most problems. Their use was discontinued some twenty years earlier when it was feared the sophistication of Leutra weapons and surveillance technology might result in confirmed detection or capture of one. The much smaller, faster, and more maneuverable scout craft continued in use. Her reflections of the past completed, Lyriel set to work on another possible answer to their dilemma.

When ULDI reminded her at 0920, she knew she had less than twenty minutes to gather her thoughts for the meeting. She expanded on an unusual idea that passed through her mind on several different earlier occasions. Pleased with the results when it was provisionally approved by ULDI's legal program, she thought. *This might work.* Usually quite punctual, the other four would begin returning about 0935.

Lyriel commanded the computer, "ULDI! Clear the display. Access none of my latest Leutra project without my voice activation." She was prepared. If any new proposal seemed superior to hers, she would not reveal her work, but support one of the others. Smiling to herself, she thought, *There are some privileges of command.*

As usual, the rumble of Stagus's voice announced his arrival before the door opened. "I still say the only thing that will work is to change their culture. Any plague or other partial annihilation will only put off the inevitable, and we're running out of time," Stagus argued as he and Shremon entered.

"Let's see what the rest think," Shremon replied. "Your idea will not pass the legality test." A moment's delay and she took her seat, the other two entered and seated themselves without comment. Lyriel noticed Llalimeno's head was quite blue.

Lyriel rose and faced the group. "I'm sure you've transferred your work to ULDI, but let's hear verbal reports first. Shremon, from what I heard as you and Stagus entered, I take it you did not come to an agreement. What a surprise," she remarked sarcastically. "Tell us what you think."

Shremon seldom agreed with Stagus, or anyone else for that matter, including her fellow Fallon. "The Thrack has no idea how hard it would be to change their culture, even if we could find a way to skirt the legalities of such action. ULDI! Display SF22 statistics."

A table of items and related numbers appeared on the screen.

"This list of Leutra diseases includes all those the Leutra currently hold in their arsenals of biological weapons. It lists the lethal effectiveness as a percent of the total population, the probable rate of growth for outbreaks under several conditions, and the probable time before the Leutra could stop an outbreak. After selection of the best prospect, we could modify some of our supply of the selected organism and use them to reduce the population to an acceptable level. This would buy us time to find a way to get them to regulate their own population and restore balance to the planet's life-forms. A reduction to about half a billion Leutra would seem most practical."

Lyriel commanded, "ULDI! What is the probable length of time after activation of SF22 that the Leutra would develop hyper–space capabilities?"

ULDI answered while displaying the same information at the bottom of the display. "Roughly two hundred standard years if the disease terminated scientists likely to be in that field in the same proportion as the general populace. It is more likely that particular group's termination rate would be closer to one third of that of the general populace resulting in a projection of only eighty standard years."

"That's far too short a time for us to do anything indirectly about their culture," Stagus commented, sneering. "If we create a controlled decrease in their food supply, the resulting starvation would reduce the population substantially and maybe teach them how important it is to control their numbers. Combine that with efforts to directly change their culture, so they themselves would stabilize the population at some desired level where the balance of life-forms on the planet would be more stable and in accord with Union standards."

"Your interference in their cultures would not pass a legality test," Shremon countered.

"Let's find out how illegal it is. Stagus, ask ULDI the legality question," Lyriel requested.

Stagus was hesitant, but finally did as he was asked. "ULDI! Test the legality of ST279."

ULDI replied, "In its present form, ST279 is too vague for a legal opinion. It is necessary for you to define the level and method of changes to their cultures to obtain an opinion."

"Damned, stupid machine!" Stagus muttered. "I thought maybe working together we could come up with an acceptable method."

"Possibly," Lyriel replied. "ULDI! How long would a major culture change, as proposed in ST279, take to be accomplished by the Leutra, given we found an acceptable method to trigger it?"

"Approximately two hundred standard years."

"ULDI! Could we reduce that to fifty standard years?" Stagus asked.

In a few minutes, ULDI replied, "No answer to that question is possible without additional data about the method and application."

"You knew that was coming," Lyriel commented. She looked at Farcos and Llalimeno. "How about you? Did you come to any consensus?"

Nodding to Llalimeno, Farcos stood up to announce their results. "We feel another genetic tweak to the Leutra would be the best solution. We could introduce a modified virus or bacteria that would reduce their numbers to an acceptable level while modifying the remaining Leutra to more tolerable levels of reproductive drive. I believe it is a hormone they call testosterone that drives their reproductive excesses. A substantial reduction in their ability to produce testosterone should have the desired effect. It would have the side effect of modifying their aggressive nature as well."

Stagus butted in. "That would never pass the legality test!"

Llalimeno, his head only slightly bluer than usual, sneered at Stagus, "Is the Thrack substituting his legal expertise for ULDI's? ULDI! Please provide an opinion of the legality and viability of FL229."

Farcos and Llalimeno both smiled in satisfaction as ULDI reported, "FL229 in its present form is barely legal. I would need more details to be certain. Genetic manipulation is not an exact science and never will be. The results of this procedure could result in the extinction of the Leutra."

Noting the dissatisfaction of Stagus and Shremon, Lyriel stood and addressed them all quite formally. "You have heard three proposals, all of which use various methods to reduce their numbers. One proposal would use genetic modification to reduce their reproductive drive, a questionable and risky process. According to ULDI, each one has questionable legality as proposed. Now, before we discuss and vote, I would like to present a proposal of my own."

Stagus stood angrily. "Don't the articles of our charter specifically state a leader cannot make a proposal? I don't see how we can legally listen to a proposal from you."

Lyriel smiled as she gazed patronizingly at Stagus. "ULDI! Under what circumstance is it permissible for a leader to make a proposal? Please display all references."

"Should the members of the New Life Project fail to agree upon the solution to a problem, and be hopelessly deadlocked, the leader may make a proposal. Article four, section three," ULDI spat out, displaying the reference on the screen.

Greatly chagrined, Stagus muttered a few expletives as he sat down, then remarked sarcastically, "All right, let's hear this proposal from our glorious leader."

"Thank you, Stagus," Lyriel replied, ignoring the sarcasm. "I propose we give them a history book."

There was a long silence as the four displayed baffled looks. Llalimeno spoke, "A history book? What kind of history book?"

"That's crazy! How could a book accomplish anything?" Farcos asked with a still puzzled demeanor.

Lyriel laughed. "Let me explain my proposal. We have the complete records of this planet since the first promising creatures evolved. ULDI could convert those records using the languages and style of the Leutra into a book entitled *The True History of Planet Earth*. The history could be extrapolated through their extinction. We could plant it in the form of a manuscript by a famous, but deceased writer. ULDI could create such a book using the literary style of the famous author. We could then TT it to an appropriate place where it would be found and then published."

Stagus was soon back complaining. "That's ridiculous. How could such a book have any effect? The Leutra would consider it fiction, ignoring the obvious."

Llalimeno slithered to a stand, his visage a pale blue indicating satisfaction as he stabbed at Stagus verbally. "I can't understand how an individual with so little insight or imagination could become a member of this project. I can see how such a book could be a truly innovative solution. I'm certain we could devise a way to word it so it would be believed. My only question is, would it have the desired effect?"

Lyriel had anticipated this type of question. "ULDI! Please report on the efficacy and legality of proposal LY83."

"The book proposed in LY38 is quite legal as described. Since I would be creating the book, the legality of the entire proposal would be assured. Analysis of the psychological effect and resulting action of the Leutra show a probable success of 70 to 90 percent."

Shremon stood and questioned, "How can we ensure the book's message will be taken seriously?"

"You all know about Supernova EMX356 that caused a major evacuation in nearby sector eight of the galaxy," Lyriel reported to heads nodding in agreement. "Its light will arrive here in about two years. We will place the precise location, date, and time of the explosion in the book. That, together with the many other confirmable facts, should make them patently aware of its accuracy. Then if they don't change their current course and control their population, we will have to terminate them and start over."

A four-hour deliberation of the details continued during which time all came to agreement except Stagus who doggedly stood his ground.

Finally, he struggled to his feet, hands held high in resignation and spoke softly, almost apologetically. "I hate to admit it, but that may be the only realistic, workable option. Let's start on it right away. Remember, the book is Lyriel's idea and should be recorded in the record books as her decision. If it works, she'll get the credit."

Lyriel smiled triumphantly at Stagus' desperate attempt not to appear totally defeated. "And if it doesn't, I'll get the blame, right?"

On Prayer (From the Sermon on the Mount)

[5]And when you pray, do not be like the hypocrites, for they love to pray standing in the synagogues and on the street corners to be seen by men. I tell you the truth; they have received their reward in full.

[6] But when you pray, go into your room, close the door and pray to your Father, who is unseen. Then your Father, who sees what is done in secret, will reward you.

[7] And when you pray, do not keep on babbling like pagans, for they think they will be heard because of their many words.

[8] Do not be like them, for your Father knows what you need before you ask him.

—Matthew 6:5–8 NIV

He who knows nothing, loves nothing.

He who can do nothing, understands nothing.

He who understands nothing is worthless.

But, he who understands also loves, notices, sees . . .

The more knowledge is inherent in a thing, the greater the love.

Anyone who imagines that all fruits ripen at the same time as

    strawberries, knows nothing about grapes.

—Paracelsus

Add a few drops of venom to a half truth and you have an absolute truth.

—Eric Hoffer

One Man's Opinion on Being a Christian

To follow the teachings of Jesus does not mean putting on the trappings of a Christian to show others. Nor does it demand that others practice Christianity the same way we do. Jesus never forced a single person to follow him. He left that choice up to us. He never demanded we look like him, dress like him, speak the same language as him, eat the same food as him, or bow down and worship him!

He never had a church or a parish. He asked only that we believe in him as the son of God and let our hearts follow his. As a boy of twelve, he questioned the priests in the synagogue. Should we not then, in following his example, question authorities of the church, if only to keep them on their toes?

Was not "Go now and sin no more!" his only command after forgiving? How about "Let he who is without sin cast the first stone!"

Is not this a charge for us to refrain from condemning others in our own lives? In washing the feet of the disciples at the Last Supper, did he not demonstrate the ultimate equality of the leader as a servant to those he leads?

Jesus showed us that by answering to ourselves in understanding, we would be serving him. Being a Christian is not a goal to be reached but a searching and striving within each Christian to be more *Christlike* each day of our lives—a private, personal drive within our hearts and souls. To be a Christian is always to lead by example as he did.

—*Howard Johnson, March 14, 1999*

Of all Life's difficulties, I have found it hardest to compel myself to recognize and concede a woman's right to meet me on even terms. But it seems equally hard for a woman to understand my attitude. No more than all the priests, philosophers and poets do I know what love is. Unlike many of them, I am unwilling to pretend that I do know. Neither do I know what life is. But it seems to me that if love or life lack dignity, neither the one nor the other is worth the sacrifice of half a moment's thought.

—*Talbot Mundy in* Purple Pirate

Faith without credulity, Conviction without bigotry, Charity without condescension, Courage without pugnacity, Self-respect without vanity, Humility without obsequiousness, Love of humanity without sentimentality, Meekness with power.

—*Charles Evans Hughes, 1862–1948*

A Matter of Dedication

Onas awoke to the warmth of sunlight on his face and animal noises some distance away. He was half-hanging, half-lying, and almost upside down in a tree, about twenty feet above the ground. He had no memory of how he had gotten there, but imagined it was painful. Trailing away from him up into the higher branches were several parallel lines of nanocord, and far above him in the forest canopy, the remnants of his gliderchute.

A drop of sweat formed at his chin and ran the length of his jaw toward his ear. His right cheek burned, which was not a good sign, and his right eye was half-stuck shut. His right foot was tangled in the cord, and when he exerted himself to free his leg, an intense pain shot up his back and nearly caused him to black out. It was all coming back to him: the sharp crack, the look up at the collapsing wing, and then his GC folded and helicoptered him down to a soft crash into the forest. He was at least two miles from the long river sandbar he and Eyalon were supposed to land on and set up the geo-research station.

A sudden lurch downward and Onas realized the GC was beginning to slip from its hold in the branches of the canopy. He watched fascinated and unable to do anything as the branch holding the GC bent and then finally broke, dropping him the last few feet to the ground. Before he could move, the GC broke free and headed straight down at him from at least a hundred feet up. He raised his arms instinctively to ward off the blow and watched as the broken wing caught the air and spun away from him at the last minute.

Damn! That was close! he remarked to himself as he tried to get up. When it didn't hurt too much, he rolled over and got to his hands and knees. The GC wreckage was right in front of him and what he saw was a shocker. The main composite member had been cut apart neatly, like with a knife. The secondary member had unexploded red primer cord wrapped around it. *Some son of a bitch tried to kill me!* He thought as he traced the primer cord to a tiny device taped to the composite brace. About three inches of the broken primer cord dangled nearby, broken away from the switch before it could be fired. That had doubtless saved his life. The device was a simple pressure switch set to fire the primer cord well below the drop height and at least half a mile above the jungle. If all the primer cord had fired, the GC would have blown apart and he would have plunged to his death. The loud crack was the primer cord going off and cutting the main member in half. All things considered, the gods had been kind to Onas.

He sat there for a while amidst his scattered test equipment and tried to decide who would want him dead. Mentally he replayed the last crew meeting aboard Mother, looking for clues. Captain Fogarty, the flight commander in charge of everything except the research station itself, was nearing retirement, and their relationship had been jovial from the start. Kropa, the young flight engineer and second in command of the ship, was on his first deep-space assignment. Reserved and seeming a bit self-absorbed, he still didn't impress Onas as the kind to indulge in any intrigue. He was too intent on furthering his career at this point. Greg, the data manager, about fifty, was rather a geeky, reserved man. Like most people who manage and record numbers, he could be curt in conversation and strongly opinionated when on a subject he knew. He knew numbers and data tracking. Adriana, the assistant data tracker, was not friendly with anyone. A plain and introverted woman of about thirty-five, she rarely spoke to anyone unless it was absolutely necessary. She made it plain that anything other than business in which she was involved was strictly off-limits. During the meeting, her only participation was to ask for direct authority to download and record all data from the research station once it was in place. At that meeting, everyone else had lots to say, even Greg.

Eyalon was the only one who had crossed swords with Onas. They had often been at odds since the project began. Second in command of the research station to Onas, Eyalon was overtly envious. They had several clashes over minor things in the configuration and operation of the research station. The last was an angry exchange about the division of actions and responsibilities during the two-month operation of the station on the sandbar. It ended with Onas putting his leadership stamp on the situation by telling Eyalon, "That's the way it is going to be." Eyalon stomped out of the meeting, grumbling unintelligibly. That was less than two hours before their scheduled drop. There were two more in the crew of eight, Salus, a grumpy old guy in charge of ship maintenance and Pirie, the steward, cook, and comedian of the group, also handled communications. Pirie always had something to say, usually a joke, but his cooking was definitely first-class. He didn't get on too well with Salus, but then Salus didn't get on too well with anyone.

Onas knew Eyalon had the time, the knowledge, and maybe the anger to rig the GC for his demise. Since neither he nor Eyalon knew which GC they would be taking until drop time, he could have rigged both GCs and easily disabled the one he was using long before the pressure switch fired the primer cord. Onas thought since Eyalon needed to retrieve his part of the equipment to set up the geo-station, he was probably looking for his body right now.

Before standing, Onas took out the emergency medical kit and applied self-sealing aid packs to his hip and his cheek. A cold crush-pack relieved his swollen right eye, but it would be a while before he would be able to see clearly. When standing didn't bring on any searing pain, he decided he was okay to gather the equipment for the geo-station and head for the sandbar.

He had been walking for about an hour when he saw the unmistakable yellow and red of Eyalon's GC wing on the ground up ahead. Before he got to the wing, he had to take back his suspicions about Eyalon whose crumpled body lay on the ground still attached by nanocord to the remains of the GC. It was obvious from the wreckage that the primer cord on his GC had all fired and ripped it apart.

"Sorry for the bad things I thought about you, old man," he said quietly. "Whoever did you is still alive and aboard Mother. They also think I'm dead, and that's to our advantage. I promise to make that bastard pay."

Onas changed plans. He began setting up the research station and rain canopy on the bank of the river under the trees, not out on the sandbar. He wasn't about to let anyone on Mother know he was still alive. He could conduct the experiments and take all the readings just as well in safety from prying eyes. Onas grinned as he mused, *The first com reports are due in two hours, before sunset. I wonder what will happen when their call goes unanswered.*

Onas began thinking about the planet and the project. Raza Three was an unusual planet. About 20 percent larger than Earth, it nonetheless had only about 80 percent of the Earth's mass and gravity. This was because it had a tiny iron core inside a huge mass of much lighter rocky material. There was no evidence of tectonic movements of the surface. It was smooth and quite level. About 90 percent of the surface was covered with a shallow ocean at most a few hundred feet deep. The land was flat as well, and because of the warm temperatures, it was very wet. Broken rain clouds moved, constantly bringing alternating rain and sunshine in irregular periods. It rained constantly at the highest elevations—about five hundred feet above sea level. The only things that sculpted the landscape were huge slow-moving rivers running from the highlands to the sea. Their flood plains were the only land not covered with a dense jungle canopy of trees. Virtually no sunlight reached the jungle floor, so it was smooth and easily traversed on foot.

Raza three rotated once every twenty-seven hours and thirteen minutes, approximately. For this reason, the program clock reset every twenty-seven hours and thirteen minutes at about midnight, Raza 3 time. This kept ground station time in sync with the planet's natural rhythm. The ship's clock remained on Earth time, so there were two reference clocks on the bridge, one for each kind of time. The planet had a huge moon about a third its size. The two rotated around a point somewhere between them but much closer to Raza three. The research project was to determine if gravitational distortion—tides in the rocks—was generating the heat that kept the planet warm. It was much warmer than it should have been, considering its atmosphere, surface conditions, and distance from the star, Raza.

Its atmosphere, twice the depth of Earth's, held a much higher percentage of carbon dioxide, 1.5%, and oxygen, 23.2%. Nitrogen, argon, and the other rarer gases were each a lower percentage of the total than Earth's. There was also a measurable portion of methane. The carbon dioxide was strong enough to make it noticeable with a slight, sharp stinging sensation when one breathed in. The surface air pressure was a bit more than Earth's at sea level. He had to know all this to properly set up the instruments. By the time for check in,

Onas had everything up and running and plugged into the data storage banks. He did not connect the data relay as that would have given him away.

Suddenly the speaker on the com unit barked out, "Baby one, are you there? This is Mother. Come in."

The message was repeated several times, each repeat a bit more urgent than the last. The voice on the other end was that of Pirie, the communications guy. Finally, almost pleading, he said, "You guys aren't fooling around, are you? Please respond."

"Onas? . . . Eyalon? . . . This is your captain speaking. Report back . . . now!"

Onas wished he could have seen their faces at that moment. The guilty party would have stood out like a neon sign on a dark night. They would have to send someone down to find out what was going on, but that couldn't be done until morning. Onas wondered if the killer would be the one to come down. He settled down in his sleeper for the night, knowing he would be ready in the morning.

Almost hourly through the night, the com unit broke the silence with, "Baby 1, are you okay? This is Mother. Please respond." Onas couldn't shut it off as that would be a dead giveaway that someone was alive. He did turn the volume all the way down.

At eight in the morning, the message changed. "We're dropping out of orbit. We will fly by and release a rescue party to see what's been going on. He should be on the sandbar in about two hours. Make sure your com units are on so we can find you." Onas wondered if this flight would be blown apart as were the first two. Only now, everyone would be watching. Also, this would be a military GC, not a civilian one and launched from a secure spot on the ship. He checked his com unit, carried it back into the forest, and set it on the ground some distance away. He wondered about the rescue mission. Would the killer be on it? Surely they would bring a new pickup rig, balloon and all, to lift the nanocord to where Mother could catch it as she flew by and lift whatever was attached to the end of the cord up into her belly as she flew away.

He camouflaged the research setup with branches and leaves as best he could and waited, hidden from sight in a small hollow of an old tree stump. He watched as Mother flew slowly by, wings fully extended at about twenty thou and released the GC. He followed the mottled green glider as it circled and descended to the sandbar. Who was piloting the craft but Lieutenant by-the-book Kropa himself. Well, of course. That would be his job. He tied down the GC and spoke on his com unit. It came through softly, but clearly on his unit, "I'm on the sandbar, and there is absolutely nothing here. What do I do now?"

"Start a search pattern of semicircles on the windward side of the river, you idiot, just like I explained before we dropped you." Captain Fogarty always said it like it was.

"Yes, sir!" Kropa clipped off as he turned and waded through the shallow river to the shore about two hundred yards downstream from where the setup was hidden.

After about forty minutes, his voice came on the com a bit unsteady. "I found Eyalon, sir. He's dead!"

"Dead? Where? How'd it happen?" Fogarty was obviously quite shocked, at least as shocked as one with so many years of military service can become.

"I don't know, sir! It looks like he crashed into the treetops and fell to his death from there. It's a hundred-foot drop at least."

"Is there anything strange or out of place at the crash site?"

"I'm examining the wreckage right now. I'll send images . . . It looks awfully broken up."

Damn! Please don't look at those broken members too closely. I thought.

"He must have hit the trees going fast," Kropa told him. "I can't believe how broken up the GC is."

As Onas was congratulating myself, Kropa added, "There is something strange though."

"What's that?" the captain queried.

"It's just that I can't find the research equipment, or his com unit. They are all gone."

Damn! Damn! Damn! Onas cursed under his breath.

"I know it was attached to his GC right on the main member. Wait a minute. Most of that piece was destroyed in the crash. That stuff's likely to be anywhere within a few hundred feet."

"Well, find it. Damn it, and let me know when you do."

"Yes, sir!"

Now much relieved, Onas worked his way to near where Kropa stood examining the wreckage. Setting his LK on stun, he placed his finger on his lips to indicate silence and stepped into view, leveling the weapon. Kropa froze, started to speak, and then stopped as Onas waved his weapon at his lips with the universal sign for silence.

"Turn off the mike on your com unit—carefully and moving slowly," Onas whispered.

A slight click on his unit and Onas knew he had complied. He then stepped over to Kropa and relieved him of both of his weapons, holstered his LK, and said, "Welcome to the deadly forest."

"What in hell is going on here?" Kropa asked when he felt free to speak.

"That's what I would like to know. Let me tell you what I do know."

As he led Kropa back to the setup, Onas explained most of what happened. When he finished, Kropa sat dumbfounded.

"Who in hell would want both of you dead?"

"That's precisely what I'd like to know, and I hope it isn't you. Incidentally, you'd better call in and report you found the missing equipment in working order and my dead body like Eyalon's. I don't want anyone on Mother to know I'm alive yet. That would be too helpful to our killer. So far it looks like an accident. Only you, I, and the killer know otherwise. That should be a great help in catching him."

After Kropa's report, the captain asked him, "Do you think you could set up that station and take those readings? The instructions are all there in the computer."

"I don't know, sir." After Onas' emphatic affirmative head shaking, he added, "But I'd like to give it a try."

"Good boy!" Onas muttered softly.

"I'll give you all the help I can from here, Lieutenant. I'm sure we can pull it off and make this mission a success. I'd hate to lose two of my crew and go home with nothing for it," Captain Fogarty said before clicking off the com unit.

Kropa was at least sharp enough to understand the realities of the situation. Now they had to take and record all the readings while finding and catching a murderer who, for all, they knew was fully capable of killing both of them, maybe by leaving them to starve to death.

At this point Onas returned the weapons to Kropa who looked at them curiously then to him, and said, "Sir?"

"Hell, Kropa, that wasn't being too risky. I shorted out the charge on both weapons before giving them to you. If you were the killer, you would have tried to use at least one of them and I would have known."

"Maybe I suspected that and am waiting for them to be recharged to use later," he remarked with a broad grin.

"Not a chance! Your eyes would have betrayed you to this old one-time psychologist. I watched them when I handed you your weapons. I'd have known instantly. Let's give Eyalon a proper burial. One due a dedicated man killed doing his job."

Kropa heaved a sigh of relief as they returned from Eyalon's burial. Onas was beginning to genuinely like this young man. Something about him struck a chord.

"How about we do a little rundown on the five remaining crew? Let's rate them as possible murderers and look for motive," Onas suggested.

An hour later, they had made little progress but had the following list of facts:

1. It was not personal, but rather was directed at the project. This was quite obvious from the fact that both science officers were to be killed.

2. Captain Fogarty was placed at the bottom of the suspect list for lack of motive. The other four were equal possibilities, but nowhere was there a shred of evidence as to a motive.

3. All had roughly equal ability to obtain the primer cord, pressure gauge, and tape used on both GCs.

4. All had equal access to the staging area where the GCs were fixed to crash and kill their fliers.

5. No one of the suspects knew Onas was alive. That was their greatest asset.

"It's not much to go on, sir," Kropa said, looking dejected.

"No, but it is a start. Now we'll have to develop a plan. We scheduled almost sixty days to finish the project and can use all that time to do that **and** find our killer."

✳   ✳   ✳

The first few weeks went without incident. When it came time to send the collected data, Onas coached Kropa. "You'll have to convert the data to a transmission format on your own. If I did it, I'm sure those data geeks would realize it was not the work of a junior military officer."

"How am I going to do that?"

"Tell me what you're doing and I'll let you know if it will be okay."

"It seems kinda like an inventory report. I've done lots of those."

"You'll do fine." Onas smiled as he kept to himself that he was storing all the raw data on digicards just in case. If one or both of the data geeks wanted to sabotage the project, he would have a backup.

✳   ✳   ✳

The end of the seventh week, Onas called Kropa over. "We're almost done here, so we'd better get things ready for pickup. It's fortunate that the balloon pickup systems on the badly damaged GCs were intact. That means we will have to get Mother to make three runs, each one protected against sabotage. As we decided, you'll go up on the first lift. It's definitely the safest. You'll tell them you are sending the instruments up first along with some fragile samples. Call for the pickup now."

"Won't they be suspicious when I show up in place of the equipment?" Kropa asked as he contacted Mother on the com unit.

"I'm counting on you disrupting their plans. If we sent the equipment up first I'm sure there would be a fatal failure on your lift. Make sure you note as much as you can how everyone reacts. Tell them you have to make a military report to Captain Fogarty and then get Fogarty to take you to his cabin for a private talk. You know what to tell him."

The com unit barked, "Are you ready for pickup?"

"Soon as I get the balloon up," Kropa replied. "This will be the equipment and some fragile samples, so treat them gently."

"We'll be there in about forty minutes. We're already out of orbit and flying."

"Wasn't that Greg, the data geek?" Kropa asked.

"Wonder why he's manning the com?" Onas questioned, "I thought Pirie would be doing that."

"He should be, especially during flight operations. I hope this isn't an indicator of trouble."

"Too late to worry about that now. Be sure you get to the captain as soon as you're aboard."

"What about the others while I'm talking to the captain?"

"There's not much they could do at that point. Fogarty will be conducting a wide circle to make the second pickup pass. Ask him to secure the rest of the crew in quarters so you and he can make the pickups without interruption. We don't want any of them to have access to any part of the pickup system or loading bay until I'm aboard."

With the balloon carrying the pickup line high above him, Kropa got ready to crouch into lift position in the pickup capsule while Onas stood behind the equipment, ready to duck under the cover as Mother flew by.

As the hum of Mother's air drives picked up, she showed up above the horizon over the river. As Kropa crouched for pickup, Onas shouted, "Pray man! Pray the lift gets you to Mother."

The catcher fork extending from Mother's belly picked up the balloon line which stretched, drew tight, and then snatched the capsule containing Kropa up into the air to be retrieved by the recovery winch. Onas was pleased to see the capsule taken aboard without incident before Mother flew out of sight.

Aboard Mother, Greg and Adriana were manning the retrieval equipment. When Kropa stepped out of the capsule, both registered extreme surprise.

"I thought you were sending the equipment up first," Greg remarked. "Why the switch?"

"Last-minute change of plans," Kropa reported. "Right now I have to report to Captain Fogarty. Military protocol, you know."

"This is a scientific expedition, Kropa," Greg remarked. "First order of business is the data. Where is it?"

"Coming on the next pickup."

"How can there be another pickup? You're up here!" Adriana asked in her most sarcastic tone of voice.

"I rigged the other pickup balloons to deploy as soon as the previous one is picked up. It was quite simple. Now I must report to the captain." That said, he stepped into the lift before they could complain and headed for the bridge.

Captain Fogarty was incredulous at the tale Kropa unfolded. "That Onas is both lucky and resourceful. I don't know that I can confine everyone to quarters without a known emergency. In the mean time, let's get Mother into another pickup turn."

"Why were the two data processors manning the catch lift? Where is the rest of the crew?" Kropa was puzzled.

"They offered to do it, and I saw no reason not to allow it. Now it is an obviously different situation. My bet is on those two as the culprits. That Adriana will do about anything Greg orders, so he's got to be the man behind the plot. I'd like to know what it's all about. Makes no sense to me."

"Me either, but Eyalon's death was definitely murder, so it must be serious."

"We'll be lined up to recover the equipment in about ten minutes. I'll send Salus down with you to the recovery bay and try locking the others in quarters. Get moving!"

"Yes, sir!"

When Salus arrived at the bay, they were about two minutes out. Kropa moved one of the mobile cargo cranes against the lift door just in case.

"What the hell's that about?" Salus asked.

"We don't need any unwanted company. I'll explain after this pickup."

"Okay, Lieutenant."

"Drop the catcher now," Kropa ordered.

Salus pulled down the lever and the winch lowered the catcher.

Kropa tapped his com unit and said, "Captain, the catcher is down and locked."

Some ten minutes later, the captain said, "Got it! . . . Raise it up . . . I'll start a new circle. The third capsule should be picked up in about twenty minutes."

It took about five minutes for the winch to bring the equipment capsule aboard. As soon as it was tied down, Salus lowered the pickup cable catch for the next pass.

Once more Kropa used his com to speak to the captain. "The catcher is down and locked."

And again after ten minutes, the captain said, "Got him! Pull him aboard. Then all of you report to the bridge. We still have a serious problem to deal with."

"Roger, Captain. We'll have him aboard in about three minutes."

Before Onas was up, the door to the lift opened, and Greg and Adriana tried to move the crane that blocked them. Somehow they managed to release the lock on the crane wheels, push the crane aside, and step into the recovery bay.

"What's going on here?" Greg asked curtly. "Why were you trying to keep us out?"

"Captain's orders!" Kropa answered lamely, positioning himself between Greg and the winch controls as the two moved between him and the lift winch. "This is now a military project and you are to return to your quarters."

Greg replied, "Not while I'm here. This is a scientific project, and in the absence of the leader and his second in command, I am in charge."

Kropa tapped his com unit on and hoped Onas could hear him. "Greg, you are no longer in charge. There's been a murder, and until that is solved, military law prevails and you are under the captain's command."

With that Adriana pulled out a Galbo blaster and leveled it at Kropa.

Greg looked surprised. "Adriana! Put that away! That's a dangerous weapon."

"Shut up, Greg. This is a whole lot bigger than any of your petty little data thefts. I'm running this show and don't you forget it."

Kropa inched his hand toward his LK holster, but she could cut him in two before he could raise it, and he was quite sure she would. "What the hell are you two up to?" he asked. Behind them, the capsule holding Onas was coming aboard.

"Saving our planet," Adriana shouted. "Saving our planet from alien invasion. Our organization is dedicated to preventing any material from any alien planet from reaching Earth. My assignment was to scuttle this project by any means possible."

"What the hell are you talking about?" Greg asked, a bewildered look on his face.

Salus grabbed for his weapon, but before he could aim it, Adriana cut him in half with a blast from her Galbo. Kropa took this opportunity to dive behind a bulkhead. The last thing he saw before he dropped to the floor was Onas stepping out of the capsule, holding his LK at the ready.

Onas stepped out of the capsule as Salus was blown apart. He was out in the open with no cover nearby. He hit both data geeks with a wide spread from his LK, knowing that would stun them for only a moment. In that moment, he ran and dove behind the loading dock bulkhead, putting three feet of steel between he and the deadly Galbo. Finding one of the round barrel covers, he hurled it toward the opening in the bulkhead where it was vaporized by a blast from Adriana's Galbo. In that same instant, Kropa rose and knocked Adriana down with his LK. Her Galbo clattered to the floor. Before he could get off another shot, she rolled over, grabbed the Galbo, and grabbed Greg, holding the Galbo against his neck.

"Drop your weapons and step out or I'll blast Greg," she shouted.

Onas called out, "We're not that stupid, you damned bitch. Go ahead, kill him. While you're doing that, we'll both hit you with our LKs—at full power."

After a short silence, Adriana began moving toward the door to the lift. Kropa was out of her line of sight, and Onas couldn't fire without hitting Greg. He tapped his com. "Captain! Adriana killed Salus and is using Greg as a shield so she can get to the lift. She's not inside yet. Can you do a complete lockdown—right now?"

Almost instantly the lock down siren went off, and all doors locked and the lift was immobilized. Unfortunately Adriana and Greg got inside the lift before lock down occurred. The lift doors were not blaster proof. Thinking quickly, Kropa rolled a heavy mobile cargo crane up against the lift door and locked it in place. It left enough room for a thin person to squeeze between the crane and the door frame. He positioned himself right beside the door.

Onas ran over to the other side of the lift door, took the same position there, and asked, "Captain, reverse the lock down—right now and be prepared to lock it again on my word. Now!"

As the door opened, Adriana pushed Greg out in front of her. Onas shouted "now!" grabbed Greg, and jerked him through the narrow space. Adriana burned a hole in the crane, but missed Greg then pulled back inside the lift. The door closed with her inside.

"Captain? I got Greg, and Adriana's back locked inside the lift."

Behind the crane, the door to the lift began to turn red, then yellow, then almost white. Adriana's Galbo was at work. Then the door melted away. A badly burned Adriana fell through the doorway, incinerated by the intense heat from the Galbo in such a confined space. She was dead before she hit the floor.

A few minutes later they informed the captain what happened. They heard the captain say resignedly, "Use the walkway and both of you get up here right away and tell me what in the hell is going on. I hate being totally ignorant about what's happening on my ship."

Onas replied, "Right away, Captain, but first let me say you have one helluva second in command. I'd recommend a promotion for this man, and I plan to put that in writing."

Kropa blushed!

❖   ❖   ❖

Enlightened people seldom or never possess a sense of responsibility

—*George Orwell*

God sends children for another purpose
    than merely to keep up the race;
To enlarge our hearts, and to make us unselfish
    and full of kindly sympathies and affections;
To give our souls higher aims;
To call out all our facilities to extended enterprise and exertion;
And to bring round our firesides bright faces,
    happy smiles and tender hearts.

—*Mary Howitt, 1799–1888*

In children we have an innocent audience not yet hardened and brutalized and made cynical. They look to us trustingly for information and enchantment. How very few of us are worthy of such trust.

—*Sterling North*

As for me, I am a mystic, not denying what I merely do not like or do not understand, nor claiming absoluteness for the truth I think I know. And I believe, and enjoy believing, that a greater mystery than human mind can know, selects and sets us amid flames of love and hate, wherein we forge new weapons and for them new uses, and for our souls new destiny.

—*Talbot Mundy in* Purple Pirate

More tears are shed over answered prayers than unanswered ones.

—*St. Teresa of Avila*

The desire to transcend the human condition is an invitation to tyranny.

—*Gertrude Himmelfarb*

Purple World

The change was as unnerving as it was subtle. Each time she looked away and back, she knew her view had somehow changed, but she couldn't figure out how. Caroloona checked her pouch. *Empty! Blast it!* With urgent need, she scanned around her. Where was Jeff when she needed him the most? Then she realized what had been changing. Her gaze rested on a dark form that moved only when she looked away. The only way she knew it had moved was that each time she looked away and then back the form was in a slightly different position against the background of purple vegetation. Still, its shape was hard to discern, and that was frightening. That it was absolutely immobile when she looked at it was quite disconcerting. Those who indoctrinated her and Jeff for this expedition had no knowledge of animal life on Koola except they suspected it had evolved from the same root as animals on Earth. That meant plant eaters and carnivores—maybe intelligent life, but no technology. Of that they seemed sure.

The deep maroon-purple of the vegetation on Koola combined with the long slender shape of the huge leaves gave forest vistas a different look from Earth. It also made the forest dark, in the midday light of Koola's sun. It was difficult to believe Koola's plant life had evolved from the same plants as the green ones on Earth and used the same photosynthesis. Then she remembered the oxalis hanging in the planter at her mother's. That plant was the same dark purple and had tiny white flowers. The colors of the forest flowers amazed her. White was common, but so was pale green, dark blue, dark yellow, and blood orange. Virtually every imaginable hue was present among them. Some of the flowers were huge and were suspected of being carnivorous. That dark form was no plant as it was definitely mobile and headed toward her position. *Why couldn't she ever see it move?*

Where in hell is Jeff? He'd been gone far too long and was almost an hour overdue on his com check in. She popped her com for him several times and received no response—a bad sign. Then there was her missing Galbo blaster. It had been nestled between two meal packets in her now-empty pouch. It and the food packets were missing. How could that be? She did not remember doing anything with the Galbo and the spare meal packets when she took out the food for her and Jeff to eat as they rested during their first foray into the forest.

They had secured the Gonga III, their explorer shuttle, and walked about twelve kilometers due planet north. *Could she have set them aside while they ate and forgotten to put her weapon*

169

and remaining food packets back in her pouch? That was extremely unlikely, but possible. Her food and only weapon for protection had vanished, and she was alone on an unknown planet with an unknown creature of some sort approaching with unknown intent and purpose. She checked the autonav unit and tried to bring up where they had stopped to eat so she could backtrack to find the missing items. The autonav unit didn't seem to be working properly and would not provide the location.

Suddenly a flickering pale purple light emanated from the form which she guessed was about fifteen meters from her position. An answering flicker came from far to her left where she saw another form. As she watched the flickering appeared then ceased, first from the form nearby then from the other. She noticed subtle color changes in the flickers. She guessed these were intelligent life-forms communicating. She wondered, *How intelligent are they, and are they hostile or friendly?*

A bright purple flash exploded in her eyes, silently. She was transfixed as the light held her gaze, and she could neither shut her eyes nor look away. Her mind became confused as memories, pictures, and a few sounds from her past life flashed through her consciousness then vanished. After what seemed like an hour, three words started and then kept repeating in her head, *No hurt, learning.* They repeated several times, *No hurt, learning. No hurt, learning.* The words were clear enough, hesitant at first, but definite and clear by the third repeat. *Learning what?* Caroloona thought out loud. The answering words came into her head clear as a bell, *Communicate you. Learning language. Know soon.*

As suddenly as it had begun, the purple light went out. It took a moment for her eyes to adjust back to the darkness of the forest floor. As they did, Caroloona realized the two forms were standing directly in front of her. They did not appear to be threatening in any way, so she assessed them visually. Four long legs topped with a short heavy body, they were unlike any creature she had ever learned about. The body and legs reminded her a bit of a small, black, smooth-skinned giraffe. About a foot taller than she, the height difference was all in the long legs. The head and other appendages were alien. Two "arms" were mounted, one on each side of the body and resembled the legs much as our arms do our legs. The ends of the arms were different. Their "hands" were seven "tentacles" ranging in size from about ten inches to no more than three. The head was larger and more oblong than ours and topped a short, thick, flexible neck. Between the wide-set eyes was a round organ about two inches in diameter—the light emitter. Beneath the eyes and centered, were what must have been nostrils, pairs of small holes surrounded by cup-shaped depressions. The mouth was a large, thin slit at the forward end of what could be called a blunt snout, much like that of a frog. There were no other openings on the snout.

There was a long silence before Caroloona blurted out in frustration, "What do you want?"

A brief flicker from the light organ from one of them and the words, *Come with us. Your friend is hurt,* came into her mind much like a recalled memory. This was going to be interesting—and a bit frightening. As they turned to lead her, she noticed a third member of the group, a tiny replica of the two, no more than eighteen inches tall with short legs. The tiny one scampered ahead, leading the way. "Search and discover" had been the bywords of this expedition, and she was doing so. Unfortunately it was not on her terms, but on those of her newfound "friends."

They walked approximately planet west for more than an hour in complete silence. The silence! That's what was so different. There were no animal sounds of any kinds, nor did she see any other creatures—ever! Only the deep purple vegetation broken occasionally with bright flowers. Then she touched the trunk of one of the trees as they passed. It was soft, smooth, and wet. She looked at her hand, and it was purple where the wetness clung to her hand. As she looked at her hand, now itching, one of her companions pulled a yellow cloth or skin from somewhere and proceeded to wipe the purple from her hand thoroughly. A slight flicker from its light organ and the words *bad, poison*, and *do not touch* came quite clearly from her memory. It pointed to several other trees nearby with the slick, wet purple trunks like the one she had touched.

"Okay! So I'll not touch them anymore," she said out loud. *Correct!* was its immediate unspoken reply. After several of these strange conversations, she realized that only when she could see their light organ did the words come into or from her memory. Clearly their communication was coming from that organ and reaching her mind through her eyes. All she had to do to stop information transfer was to close her eyes. That they had gained access to her memory and thoughts via her eyes also became obvious.

She became aware of a soft *clicking* in her head as she walked. It was getting steadily louder as they approached a wide clearing. When she shut her eyes, the clicking stopped. One of the creatures turned toward her, its light organ flickering. *Danger ahead. Stay close to us,* popped into her thoughts. The two drew close together on either side of her, and the tiny one got in between the front legs of one of them as they approached the clearing.

Suddenly the clicking stopped, and an actual whirring sound came from above in the clearing. As she looked up, bright red flashes and streaks tore through the air at and from about half a dozen flying creatures descending toward them. It was over in a few seconds as the six bodies plummeted to the ground with resounding "thunks" and lay still. Not unlike her friends, they were the size of a large dog, but with wings in place of the *arms* and front pair of legs. The rear pair of legs ended in the same type of tentacles as the "hands" of the others. There were also three small tentacles at the end of each wing. One of the huge open mouths looked amazingly similar to the mouth of terrestrial sharks. *Parallel evolution,* she thought.

Immediately into her mind came, *Carnivores. Dangerous, but not intelligent. Safe now.* She wondered if that large slit of a mouth of her friends was armed with equipment similar to the flying creatures.

As they walked past the dead creatures, rustling in the debris on the forest floor near one of the bodies caught her attention. It was a fat, snakelike creature, about a foot long and its business end was obviously trying to bite its way into the dead animal.

Scavenger, no danger. She learned. *It will grow to many times its length while eating its way through the dead.*

Name? she asked mentally.

We use no names as in your language. Evolutionary classification replaces names. Very complex. May be impossible to convert to your understanding.

"How do I call you?"

After a long pause, *Call me Blue, my friend Red, and the little one Black. Not accurate, but it should fill your need.*

It was obvious at least Blue had learned a great deal about our language and was able to convey complex meanings and concepts. I still wondered what their mouth slits hid. Were they also carnivores?

Like you, we are omnivores. Told me my thoughts had been read through my open eyes. It was obvious they knew a whole lot more about me than I did about them.

We came upon a large clearing. It appeared to be several kilometers wide in each direction. In the center stood a large building seemingly made of grey stone with many large glass windows. We were quite obviously headed for the building. "What is that building?"

Hospital in your language. Your friend is there.

"How badly is he hurt?"

That's why we need you. Your anatomy is quite unfamiliar.

"How did you find him?"

He must have tried to climb one of our "trees." The "tree" didn't like it and threw him out. We "heard" the EMF disturbance and arrived in time to prevent the flying carnivores from attacking him. They would have received a nasty surprise. We believe your flesh to be toxic to all life here. Though the chemistry is similar, the subtle differences are quite major.

"So we're toxic to you guys. That's comforting."

But you would be dead before the carnivores realized that.

Small comfort, that. I thought as we entered the building through a large transparent door.

Your friend is down this way came as we headed down a short hall and into a small, brightly lit room. Jeff lay there on a soft table, a bit bloody, but still conscious.

"Jeff, where do you hurt?"

"Everywhere, damn it! These geeks weren't gentle moving me, and I'd guess I've a broken leg and arm. They wouldn't do a damned thing to help me."

"No? They brought you to this hospital and brought me here to help. I'd say they saved your life. Let's take a look at your arm and leg."

A cursory examination and I determined his leg and arm were bruised, but not broken. "Hell! You're sore from bruises. Nothing's broken. A little painkiller, some rest, and you'll be good as new." I gave him a couple of pills from my aid kit and dressed his small, but numerous wounds. I massaged his arms and legs with painkilling cream for about fifteen minutes. He then seemed much better.

"There! . . . You should be feeling a bit better."

"Easy for you to say. How are we going to get away from these four-legged monstrosities and back to the shuttle craft?"

"First of all, they are not monstrosities. They are intelligent beings who are taking great risks to help us."

"How can you say that? They don't talk."

Blue's words came through to my mind loud and clear. *He has resisted all our attempts at communication. We were quite surprised when you responded so well to our efforts. It appears your "brains" are, as you say, wired quite differently from each other. He reacted quite differently from you. We didn't learn much from him.*

"Too much testosterone addles the brain, I guess. But it sure helps out in the quick decision-making process."

"Who the hell are you talking to?" Jeff commented.

"Blue here was telling me our brains are quite different."

"Blue? Who the hell is Blue, and how is he telling you about brains anyway?"

"Blue is this one right here, and he reads your mind through your eyes with light. I haven't a clue how, but he does. He also makes words and thoughts appear in my mind—almost like remembering, but different. We communicate in English!"

"That's bullshit!

"Incidentally, how is it that you are able to communicate so well with me, yet can't with him? You seem to have an excellent grasp of our language."

I believe that with the knowledge we gained from you, we could now reach his mind, provided he doesn't fight our efforts. See if you can get him to relax, look at my face, and try understanding what he remembers. That's the best way I can describe it.

"Who in hell are you talking to? Your words make no sense to me."

"Look directly at the organ in the center of his face, relax, and see what you remember."

"That's nonsense!"

"Just do it! That's an order!"

Greetings, Jeff. Your commanding officer knows what she's talking about.

"Wow! Wow!" was all Jeff could manage. "I heard or should I say, thought all those crazy thoughts before. Never realized where they were coming from."

"It will take a while to learn how, but it will become almost like talking."

"Great! We can have them help us get back to the shuttle so we can get the hell off this purple hell."

Caroloona turned to Blue. "Will you help us? I need to find a tool and some food I lost. It's important. Without that food, we won't last more than three days."

With that Blue retrieved several items from somewhere on his person and showed them to her. He held the Galbo and the two food packets. *You may have the food packets, but the weapon we will keep* came through clearly.

Caroloona closed her eyes and began thinking furiously. "Jeff! Don't think or question, but close your eyes and tell me when they're closed."

"Okay! They're closed."

"As long as our eyes are open, they can read our thoughts. I don't believe their hearing is well developed, so they are unable to detect our speech or read our thoughts as long as our eyes are closed."

"Wow! I'll keep them closed."

"Good! I learned that they somehow took our food packets and my Galbo back when we stopped to eat. As you could see, they returned the food to me but will not return the Galbo. I'm beginning to smell a rat. It could be a security measure on their part, or it could indicate something a bit more sinister."

"Oh? What would that be?"

"I'm not sure, but I am going to open my eyes and avoid thinking about it while I inquire a bit more. Keep yours closed."

"How about using the words *shut down* whenever either of us wants to talk to the other in private? That should keep them out of our heads."

"Good idea! I'm opening my eyes." I continued talking as if none of my secretive words had been said. "Why won't you let me have my weapon back?"

"Mine too!" chimed in Jeff.

*We allow no weapons other than those that have evolved. In your home worlds, you developed technology outside of yourself. We **evolved** our technology.*

"Then we'll have to get back to our shuttle craft immediately." Caroloona said to Jeff.

"We'd better get started right away. It will take us several hours to get there," Jeff commented.

We cannot let you leave came through clearly to both of them.

"I'm sorry, Blue, but we have to get back to our food source or we'll die of starvation. Your water tests fine for us, but the local plant and animal life is definitely not compatible with our chemistry."

We cannot let you leave came through clearly once more to both of them.

"Sounds as though he's stuck in a verbal loop," Jeff remarked and then said, "Shut down."

She closed her eyes and said, "What's up?"

"I realized we each have a deadly weapon we can use if necessary."

"What's that?"

"You still have that machete strapped to your leg, don't you?"

"Damn! I never thought of them as a weapon, but they are good ones. They must not view them as weapons either, or they'd be gone like our Galbos. I'm going to open my eyes and push them to see what happens."

With that she opened her eyes and said, "What will you do if we try to leave?"

Her thoughts caused Blue to stand alert and soon Red was in the room with them. *Restrain you* was his answer.

"What if we fight?"

You will be restrained.

"How will you restrain us?"

With light energy from our light organs. You saw what they can do when we were attacked by those carnivores we encountered.

"In other words, we're dead! Shut down!"

She spoke quickly and emphatically to Jeff. "Since we're going to die anyway, let's do some damage. Offense is the best defense. Loosen your machete and get it ready for use. Do you still have your UV-block sun visor?"

"I sure do."

"Then put it on now. Then when I say go, you take the one on the right closest to you and I'll take old Blue. I'd go for a chop right through that light organ. Hit its head with all your strength. Then let's head out the same way we came in. I still have my com unit, and with that, we should be able to find our way back to old Gonga baby. Hopefully, they haven't set a guard around it. No need with both of us in custody."

She found her UV visor and put it on. "Is your visor on?"

"Check!"

"Ready, go!"

The attack caught the two off guard. It was like there was no bone in their heads, necks, or upper bodies, and they were split in two down to where their arms attached. They dropped like stones. She and Jeff bolted for the doorway, ran down the short hallway and out the entrance. They hadn't seen another Koolan anywhere.

"That was way too easy! Something's not as we think it is," Jeff muttered as he ran.

"I hope you're wrong!" Caroloona replied. "My com unit shows Gonga to be about five kilometers that way." She pointed across the clearing to a spot at the edge of the forest.

As they entered the forest, she warned, "Don't touch any of those trees with the slick, wet trunks, they're deadly. Most of them seem to be near the clearing."

"Gotcha, boss!"

They ran until they were exhausted. "My com unit says one more kilometer. We should be blasting off this purple pit in less than an hour. Thankfully, the higher percentage of oxygen in Koola's atmosphere makes catching our breath a short experience. Let's take a five-minute rest and approach Gonga from downwind. I sense trouble when we get there."

Jeff started to lean against a tree and realized it was one of the wet poison ones. "Wow! That was close." He said as he moved away. A familiar whirring sound came from above.

"Nasty big carnivores coming. Use your machete!" she screamed. Again, their large knives sliced through the nasty critters like a hot knife through butter. In five minutes, it was all over. Five mutilated bodies lay on the ground, and the rest had retreated out of sight and sound.

"Well, how's that for a nice rest? Let's head for the shuttle before we get into any more trouble."

In about half an hour, they were in the clearing where the shuttle craft, Gonga III, awaited them. Unfortunately it appeared that a small forest of young poison trees had grown up around it.

"How in hell are we gonna get through that?" Caroloona commented. "Those poison trees aren't there by accident. Our friends must have planted them shortly after we arrived."

"Machete time," Jeff joked.

"We'll have to be careful not to let any of those branches touch us as they fall."

"Our hazard protection gear is in Gonga, but I happened to stash the gloves in my pouch," he said, reaching for them and then displaying them gleefully. "These should protect me, so I'll have at 'em. You stand back just in case."

After a few minutes of cutting the two-meter tall trees so they wouldn't strike him when they fell, it became apparent it would take hours to cut a path to Gonga.

"I wonder if these suckers will burn."

That said, Jeff took out an igniter pack, crushed it to start the reaction, and threw it into the pile of trees he had cut down. The flames grew so quickly it surprised them both.

"That slick material on the trees must be an organic oil of some sort. Those trees burn like they were doused in kerosene."

The conflagration grew rapidly, and they both drew back, driven away by the intense heat of the fire.

"Great idea, Jeff. I only hope Gonga doesn't get so hot we can't touch the hatch to get in."

It took two hours for the fire to burn out and the hatch door to cool so they could enter Gonga. Once inside they ran the prep sequence for takeoff, and in another ten minutes, Gonga was heading back to Big Mom. Their ship, affectionately called Big Mom, had been orbiting Koola since they left in the shuttle to explore. It would take them at least three hours to rendezvous, so they settled back for a well-earned rest. As they did so, Jeff switched on the spec display to check fuel and rendezvous time.

"Uh-oh!" he exclaimed as he viewed the spec report. "We have 120 kilograms more weight on board than we should. Where the hell is all that mass?"

"Could it be something stuck to the hull?" Caroloona wondered aloud.

"I don't see how that could be. I didn't see anything when I did the preflight."

"We'd better check the cargo compartment."

A short weightless traverse to the cargo door, and she grasped the handle. "The damned handle is stuck."

"Try the electronic release."

"Nothing!"

"Let's take a look with the cargo camera."

As the display came on, they were greeted with the sight of one of the Koolans like Blue grasping the inside of the hatch with its hands, and with its four feet planted securely against the surrounding bulkhead.

"We have an unwelcome stowaway. We don't want him aboard Big Mom. How are we going to handle that?"

"Well, we have three hours to figure that out."

About an hour into the flight, Jeff asked, "I know all cargo hatches can be opened from inside, but couldn't we weld that one shut from outside? Then we could leave him in there, and let the experts deal with him when we get back to Earth."

"I don't see why not. Of course, we'd have to drop the shuttle at the decon base on the moon, but that's no problem."

"There's a welder among the repair tools in the tool compartment. It's a small one, but it might do the job."

"Do you suppose he'd live through the trip? I wonder if he brought any food?"

"Well, if he didn't, we may have a pile of goo when we get there."

Some things didn't add up to Caroloona, and the more she thought about it, the more it troubled her. *Why had our escape been so easy? It seemed the Koolans may have wanted us to escape, but why? And why would one of them stow away on the shuttle since that meant almost certain death?*

She shared her concerns with Jeff. "I think we should send the shuttle back down to Koola as soon as we get to Big Mom. Then we won't have to worry about contamination."

"Old Pam will be mighty pissed if we do. That's a valuable piece of equipment to toss away."

"Yes, but that's the only sure way we have of preventing contamination of any kind."

"How long would it take to get a permission message through on the Q-com?"

"I don't think we have enough time. At this distance, it would take about a week for a two-way response—three days for the questions and three more for the answer, if we get one."

"Why don't we weld the door and decide what to do about our passenger when we get home? We could always dump the shuttle into a sun trajectory before we set down."

"That'll work. See if you can get that door welded before we rendezvous with Big Mom."

"You de boss!" Jeff said with a grin as he headed for the tool locker.

They outlined the situation to mission head, Pam, when they arrived in Earth orbit about two months later. Her decision? Drop the shuttle at the moon decon base. They could wait there until the decontamination squad did their job and then come home with the shuttle only slightly damaged. Pam was a stickler for not wasting anything that cost money. The decon squad took a full week to do their job.

"She's all yours. Clean as a whistle," Mac, the head of the decon squad, reported. "We found no sign of the critter you said was in there. All we found was about 120 kilograms of dark sand spread all over the cargo compartment."

"Sand? What kind of sand?" Caroloona asked.

"Our guys have it all in a sealed decon isolation pack. We should have a complete analysis for you in a day or so."

"I'd be careful of that stuff," Caroloona warned. "Koola was a weird place with some strange life-forms. Are you sure you got every grain of that sand?"

"Every single one! We removed everything but the paint on the storage shelves. Standard decon procedures."

"Paint? What paint?"

"Just the standard paint used on most storage racks. It's a thick plastic coating, hard and tough and contamination proof."

"I hope you got everything. Be sure to let me know as soon as you analyze that sand."

"The weight of the empty shuttle matches its exact tare weight to a hundredth of a kilogram, so we know we got everything. We are thorough."

By the time Big Mom took them home, they had been gone for almost two years. There were lots of records to go through, so they stayed in Big Mom for eight more days while all the samples went through standard cataloging and storage procedures. Funny thing about that sand. Decon reported it analyzed as organics with some unusual silicon compounds. The day before they were to head for home, the two researchers went into Gonga III to pick up the hard copy of the trip log and check for anything they may have left behind. Caroloona was in the cargo compartment when she realized something was wrong.

"Jeff! Get security in here instantly!" she shouted over the com unit.

"What's wrong?' he asked as he rushed into the compartment."

"Did you call security?"

"On their way! What's cooking?"

"Look around? See anything wrong?"

Jeff looked all around and shook his head. "I don't see any problem."

"The shelves."

"Clean metal shelves are all I see."

"Where's the paint? That heavy plastic paint. I thought it strange when decon mentioned it."

"Damn!"

"Damn is an understatement. We've been invaded."

Epilog

For ten years, there was no sign of any Koolan life. Security decided it must have perished in an environment that was chemically hostile. Then reports began coming in about strange purple flashes in the night. It took ten more years for expanding Koolan life to reproduce enough from the "seed" that had been the "paint" on the shuttle shelves to come out of hiding. The effects of Koolan micro biota were devastating to virtually all Earth life. In another decade larger, Koolan life-forms took control of the Earth. While the Koolan life-forms were poisonous to Earth life, they easily digested Earth life as nourishment. In another three years, purple vegetation had largely replaced green as the alien life overwhelmed the native green plants. Evolution trumped technology. About fifty more years and all Earth life had been eradicated in the now-purple world.

If there were no God, there would be no atheists.

—*G. K. Chesterton*

The Hygienist

I later found out her name was Barbara LeFang. I met her last week in—well, you know—one of those bars. Business had been great, and I wanted to celebrate. Honestly, I had only had a couple of drinks when she walked by my table. Little did I know she would drastically alter my life when she sat down at the bar near me.

Most of the women in this remote outpost on Apodia 5 were missing a tooth or two from fights in bars like this one. I oughta know, I'm the only dentist in this section of the galaxy, and I've seen some horrible mouths—especially on women. Why women get into so many bar fights here is a mystery, but they do, at least ten times as many as men, according to the stats. The news junkies here are big on stats. Go figure!

And yes, I was celebrating. Lately I'd had a run of major rebuilds on mouths busted up in local bar fights, and my cash box was overflowing. Yes, cash! Mostly hard cash at that. No CCs or DCs here, no credit or credits of any kind and no checks or IOUs. Paper money, registered Centars, are discounted by at least 50 percent. Coin of the realm, gold, platinum, or diamonds for major purchases is the rule. Sure, it's a bit archaic, but this far from civilization, no one trusts anyone. Everyone carries a Waxtal analyzer to check on coin quality and value, a small price to pay to prevent being cheated.

My little sign, "Dentist, Reasonable Fees," sat on the table in front of me. It was about the best way I had found to advertise my services. After I saw her eye my sign, she took her fresh drink from the bar and ambled over to my table. Barbara was different. Her revealing blouse competed for my focus with her perfectly formed face framed with lots of tousled black hair in that just-hopped-out-of-bed look. I noticed her pearly whites were neat and straight.

"May I join you?"

"Of course. Any particular reason?"

"I saw your sign and thought you might be interested in what I do."

"Yea? What's that?"

"Let's say I'm a dental hygienist looking for a dentist to hire me."

"A dental hygienist? Lady, most of my clients are lucky to have teeth, especially the females. About all I do is rebuild busted mouths. My operating room hasn't seen a prophy in fifteen years."

"I'm a special kind of dental hygienist. Do preventive care, you might say. I can help your *clients* keep their teeth, bright and straight in the first place."

"If my clients don't lose their teeth, I'm out of business. Right off the top, I don't like your preventive care."

"How much do you think you could charge if you promised—no, guaranteed their teeth would be perfect permanently, and you would never have to rebuild their mouth again?"

"Aw, c'mon. No one would believe that."

"Even if we could prove it to them . . . convincingly?"

"How in hell could you do that?"

"Easily! I can do it right away if you're game."

"Lady, I have no idea what kind of con you're trying to pull, but I'm not buying."

"Even if I took one of your patients and made their smile beautiful and indestructible?"

By this time I was getting tired of her little game, yet I was still curious. "What's it gonna cost me?"

"You provide me with your worst patient in your office, and I will do a single demo for absolutely nothing. Free!"

"If I fall for it . . . and if it actually works?"

"I want 50 percent of all your fees my services provide."

I couldn't believe I was going to let her work on one of my patients. "Okay! I'll let you do a demonstration, and if I like it, we have a deal."

"How soon can we get started?"

"How about tomorrow morning at nine? I have a patient in mind, Lowiece Grenby. She has already had three rebuilds and needs a fourth scheduled for that time. She only has about five real teeth in her mouth. The rest are all bridges and crowns, a shaky situation."

"I'll be at your office at seven as it will take me some time to set up my equipment. Is that okay?"

"No problem!"

I'm not a bad looking guy, but when I propositioned her she laughed. It didn't take long for me to find out she liked sex, but with the same gender I did so it was back to the dental business. As soon as we agreed, she left. I was wondering what kind of scam I had gotten myself into. I had to find out. Curiosity killed more than cats.

Next morning at seven, she was at my office with a small truck and a large black piece of equipment that looked ominous. It took us fifteen minutes to muscle it into the building and into my spare treatment room.

"What do we do now? Where's the power cord?"

"It has its own power supply. Doesn't need a cord."

"Okay! How do we work this thing?"

"We don't! . . . I do!"

"Okay! So how do you work it?"

When she opened the front bottom-hinged panel, a chair appeared. There was also a gadget that looked like a space helmet with a large cable attached to the rear of the device. The cable led to a box under the chair.

"Now! All we need is your patient."

"I figure the helmet thing goes over the patient's head, but what's inside?"

She picked it up, turned it over, and pulled out what looked like one of those form-fitting mouth guards attached to the back of the helmet with another stout cable. She smiled as she told me, "This is the business end of my system. The patient places it in their mouth and holds it firmly. I place the helmet over their head, fasten the straps, and turn it on. In about half an hour, they have a nice new set of indestructible teeth in place."

"That's all there is to it? You've got to be kidding."

"That's it! . . . Job over . . . I get paid. Once this demo is finished, and I use my machine . . . I get paid. Every time."

"How do I know it will work?"

"Simple! . . . Try to damage or remove one of her new teeth."

"Ha! One crunch with my forceps, and your job will be wrecked."

"Ha yourself. More likely your forceps will be wrecked."

"We'll find out soon enough. Ms. Grenby should be here shortly."

I introduced Barbara to Ms. Grenby as my new *hygienist*. Lowiece's answering smile showed that about half of her teeth were missing. When we ushered her into the room, she looked a bit apprehensively at the strange chair we asked her to occupy.

"Don't worry! It's a new technique I'm sure you will like . . . a lot!" I assured her.

Barbara explained, "Here! Put this piece in your mouth and bite down hard. It will reshape to fit your mouth and existing teeth. And don't worry! This will not hurt a bit . . . Really!"

"Go ahead, Lowiece! It's okay!"

Once she bit down on the mouth piece, she relaxed noticeably. She did not move or utter a sound as Barbara placed the helmet over her head and fastened the straps. She sat motionless and uncomplaining, without exhibiting her usual stream of nonstop questions and idle chatter.

"The system uses a powerful tranquilizing system," Barbara explained as she stepped back. "Now, let's leave the room. Consider that as an X-ray machine and stay away from it as it does its job."

"What about Lowiece? Isn't it dangerous to her?"

"Like X-rays, emanations from the machine are perfectly harmless at low doses, but can accumulate and do considerable damage during repeated and long-term exposure."

"That I can understand, but what kind of emanations?"

"If you were a particle physicist, I would still have a hard time explaining it to you. Let's say it's doing some serious quantum gyrations and leave it at that. It's called quantum repolarization."

With that, I began to worry a bit. What in hell was I putting my patient through?

"You're sure, absolutely sure, that this will not harm Ms. Grenby?"

"Not in the least. In fact, she will feel better than she has in her entire life."

"I'm still quite skeptical, you know."

"In about half an hour, you will be both surprised and reassured. Why don't we just sit back and relax 'til it's over? Just remember our bargain."

The emphasis she placed on those last words concerned me a bit. In fact, I was getting more nervous as each minute ticked by. By the time the treatment was over I was almost a basket case.

As Barbara headed for the patient, she grinned broadly and remarked "Bargain time!" as she pressed the remote control and turned the machine off.

Lowiece began to move in the chair as Barbara lifted the helmet off. She was bright eyed and animated as the cable was unplugged from the mouthpiece.

Barbara explained, "The mouthpiece has formed closely around her teeth as you can see. I will peel it off, and you will see a new and indestructible set of teeth."

I was amazed as was Lowiece when she looked in the mirror. "I can't believe it!" she exclaimed with a broad smile. "My teeth have never looked so great!"

I shared several minutes of disbelief with Ms. Granby before I checked her mouth. It was flawless.

"Now for the acid test," Barbara said, smiling as she led us into the other treatment room and motioned for Lowiece to sit in the chair. "Dr. Dunning will demonstrate the indestructibility of your teeth."

"What do you want me to do?"

"Forceps, you idiot. Try to pull one of her teeth."

"I couldn't do that."

"You will, or I will. Take your pick. I said this was a demo, so try to prove me wrong."

Reluctantly I took out a set of forceps and proceeded to try to remove an upper central. I was gentle at first, but when Lowiece said she was feeling no pain, I tried harder and harder. I have done literally thousands of extractions, and when I braced my other arm against her forehead and wrenched with all my might, not only did it not budge, but Lowiece said it didn't hurt a bit.

"Now, do you believe me?" Barbara said with a smile. "Let me show you something."

With that she took the forceps from my hand and belted Lowiece directly in the mouth with all the force in her considerably strong arms. Other than her head snapping back a bit from the blow, and Barbara's hand holding the forceps bouncing back like a rubber ball, it was as if nothing happened. No blood! No broken teeth! No pain! This was far more than I had bargained for. Lowiece was delighted as she began realizing what this meant.

"Wow! Doc, your new machine is a marvel. I can't wait to tell everyone of my friends. My enemies will find out about it soon enough."

"Now, Lowiece, don't be too hasty," I said, sounding like an old-school teacher.

"It will take some time for you to get used to your new mouth, so be cautious," Barbara warned.

Several days later and after we had *repolarized* numerous other patients, we learned that Lowiece had nearly killed a woman in a bar fight when the woman hit her in the mouth with a beer bottle. The news announcer expressed amazement at her lack of injury from the bottle and the fight that ensued. When interviewed, Lowiece gave me a plug.

With that publicity, my practice grew by leaps and bounds. Although I raised my fees to triple what I had been charging, my office was overwhelmed with new patients. Or should I say Barbara was overwhelmed. All I had to do was rake in the loot. When a patient asked her last name, I realized I had never done so. I was not surprised by LeFang, the big tooth, as she made it big in the tooth business.

*   *   *

Six months later, most of the personnel at the outpost on Apodia 5, about nine hundred individuals, had been made indestructible. Oh yes, once treated, their entire bodies became indestructible, not just their mouths. I, of course, was among those treated. Barbara finally admitted to me that her machine not only regenerated mouth and tooth tissue, nerves, dentin, enamel, and all. It also changed the atomic structure of each atom in the body. *Quantum repolarization,* she called it. Roughly the energy expended in a blow is reversed or *bounced* back into the matter of the striking object. That's about all I could fathom, but the result is obvious—indestructibility.

Then the ship arrived. It was a small ship, with room for no more than a few hundred individuals and a crew of five. When they disembarked, we received a shock. The five armed crew members reported to Barbara, saluted (at least that's what it looked like) and proceeded with her to Outpost Commander Quelter's quarters. We were all escorted by the armed crew to the Commander's quarters for an announcement. The announcement was made by Commander Quelter, who was obviously under duress.

"To every member of Apodia 5. All but a few of us have been *repolarized* as you know. The rest will soon be given the same treatment. I have been informed that in spite of what it may seem, we are not indestructible. I repeat, we are not indestructible. The weapon each of the crew of the ship *Freedom* holds will instantly depolarize anyone struck by its Q-ray. The repaired parts of our bodies will remain, but they will no longer be indestructible. We are all to be trained as an armed force that will, according to the crew, begin systematically conquering the entire galaxy. Those who are willing to join this force will be accepted as comrades. Those who do not will be eliminated. Do I make myself clear?"

Commander Quelter was visibly shaken as he spoke. The resounding yes throughout the compound was obviously driven by fear, not enthusiasm.

"Great!" Quelter continued. "About half of us will board *Freedom* in ten hours. The rest will remain here for the second ship which will arrive a few days after the first one leaves. Don't take anything with you as everything needed will be provided aboard ship. Now, those of you who are notified to be in the first group, do what you will, but be at the ship at 0200 sharp. We intend to leave at 0300 on the dot. See you aboard."

It took me almost an hour, but I managed to get to talk to Barbara.

"Can you tell me what this is all about?"

"Well, I suppose since you were so helpful, I'll give you a special job. You can repolarize the remaining members of the outpost. That means you'll be on the second ship."

"But why? What's it all about, conquering the entire galaxy?"

"Doc, don't concern yourself about it. Being indestructible, I doubt we'll see much real combat. It should be a piece of cake."

"But your guys have those depolarizing weapons. Suppose some others have them as well?"

"Impossible! I've got to go. Conquests to plan and all that rot. You understand."

"I guess," was all I got out before she turned and hurried off toward the commander's building.

By the time I had finished *depolarizing* the rest of our little army—these were for free—it was a half hour before *Freedom* was to lift off. I walked around my office and looked longingly one final time at the store of precious metals in my safe. *I could have lived like a king*, passed through my wistful mind. *Oh well, easy come, easy go.* I locked my safe—yea, stupid, useless effort that it was—and stepped out my door for the last time to head for the launch pad where I was to board the second ship with the remainder of our group. As I walked down the deserted street, a shadow fell over me. It was far too early for the second ship, and the first one wasn't to leave for at least fifteen minutes, but that shadow was definitely caused by something big—something very big.

One blinding, silent red flash then another, and I stumbled to the ground. My knee started bleeding. For me at least, physical normalcy had returned.

✸　　✸　　✸

It was about six months later, and I was entertaining friends at that same bar—I was the new owner, all that gold and platinum you know. We were discussing how wrong Barbara had been about the invincibility of depolarization. It seems the commander of the federation starship was an old girlfriend Barbara had dumped. Hell hath no fury, remember? The crew of the federation starship turned us all back to normal, destroyed the *repolarization* machine, and hauled Barbara and her buddies off to the hoosegow—all without a single fatality. I defended my being the only one to have profited from the venture by reminding them that the whole incident had put Apodia 5 on the galactic map. The resulting inundation of tourists meant huge profits for the locals. We even got a federation bank and useable credit cards. The bar fights between women ceased, and the new dentist—I sold him my practice—wasn't making out so well. Civilization had tamed Apodia 5.

There went the neighborhood.

❖　　❖　　❖

A trained flea can be taught to do most the things a congressman does.

—*Mark Twain*

The Bridge Builder

An old man, going a lone highway,
        Came, at evening, cold and gray,
To a chasm, vast and deep and wide,
        Through which was flowing a sullen tide.
The old man crossed in the twilight dim:
        The sullen stream had no fears for him.
But he turned, when safe on the other side.
        And built a bridge to span the tide.
"Old man," said a fellow pilgrim, near,
        "You are wasting strength with building here;
Your journey will end with the ending day;
        You never again must pass this way;
You have crossed the chasm deep and wide,
        Why build you the bridge at eventide?"
The builder lifted his old gray head:
        "Good friend, in the path that I have come," he said,
"There followeth after me today
        A youth, whose feet must pass this way.
This chasm, that has been naught to me
        To that fair-haired youth may a pitfall be.
He, too, must cross in the twilight dim;
        Good friend, I am building the bridge for him."

—*Will Allen Dromgoole*

True value in life lies not in finding that which we like, but in liking what we find. The secret of happiness is not in doing that which we like, but in liking that which we must do.

—*Howard Johnson, December 1980*

Every former protester I know passionately defends the actions of the 1960s and early '70s as exercising our First-Amendment right to criticize government policies. None seems to have read the First Amendment to the end where it speaks about peaceably to assemble, and to petition the government. More importantly, even in their advanced years, many seem incapable to confront the reality of having served the interest of America's enemies.

—*Balint Vazsonyi*

He who has begun, has the work half done.

—*Horace, 65–8 BC*

A Repair Mission

33.64.21.0600 (Stardate/time)

Captain Woolgah headed his starship, the Gelwah, away from the Vega sector at maximum welt. A sudden peace following millennia of constant warfare with the Scentar caused Captain Woolgah to struggle with his new peacetime role. The peace treaty between the Segwah and the Scentar had held for eight years after millennia of bitter and deadly fighting. Negotiated cleverly by his human friend Draxel Syl, the peace was extremely beneficial for all parties.

33.64.23.0600 - Two days at max welt speed behind them, first officer Jemrah reported, "Sir, there is a Scentar ship following us. It left from near the Vega system about twelve hours after we left. At first they were vectored substantially away from our path, but recently they shifted heading and are now on an intercept course."

"Have you identified the ship?"

"No sir, they are running without a transponder or ID code. According to their Iway pattern, their ship is a Delfro class warship."

"Their ship might be faster, but the Gelwah could easily out maneuver it in virtually any combat situation. I wonder why they are chasing us? I'd like to talk to them as soon as they are within hailing distance. Set navigation control for course change of one hundred-twenty degree Y, forty degree Z, and 0 degree X. Then be prepared to execute."

"Yes sir!"

A few moments later Jemrah said, "I estimate hailing and active combat contact in about three hours."

"Hail them as soon as they are within range and be prepared to make that course change on my command. I don't want to give them much of a chance to fire on us just in case."

33.64.23.0900 - "Hailing, Captain."

A minute later the Captain said, "They don't seem to want to answer our hail. Change course—now!"

"Done Captain."

A few minutes after the course change Jemrah remarked, "It looks as though our move surprised them. I estimate it will take them thirty to forty hours to get back within combat range again."

"Thank you, Jemrah. We will continue on course. That will take us to eighty degrees Y, sixty degrees Z, and ninety degrees X. Then in fifteen minutes, reverse the procedure minus five degrees in each plane."

33.64.24.1500 - Jemrah reports. "Sir, the Scentar ship will be coming within combat distance in about fifteen minutes."

"This time, on my command, repeat the last maneuver with a plus five degree correction instead of minus five. Then, in five minutes, put us back on our original course."

"May I ask why the same maneuver we did the last time?"

"That's because they won't expect it. They will be set up for the opposite maneuver and overrun our position by several hours. By the time they figure out where we've gone we will again be at least thirty hours beyond them."

"Do you want me to try hailing them when they come close enough?"

"Not this time. They didn't respond last time so let's see if they will hail us if we remain silent."

"Yes Captain. Uh, May I speak freely sir?"

"Certainly! What's on your mind my friend?"

"I've been considering our situation——this critical mission to try to stabilize a growing tear in the space time continuum. This tear was started by crude gravity wave experiments conducted by humans."

"Yes, Jemrah, what are you getting at?"

"The Segwah technology in the form of our new gravity-based defensive energy shield looks like the only system available that can repair that tear and stop its propagation, correct?"

"Damn! Jemrah. You know that is an unproven theory. The Scentar, with their understanding of inter dimensional rifts, admit they are guessing. They do know that if that rift continues to propagate at an accelerating rate the Scentar universe and the human universe will eventually fold into each other through the rift and be annihilated. Our mission is to try to repair the rift and stop any possible collapse."

"I realize that, Captain, but such being true, it escapes my understanding why a Scentar ship would follow us in such a threatening manner."

"Mine as well, Jemrah, but so far they have only been a nuisance. The times they were within combat range they did not fire anything, and they could have. We must make certain they do not get within combat range again until we've placed and engaged the shield."

"If they are desperate enough they might portal to where the rift is and wait for us there in spite of the risk. They must know as we do that using a portal near the rift could trigger the same gravitational collapse we are trying to prevent."

"Surely they're not that stupid. Should that happen we would all be annihilated. I wish I knew what their game was."

Jemrah, first officer of the starship Gelwah ex-captain of the starship Remlah had been demoted when he made a foolish move in a fleet battle with Scentar starships almost twenty years ago. His ship, the Remlah was destroyed, he lost half of his crew, and he was disgraced. Captain Grala Woolgah took him on as first officer, the position he held earlier. Since that time, Jemrah distinguished himself in a fierce battle with renegade Segwah on the Scentar base on Vega Five. At great personal risk, he had attacked a large group of renegade warriors trying to assassinate two humans and restart the war between the Scentar and Segwah. For this action, he was awarded the highest military honor, the order of Cheemah, reinstated in the military, and offered a new ship as captain. He refused the commission and chose to stay as Captain Woolgah's first officer.

33.64.24.1515 - "Captain! The maneuver is complete. The Scentar ship is far off our course. It will take them at least 30 hours to catch us."

"Thank you, Jemrah. One more course change to lose them again, and we will almost be there. We should be in position to deploy the shield in about forty-two hours."

"Sir! I hope my gravity shield theory is correct, and the shield stops the rift when we operate it? My original idea was based on my understanding of standing gravity waves used to generate and maintain the shield. That seemed logical to me. You know my gravity wave expertise is self acquired."

"Jemrah, you know much more about that than anyone else I know of. All I know is that the Scentar scientists confirmed your basic idea. They think that it will reverse the tear if we operate the shield at the end of the tear where it is coming apart. They told me the tear had much in common with unzipping a zipper, and that use of the shield would be like zipping it back up. That is, of course, a loose example. Unfortunately, if it doesn't work or if we miss the targeted area, the Gelwah with us in it will not survive."

"Then we had better not miss."

Captain Woolgah smiled. "That, my good friend, is imperative. Now let's check the shield generators to make sure they will be fully functional when we need them."

33.64.25.2209 - Jemrah shouted, "Captain! The Scentar ship is within half an hour of reaching active combat contact. But that's impossible with rational action."

"They must have used their portal. That's the only way they could have reached their present position so quickly."

"We're lucky that didn't trigger the collapse."

"They must be in maximum welt drive. An instant before they reach combat range, reverse course, and they will blast right past us. When they do, reset course for the end of the rift. That should give us the ten hours we need to reach the end of the rift using welt drive. Once we operate the shield they won't be able to touch us."

"Yes Captain! Course changes entered as ordered."

As the Scentar ship reached combat range, Jemrah executed the changes, and the ship flew past them as expected.

33.64.26.0802 - Captain Woolgah from the bridge. "Jemrah! Cut the engines, and prepare to engage the shield generators in seven minutes. At that time we will be at the optimum point to engage the shield around the rift."

"Yes sir!"and a few seconds later, "Captain, the Scentar ship is directly in our path. Four minutes to impact."

Suddenly the com system lit up with incoming on screen. "Captain Woolgah, this is Captain Raoul Saras of the Scentar starship Intreba. You are ordered to stand down. Abort the mission. I repeat. Abort the mission."

"And on whose authority do I abort this mission?" As he spoke he signaled Officer Jemrah to be ready to activate the shield control.

"On my authority, and because we have massive disrupters powered and trained on your ship. If you do not do so we will obliterate you."

"Give me two minutes to shut down the deployment series. It will take that long."

"Granted. But not a second longer, Captain."

With that Captain Woolgah shut down the com unit and issued rapid fire orders to Jemrah. "Set sensors to activate the shield generator the instant they fire anything, not before. Set the envelope to include the Gelwah, no more. The shield should not only stop anything they fire at us, but should brush their ship aside like a toy when we impact them. They've never faced our shield before and don't realize what it can do. Any kind of energy poured into the shield is absorbed by the shield and strengthens it. When we hit them, they will bounce off the shield like a rubber ball. I doubt anyone on their ship will survive."

"Yes sir!" Jemrah said with a smile. "What about the rift?"

"As soon as you know we have cleared their ship, shut down the shield and reset it to the original pattern. We should be in contact with the rift in less than five minutes."

"They're firing sir." Jemrah announced as the sensors activated the shield.

The screens went black, No impact was felt, and there was no sound for two minutes.

Then Jemrah reported, "The shield is off and reset as you requested, sir. We are fifty seconds from optimum position."

"Great work, Jemrah. I see no sign of the Scentar ship. I wonder what happened to it?"

"Prepare for shield activation, Captain." Jemrah said as he brought the coasting ship to a halt with a retro burst and in line with the end of the rift.

Once more the screens went black as Jemrah activated the shields. "Now what do we do, Captain?" Jemrah said as he looked at the Captain with a big grin on his face.

"You know what we do. We park here for fifty hours and hope it works. We have enough fuel to power the shield for at least seventy more hours, but you estimate fifty will definitely do the job." Captain Wòolgah said, grinning. "I don't know about you my friend, but I am going to take this opportunity to catch up on my sleep. I suggest you do the same. We've earned our rest, and we can't go anywhere or do anything anyway."

33.64.28.1009 - Captain Woolgah and First Officer Jemrah finished their two hour workout on the exercise machine and stepped off the platform. Then they showered, dressed, and walked together toward the bridge.

"Jemrah! It's about time. If the shield hasn't done its job by now it never will. Let's power it down and see where and when we are."

"When? Captain?"

"Yes, my friend, when. We have been dealing with unknowns here, unknowns based on sketchy theories of a bunch of scientists who deal with gravity and gravitational rifts using their expertise and super computer simulations. We're the ones who risk life and limb to test those theories in reality. I'm betting we will be in for a few surprises when we drop the shield."

"Is that why you insisted the entire crew stay behind?"

"No need to risk lives unnecessarily. You and I are enough crew to handle this mission. I thank you for volunteering."

"Every man aboard ship volunteered. They are quite loyal. Thank you for choosing me. It is a great honor."

"The honor is mine to have such a man as yourself by my side."

"Thank you sir."

"Here we are. Let's see what happened. Shut down the shield."

"Yes sir!"

In seconds the shield was gone. All they could see in the display was blackness. Not a single light, star, galaxy or other visible object could be seen.

"Well, Jemrah, where could we be? Activate the position search system."

Several tense moments of waiting and the system reported "location unavailable."

"Captain. I don't like the sound of that."

"I don't either. Prepare the basic portal for transport to a known position in the human universe. That is where we were."

"Yes sir!" A few minutes at his console and Jemrah reported, "Ready sir."

"Execute."

The blue glow gradually engulfed the bridge then died away. It was not a good sign.

"Well, Jemrah. It looks as if the portal couldn't find the destination coordinates. At least we know we are not in the human universe. Let's try our own. Set a search for home base."

After a few minutes Jemrah reported, "Ready sir."

"Do it!"

The blue glow grew bright in the viewer and then died away.

"That leaves only the Scentar universe. I hope this is it."

Once more the blue glow intensified then died away.

"Damn it, Jemrah, we seem to be outside of any known universe. We may have to use the universe portal, and I don't want to use the power that will require. Check our power reserves."

A few moments at the console and Jemrah reported. "Welt drive power is down only 20 percent, sir, steady at 60 percent. We used much less than we thought we would to generate the shield. Portal power is maximum since we haven't used the portals. System power is only 22 percent. Weapon power was never used and is at max."

"That's good news. Check the transfer conduits to make sure we can transfer power if necessary."

"Yes sir."

It took Jemrah no more than ten minutes to set up the tests, execute them and report. "All tests are positive, sir."

"Good. Let's run a forward impact weapons fire test on visual with trajectory analysis."

"Sir? The purpose?"

"I know it's a strange request, but I want to check out one of my suspicions. Fire as soon as you can."

Within seconds Jemrah fired the forward impact guns with tracer rounds. The tracers did not go straight, but curved sharply up and to port, and went out of sight.

Captain Woolgah barked an order. "Set courses for a 90.30.150 turns and give me full Welt drive as quickly as possible. We're in the grip of a massive gravitational force and quite close to its event horizon."

There was a noticeable jolt of the ship as the welt engines powered up. This was followed by surging vibrations. The ship groaned and creaked as if it were being twisted and stressed physically. He watched the display for any sign of light.

"Repeat the forward weapons test, now!"

This time the tracers made a much wider curve directly upward.

"Change course minus 45.00.00 and maintain full welt."

"Yes sir! - - - Done!"

"Repeat the weapons test."

"Done."

This time the tracers went straight ahead with no curvature until they burned out.

"Thank you, Captain. That was close. We were about to fall into some kind of event horizon, weren't we?"

"Right you are my friend. Look at the display, the view outside. That's a beautiful sight, all those points of light. Let's do a position check."

"Yes sir." Jemrah said with a smile, "Yes sir indeed."

A few motions at the console, and Jemrah reported, "It still indicates the location is unavailable, sir."

"All right my friend, let's repeat the same jump sequences as before with the portal and see what that gets us. Before we do that, shut down the welt engines. No sense in wasting fuel."

"Yes sir."

Engines were cut off and the sequences run. The result was the same as before.

Jemrah looked at his Captain and said, "We must be in another universe in a still different dimension from the three we know of."

"More likely we are still in the human universe but at another, later time."

"Of course. The result would be the same, wouldn't it?"

"Yes! Out of curiosity, do a search for known objects, million kilometer range. We might pick up something."

Jemrah executed the search command and the image of a ship appeared on the screen. It was the Intreba and she looked distorted.

"Distance?"

"About fifty thousand kilometers, Captain, and we are approaching. She's within hailing distance."

"I doubt anyone is alive, but hail them anyway, Jemrah."

"Yes sir."

The display flickered, but there was no answer.

"She's derelict, sir. Looks to be badly damaged."

"Yes, Jemrah she's quite misshapen, and look at all those pok marks on her outer hull. She looks like she's been here a long time and received many meteor strikes after her impact shields quit working."

"Captain, it would have taken more than a hundred years to have so many meteor impacts. I think you were right about another time."

"All Scentar ships carry flight recorders as do ours. Those recorders also have permanent atomic clocks that run virtually forever. I wonder if we could find theirs and remove it?"

"Sir, if we could get that recorder it would tell us a lot about what happened and maybe where we are."

"Our current trajectory will take us within about ten-thousand kilometers. When we reach that position, set a corrected course to bring us alongside Intreba. Find out how long that will take."

Jemrah adjusted the controls and did a few calculations. "At minimum power use it will take about five hours. I assume you don't want to use any more power than necessary."

"Correct. Until we know more about our situation, we have a lot more time to spare than power. We don't know when or how we might get home."

33.64.28.1525 - "Captain, We are in sync with Intreba about a quarter of a kilometer off her starboard bow."

"Thank you, Jemrah. Well done. Let's see if that atomic clock is still working. Set the energy scan for the lowest emission and scan Intreba. If we find anything, we'll then have to figure out how to get it."

"Sir, there are three small energy sources that showed on the scan. From its emission signature I know that the forward one is the atomic clock. I have no idea what the others are. There are no life signs, and there is no air in the ship, only the vacuum of outer space."

"Is there a hatch forward near the clock?"

"Better than that, sir. The flight recorder is in its own easy access tube that can be opened from the outside. The Scentar at Vega Five showed me when I took a tour of their ship. I can put on a vacuum suit and take a scooter over to get the recorder. I'll need a Darium cutter to open the tube because we don't have a key."

"Do it! I'll keep watch in case something goes wrong. Check to make sure our com units are working before you leave the airlock."

"On my way, Captain."

As Jemrah prepared to go get the flight recorder, Captain Woolgah started searching Scentar ship schematics to try and find out what the other small sources of energy could be. He had examined energy sources on three Scentar ship plans when Jemrah clicked his com unit.

"All set, Captain. The air lock is open, and I'm ready to go."

"Good luck my friend. Don't take any chances. I can't afford to have you injured. Get back to the ship the instant something doesn't seem right."

"Yes, Captain. I'm out of the airlock and on my way."

Captain Woolgah went back to his searching for energy sources. Every few minutes he looked at the display focused on Jemrah and the scooter. The bright white flashes of the Darium cutter told him Jemrah had found the flight recorder tube and was working to extract it. About an hour later the com unit signaled.

"I've reached the recorder." Jemrah reported. "It's going to take a while to cut it out. Both the outer and inner hull shells are crushed in, the tube is badly mangled, and the recorder itself is jammed into the bulkhead it was attached to. I would say I'm about half way finished with the cutting. That means another hour at least to get it free and carry it back."

"No hurry, Jemrah. Make clean cuts and don't leave any jagged edges that could catch and puncture your vacuum suit. While you are doing that, I'm searching Scentar ship plans to try to find those other energy sources. They have to be significant sources or they wouldn't still be active."

"Yes captain. I'll let you know when I have the recorder free."

With that Captain Woolgah went back to studying Scentar ship plans. True to his estimate, Jemrah called about an hour later.

"Captain, I'm about to cut what seems to be the main power cable to the recorder. I don't understand why it is so thick, about the size of my wrist. Once I cut through that, the recorder is easily removed."

"Jemrah, be careful. I know no reason why there should be any power cable to the recorder. They always have their own separate power source inside the box. Maybe it's a security lock or hold down. Wait a minute. Here is a diagram of a flight recorder installation just as you describe. I think I also discovered those two power sources."

"Captain, I'm cutting that cable. It's hollow."

Immediately the display of the energy sensor lit up with a brilliant display. At the same instant, the Captain realized what they were.

"Fusion Bombs! Jemrah get out of there!" He shouted over the com unit.

As he watched the display, he saw Jemrah mount the scooter and head back toward the Gelwah. In one fluid movement the captain set the shield to include Jemrah and turned it on. In the instant before the shield was deployed, he saw the hull of the Intreba turn white hot.

"Jemrah!" the captain screamed into the com unit. There was no answer. As the captain bolted for the door to head for the airlock, his com unit popped, and Jemrah's calm voice stopped the him in his tracks.

"Yes Captain? What happened? I was half way to the airlock when everything turned dark. I had a hard time locating the com unit in the darkness. Could you turn on the airlock navigation lights so I can get back into the ship?"

Captain Woolgah slumped into the nearest operations chair on the bridge in immense soul-wrenching relief. "I'm sorry, Jemrah. The lights are on. Please get aboard and come directly to the bridge. We were both extremely lucky."

"Yes? How's that, sir? And why did you turn on the shield. I assume that's what happened."

"I'll explain when you get to the bridge."

Half an hour later Jemrah walked onto the bridge carrying a large metal cylinder with at least a meter of large diameter tube hanging out of one end.

"What happened, captain? I had started the scooter back toward the ship when everything went black. Why did you turn on the shield?"

"Those two energy sources I couldn't identify? They were two fusion bombs connected to the flight recorder with a wave guide trigger. They used a similar long-lived atomic power source like the one in the flight recorder. When you cut that waveguide, it triggered the bombs to go off. I realized what was happening when the energy sensor display lit up when the firing mechanisms energized and set the bombs to explode. I tried to tell you what was happening, but apparently your com unit was turned off. As soon as you were clear of the Intrepid I set the shield to include the space you were in and turned it on."

"Was it that close?"

"Here's how close it was. I saw the hull of the Intrepid turn white hot no more than a nanosecond before the shield deployed. Which reminds me, Why didn't you respond when I shouted ***fusion bombs*** over the com system?"

"I had to turn off and stow the com system in order to get the recorder out of the ship, it was that tight. As soon as I boarded the scooter I turned it on and heard you screaming for me. I was only a few meters away from the Intreba when all light disappeared."

"That was way too close. I'll wager we benefitted from those bombs when their energy fed the shield and filled storage."

"I'll check power storage." Jemrah punched a query into his console. "You were right. All energy storage systems report full."

"Good! Turn off the shield and then let's examine that flight recorder clock and see what it tells us."

"Damn! Captain, look at the display. All those incandescent objects must be the remains of the Intreba. They're all rapidly receding from us."

"The advantage of being at the center of a huge explosion while protected from it."

33.64.28.1735 - It took almost two hours to remove the flight recorder from its heavy protective cylinder. The record showed that the ship had jumped three times since leaving Vega Five. It had then sustained a sudden acceleration of almost forty G's. There was nothing after that. The recording continued until the space filled. No living being could have survived that sudden acceleration.

Captain Woolgah looked up at Jemrah. "Whatever those Scentar wanted, whatever their mission or purpose, we will never know. The impact with our shield killed all of them instantly. Whatever records they had were incinerated when those bombs went off. There's another shocker. The atomic clock shows we have jumped 126 years into the future. That's why the locator can't tell us where we are. The data base of celestial objects is that much out of date. Those objects have all moved."

Jemrah entered some information on the console and queried the locator once more. "I've updated the data base with the known movements of those baseline objects. That should provide us with a fix. There! At last we know where we are."

"And when!" the Captain added. "We're slightly more than ten million kilometers away from where we sealed that tear. We're also about fifty days from the current position of Vega Five"

"Do a dimensional gravity scan to see if the tear is there."

After a few moments, Jemrah said, "No sign of any disturbance, sir."

"I guess that means we should head back to the base on Vega Five. Set the return heading to the new coordinates, and let's go. I'm wondering what we'll find after being gone more than a hundred years. It should be interesting."

Jemrah grinned. "At least we'll have a clean slate. No one there will know us."

33.66.18.1735 - Approaching Vega Five orbit, Captain Woolgah speaks. "Jemrah, open up a direct channel and see what we can find out."

"Yes sir, Captain."

The display held a background full screen shot of Vega Five from the Gelwah forward cameras.

"There seems to be no response to our signal. Send out a frequency pattern call and zoom the camera in close on the base location."

"Yes, Captain." The display filled with an expanse of green vegetation with a huge, black tower. Atop the tower was a fairly detailed model of a space ship. "What the hell is that?"

"Unless I miss my guess, that is a model of our ship, the Gelwah. Are you getting any confirmation of our signals?"

"Not yet, sir. Wait, there's something digital coming in. It's in human English requesting we identify ourselves. They do not recognize our ID signal."

"Switch the display and communications to their frequency."

A disturbed and obviously human face displayed on the screen. "Please identify yourself. Your ID signal is unknown to us."

"This is Captain Grala Woolgah of the Segwah starship Gelwah. Aboard with me is First Officer Der Jemrah. There is no one else on board. We left the Scentar base on this planet more than a hundred years ago on a critical mission which has been accomplished."

"One moment, please" was followed by a long silence as the face on the screen looked off to his left. At least fifteen minutes later the face on the screen turned to the camera. "Please enter security orbit D for David and wait for further instructions."

The Captain smiled. "If you will provide me with the vector and coordinates for security orbit D, I will comply. Who are you, please?"

"One moment, Captain. We're not accustomed to providing such information as it should be in your ship's navigation data banks. My name is Charles Sung. I'm a security officer on Vega Five."

"Our navigation data banks have not been updated in more than a hundred years so it's no wonder they are out of date. If you put our ship on visual, you will note it looks exactly like the one atop that huge black tower near your base. That is no coincidence. Please provide us with the requested navigation guides so we can comply with your request."

Suddenly the display went blank for a few seconds and then a different and smiling face appeared. "I am Grace Shelbourne, governor of the Vega Five territory, the oldest human outpost in the galaxy. I want to take this opportunity to welcome you back, Captain Woolgah, and to apologize for your earlier treatment. Your return is miraculous and quite unexpected."

"Thank you, Governor Shelbourne. It is good to be back."

"Is your ship capable of landing on the surface without assistance?"

"Tell us where we can land, and we will do so unassisted."

It took less than an hour to land the Gelwah on the broad grassy area near the tower and next to the road that circled it about a third of a kilometer from the tower itself. It took another hour to shut down all flight equipment and prepare the Gelwah for ground access and control before they could exit for any length of time.

By the time they lowered the entrance ramp a large crowd had gathered from all directions, running across the grass and along the road. A military detachment had surrounded the legs of the Gelwah and held the crowd back. As the entrance ramp lowered from the nose of the ship a group of small busses or "porters" pulled down the road, through the crowd, and stopped nearby. Two women and three men stepped out of the first bus and walked over to the staircase as Captain Woolgah and First Officer Jemrah came down the ramp. Media cameramen and reporters tumbled out of the other busses and literally fell over each other getting positioned and set up. It was a historic occasion.

As the two Segwah officers stepped on the grass, the taller of the two women spoke. "As Governor of Vega Five it my great honor to welcome two historic figures from the past. But for their courage and daring in a successful mission of great danger, none of us or our universe would exist. Welcome, Captain Woolgah and First Officer Jemrah. Welcome back to your adopted home. We all read, in our history books, about two dedicated soldiers, Segwah soldiers, who risked their ship and their lives to save both the Scentar and Human universes from annihilation. All the more amazing is that their home universe was not threatened, and they could have gone back there and lived out their lives without danger."

With that the crowd applauded and stomped their feet in approval. They also called for Captain Woolgah to speak.

The Captain looked at his First Officer, smiled and addressed the crowd. "Madam Governor, citizens of Vega Five: believe me when I say we are extremely pleased to be here. We are especially pleased that you chose the Segwah foot stomp to honor us. According to our clocks, it was sixty-six days ago we left this same spot on our mission. During that same time, you have experienced more than a hundred years. That will take some getting used to on our part. I'm sure we will have much to share with each other in the future. Right now we are exhausted from the rigors of flight deceleration and orbital maneuvering."

After much hand shaking and greeting of members of the crowd who responded with a few Segwah stomps, Governor Shelbourne turned to the two travelers with a big smile. "We knew you would be exhausted from your ordeal so we have arranged for you to stay in the VIP suite at our finest hotel. We will hold any debriefing and celebration until you are refreshed and ready to enjoy our hospitality."

At this time Captain Woolgah took and raised the hand of his First Officer and said, "We may be tired, but before we go to the hotel, I want to say something important about First Officer Jemrah, something the history books may not have mentioned. I publicly acknowledge that First Officer Jemrah was the one who conceived the possibility that our gravity-powered shield might repair the damaged space/time fabric. He conceived of this entire mission. True, Scentar theoretical physicists confirmed Jemrah's concept and calculated the details of the mission, but this man initiated the entire project. That such a man would choose to remain my first officer when offered a command of his own was the greatest compliment, honor and reward I have ever received. No man had a better or truer friend."

Jemrah looked a bit embarrassed. "Thank you, Captain. To hear such words from you brings me great joy. You will always be my friend, my mentor, and my captain. I'm not good with words, but I have to say we have both lost all of our friends, save each other. True, we will make new friends, but to lose all one's friends and family in two months brings great sadness to my heart, and I'm sure to my captain's as well. Thank you all for being so kind."

Governor Shelbourne responded. "There have been momentous new revelations today, happy and pleasant revelations. Let's get you two travelers to your hotel and some rest. This will also give us the opportunity to prepare a proper welcome along with a few surprises for you."

A few more handshakes, cheers, and foot stomps, and the group headed off to the hotel in the porters.

It was afternoon about a week later when Governor Shelbourne and a larger group of officials picked them up in several porters. They entered the porter with the governor and her aide. The governor spoke. "Now we would like to take you to a reception in your honor. It is being held at our life research facility which incidentally was funded by generous gifts from Humans, Scentar and Segwah alike in memory of the two brave adventurers who saved two universes. The full name of the building is, The Woolgah-Jemrah Life Science Research Facility."

"Thank you for the opportunity to rest. I assure you Jemrah and I appreciated your thoughtfulness." Captain Woolgah said. "It is wonderful to have the opportunity to meet new people and make new friends."

The Governor responded, "Do you remember the friends who said good-bye when you left? Our historical records indicated there were many at your departure. Those who were there to send you off included the crew of your ship and a wedding party gathered for the marriage of two close human friends."

"Of course we do. We will miss each and every one of them." Captain Woolgah replied as the porter pulled up to the Life Sciences Research building.

As they walked inside, Governor Shelbourne guided them to a large reception room where the trappings of a major reception celebration were laid out, and a large number of people were gathered. The Governor walked to the podium to address the crowd.

"Welcome ladies and gentlemen and thanks for coming to celebrate the miraculous return of two of our most cherished heroes."

The governor continued recounting their departure, realization, and acknowledgment that their efforts had succeeded, and the long wait not knowing when or if they would return. She then described the purpose and mission of the research facility.

"Some important and successful research has been accomplished in this establishment and the fine people who worked here over the years. One major accomplishment happened after our travelers left on their mission. This success was in the Life Stasis project. This project was to develop the means to suspend life activity for long periods, mainly for those with incurable, fatal diseases. It was hoped that once a treatment was perfected, those suspended could be awakened and cured. I would like to introduce a group that chose to be suspended, not for health reasons, but for another that should be obvious. Let me introduce this group who went through the process of rejuvenation and were awakened during the last week. This was done carefully so they could be here at this reception today."

At that point the doors at the end of the room opened, and a group of people walked into the room. The Governor announced, "Captain Woolgah and First Officer Jemrah, I give you your friends, Draxel and Maria Syl, the newlyweds, and most of your original crew of the Gelwah. These loyal friends and comrades chose to go into stasis to wait for your return."

Is the music a tool of the performer?
 Or is the performer a tool of the music?
Is the message the tool of the author?
 Or is the author a tool of the message?

—*Howard Johnson, 2001*

People unfit for freedom—who cannot do much with it—are hungry for power. The desire for freedom is an attribute of a ***have*** type of self. It says: leave me alone and I shall grow, learn, and realize my capacities. The desire for power is basically an attribute of a ***have not*** type of self.

—*Eric Hoffer*

The New Job

Mary lifted the pen from the old desk. Sounds of sirens came through the open window. *Had David called the police to try and stop me,* she wondered? *Would I have time to finish writing the goodbye note? Would anyone believe the contents of the note?*

Aerlo waited in the bedroom beside the open door of the stargate, an aperture that occupied much of the wall where her plasma TV hung. "Hurry!" he urged, knowing that the siren might mean trouble.

"I have told my son I am leaving and why. He doesn't believe me, of course, and thinks I am losing it. Hopefully he'll believe what I said about not coming back and that he should not consider searching for me. It will only take a minute to finish writing the note," she called from the study.

"We must be through the gate and close it before anyone enters the room."

Mary's thoughts ran through the startling events that had begun three hours earlier.

She awoke at six and turned on the TV to view the morning news, her normal routine. As she was thinking out the breakfast she would soon prepare for her and her son, David, the news program was unexpectedly interrupted. A strange man's face appeared on the screen and spoke directly to her.

"Mary Carlisle, you have been chosen for a special privilege. I am speaking directly to you, and you can answer as if I were there in the room with you."

A bit startled, Mary sat upright and clutched the covers to her.

"Can you . . . see me?" She asked haltingly.

"The same as if I were there in the room with you," the man answered.

"What is this? . . . How? . . . Are you serious?" She asked as she arranged the covers closer.

"My name is Aerlo, and I am speaking to you from a craft that is orbiting your planet. I came here from another planet in your galaxy. I am quite familiar with your Earth, having spent several years among you during my training. My species is closely related to yours. We are virtually another race of *Homo Sapiens*. In fact, both of our species came from the same ancestors. Please do not be frightened as I will not harm you in any way. Ask any question you wish, and I will reply truthfully."

"I don't know what to say . . . Why are you speaking to me? What do you want?"

"In a way, I am offering you a job. It will be yours to accept or reject. Should you reject my offer, I will leave and not trouble you again."

"A job? What kind of job? I already have a fantastic job I love."

"That is precisely why we would like to hire you, your expertise on cetaceans."

"I'm a marine biologist—a specialist on cetaceans. I work with them every day here at the Oceanic Research Center. It's the foremost research installation of its kind in the world."

"That's precisely why we want you. We have an interesting research project involving cetaceans on another planet quite far from here."

"This isn't some kind of weird promotion, is it? You can't be for real. How do I know you aren't some con artist? It is hard for me to believe you."

"Tell me what I must do to convince you this is all true and proper. I will do whatever you say unless you ask something impossible or that will reveal my presence to others."

"Anything?"

"Anything!"

"Let me think."

"Mom!" called a voice from elsewhere in the house. "Who are you talking to?"

"The TV news. You know I always talk back when they say something I don't like." And as an aside to Aerlo, "That's my son, David. He's staying with me while his group is trying to free some beached whales nearby." It was a lame explanation to David, but it worked.

"Well, I have to get to the beach in less than an hour, so if you're going to fix breakfast, please hurry up," came from the far end of the hall.

"Attend to your son. I can return after he has gone," Aerlo said.

"I am not due at the center until noon, so can we talk in, say, an hour?"

"That's okay with me. Incidentally, if you turn off the TV, it will cut off our connection. I will not be able to contact you, so be sure it is on in one hour. Also, I have only about three hours before my departure window closes. I must leave by then and will be unable to return here for about eight years."

"I can't leave in three hours, that's impossible!"

"Wait until you have heard what I have to say, please."

"Okay! In an hour then. Now I will turn you off so I can dress."

"Goodbye!"

During breakfast, she told her son nothing about the visit, only that a new job opportunity had come up. "I have to make a decision now and will leave before noon if I accept."

"That's unbelievable!" David remarked. "Are you crazy? Who are they? Why you? And why so suddenly?"

"Whoa there, son! One question at a time. No, I'm not crazy. Besides, I am already eligible for retirement, and you are my logical successor. You've been with me at the center for fifteen years and know more about what I have been doing than anyone. You could take over in a heartbeat. We've already discussed that."

David looked almost in a panic. "Can't this hold until after I finish at the beach? We should be finished by dark. Why don't we discuss it then?"

"If—and that's a big if—if I take the job, I'll be gone before noon. Also, I'll be out of the country and unable to communicate with you for some time, so if I decide to go, please don't worry."

"Mom, I'm going to cancel my work at the beach, stay with you, and find out what this is all about. I'll call Denise, and she can take over for me."

"No, David, you'll do no such thing. I insist! That project is quite important and will not wait for another time. Denise is smart, but she hasn't done this before. Those beached whales will be dead by nightfall, and that must not happen."

"But, Mom!"

"Hush! If you won't talk sensibly and rationally, I'll end this conversation and run you out of the house."

"Look! Who's talking about sense and rationality? You sure don't make either."

"Please, David. I have been searching for something different to do ever since your father died six years ago, and I think I have found it. Let me do something crazy with my life if I decide to. Now, go take care of those poor beached creatures."

"I still think it's crazy, but have it your way. Call me on my cell phone when you decide."

With a shake of his head, he hugged her and left with the admonition, "Don't do anything stupid, please!"

"I assure you that I won't."

She cleaned up the dishes and headed back to the bedroom to turn on the TV. The eight o'clock news was starting. Before it had been on five minutes, the screen went blank, and once more Aerlo appeared.

"Have you decided on a test of my veracity?"

"How about showing me pictures of your ship, inside and out, your home planet, and the place where the research project is located."

"Your wish is my command."

With that, a picture of a large vessel appeared. It was sitting on what was obviously a launch and retrieval platform with open countryside in the background. People and loading carts or carriers were moving in and out of numerous hatches on several levels.

Aerlo narrated, "That's the *Curex* as she was being loaded for this trip. She carries a crew of sixty and provisions for trips of up to four million light years. Right now she's orbiting Earth about six hundred of your miles above the surface. EMF shields are on, so she is invisible to light, radio, radar, and any other kind of EMF radiation sensors or detectors."

The picture changed to what was obviously living quarters not too unlike the cabins on cruise ships Mary had been on, but a bit larger.

"Nice room," she commented.

"There's one like it waiting for you," he added. "Both the bridge and the residence areas have artificial gravity systems and inertia neutralizers as do all passageways and lifts."

"How do I know those aren't fakes?"

"Would you like a personal tour?"

"Come on."

"Seriously, I can transport you there in a second through this portal."

With that, an opening appeared in the wall near the TV screen. It looked like an open window to the ship's interior, a salon or recreation area. Aerlo stood on the other side.

"My god!"

"No, just advanced technology. Your own technology is not many years behind all of this."

"How do I know you won't whisk me away once I'm aboard?"

"That's a risk you'll have to take unless . . . wait a minute." He walked away from the window and reappeared carrying a small furry animal about the size of a squirrel, but with no tail.

"This is an oolabit. It's what you would call a pet, a small version of, say, a cat on your planet. Take her! She won't bite."

Mary held the little critter which curled around her hand and held on softly with padded feet, but no claws. "She's cute and cuddly."

"Does that convince you that I am from another planet?"

"Absolutely! How about that tour."

Aerlo helped her step through the portal into another world . . . literally.

"It's lovely," she remarked as she walked through the room toward the windows in the side. What Mary saw through that window blew her mind. There was her beautiful blue Earth from six hundred miles up. "That is spectacular!"

"A nice view, yes. Any more tests?"

"Tell me about this job. I'd like to see pictures of the lab and where the animals are kept."

"Don't be shocked please, but your lab is an entire planet with deep oceans."

"An entire planet?"

"Yes! It's called Stentor Seven and is a planet orbiting the star Stentor near the star you call Vega. We call this area the Vegan sector of the galaxy. It's an artificially *enlivened* planet about halfway in size between your Earth and its moon. All the water, the ocean-river systems, the plants, and animals have been transported to the planet to create what you might call a recreational world—a place for R and R as your people say."

"Are you serious?"

"Very! We want to transfer some cetaceans from your planet to Stentor Seven, and would need to provide them with an environment fully stocked with appropriate food and support systems so they would live as naturally as they do here."

"That's a tall order!"

"And one we know quite well. You have the mental tools and experience to accomplish what we need—tools which we lack."

"All the records of my work, my research, are in the center's computer system . . ."

Aerlo interrupted. "And in our data banks as well."

"Then why do you need me?"

"For the immense store of knowledge and experience in your brain that is not in those data banks."

"You think of everything, don't you?"

"We try. Let's get you back to your home so you can make your decision in a free environment—and soon, I hope."

This is incredible! Mary thought as she headed back to the portal with Aerlo. "I'd like to do it, but there are so many unasked questions in my head. Will I ever be able to come back here?"

"Not for slightly more than eight years. Space travel over such long distances is only possible under certain conditions and at certain times. That's a limit our scientists and engineers are trying to overcome. The temporary window we are experiencing lasts only a few days and ends in a bit more than one of your hours. We have no time to waste."

"You mean that if I go, I'll only have a few days after eight years?"

"No! This window was a short one. Some of them are much longer. The one open about five years ago was almost six months. That's when I trained for this project. The one in eight years is at least a year, maybe a bit longer. I'd have to look it up to be exact."

As they stepped through the portal into her bedroom, Mary decided she was going to go. "What do I have to do if I decide yes?"

"Just step aboard. Everything else has already been taken care of. Your clothes, books, medications, records, everything. We've prepared a cover story and arranged for David to have legal access to all of your records and property."

"Everything?"

"You will have veto power if there is anything about David's legal access that makes you nervous."

"No! David can be trusted, but the thought of eight years with no communication is disturbing."

"I don't mean to press you, but we have less than an hour before I must leave, with or without you."

Mary turned and walked into the study and sat down at her old desk. "I'll do it! All I have to do is write a letter of resignation to the center and recommend David as my replacement. I've already talked with the board about that, and they are in agreement. What can I say to all my friends and other relatives? I have such a short time."

"We have a contact and control person here who will be able to help David handle all that for you. His Earth name is Ralph Gora, and you and David already know him."

"Dr. Gora? Our dentist? He's been our dentist for more than fifteen years."

"He was my contact when I trained and has committed to living his life here on Earth with his family. He is a good one."

"You seem to have thought of everything."

"We've been planning this for at least twenty years, but had to fill in all the blanks."

"What would have happened if I had said no?"

"It would have set our project back the eight years until the next window. By that time, we would have lined up several other prospects and had a full year to find one. Incidentally, that's when we plan to pick up the whales and dolphins."

"No animals for eight years? What will I be doing?"

"You will have those eight years to prepare their home. At least to design their home and prepare to transfer all the required life forms. We are carrying a rather large amount of seawater from several of your oceans where the whales and dolphins roam. Those will provide at least part of the required biota for designing Stentor Seven's oceans."

"You do plan well. I hope you haven't missed anything."

"If we have, it will be eight years before we can correct it."

With that, Mary sat down to tell David the story of her new adventure. When Aerlo suggested she tell him the truth, Mary replied, "He'll never believe it."

She was almost finished when the sirens startled her. When they stopped in front of her house, Mary panicked. "That boy of mine must have had second thoughts and called the authorities."

"We must leave now!"

"As soon as I finish this last sentence for David. It will take them some time to get in if I answer and say I'm dressing and will soon be at the door." As she finished, she walked into her bedroom and was greeted with a room empty except for her desk, chair, and the TV hanging on the wall.

"We are taking all your things as I explained earlier. My people came and moved them into your quarters on the ship while you were writing your note. Your house is virtually empty."

"My god, you move fast!"

"Have to. We have little time to spare."

As he said that, the doorbell rang and the intercom sprung to life. "Mrs. Carlisle? Are you all right?"

"Quite all right, thank you. What do you want?"

"Your son, David, asked us to check on you. Would you please come to the door?"

"Wait 'til I put on my robe."

Two men climbed through the portal and headed for the study. "I assumed you would like to have your desk and chair?" Aerlo said with a broad smile. "Please follow me through the portal while they get the desk."

"What about the note?"

"David will find it on the floor where your desk once stood. Now, let's hurry."

After Mary had stepped through the portal, the men came through with her desk, chair, lamp, and remaining items from her study. As they did, she heard a loud crash. The police were coming in.

The sounds of their shouts were silenced as Aerlo turned off the portal.

"All they'll find is a plasma screen TV hanging on the wall. Let's head for the Vegan system."

"Well, Mary Carlisle is off on a wild adventure. I certainly hope this was no mistake."

"I think you will find it more rewarding than you can imagine. Stentor Seven is a marvelous place."

The Ultranet

*T*his introduction is information describing some new ideas and technology that will soon affect much of the world. The devices and concepts described are quite startling, but they are all in use or in the R&D stages as this is written. Many will be available surprisingly soon along with the worldwide systems they serve and require. Understandably, this may seem a bit fanciful to those an arm's length from digital technology, but wouldn't the Internet have seemed so in 1980, a few years ago?

RFID or Radio Frequency ID, tags identify a great many details about the package or item to which they are attached. They are read by **RFID** readers (handheld computers) that can extract the information, pages of information. These tags have been in use for several years! They are one reason for the success of marketers like Wal-Mart.

The **FMID**, or FM Input Device, is already a reality. The Blackberry, iphone, and ipad are handheld wireless devices that can access the Internet and so much more. They have been popular for several years. One version can be used as a wireless access device for a computer and the Internet as well as a cell phone and digital camera. With available verbal commands, an advanced **FMID** would combine all these features and also replace the mouse and keyboard.

PUAIs, or personal universal access interface devices, will replace FMIDs in the near future. They look like thick credit cards with a viewing screen filling one side. They have no buttons and require no manual entries. All commands are voice or mind actuated and keyed to the owner. Right now **PUAIs** are merely wireless interface devices for the Internet with far more power and functionality than complete computer systems of a few years ago. They carry no programs or data since that is all stored in multiple digital storage banks serving the Internet. **PUAIs** can also serve as credit or debit cards and as a telecom system. They can hold personal ID information, including an ID code that positively identifies the owner including a picture, fingerprints, and an iris scan. They also carry a GPS device so they can be positively located. They contain an internal power source that needs recharging only once every few months, and they are always on and active. Any function that needs to be turned off for any reason can be turned off by a simple mental command as can the entire system. The GPS signal cannot.

*The **IBI**, or intra brain interface, is being researched and tested at several universities. Types of IBI are being developed by our military for pilots to use to control their aircraft. These devices could replace other access devices in the next ten or twenty years. The IBI, a link between the brain and the owner's **PUAI**, will be part of the next quantum leap for the digital world. Look at the screen and a cursor appears and moves wherever the eyes focused. In discussion mode, whatever anyone says comes out from speakers at the workstation of each member of the discussion group or class. Sometimes those discussions become quite animated. All the subtle and not so subtle actions of any group e-meeting, e-class, or e-conference will be available to every member of the group attending the e-meeting. This will be accomplished either directly by the **PUAI** or by its being connected to a screen and speakers.*

*In text mode, words or commands, spoken or mentally imaged, are digitized and converted to text on the screen in the format of the program being used. Graphics entry, a bit more complicated, requires some fairly intense training. The entry of programming code is almost as simple as spoken or mental, **SM**, commands, converting directly into code. Editing is quite simple and direct using SM commands. The **IBI** device is a tiny chip or several chips inserted surgically under the scalp and near the speech, visual, and other thought centers of the brain. The chips pick up brain activity and turn it into digital input read directly by the individual's personal universal access interface device - **PUAI**.*

These "Star Trek" technologies either are a reality already or will be in a few years. Fact seems to be outrunning fiction. The story that follows is an extrapolation of the preceding—it is science fiction. The same thing has happened many times before. Today's science fiction is tomorrow's real science and technology. The time span for that conversion is shrinking from centuries to years to months to weeks to days.

The story begins:

I am Leon Moon of Fort Wayne, Indiana, the first member of a training research group here at CDI Fort Wayne. Mara Singleton of CDI Mexico City was assigned as the second member of my group, and Vivek Piloto of CDI Rome was the third. Brought up on a Hoosier farm, I dove into the exploding world of Internet technology like a hog after a bucket of Indiana corn. We are IBI development engineers at CDI currently assigned to develop a training program for our own people to use the latest development from CDI, the Intra Brain Interface, or IBI.

The IBI is a set of small chips, highly miniaturized and compacted digital devices combined into a high-speed, wireless interface between a person's brain activities and existing Personal Universal Access Interface devices, or PUAIs, pronounced "pooies" by all in the trade.

Our introduction to each other was by Gregory Stilling at a meeting in the Inet conference room. Stilling, a stuffy, balding, slightly pudgy man, is vice president of the IC technology section here in Fort Wayne. It was eight-thirty in the morning, local time, when we met. We often speculated that Stilling either inherited a major portion of the local organization or had something on someone high up in CDI. He never seemed to say anything intelligent or be doing anything productive we knew about. However, whenever Chinese big shots visited, the ones from China especially, Gregory squired them around, and they seemed to like him a lot. CDI or China Data Industry is headquartered in Dalian, China and has hundreds of research centers around the globe. Fort Wayne is one of eight in the United States. Like the other seven, it is located in a small city not known for high technology or digital expertise. For some reason, the Chinese think it safer, less costly, and just as efficient.

Stilling made sure all three of us were on line, then launched his inspiring introductions. "This is . . ." until all three of us had verbally shaken hands via the conference room TC setup. "Now that introductions are over, let us begin."

I was already bored. I looked the other two over. We had no chance to speak to each other beyond the short greetings of introduction before Stilling launched his inspiring lecture. Vivek looked like the typical Italian geek—tall, skinny, and awkward—but don't all geeks look like that? On the other hand, Mara looked to be a real dish, so I studied her carefully. It was hard to tell how tall she was. TC screens often distort the vertical. She looked good to me, but not the least bit Mexican.

What was she doing in Mexico? I wondered. *She looks more like a Swede or Hungarian than a Mexican? Singleton, that certainly wasn't a Mexican name.*

"Mr. Moon? . . . Mr. Moon?" It finally registered that those words were coming from Stilling.

"Yes!" I answered, pretending to know what was happening and not appear startled.

"You agree then to being group leader for this project?"

Damn! I thought. *I should have paid some attention to what Stilling had been saying.* All three faces stared directly at me, and I didn't feel any warm fuzzies.

"Yes, Mr. Stilling. I'll be pleased to be group leader." I had put on my best loyal-and-cooperative-employee attitude along with a friendly smile.

Stilling didn't seem too happy at this point. "For a moment I thought you weren't on the same page with the rest of us. I seem to have been mistaken."

Had I missed something? I was suspicious. I must have missed something, something important at that. No problem, I'd pick it up on the transcript after the meeting. Stilling always provided transcripts.

Stilling continued, "Ms. Singleton, you will handle the editing of all reports from Mr. Moon. Mr. Piloto, you will be the coordinator of all data for the group."

As they both agreed, I knew I had missed something important. *Damn!* I said to myself once more. *Damn! I'd best look that transcript over ASAP!*

"Mr. Moon will create a complete outline of the work to be done by the group and provide you with a finished copy by tomorrow at 2200 GMT. Is that all right with you, Mr. Moon, or maybe you will need a bit more time?" There was a sarcastic smile on Stilling's face as he finished.

I knew something was amiss, so I fired back my cleverest response. "Let me look over the scope of the project thoroughly. I'll let you know if I will need more time."

It was not the response Stilling had hoped for. "I hope you will not disappoint us, Mr. Moon."

"I'll find a way not to do so," I answered quite honestly as I picked up the card with the project code he left me. That pompous ass was not going to trap me.

As soon as the meeting was over, I headed for my office to examine the transcript. It was then I realized he left no transcript of the meeting. *Shit! That bastard set me up. I'll download the meeting record, and it will all be there.* I had started to look for it when my phone clicked. I looked at the screen of my PUAI and saw the lovely Ms. Singleton's smile. *What's this all about?* I wondered.

"Hello, Ms. Singleton. What can I do for you?"

"Mr. Moon! I don't know what kind of problem you have with Stilling, but get over it or we'll all look bad. I can't afford to appear to look even a tiny bit bad on this project. Also, call me Mara. I'm not one for formalities."

Obviously Ms. Singleton, in addition to being informal, was not one to mince words or practice diplomacy. "Okay, Mara. Call me Leon, or Moon for that matter. I'm no stickler on formality either, not like old Stilling. Incidentally, I don't have any real problem with him. It's just that he's so stuffy."

"Unless you want to have a problem with me, try to treat him with a bit more respect. He may be stuffy, but his standing with CDI bigwigs trumps his personality."

"Okay! You made your point. I'm sure that's not what you want to talk to me about, so get on with it. I have a ton of work to do on this outline, and Stilling's information is fairly sketchy. At least the part I've been able to check out thus far."

"Here's a question for you: How long have your implants been in place, and what version do you have?"

"Getting rather personal, aren't you?"

"Grow up, Moon, we're both at the same clearance level and working on the same project. Mine have been in for fifteen months and have been updated to version 1.02."

"Sorry, Mara. I have a hard time not pulling your chain. I'll try being a bit more serious if you will try lightening up a bit. Mine have been operational for almost eighteen months. I'm up to revision 1.01. I'm supposed to get the 1.03 upgrade tomorrow. I didn't see the need to upgrade to 1.02 when the other one was almost ready."

"I guess I did come off a bit caustic, but I'm bucking for a big promotion and need this little project to go extremely well. What I wanted to confer with you about is something I've been experiencing lately. It's been happening since I was updated to 1.02."

"Oh! What's that?"

"I seem to be occasionally having some unusual interference between my IBI and PUAI. It's as though I am having hallucinations—at least they seem like hallucinations—both visual and verbal. I find at times I am *remembering* things I know are not in my memory. I thought maybe that since you worked on the revisions, you might be able to help me out."

"I haven't a clue! Never ran into anything like that in any of our tests and research on that upgrade or on the whole project for that matter. Maybe it's a chip failure, but that's extremely rare anymore. How many have had that upgrade? Do you know?"

"I'm the only one here in Mexico. I don't know about the rest of CDI. I thought you would know, having worked on it."

"Nope! Not my concern. I leave that up to the statisticians. But there has to be some kind of record."

"I couldn't find any. Lots of info about the initial few hundred insertions, but nothing about the upgrades."

"I guess no one thinks that is important. Those first two upgrades were minor changes. This new one, 1.03, is a significant advance and is a lot more sensitive. I'm supposed to conduct a series of tests as soon as the four of us here are upgraded, and that is scheduled for tomorrow. We also have scheduled installation of more than two thousand IBIs here at the university during the next few weeks. If I'm not mistaken, CDI has several million IBIs ready to go. They are waiting for the okay to be given as soon as our training program is ready. The training program will be made available to the public so all those customers can use them. It's like teaching people to walk or talk, and I'm still learning. My guess is that will be a never-ending project."

"Let me know how it goes with the upgrade. I can't help but think the IBI/PUAI system is doing something it was never intended to do. I hope it's nothing harmful."

"Likewise! In fact, CDI is planning on selling two billion of these things in the next few years, so there had better not be any problems."

"That's for certain! There's a lot of money, betting these will be a fantastic success. There will be hell to pay if they aren't for any reason."

"You can say that again . . . I'd best be getting on with this project outline, or Stilling won't be happy. I'll say goodbye. See you later."

"Okay! Just don't forget to let me know about 1.03."

"Willco! Out!"

The screen went blank, and I went to work on the project outline.

Two days later, after my update to version 1.03, I knew exactly what Mara had been describing. Fleeting thoughts and images, unfamiliar ones, came into my head and then were gone. It happened when I was relaxing, not when I was reading, talking, or concentrating on any mental task. It became obvious that concentrating on moving a cursor, writing, drawing a graphic, or accessing any program or data, swept these mental aberrations away.

✹   ✹   ✹

By the end of the six months it took to complete our project, Mara and I became good friends as well as colleagues. She also received the next upgrade, and we both continued having the strange images and thoughts. It didn't take us long to learn to push them out of our minds like unwanted thoughts.

✹   ✹   ✹

During the next few months, CDI delivered hundreds of millions of IBIs. They had become the standard system for interacting with the Internet—fast, accurate, easy, inexpensive, and quite reliable. CDI was unable to keep up with the demand and waiting time for customers grew to months.

During this time, something unprecedented happened.

Suddenly, out of nowhere it was there. It took some time before anyone realized what was happening. The reality sneaked up on everyone. It was a wholly new concept, entity, system, whatever one could think of to call it. One day it didn't exist, the next it did. Talk about a paradigm shift, this was the paradigm shift of all paradigm shifts—and no one created it. Hell, no one knew what or how it was, just that it was. With this sudden appearance, everything changed—instantly.

✹   ✹   ✹

My first inkling of this happened one day when I was thinking about contacting Mara, her voice was in my head. *Moon, is that you?* came through clearly with a strong feeling of amazement.

The thought and feeling of incredulity coursed through my mind. It was a strange experience. *What? . . . How? . . . Mara?* The words crossed through my mind and were soon answered.

Am I hallucinating, or is that really you? came into my mind directly from Mara—of that there was no doubt.

I hear you in my head loud and clear. That is, it seems I hear you, but how can that be? Mara's thoughts as well as her disbelief rang in my head.

Suddenly, from knowledge I had from my work on the IBI, I began to realize what was happening—what those "hallucinations" were. It was an awesome realization. *Mara? Hear and listen to my thoughts. I think I know what is happening, and now you do as well.*

I'm listening, came into my head crystal clear.

You do know that the IBI/PUAI system is a bidirectional communication system?

Yes, I'm beginning to understand. Your thoughts are coming to me clear as a bell. Somehow your thought patterns are being picked up by your IBI, transferred to your PUAI, to the Internet, to my PUAI, to my IBI, and thence directly into my thoughts, virtually instantly. It's digital telepathy on a grand scale.

Suddenly we realized we were thinking together. It was so much faster and clearer than talking. Then came the visual—I could see clearly what Mara's eyes saw and she saw what mine saw. Our minds were linked.

This may pose some problems we'll have to learn how to handle soon, I thought.

It certainly does. I can think of one glaring example. I must close my eyes or turn off my PUAI when I'm getting dressed or looking in my mirror.

Never thought of that. There goes privacy. Hell, we both know what we think of each other, and I like what you think.

I do too. That sure cut out a whole lot of thoughtful speculation and planning, didn't it? . . . You're right. We will have to try to get together soon. Of course, we're thinking about when and how.

Mara, this is going to take a whole lot of learning—other minds, strangers, people we don't like as well as those we do, may have access to our thoughts unless we can learn to block them. It's getting quite scary. Yes, I can think that again . . .

Then Mara contributed an insightful thought, *Our two minds seem almost to be as one, communicating is almost exactly like remembering. That is going to take a great deal of adapting and learning.*

In the ensuing months, we learned a great deal about how to communicate over what became known and came to be called the "Ultranet." Our plans to meet brought on growing and shared excitement.

Mara! My lovely Mara. Where in this electronic maelstrom are you? buzzed in my mind as I tried finding hers. She couldn't be asleep at this time of day—not even dozing. After several months of Ultranet thought sharing, I was finally going to meet her in person. Our lunch date at the CDI salad bar was set for eleven-thirty, two hours away, but why couldn't I find her? For some reason she was off-u.

Starting Ultranet contact is much like trying to remember a name or recognize a face: trying to make a connection with someone, *remembering* their pattern, connecting with that one thought process among the billions out there. There are hazy images, mumbled words, whispers, all in the mind. Unintelligible visual patterns like remembering the face of someone you met in the past. They all flash through your mind like experiencing a changing mental collage. It is hard to describe as every part is a new and unstudied experience for anyone who goes, *on-u.* Using the Ultranet, being *on-u*, is almost like silently talking with yourself, but you know it's another person, not you. You also know with whom it is you are connected.

This direct mind-to-mind connection ability was dubbed the *Ultranet* by some unknown person soon after its reality was recognized. It was nothing like the telepathy of science fiction and charlatans. It was sometimes hard to do. After a year of learning this new means of communication, most people had just scratched the surface. Being specialists in developing the intra brain interface, Mara and I were in the forefront of that learning process. We continued working closely together at CDI. Understanding this amazing thing that had happened was a daunting challenge. We were collaborating on trying to learn its use and capabilities and teach others—she in Mexico City and me in Fort Wayne. We were among the early ones to recognize the existence of the unimaginable soon after IBIs came into common use. And we were among the few who had a fairly good idea of how and why it worked.

During the months of working together, our feeling for each other grew from that first sharing of thoughts. We realized we had a special personal *connection.* There was no point in denying it as we each had access to the other's innermost thoughts and feelings.

There you are! Mara cued in that sweet, warm tone of thought that always melted me away. *I accidently shut down my PUAI while searching for lip gloss in my purse. Sorry about that. I knew you must have cued me when I realized I had shut it down.*

When I cued and you weren't there, I wondered if you had gone to sleep on the plane, I responded.

I cannot sleep on planes. Too much mental noise with all those high-profile minds working in such tight quarters. Evidently lots of people haven't learned an effective blocking technique as yet. Many never learn to turn their PUAIs off, she cued as she exited the plane.

I suppose they don't think anyone can get into their thoughts without their noticing. I could make out the hairy neck of a short, fat man walking in front of her. *That guy in front of you sure has a hairy neck,* I cued and laughed out loud.

Yeah! So I noticed. The view shifted to the floor and her carry-on case as she checked it out.

When those hazy visual images from another's eyes first registered on my mind, and I realized what they were, it blew me away. It was unnerving. *Remembering* in visual context what another person was seeing and sorting a single *vision* from the multitude of competing mental images took a long time to learn.

My first clear picture of Mara from her eyes happened about a month after we started Ultranet communicating together. The vision came clear as can be as she stepped out of the shower in front of her full-length mirror. We were both startled, and she immediately closed her eyes—show over. She gave me a hard time over that, but realized it was unintentional. Though we were several thousand miles apart at the time, we could easily have been in adjacent rooms. While *on-u*, one learns not to look at what one doesn't want another to see. Of course, going *off-u* by shutting down one's PUAI makes a complete disconnect, so there is that option.

After eight months of trying, Mara managed to wrangle a temporary assignment to CDI Fort Wayne from her boss at CDI Mexico City. A year working closely together long distance almost every day, and we were soon going to be together for real for the first time off-u. We were both so excited as the last minutes ticked by. CDI Mexico City was finally going to meet CDI Fort Wayne, in person.

It was late June when *I see you* rang through my mind from hers as her rental car pulled up to the door where I was to meet her. *You're much taller than I saw you to be in the conferences* followed.

I opened the passenger door to get in and direct her to the parking lot. I was not prepared for my reaction when we met in person. "You're much more beautiful in person." I said out loud without exaggeration. After one long awkward moment, I began directing her to the parking lot. "Wow! . . . Wow," was about all I could get my mouth to say. At the same time, our minds were speaking in an entirely different language and on an entirely different level.

This was a strange situation for both of us. Our personal, direct visual and vocal contact was as awkward as any first meeting of strangers. Our continuing intimate mental contact did not affect our new, physical meeting. We were, in effect, on a first date—awkward. We were two different couples, one with intimate and close personal contact, the other, two strangers who met in person after having corresponded for some time over long distances. In person we were both shy and reserved.

"I can't believe we have such difficulty talking," Mara said.

"Me too," I replied.

At the same time our minds were locked in a battle. *Why can't we verbalize? Should we kiss? I'd love to hold you,* flicked through my mind. I didn't know if they were my thoughts or Mara's. They were almost in images rather than words—like memories. It's quite difficult to describe.

How do we handle this? Our unison thought was like two people saying the same thing at the same time, only more intense. The disconnect between our thoughts and our physical actions and voices was utterly amazing—weird.

Our thoughts suddenly came virtually in unison, *Shut off our PUAIs!*

The resulting mental quiet was wonderfully relaxing, quite similar to turning off a too-loud radio or blaring TV. That was the last time we ever let our PUAIs remain on when we were together. This worked beautifully; our minds may have communed when we were apart, but were a strong source of distraction when we were together. We were now together in new and unknown territory. It was strange.

We had to get to know each other the old-fashioned way. Our work together in the electronic conferences did not prepare us for direct contact. All of our senses were in play, not just sight and sound. The subtleties of face-to-face contact: the interplay of all of the senses: the power of direct three-dimensional eye contact: and especially the actual presence with all those powerful pheremones in play, overpowered the mental aspects of the familiar Ultranet memories, images, and sounds. This was heady stuff. Our Ultranet communion had only intensified all of these sensory experiences and emotional connections as they occurred.

We both remained silent as I guided her car to a spot in the parking lot. As soon as we got out of the car, Mara walked up to me, smiled, stuck out her hand and said, "It's nice to meet you, Mr. Moon. I'm Mara Singleton from Mexico City."

The ice was broken. We stood laughing almost to the point of tears. It was a special moment. I took her hand in mine and as formally as possible said, "Pleased to meet you, Ms. Singleton," whereupon we had another laugh.

The bright dance of her eyes could only be appreciated in person. "I'm glad we thought to turn off our PUAIs. We are now in a much more real situation. I did not expect this kind of first meeting."

"Nor I," Mara said as we walked hand in hand toward the lunchroom across the parking lot deserted of other pedestrians.

At the sight of someone exiting the building, we self-consciously dropped each other's hand. We were truly like two kids on a first date—self-conscious and unsure of what to do or how to act.

By the time we had finished our meal from the salad bar, we were much more relaxed and talking quite freely. "I think I'm beginning to like you, Moon." Mara's grin and downcast eyes when she said it indicated the ice was not only broken but rapidly melting away.

My response, "I know I really like you—in fact, all those thoughts and feelings we shared are running through my mind—WOW," was followed by a meaningful silence.

✳   ✳   ✳

It was late Friday afternoon when we finished our tour of CDI and the meeting with Stilling. We scheduled her visit to start on a Friday so we could have the weekend together free of the concerns of our professional lives. That evening, after she checked in at her hotel and freshened up, I led her to the Riverwalk.

"It's quite beautiful, cool, and friendly." Mara commented as we wound our way along the shore of the river past shops, art displays, numerous restaurants, clubs, a few small theaters, and a movie house, all under the early evening sun.

"This is one of my favorite spots in town," I explained. "When I was little, this was a stinking, dirty river lined with old buildings—mostly abandoned warehouses and manufacturing plants. It was not a pleasant place. During the economic recession of 2008 to 2015, Chinese-American companies made a huge investment in the city, putting lots of people to work. One of their projects resulted in this beautiful Riverwalk in the downtown of the city. The river was cleaned up, most of the old buildings were leveled, and what you see came to be. It took four years and lots of investment to get it ready. Since then it has grown tremendously and repaid the investors many times over."

"What's that?" Mara asked as a small tour boat rounded the nearby bend and headed toward us with about forty people on board.

"That's a boat we can ride if you like. It takes about an hour to go around the city loop of the river all along the Riverwalk."

"I'd love to."

"Done! But not until after we eat. I'm starved, and the Riverwalk Grille is right there ahead. They have some of the best Italian food in town."

"How do you know I like Italian food?"

"I'm a mind reader," I answered with a broad grin.

"I forgot we already shared that information."

After dinner—ahi tuna, a green salad, focaccia bread, all topped off with freshly made cannoli—Mara was emphatic. "That was one of the best meals I have ever had. The tuna was absolutely fantastic."

"No disagreement from me. Now, how about that boat ride?"

By the time the boat tour was ending, it was almost nine-thirty. The sun had slipped below the horizon. "It is so beautiful," Mara said, almost misty-eyed. "The clear water, fresh smells, orange sky, and fireflies everywhere. It's hard to believe we are in the center of a city and a small one at that. It's much like Venice, but without the unpleasant smells."

"You've been to Venice?"

"Went there during a visit to Italy with my parents when I was about twelve."

"Awesome! Then you've traveled quite a bit?"

"With my parents when I was quite young."

"That's right. I seem to recall your father was a diplomat of some sort. One that moved around quite a bit."

"Still is!"

The slight bump as the boat nudged the landing dock interrupted our conversation. We joined the passengers as they exited the boat and dispersed down the Riverwalk on their separate ways.

"You've had a long day, starting out in Mexico City. I suppose you'd like to get some sleep?"

"Not quite yet. I'd like to walk some more if you don't mind, but thanks for your consideration."

A few minutes later when we walked through a particularly dark arch of trees and bushes, Mara turned, grabbed my hands and turned her face up. "Kiss me right now before I drag you off into the bushes and attack you."

After a long silence interrupted only by heavy breathing, my only comment was, "WOW! . . . WOW!" All those close-shared thoughts and feelings of the last year overpowered us as they rushed into our consciousness. We were definitely in love—a state I had never experienced so intensely before.

We turned and purposefully started for her hotel. Anticipation and swelling passion soon overcame us. Our steps quickened so that we were running by the time we turned onto the street where her hotel waited. We barely controlled our wild physical contacts as the elevator lifted us toward her floor. Mara could hardly get the key card into its slot, her hands were shaking so. The door closing, the total darkness of her room, and her body hard against mine, was the last thing I remembered.

✳   ✳   ✳

I was laying face down on something hard, rough, and cold. When I tried to push myself up, I found my face was glued to the concrete by dried blood. Pain blasted through my face when I tried to pull free. I struggled to open my eyes but could see nothing in the pitch blackness. Something hard and heavy lay on top of me. I freed my left hand from beneath my body and dug in my pocket for the knife that was always there. Cutting at the dried blood to free my cheek took forever. My face must have been in that position for many hours for the blood to have dried so hard. My mind snapped into reality and screamed for answers with a mix of anger, wonder and fear. Mara! What had happened to my Mara?

My face freed from the grip of the dried blood, I tried once more to get up. The heavy thing that weighed me down finally fell over with a bang, and I arose to my knees. On the surface, a few feet away was a thin slice of light. It was coming under the bottom of a large door, an overhead. Groping my way toward the faint light, I felt for the vertical surface above and found the narrow metal strips of a commercial truck door or loading dock. I tried standing, but dizziness and pain in my head kept me on all fours. I followed the door to its right side and felt for any lever, switch, or other opening device. Nothing!

Once more I tried to stand, this time I succeeded in spite of the pain and dizziness. I felt around on the wall for something to open the door. A switch! I flipped it and was assailed with light as the bare, overhead lightbulb illuminated what I decided was a small warehouse. There were no windows or man doors in the grey room that was my prison. The only way in or out was the overhead door. Then I knew where I was. It was an empty storage space, the kind people rent to store cars and other belongings when they don't have space available in their homes. On the left side of the door, I noticed a red handle—it was an emergency door opener to be used by anyone stupid enough to close the door while inside.

One pull down on the lever and the door raised about an inch, letting more light under the door. It was daylight outside. Another crank was unsuccessful. It was apparent the door had been locked or at least latched on the outside. I looked about for something to try to pry up the door. There was nothing but the heavy door that had been on top of me. Other than that door, the room was clean. There was not so much as a splinter of wood or metal. There didn't appear to be anything I could tear off the walls, floor, or ceiling I could use on the door. Then I heard the sound of a motor. Were my attackers coming to get me? Maybe to beat and question me? If they planned to kill me, they could have done that earlier. What was this all about? I reached for my PUAI to contact the world outside and brought a smashed and useless mass of crumpled electronics from my pocket. My attackers, whomever they might be, had been quite thorough.

Men's voices sounded outside. I began kicking the door to get their attention and shouting for help between kicks.

"Someone's in there," said one startled voice.

"I'm trapped in here!" I shouted. "Can you open the door?"

"How'd you get in there with a locked door?"

"I was attacked, knocked out, and dumped in here last night. I have no idea who did it or why. Is the door locked out there?"

"With a big padlock," was the reply.

"Call the police, please," I pleaded. "I've been injured—lost some blood and don't know how badly I'm hurt."

The sound of a phone call to 911 was a great relief and comfort.

An hour later I was being interviewed by Detective Karl Reston from a hospital bed in Lutheran Hospital, my head cleaned and bandaged and a few other cuts and abrasions patched up.

"You say you were attacked in a room in the Regency Hotel?"

"Yes, and I am worried about the lady, Mara Singleton, who was with me."

"Maybe she was a hooker, and you were set up for a robbery. That's not likely here, but it has happened."

"Impossible! Mara is a scientist at CDI. She's a colleague and a longtime friend. You must find her."

"We have officers checking your story at the Regency right now. They'll report to me here as soon as they find anything. The Regency is a first-class hotel. We rarely have any problems there, especially the kind you told us about. Hang on. I'm getting a call."

I watched as Officer Reston conferred with the men at the hotel. He was using the old cell phone system. IBIs were not yet in use by the police.

"They report there is no record of a Mara Singleton ever checking in. They looked at the room that you claim she was in, and it's quite clean. It hasn't been occupied for at least two days. How do you explain that?"

"I can't. She arrived at the airport yesterday morning after a flight from Mexico City. She rented a Toyota from AVIS at the airport, came to CDI where we had lunch. After we left CDI, I went with her to the Regency and stayed with her as she checked in. I went up to her room with her and watched as she unpacked her things and placed them in drawers. She was planning to stay at the hotel for at least three weeks. Our company made all the arrangements."

After another lengthy cell phone exchange, Reston came back. "Everything you say—at least most of it—has been confirmed. CDI did make that reservation, and it was for one Mara Singleton. She was on the plane from Mexico City, and she did rent a Toyota from AVIS. They reported she changed her plans, returned the rental car last night, and caught an early morning flight back to Mexico City. She spent the night at a hotel near the airport."

"That's not possible! Something is wrong here. I'm a serious-minded engineer and have been working at CDI for five years. I am not subject to hallucinations or flights of imagination. My record is impeccable. I know what I described to you happened."

"Well, Mr. Moon, that may all be true, but we are bound by facts we can verify, and those facts do not confirm your story. Fact is, they quite effectively refute it."

"Well, someone is messing with all of us big time, and I'd like to know why. As a matter of fact, I can quite accurately verify my story. I need another PUAI to replace the one my attacker demolished, and I can download records of most of what I told you." As I said it, I realized I could do no such thing. Both Mara's and my PUAIs had been turned off since shortly after she arrived. There would be no data recorded from that time on. The only data available would merely confirm the records Reston reported.

"I suggest you get some rest and take time to heal. That was one nasty knock on the head you took, and it may have muddled your memory. In the mean time, we'll get hold of your Ms. Singleton in Mexico City and listen to her story. I've got to get back to work." As he walked out the door, Reston turned and faced me, a small grin softened his stony demeanor. "I don't think you're crazy. There is certainly more to this than appears on the surface. There is one powerful fact that adds credibility to what you told me."

"What's that?" I asked a bit sarcastically.

"That knock on the head you received, and you in that locked storage room. That's quite real and obvious. You didn't imagine that. So don't think my mind is closed on the subject, or that I don't believe you. Give me some more evidence. I'll listen."

My opinion of Officer Reston took a huge leap upward. That was the most encouraging thing that happened since the blow. Now I had work to do. The nurse informed me I was to remain hospitalized for the weekend and would be released on Monday. I had other plans. It took me thirty minutes to find my clothes, get dressed, and sneak down the service stairs to the first floor. I squared my shoulders and marched straight across the lobby and out into the parking lot, hoping my bandaged head and bloody clothes would not give me away. When I walked back to the Regency to retrieve my car from the parking lot, I walked through their lobby to see if any of the people I remembered from yesterday were there. No luck!

As I reached my car, I realized there were no keys in my pocket. A quick reach under the side and I retrieved the magnetic box containing spare keys to both my car and my apartment. I mentally patted myself on the back for such genius and headed for home. After changing my bloody clothes and checking my bandages, I grabbed the keys to my lab and headed for CDI.

Once inside my lab, I grabbed my spare PUAI from its charger on my desk and turned it on. *Mara! Mara!* My mind cued, searching for hers. Nothing! Wherever she was, she was off-u. I spent the rest of the day trying to make sense of what was going on with little success. Hell, I didn't even eat, I was so upset.

It was early evening when *Moon! Are you there?* came into my head. Another mind was on-u and linked. It was Vivek, the Italian member of our training group from a year ago. I was not in the mood for tech talk, but tried to keep that thought out of my mind. Piloto was a bit sensitive and easily hurt, even by what I considered the most innocent of thoughts.

What's up, man? I haven't heard a peep from you for months.

I've been doing training on advanced uses of the IBI since we completed our project. They made me head of our training program, not that I wanted it, but it does carry a nice increase in pay.

Congrats! You deserved a break. We both know how hard you worked on that project.

Thanks, I appreciate your thoughts, but that's not what I contacted you about.

As his thoughts flooded my mind, I was astonished. *I realize that now. It's Mara? She left CDI and went to work for Solomon Rachid in Pakistan. I find that hard to believe.*

Some rapid mental communication and my recent experience with Mara, the knock on the head, his source of the information on Mara, and our speculations about what this all meant was shared. We also shared that something was definitely wrong. We spent the better part of an hour linked mentally, sharing and speculating before we broke our contact promising to keep in touch.

Reflecting on what I learned from Vivek was sobering. I downloaded the TV news report from Pakistan, less than an hour old. There was Mara, with Rachid beside her, announcing she had left CDI and was joining Rachid's research group to further development of advanced IBI communication devices. She was definitely not herself. She appeared listless, heavily sedated and, I thought, a bit roughed up. Her speech was slow, deliberate, and brief. I knew she was not there of her own volition. Rachid, a known radical and enemy of all progressive nations, hated the Chinese and Americans and could never have gotten Mara to go to work for him willingly.

My next call, not IBI but cellular, was to my friend, Leo, at the Riverwalk Grille. After a few greetings, Leo saved me when he asked, "Who was that gorgeous little lady you had with you last night? She is sure something."

"Leo, you have no idea how great it is to hear you say that. Someday I'll tell you why, but for now, please know I will be forever grateful for your words."

"Moon? I always thought you were a bit strange. That confirms it." I could tell he was smiling and pulling my chain. "I also noticed you two seemed to have something going between you. Am I right?"

"Right on, Leo. You are very observant. Do me a favor and keep that under your hat, will you? Don't let it be known to anyone else—that's anyone, please!"

"Mum's the word, friend. You can count on old Leo."

"There is one person you can share that with—if he ever asks about it."

"Who's that?"

"Lieutenant Reston, a Fort Wayne Police detective. Feel free to tell him about what you saw, but do so only in private."

"I know Reston. He brings his wife here frequently. He seems to be an okay guy."

"That he is, my friend. I'd better get back to work."

Reston! I thought. *I must talk to him and soon!* With that, I cued a call to Reston's home number and waited.

"Okay, Moon. What is it? Aren't you supposed to be in the hospital?" Reston's voice was definitely in the bothered state.

"I think you had better come to the lab and talk to me. I not only have a reliable witness to confirm my story but have a good idea what is going on."

"This had better be good, Moon. Don't you realize it's after ten on a Saturday night? I don't like wild-goose chases and this certainly smells like one to me."

My emphatic "It's not!"

This was greeted by a loud "Shit!" followed by "Okay! I'll give you a listen, but it had better be good."

In half an hour, Reston sat with me at my desk at CDI. After a short phone conversation with Leo and a view of the news reports about Mara, Reston was convinced and definitely on my side. He dispatched crime scene investigators to the hotel, AVIS, and the airport with instructions to "lean as hard as is necessary on anyone who might know anything about this little caper."

"Isn't this federal?" I asked. "Aren't the feds always brought in on kidnapping?"

"True enough, but first of all, there were several crimes committed in my jurisdiction, or have you forgotten the knot on your head?"

"Not at all. It still throbs."

"Okay then. Second, we don't have enough evidence that a kidnaping took place, and until that happens, I have no authority to call in the feds. As soon as we have such evidence, I'll call them in personally. And third, this may be an international situation where I can't be of much help. Hell! At this point I wouldn't know who to notify. Would it be the United States, Mexico, or Pakistan?"

"I see what you mean."

"So let's deal with what we have here and now—the things we can legally investigate."

Reston's phone clicked as he was talking.

"Good work, Ramon. Don't leave until you have questioned everyone possible."

Reston turned to me. "It seemed someone paid the night clerk at the hotel an enormous sum to remove all records of your friend's stay. Apparently, the lady walked out with two men while another two carried you out the service entrance, all to be forgotten and records obliterated in exchange for a large amount of money—several thousand dollars in fact. Ramon passed this information on to the boys who'll be calling on AVIS in the morning. They will use it to pry the truth out of the AVIS people."

"That sure makes me feel better. I knew I wasn't crazy."

Stilling walked in. "Moon, what are you doing here? I heard you were in the hospital, that you were mugged? Who is this?"

"This is Lieutenant Reston, a detective from the Fort Wayne Police. He's investigating the attack and the disappearance of Mara Singleton."

"Did you know she jumped ship yesterday? Went to Pakistan and joined that sleaze bag Solomon Rachid when she was supposed to be here. You helped arrange her supposed visit here, didn't you?"

"Yes!"

Stilling's face flushed with anger. "Did you know that after she left here she hopped a plane to Pakistan early the next morning?"

"She was kidnaped. She didn't go willingly," I protested.

"How could you think that? In her press conference an hour ago, she said she chose to go."

"We were together all of yesterday until someone knocked me out and dragged her away last night. Believe me, she did not go willingly."

Reston piped in, "He's right about that. We have confirmation he knows what he's talking about."

"Incredible! Incredible!" Stilling repeated. "Does anyone else at CDI know about this?"

"Not that I know of. I didn't tell anyone—other than Reston here and Vivek Piloto in Italy. Oops!"

"I'd better inform Dr. Huer in Dalian right away," Stilling said as he rushed for his office and the secure phone.

"That should stir things up a bit for our friends in Pakistan," Reston said, smiling.

"I hope it doesn't put Mara in greater danger than she is already."

Reston's response was not reassuring. "I doubt it will. But it could put international pressure on Rachid to give her back."

"That son of a bitch won't yield to international pressure. His puffed-up ego wouldn't allow it. No, he'll play this one out to the bitter end whatever his game."

"What could he want with her anyway? Do you have any idea?"

"Not at this point. I have no clue other than it must have something to do with the IBI system and our training program. What it might be specifically is anyone's guess."

"I'd better head for home. I told the missus I didn't think I would be long, and I wouldn't want to disappoint her. Besides, there's not much we can do until Monday anyway."

"Yeah, I guess you're right, I'll soon be headed for home myself."

It wasn't more than two minutes after Reston left that Stilling burst into my lab, talking excitedly. "Dr. Huer wants us to do something about this!"

"Yeah? What?"

"It's dangerous, and he won't order you to do it, but he thinks you are the only one who could pull off her rescue."

"Rescue Mara? Me?"

"That's right, you!"

"What makes that genius think I can do it, or would try to? That's a dangerous game he's talking about. Solomon Rachid is no pansy. He plays for keeps. He kills people for fun."

"Exactly!"

"What do you mean, exactly?"

"Dr. Huer says you are young, strong, clever, ambitious, and in love with Mara Singleton, strong enough motivation."

"How could he possibly know that?"

"He's the inventor of the IBI system. How could he not know a whole lot about the people working in his organization? He never for a moment believed Mara went there willingly. He understood your experience the moment I related it to him. He knows you two are in love."

"Stilling, you amaze me. I think maybe I would like to try that crazy idea. I did have some Ranger training when I was in the military and I have kept myself in fairly decent shape. How does Dr. Huer propose I win at this undertaking?"

"First of all, I know you don't care much for me. You think I'm a jerk and not very sharp."

"Stilling? Don't say those things. You've already gone way up in my estimation. I know I seriously misjudged you."

"Don't apologize, Moon. We haven't the time for anything but getting you prepared and off to Pakistan. You'll have to fly commercial, and that first flight leaves early in the morning. There's lots you don't know that you must learn in the few hours we have left."

"Oh?"

"Yes!" Stilling was emphatic, excited, and unlike any other persona of his I had experienced. "You have much to learn, including how to use the latest military IBI unit which I plan to implant in your skull in the next ten minutes."

When I caught the six o'clock plane for Detroit, I was as steeped in new information and technology as could be completed in seven hours. My spare PUAI, still working properly, was in my pocket, but that was only as a ruse. The new military system, miniaturized to the size of a short length of pencil lead, was inside my skull, shielded by a new technique to be invisible to all kinds of indirect examinations including X-ray, sound, MRI, and several other sophisticated scanning systems. It used, but did not compromise, the existing IBI system already in use. The screen for this unit was my eyes. It appeared somewhat like the visual display used by fighter jets of a few decades ago. I could focus on the display or on the usual visual field beyond.

With a thought, I could use the Ultranet normally or as a background to the MI-1 system which would override my PUAI in all situations. There was one other neat little feature. The MI-1 could communicate with all of the new IBI inserts, giving me selective access to the thoughts of those within about fifteen feet even when their PUAIs were shut off and without their knowledge. To sharpen my skills, I practiced doing this during the flights to Detroit and then New York. From New York to Pakistan, I slept most of the time, knowing that might be my only sleep for several days. My cover, sent out over a secure network operated by Rachid, was that I was coming to join Mara because I was so in love. They knew the truth about that, and we were counting on it to at least get us into his compound. From then on, I was on my own.

Oh yes. CDI equipped me with a new simple secret weapon, a pen with an air-pressure-fired bamboo dart that immobilized an average man in about two-tenths of a second. Crude but effective, it held six darts invisible to X-rays.

So equipped, I stepped off the plane in Karachi where I was arrested by Pakistan special military police. They took me to an interrogation room from hell, emptied my pockets, and took away all they found along with the small suitcase containing my clothes and toiletries. The questioning of the two black-clothed interrogators was neither gentle nor considerate. When I explained that I came to Pakistan to work for Solomon Rachid, they laughed.

"Who do you think we work for, American? We are part of Rachid's private army—his VM guard. He would like to know why you came, and we will soon be finding out."

As he pulled his arm back to strike me, another man wearing the same uniform as my tormentors came in and stopped him. "This is Mr. Leon Moon of the USA. He is to be the guest of Dr. Rachid and, as such, is not to be harmed, but protected from any who might wish him harm." Turning to me, he bowed. "I'm so sorry, Mr. Moon. My men have made a foolish mistake and will be punished accordingly. Please follow me."

What he didn't know was that I knew his thoughts as clear as a bell. He obviously was using a CDI IBI implant system. His thoughts conveyed a far different picture from his words. *Count one for the good guys*, I thought as he handed me back my belongings and led me to where I was whisked off in one of three white SUVs.

An hour's bumpy ride and we arrived at what appeared to be a large fortress built back against a sheer stone cliff in the mountains. We waited as the huge gate swung open far enough for us to drive into a large, walled-in enclosure of at least twenty acres. We drove to the front of a rather impressive stone building with many carvings of people and animals atop low walls and in front of higher walls. It all looked ancient. Once inside the large wooden doors, it was another world. It could have been the campus of modern, multistoried office buildings anywhere in the United States. The rows of windows apparently served offices and laboratories in what I realized was Rachid's main research facility. I was ushered directly into one of the smaller buildings and into a lavishly decorated reception room. We waited for but a moment until a neatly dressed woman in Western garb invited me to follow her.

"I am Rachel, Dr. Rachid's special assistant," she announced in flawless English with little accent as she motioned me toward a large ornate brass door which slid silently into the wall. "Dr. Rachid will see you now."

The room beyond the door was stark white, including the furniture, drapes, and carpet. Behind a rather modest white desk sat Solomon Rachid in a white lab smock over white pants and shirt. Seated beside his desk was Mara, also dressed completely in white.

"Come in, Mr. Moon, and welcome to our research facility," Rachid said quietly with a smooth, almost liquid voice. "I believe you know Ms. Singleton. How fortunate for us you decided to join our group. That is why you are here, I am told."

The MI-1 gave him a perfect reading of Mara's thoughts, but Rachid's were somehow blocked. Perhaps he did not have IBI implants although that seemed unlikely. Mara's mind fairly screamed at him in alarm. He could not link with her since he kept his PUAI turned off, closing his mind to unwanted intrusions.

"Yes, I thought since Mara chose to join you, I would like to do so as well, depending on your offer of course."

"Of course, of course," came from his narrow lips with a tone of sarcasm. "I understand you two are in love. The loyal lover coming to protect his loved one. How touching. Foolish, but quite admirable."

"Moon, you shouldn't have come. This madman now has us both," Mara almost cried as she said it. Her eyes were so terribly sad and her thoughts were more sad and terrified.

"Why is my coming so foolish? I have much to offer. You could consider it a ransom to recover, my dear lady. You should consider it."

"Why, Moon? I have you both. I see no compelling reason to release either of you for any reason. I hold two of the top IBI people in the world. Either of you would do my bidding to keep the other alive."

"True, but what if I told you that if one of us dies, the other will die instantly as well."

A sudden jolt of intense mental activity told me I had struck a nerve. This unblocked his mind to mine for a few seconds, long enough for me to understand what he planned and several other interesting items. It was obvious he knew nothing about the MI-1. He thought my mind was blocked because my PUAI was turned off. Advantage mine.

"And how do you intend to carry out such a feat?"

"That's for me to know and you not to understand."

"A foolish statement, young man. Perhaps I will have you incinerated and see what happens."

"You'll not do that until you know why I'm here and what I intend to do."

"How can you be so sure? This little game does intrigue me, so perhaps I will let you live, at least for a few days while my associates extract what they can from you. Then I **will** incinerate you."

I had enough. I leaned over his desk, spoke sharply as to emphasize a point. I hit him dead center in the forehead with one of the bamboo darts from my pen directly from my pocket. He folded like a wilted plant. I pulled the dart from his forehead, grabbed his ID badge, and called for Rachel.

As she entered, I pointed to Rachid and said, "You'd better call a doctor. Dr. Rachid has collapsed from some sort of attack."

Her mind revealed a mixture of thoughts that I had killed him and genuine concern that what I said was true. When she found a strong and regular pulse, she was relieved and decided indeed to call for a doctor. She was doubly reassured when neither of us made a move to leave.

"What should we do?" I asked. "Can we help?"

"Go to your quarters and wait for further instructions. Here's the doctor. Go! Go!" Rachel was quite confused, and her thoughts betrayed her.

"Let's go," I urged Mara as I took her hand and headed for the door. "Where are your quarters?" I asked as quietly as possible.

"Down this way! But why do you want to go there? If you're thinking about escape, that's the wrong place to go."

"Walk naturally. I can't explain now, and turning on my PUAI would be a disaster, so trust me. I know what I'm doing."

"All right, Moon. And I do trust you."

"Take out your PUAI and drop it on the floor, now!"

As soon as it hit the floor, I ground it into a pulp with my heel. "Now they won't be able to track us. The GPS in my PUAI has been disabled, so they can't track me either."

"They have cameras all over the place. They'll still know where we are."

"I'm counting on that. I know there is a heliport near here. Do you know where it is and how we can get to it?"

"It's right at the end of the hall where my room is located."

"Great! That will make it easy."

"There are several guards always on duty, and the helicopter is locked down. It is released only when Rachid is planning to use it."

"He plans to use it in about ten minutes, or did before I zapped him."

"How do you know that?"

"I'll explain later. Here, pin this on under your blouse."

"What's that?"

"It's Rachid's ID badge. I yanked it off his coat as soon as he collapsed. I'm counting on it to get us to that helicopter."

"How do you expect to fly it? Do you know how to fly one?"

"Flew several different types when I was in the military. This one can't be too difficult. But first we need to get inside your room."

"Why so?"

"We leave your ID badge there, and they'll think that's where you are. I doubt they will check. Rachid's ID should take care of getting you and me to the chopper. After that, it's prayer time."

As soon as we entered Mara's room, she dumped her ID badge on the bed, grabbed a long white coat, and we headed for the door. As we closed it behind us, we heard the announcement. "All guests, please stay in your quarters until notified it is safe to move about. There may have been a security breach. Any guest found outside their quarters before the all-clear is sounded will be terminated."

"Well, the fats in the fire for sure," I remarked as we hurried toward the heliport.

"What?"

"It's an old Hoosier saying meaning there is no turning back."

"You can say that again . . . There's the stairs to the heliport. I wonder where the guards have gone. There are always two of them by the entrance to the stairs."

"My guess is they are all up on the roof guarding the helicopter, and the door at the top of the stairs will be locked. Let's go," I said, taking her hand and sprinting to the stairway and up the stairs to the door.

"There are at least six guards on the other side of that locked door. What can we do?"

"Stand next to the opening side of the door and try not to show your face. Hopefully they will rely on their ID scanner and not check you out too thoroughly. After all, you are now the big boss."

When we were positioned by the door, I began pounding on it and shouting, "Guards! I'm here for my trip. Open the door quickly and go down to guard the bottom of the stairs. Whoever caused the alarm may try to get up here."

Immediately four of the six unlocked the door, burst through, and headed down the stairs. The ruse worked. "Now, dear one, head straight for the chopper and get inside. I'll deal with the other two guards."

By the time Mara reached the chopper, I had locked the door. One of the guards opened the door and helped Mara inside as I held my breath. As he closed the door, the other guard approached me. He never saw the tiny bamboo dart and then folded to the deck.

I called to the other guard, "Help! Your fellow here passed out." A few steps in this direction and he joined his buddy on the deck. They would be out for several hours.

Climbing into the chopper, I took a quick look around. Controls all looked fairly standard. I could fly this bird if I could get it started. As soon as I grasped the yoke, a voice startled me, "Voice print command please."

"Shit!" I uttered out loud. "This damned bird requires a voice print to start. Now what do we do." Then it hit me. Mustering all the mental power I could find, I thought clearly, *Start engines!*

Immediately the starting sequence began, and as the rotors gained speed, the voice said, "Have a successful trip!" Rachid had a mental override to the starting lock.

"Moon, you're a genius!" I remarked as we lifted off and headed for India a few hundred miles away. Mara beamed.

"We're not out of the woods yet. This chopper does at best about two hundred. That's at least an hour and a half to friendly skies. Pakistan has jets that can cover that distance in less than a third of that time. Oh yes, toss that ID out. We don't want them using that to track us."

As soon as the ID was dropped, I changed, heading far to the east. I explained to Mara, "That will take us longer to reach India, but anyone chasing us will assume we were continuing on our previous heading and spend most of their search efforts in that area. By the time they figure out what we did, we should be safe and sound." That is if the chopper itself doesn't have a GPS, a probability I didn't share with Mara.

About forty miles from safety, our luck ran out as a jet fighter blew past us at high speed. Before he could make his turn for another pass and a kill, I dropped down on the deck and began following a riverbed that snaked through the mountains toward India.

It was chancy, but not doing it was suicide. On his next pass, the jet fired a missile that missed by at least fifty feet and slammed into a nearby mountain side.

"My guess is the next missile he fires will be a heat seeker. I'll play hell avoiding that."

After two more misses, it was obvious he had no heat-seeking missiles mounted. We were in luck, at least for the moment. Seeing a narrow canyon leading off toward the south, I took a chance and dove for it. With luck it would be a shorter route to safety. Wrong choice. The canyon wandered about, narrowed dangerously, and ended at a sheer cliff. Up and over was all I could do. Our luck changed as the jet flew under our sudden change of direction and plowed directly into the cliff. At his speed, he couldn't turn as sharply as we could and paid the price. We were safe once more, at least for the moment.

Back up to higher altitude, we headed directly south toward India, less than ten miles away. Unfortunately, these jets would pursue us on into India in spite of the border. This too, I neglected to tell Mara. No need to alarm her.

My welcome words "We are now over India" were greeted with a sudden lurch as a pair of jets passed a few feet above us and hit us with their backwash. It almost turned us upside down, dropping us at least five hundred feet before I regained control.

"Another one like that and we'll be kissing those mountains below us," I warned Mara.

"Aren't we over India? Aren't they supposed to stay in Pakistan?"

"I'm afraid those pilots don't concern themselves too much with borders. I don't want to alarm you, but we don't have many options left. The terrain is flattening out, which means there are no deep canyons for us to fly in. If we get too close to the surface, they can hit us with their backwash and knock us into the ground. If we fly too high, they can take us out with missiles or those old-fashioned guns they carry.

"I'm going to fly backward for a while. Then I can see them coming and maybe avoid their attack by dropping or veering sharply. At the speeds they're flying, it's hard to change directions quickly. Watch for them, high and to the left toward the sun. That's where they're most likely to start their attack. I'll have enough of a problem keeping from flying backward into the ground, seeing as I can't see where we're going."

"There they are!" Mara exclaimed, pointing exactly where I said they would be coming from.

We watched as the tiny black dots grew larger and larger, coming straight at us. I planned to spin and dive for the ground at the last minute, or when I spotted missile exhaust trails. Unexpectedly, the jets turned and headed away from us. The reason soon showed itself as four jets with clearly Indian markings flew by about a thousand feet above us and directly toward the Pakistanis.

In less than half an hour, two Indian helicopters guided us to a safe landing at a military airport while the four jets flew past in an obvious salute maneuver. We were back among friends.

We were surprised when greeted by a large "Welcome CDI" banner above the terminal plaza entrance.

When the base commander greeted us by name, he added, "Thanks to the Ultranet and the old Internet, we have been following your exploits with interest since you took off from Rachid's compound. That was a daring rescue, Mr. Moon."

"We were lucky."

"Lucky and clever. You'll be interested to know that those were not actual Pakistani jets trying to destroy you. Rachid has been supporting a breakaway rebel military that has quite a bit of captured Pakistani equipment including four jet fighters. Your escape caused them to reveal themselves and their base. As a result, the real Pakistan military can now attack that previously unknown base and close it down. The same action also provided the Pakistan government with a valid reason to put Rachid out of business, permanently."

I looked at Mara and grinned.

Another New Serenity Prayer

May God, the order of the universe,
    grant me the serenity to understand belief,
The courage to accept truth,
    and the wisdom to know the difference.
                    —Howard Johnson, February 20, 2001

To know a person's religion we need not listen to his profession of faith but must find his brand of intolerance.

                    —Eric Hoffer

It is best to be careful what family you choose to be born into!
                    —Howard Johnson, January 1999

Beauty of face and body attracts only; it cannot hold, nor will it last for long. Beauty of heart, on the other hand, grows with time, holds people together, and brings joy to all who have the good fortune to share it. Ah! But beauty of soul, the greatest of all, makes life worth living for everyone touched by it, whether for a brief moment or a lifetime.

                    —Howard Johnson, 1974

When only cops have guns, it's called ***a police state***.

                    —Robert Heinlein

A decision is necessary only when the facts at hand do not reveal the only course to take.

—Howard Johnson, 1960

The United States will be a socialist dictatorship by 2030. At this time those leftist activists will happily extol the joys of socialism as they are being carted off to the salt mines.

—Howard Johnson, 2008

With our progress we have destroyed our only weapon against tedium: that rare weakness we call imagination.

—Oriana Fallaci

I am familiar with the arguments of priests whose truth I vigorously doubt because they take for granted claims impossible to prove. I am equally familiar with the logic that denies all speculative thought, as if a midnight to a midnight were the limit of existence, and a man no more important than a louse. I find the one as superstitious as the other, and, of the two, perhaps the priests less stupid.

—Lord Tros of Samothrace - Talbot Mundy

There's no such thing as life without bloodshed. I think the notion that the species can be improved in some way, that everyone could live in harmony, is a really dangerous idea. Those who are afflicted with this notion are the first ones to give up their souls, their freedom.

—Cormac McCarthy

A man of knowledge is free . . .He has no honor, no dignity, no family, no country,
But only life to be lived.

—Juan Matus

A man of heart is not free . . .He has honor, dignity, family, home, country,
And love of life to be enjoyed.

—Howard Johnson, 1971

The liberal Democrat party in the US has a record of failure, corruption, and greed so blatant that only an intellectual could ignore or evade it.

—Howard Johnson, 2009

Color Me Purple

Scene: a suburban bathroom in the morning. Darryl, fiftyish, is looking in the mirror and having an involved conversation with himself.

DARRYL. I'm purple! Wow! I know this is a crazy dream, but purple? A nice shade of orange wouldn't be bad or light blue or pale green, but purple, a bright purple face. It's my face all right, but purple. My hands are purple. Okay, I'll go along with the dream. Let's see. My eyes are still green, and the whites are still white. My tongue is purple, but my teeth are still white. Okay, greyish yellow, but my hair is definitely white. Ah, you handsome dude. Even purple, you're gorgeous. I wonder what my wife will think? I'll have to see if she's purple too.

(He leaves the bathroom, looks at his sleeping wife then returns, examines himself, and continues talking under his breath.)

Nope, her soft sleeping face is like always. I think I'll let her sleep 'til I figure this out. I don't seem to be dreaming, but this can't be for real. Ouch! Pinching hurts. Nothing changed, so I'm obviously not dreaming. Let's get these PJs off and see what the rest of me looks like.

Yikes! Purple chest, purple legs, purple feet, and purple—well, you know, purple everything. This can't be for real . . . can it? If I'm dreaming, it's convincing. Could I have been poisoned at that dinner last night? I feel fine, but I'm beginning to be a bit irritated. Did those guys spike my dinner with some kind of dye? I'll kill 'em if they did. I'll take a quick shower and maybe it'll wash off.

(He turns on the water, steps into the shower, and begins scrubbing vigorously.)
Damn! If it's a dye, it's a permanent dye. Waterproof too. The palms of my hands and bottoms of my feet are purple. Pale purple, but purple. Maybe it's just on the surface. I know. I'll try that abrasive loofah sponge.

(He scrubs rapidly on his arm with the loofah sponge.)
No luck. It doesn't lighten in the slightest. What will Jenny think when she wakes up? And the kids? They'll all be waking in a few minutes.

JENNY, *his wife, calls from the bedroom.* Darryl? Are you taking a shower? You took one last night, or did you forget?

DARRYL, *a bit startled.* Yeah! I musta forgot. Well, you can't be too clean. (*Then to himself*) What'll I do when she sees me? I'll bet she faints dead away. (*Now aloud to Jenny.*) Are you . . . getting up?

JENNY, *sarcastically.* Of course, idiot. In case you've forgotten, I have to get up and go to work. If you made more money, I wouldn't have to work. When are you going to ask for a raise, Darryl?

DARRYL, Not now . . . uh, not until we get the next quarterly performance review anyway. (*Then to himself*) She'll be in here in a moment, then things will get exciting.

JENNY, *walking into the bathroom.* What are you standing there naked for? At least put on your underwear.

DARRYL, *looking worried.* Uh . . . don't you see anything peculiar?

JENNY, *in a joking tone.* You're always peculiar. Especially when you're naked.

DARRYL, *to himself.* What's going on here? No reaction at all and she looked straight at me. (*Aloud*) Don't you see anything different about me?

JENNY, *genuinely concerned.* Well . . . maybe it's your potbelly. It looks a bit smaller this morning. Have you been dieting? That's wonderful. Maybe by summer you can get it down so you don't have that dunlop problem.

DARRYL, *quizzically.* Dunlop problem?

JENNY, *complaining.* Yeah! You know. Your belly dunlop over your belt. You used to have such a neat body when we first married.

DARRYL, *a bit hurt.* Is it that bad? . . . And no, I haven't been dieting.

JENNY, *lecturing.* Well yes! It's not like it used to be and you should be dieting.

DARRYL, *aggravated.* That's not what I'm asking about. (*Then to himself*) Maybe she's ignoring it in hopes it will go away. (*Aloud*) I mean my skin. Doesn't it look different?

JENNY, *in a joking tone then concerned then aggravated again.* Looks like the same old pasty skin to me. You know, you should go to one of those tanning parlors and get a tan. You look so much better when you're tan. Why are you looking at yourself so intently?

DARRYL, *defensively.* Okay! Okay!

JENNY, *questioning.* What's taking you so long to get dressed? And why are you looking at your hands that way.

DARRYL, *explaining.* I'm getting dressed, and in what way am I looking at my hands?

JENNY, *sarcastic again.* You're looking at them as if you'd never seen them before. You don't have polish on your nails, do you?

DARRYL, *matter-of-factly.* No! Of course not. They seem a bit strange to me. Like they're discolored or something.

JENNY, *concerned and in motherly tone.* Let me see . . . They don't look a bit discolored to me. You're imagining things. Why don't you go start breakfast. The kids will be up soon, and I have to leave early.

DARRYL, *patronizing.* Okay! As soon as I finish dressing. *Then to himself as he goes to the kitchen.* What the hell is wrong with my eyes? Nothing else looks purple. Not in the slightest. The hallway looks normal. The kitchen looks normal, and the appliances are still stark white. Boy, does my hand look purple against the frig.

(His son, Jerry, enters the kitchen.)
JERRY. Hi, Dad! What's up?

DARRYL, Hi, Jer.

(Jerry is fifteen and has mouth to match.)
JERRY.   What's the matter with you? You look strange.

DARRYL, *to himself.* At least he notices. *(Aloud)* How do you mean, strange. *(To himself again)* Now I'll get some answers.

JERRY, *almost concerned.* You look so worried . . . or maybe it's surprised. Like someone startled you or something.

DARRYL. Is that all?

JERRY, *still untypically concerned*. Well . . . yes. You do have a strange look on your face . . . like Mom did yesterday when I dropped her bowling ball right behind her and scared her out of her wits. She had that same look on her face. That is until she realized I was the culprit who scared her. Then she looked like she wanted to kill me.

DARRYL, *in parenting mode*. Jerry, you ought to stop scaring people like that. Some day you'll do that to the wrong person and you'll wish you hadn't.

JERRY, *back to know-it-all teen again*. Ha! What could happen?

DARRYL, *still in parenting mode*. The wrong person could get angry and beat the tar out of you. I'm surprised Mom didn't at least backhand you.

JERRY, *typically*. She wouldn't do that. It was only a joke.

DARRYL, *seriously to himself*. A joke! Maybe they're all playing a joke on me. Jerry could be the culprit, and the rest of the family could be in on it. Should I play along or call them on it now, or maybe at breakfast.

DONNA, *ten years old, enters the room*. Hi, Daddy!

DARRYL, *lovingly*. Good morning, pumpkin. How about a kiss?

DONNA, *concerned*. Sure, Daddy! (*smack*) What's wrong? Are you mad at me?

DARRYL, *sweetly*. Of course not. What made you ask such a thing?

DONNA, *almost pouty*. You look angry, that's all.

JERRY, *typically*. He's not angry, Donna, scared or surprised or somethin' like that, but not angry. He sure is out of it this morning.

DONNA, *concerned*. You and Mommy didn't have a fight, did you?

DARRYL, *solemnly*. Your Mommy and I never fight. We have discussions, that's all.

JENNY, *coming into the kitchen*. What's this about your father and me fighting? We don't fight and you know it.

DONNA, *questioning*. Then why does Dad look so strange?

JENNY, *in command mode*. He looks normal to me. You kids, set the table—now! I'm in a hurry this morning. Honey, will you get the cereal and bowls? I'll get the milk and juice. Cooperation—that will get things done in a hurry.

DARRYL, *muses to himself*. If this is a joke, it's a damned good one. They couldn't act this normally with me all purple. It's like I'm the only one that sees it. Maybe I'd better call in sick and go see a doctor. I'm apt to crack up if this continues.

JENNY, *as mother superior*. Darryl, you'd better hurry yourself. You know you have that important meeting with Mr. Herkimer at nine, and you know what a stickler he is about being on time.

DARRYL, *startled, gets up, speaks then thinks to himself*. Holy cow, thanks. I had forgotten all about that meeting. (*To himself*) That should be interesting. If I look purple to Herkimer, he'll fire me on the spot. Damn! I sure wish I knew what's going on. Let's see . . . I've got roughly an hour and a half. Unless I'm the only one who sees it, my purple face should get some interesting responses on the bus ride to work.

– ☺ –

DARRYL, *muses to himself, walks into the office and up to the reception desk*. No one seemed to notice on the bus or here in the office. I'll check in with Glynda. (*To Glynda*) Good morning, Glynda. Is Mr. Herkimer in his office? We have a meeting in about five minutes.

GLYNDA, *the master receptionist*. No, he's waiting for you in that little conference room at the end of the hall. He was there when I came in. See him there sitting with his back to the door beyond the table?

DARRYL, *seriously*. Yeah. That's funny. Not like him.

GLYNDA, *surprised*. He wouldn't turn around when he said good morning as I popped my head in to greet him. Told me to leave and not let anyone in but you.

DARRYL, *worried, then talks to himself.* I'd best be cautious when I go in. That is strange. It doesn't look good for me. This is supposed to be an important meeting and he's already acting strangely.

MR. HERKIMER, *worried.* That you, Darryl?

DARRYL, *solemnly.* Yes, sir.

MR. HERKIMER, *worried, but still the boss.* Take a seat at the table, and we'll talk.

DARRYL, *incredulously.* Like this? I mean with your back to me?

MR. HERKIMER, *definitely in command.* That's right. I'm testing out a new theory for talking with employees. It's nothing personal.

DARRYL, *to himself, surprised.* Holy cow! He doesn't realize it, but I can see the reflection of his face in the window on the other side of the room. It's as purple as mine. (*Then to Herkimer in a commanding voice*) Sir, I think you ought to take a look at me.

MR. HERKIMER, *startled.* No way! This new method doesn't allow it.

DARRYL, *straightforward.* Mr. Herkimer, my face is as purple as yours.

MR. HERKIMER, *incredulous.* What are you talking about, Darryl? Whose face is purple?

DARRYL, *straight talking.* It's true, sir. That's how I woke up this morning. With purple skin—all over. I'll come around in front of you so you can see.

MR. HERKIMER, *startled, still wary.* Damn it, Darryl. Stay right where you are.

DARRYL, *convincingly.* No one else in the family saw anything unusual at breakfast, and on the bus, no one paid any attention to me. I tell you, you look normal to everyone. Except for me that is, and I'll bet I look purple to you. Take a look. You can see my reflection in the window and I can see yours.

MR. HERKIMER, *startled and then amazed.* Well, I'll be. You're right. Come around in front of me so I can see you better.

DARRYL, *straight talk again*. Okay. What do you think is going on? (*Then to himself*) That usual florid face of his is a really bright purple with darker purple blotches where he usually has red ones. At least I'm not the only one.

MR. HERKIMER, *explaining*. I have no idea. When I got up and saw myself, it was the shock of a lifetime. Fortunately, Ethel is away, so I got dressed and drove down early before anyone else came in. You're the first person I've shown my face to. I didn't see another single soul like this, all the way downtown. Everyone seemed normal. Now you. Have any ideas?

DARRYL, *knowingly*. No, but it's obvious regular people see us as normal, so we don't have to sneak around hiding our skin.

MR. HERKIMER, *relieved*. That's a relief. It's also a great relief to realize I'm not the only one. What do you think happened to us?

DARRYL, *more confidently*. I have several ideas now that I find you have the same problem. It must have something to do with our eyes.

MR. HERKIMER. How's that possible?

DARRYL, It's only logical. We both see ourselves and each other as purple, and no one else does. It's much more likely that our eyes see us as purple than everyone else seeing us as normal when we actually are purple. The latter is highly unlikely.

MR. HERKIMER. Why don't we see some other things as purple then? If it's our eyes, surely we'd see something else as purple when it's not. Wouldn't we?

DARRYL, *now in his element, explaining*. Not necessarily, particularly if the same thing was in our skin. Sorta like things that fluoresce in ultraviolet light.

MR. HERKIMER, *impressed*. Darryl, how do you know all this? I never knew you understood chemistry or other science. You're a lot smarter than I gave you credit for being.

DARRYL, *pleased*. Thank you, sir. Now we have to figure out what the two of us have done the same that could create this effect.

MR. HERKIMER. Maybe we're not the only ones.

DARRYL, What do you mean?

MR. HERKIMER. Let's walk through the plant and see if there are any other purple people. We can round them up and get their help.

DARRYL, *smiling broadly.* Great idea.

MR. HERKIMER, *turning and heading for the outer office.* Let's go.

DARRYL, (*to himself while peering about at all the faces.*) I can't see a single purple face in the whole office. Maybe Mr. Herkimer can see some. (*To Herkimer)* There's none in the office that I can see. What about you?

MR. HERKIMER, *disappointed.* Me either. Let's walk through the plant. There may be some among the two hundred odd people on the shift.

DARRYL, *after they walked through the plant.* Do you see any? I don't see a one.

MR. HERKIMER, *turning back toward the office.* Let's check the research lab.

DARRYL, *as they walk into the lab.* Where is everyone? John, Ken, and Jan don't seem to be around.

MR. HERKIMER, *to David, the lab manager.* Marilyn isn't here either. David, where is everyone? Are you and Penny the only ones here?

DAVID, *explaining.* The four who are missing all called in sick this morning. They were the four that were working with you two on that new systemic insect repellant last week. What's going on?

MR. HERKIMER, *a light goes on in his head.* Darryl, are you thinking what I'm thinking?

DARRYL, *as things click into place.* Right with you, Mr. H. I'll bet they stayed home because they're purple too.

DAVID, *surprised.* What do you mean, They're purple too?

MR. HERKIMER, *laughing.* It's a little joke between Darryl and me.

DARRYL, *thinking quickly.* Yeah! We were wondering what it would feel like to wake up and be purple.

DAVID, *shaking his head in disbelief.* Weird, weird. You guys are really weird.

MR. HERKIMER, *as they head for the office.* Let's go call our missing lab workers and see if they can answer some of our questions.

DARRYL, *knowingly as they walk together:* I'll bet that's it. Somehow that repellant affected our eyes and our skin. I'll bet we surprise the others when we ask if they're purple. That has to be it.

– ☺ –

Darryl's bathroom some three weeks later. Darryl gazes intently at his face in the mirror, talking to himself.

DARRYL. Finally, after almost three weeks, the purple is beginning to fade. It's blotchy on my face, and my chest is almost back to normal. I don't know if I'll ever tell Jenny. She wouldn't believe me anyway.

JENNY, *walking into the bathroom.* Darryl! Are you still staring at your face in the mirror? Honestly, I don't know what's gotten into you these past few weeks. You were never so vain before.

DARRYL, *smiling broadly in satisfaction.* Sorry, honey. A few of the guys at work developed a rash because of some of the chemicals we've been using. I keep checking to see if I've got it.

JENNY, *matter-of-factly.* Well, have you?

DARRYL, *smiling and pleased with himself.* No. Everything seems to be okay.

JENNY, *relieved.* Thank goodness. I wouldn't want anything to interfere with your new job. That big raise will sure make things better around here. I wonder why Herkimer promoted you before your review. And such a big promotion.

DARRYL, *smiling broadly in satisfaction at his clever response.* I haven't the slightest idea. Maybe it's because he turned purple.

JENNY, *sarcastically.* Yeah! Right!

❧   ❧   ❧

Bloom where you are planted.

—*Unknown*

Natural science does not consist in ratifying what others have said, but in seeking the causes of phenomena.

—*Saint Albertus Magnus*

Patriotism has been called the last refuge of scoundrels, but anti-patriotism truly is much more so. Those who accuse opponents of any kind as using patriotism to hide their false intent are often the true scoundrels. False patriots are quite obvious to open minds.

—*Howard Johnson July 4, 2011*

One of My Father's Many Poems

The melancholy days are here, the saddest of the year,
When we must close the cottage and take in the pier.
The fishing tackle and the boats must be stored away
To wait the coming of spring or early summer day.
The sun is not so warm now; the breeze is very cool.
The water's getting colder in our outdoor swimming pool.
The trees have changed their colors; the leaves are coming down.
The grass is still a little green, but spots are turning brown.
But there is a certain beauty we find here in the fall.
Too bad more cannot enjoy it, even our children all.
The sky has a peculiar shade and threatens rain today,
But the sun sneaks through so often, it's really hard to say.
This summer has been wonderful, and this fall as well.
We really have enjoyed it more than we can tell.
So we are really sad when closing this lake home.
We'll come again next summer when we begin to roam.
Thank God we both enjoy good health and this kind of life,
Living close to nature and away from city strife.
For all those blessings we enjoy and for our family's love,
We will be forever thankful to our great God above.

—*Howard R. Johnson, September 8, 1954*

Two Letters

Alonzo and Stephanie were high school sweethearts in Cleveland Ohio, They were the most popular couple in their school and had many friends. So it was quite natural that they married not long after their graduation. Alonzo, who became interested in working with wood in shop class at school, started working as an apprentice carpenter to his father as soon as he became eligible for the carpenter apprentice program. Soon after he started, they were married. While he was an apprentice, Stephanie went to secretarial school and soon had a job as a secretary for a small local business. They were happy, successful and in love.

After about three years they had saved enough for a down payment for a small home near her parents in a lower middle class neighbor hood on the west side. It wasn't too long after they moved into their new house that Stephanie announced, "Guess what, Alonzo, I'm pregnant."

Alonzo was not happy. "Why did this have to happen when we've taken on the responsibility of a home?"

Stephanie's head slumped as the tears blossomed and crawled down her cheeks. "I'm sorry. I expected you to be happy at the news."

"Well - - - I am, sort of. But it's gonna make things tough around here when you have to stop workin' and we lose your paycheck. I don't know how we'll manage."

"That won't be for at least six months. We'll have to save up."

"And give up the idea of that new car we were lookin' at." He replied gloomily.

"When I told mom she said we could have the crib and baby bed my sis has in storage so we won't have to pay for that. We'll get by OK. She is also giving me her old sewing machine."

"Big deal! We still need living room furniture. That bare room always bothers me."

Stephanie smiled and her face brightened. "You'll have your carpenter ticket soon—about the time the baby gets here, then your pay alone should equal both our present incomes."

Alonzo leaned back in his chair, his face relaxing as the tension subsided. "That's right. Maybe I can get some overtime work to help out."

"Now you're thinkin'! Imagine how wonderful—our own little baby."

"I can build him some cabinets and shelves for his room, and a youth bed when he gets a bit older. Yeah! It'll be fun."

"Hold on there Mr Spade. How do you know it won't be a girl?"

"Well, I guess a girl would be OK. They're cute and pretty, but a boy—we could play ball together, do all those guy things together."

Alonzo got his way. They had a son and named him Gregory after his uncle. A year later they had a girl, Selma, and two years after that another girl named Velma. They had to wait almost five years before they could afford that "new" car and it was a used Ford station wagon with about thirty thousand miles on it.

During the next year Alonzo and his dad joined forces and started their own contracting firm, Joseph Spade and Son, and started building houses. During the next ten years the children grew and the business expanded. In fact, Alonzo and his dad became modestly wealthy. They both built rather expensive new houses in an upscale suburb and their business sported a classy new building to house their trucks and equipment.

The children did quite well in their new schools. Alonzo became quite an athlete and Velma, a bright young girl, was a top student. Things went smoothly for a while until Selma began running with the wrong crowd of wealthy trouble makers. When she was fifteen Selma became pregnant.

Her attitude, "All the other girls were doing it. So what's the big deal? I can bring it up OK."

Stephanie and Alonzo were soon arguing about whether she should have an abortion. Alonzo was for it, Stephanie against. They were greeted with a nasty response from Selma.

"I will not abort this damned kid. He's mine and nobody but me will make that decision, so shut up about it."

"How are you going to handle it? Is your mother going to have to give up her new job to stay home and bring up **your** kid?"

Stephanie bristled. "Don't you talk like that to my daughter. Of course we will help, do whatever we can. That's our grandchild you know."

Alonzo stomped out of the house and headed for the local bar, an action he would take increasingly as the tension and conflict in his home grew and expanded involving the other two children. Mother and baby moved into a spare room and stayed. The baby was a boy they named Fred after his father. The father, a senior about to graduate first promised to marry Selma and provide for the boy. That soon evaporated as he denied all responsibility and moved far away to go to college. It took a great deal of persuasion to get Selma to go back to school. Never much of a student, she went back reluctantly and did poorly.

Alonzo, now a regular at the local bar, was drinking "with his buddies" more and more. This led to arguments with both Stephanie and Greg. Both of them had to pick Alonzo up at the bar whenever he became too drunk to drive. The year Stephanie turned forty-five, she could handle no more. She threw Alonzo out of the house and filed for divorce. Alonzo was emotionally devastated as he still loved Stephanie. By the time the divorce was final Alonzo had moved into an apartment near his business. That same year Alonzo's dad, Joseph, died of a heart attack leaving the contracting business to Alonzo. It was more than he could handle so he sold it to two brothers and stayed on as a carpenter. Alonzo was fifty and a recovering alcoholic.

Stephanie tried several times to bring herself to reconcile with Alonzo. Each time she was afraid things would go back to the way they were. This was true even after she learned Alonzo was no longer drinking. She couldn't bring herself to stop hating him for what he had become. Alonzo Jr, finally reconciled with his dad who had watched his college and then professional football career from a respectful distance. Alonzo adored his children. They were reconciled before Greg retired from pro ball, so Alonzo Senior was able to attend a number of games as his son's guest. Velma, always by her mother's side, never reconciled or met with her father in spite of his frequent efforts. Selma never reconciled with anyone. Two failed marriages and drugs were defining her life.

As the years passed, Stephanie, by now a bitter woman, cultivated her anger with Alonzo, blaming him for all the family's woes. On the other hand, Alonzo's longing and love for his wife grew. Afraid of her reaction, he never contacted her though he did drive past her home frequently and went to the library where she worked to view her while remaining unseen.

Suddenly Alonzo found himself in the hospital, the victim of a serious heart attack. He was seventy. As he lay in bed he decided to write Stephanie and tell her how much he still loved her. He wrote a full page letter saying how sorry he was for what he had done and asking her forgiveness. He also begged her to come to see him before he died. His letter was mailed on a Tuesday.

The previous week Stephanie had an especially hostile meeting with Selma who was, as always, asking for money. Fresh off that encounter, Stephanie vented her pent up frustration and anger at Alonzo in a hate letter she mailed to his apartment. Friday she received Alonzo's letter which she first threw in the trash. After some time, she retrieved the letter and read it. All the hate she had cultivated for so many years left in an instant. The love she knew when they were young returned in a flash and rushed back to overcome her heart. In a gush of tears she hurried to the hospital to try to see Alonzo and retrieve her angry letter before he could read it.

Rushing up to the desk she asked breathlessly, "I'm Stephanie Spade. I learned my husband has had a heart attack and is here."

The nurse checked and announced, "Yes he is here in intensive care. Dr. Bowman will be out to see you in a few minutes."

"Can't I get to see my husband? It's important."

"Not until Dr. Bowman can speak with you. He knows you're here and will be with you in a few minutes. There's a small conference room right here across the hall. Please wait for Dr. Bowman there."

Stephanie walked across the hall and sat down in the tiny room. After about fifteen minutes Dr. Bowman entered the room.

"Mrs. Spade?"

"Yes."

"I'm Dr. Bowman from ICU. I've been caring for your husband."

"How is he? Can I see him?"

"I'm terribly sorry, but Mr. Spade passed away several hours ago."

Stephanie burst out in uncontrollable sobs and finally managed to ask, "How did it happen?"

"His heart attack was a major one. He barely made it to the hospital. He lived several days longer than we expected him to. Didn't you know he was here?"

"No, we have been divorced for more than twenty years. I received this letter from him only this afternoon and I rushed right over."

"Such a shame. He seemed such a kind gentle man, always spoke highly of you and of his children, three I believe."

"Yes, he always loved our children."

Dr. Bowman struggled for a while trying to remove something from his coat pocket. Finally he retrieved a crumpled piece of paper "Incidentally, here is a letter he received. It was clutched in his hand when we found him."

Too many people are allowed to assume a performance fantasy, never following up on many actions they endorse or initiate. They are surrounded by papers they never read, people they never really influence, and frequently live in a blizzard of involvement well beyond their ability to relate to and implement.

—Unknown

The End of the Beginning of the End??

In the beginning—or was it really the end? Maybe it was the beginning of the end, or was it the end of the beginning? Maybe it was the beginning of the end of the beginning—of the end? Damn! It's confusing. Anyway, she was there, as was he. At least—I thought he and she were there. I could smell her fragrance, or was it his breath I smelled? Of course, it could hardly be his breath since she—or was it he? —anyway, we don't take breaths. We simply breathe through our skin. Maybe it was another one like her, and him. I couldn't see a thing, but that's not surprising since I have no eyes or organs with which to see. Oh, I can sense light—and lots of other electromagnetic energy—it's that I sense it, like I do radiant heat, all over my body. I have a general idea about the direction of the source. I know to go the other way when it gets too strong. Sometimes that's hard to do.

You see, I have no legs, or arms for that matter, so I can't run away. She doesn't either, but he does get around. I'm on his and her trail again. I can taste that she and he have been here—recently. Oh the passion of it all, the wonderful feeling of touching skin against skin. We do have skin, which, in fact covers our entire bodies. We have mouths, no teeth, but mouths, soft mouths with which we ingest all that wonderful stuff. We also have tiny claws or bristles along the last third of our body. They let us hold on to our homes when creatures try to pull us out.

Then there's the touching, the wonderful sensual pleasure of merging our skin with each other and sharing bodily fluids. It is an ecstacy so delicious that we risk our lives to enjoy the sharing. Leaving the relative safety of our homes, we venture out in search of each other and that wonderful sensual, sexual sharing.

That's when the monsters often attack. Things with claws, or teeth, or large gaping mouths, or beaks, or—well, all manner of horrors they employ to snatch us up or pull us from our homes and eat us. That's why at the slightest sound, the slightest vibration, we retreat into our homes, deep into our homes where it's safe, where the claws or teeth or mouths or beaks cannot reach us. We hear the slightest sound the monsters make, even without ears. We feel the vibrations in the air and through the earth around our bodies. That's how we know the monsters are out there. When they are, we only feel safe when we are alone and deep within our homes. Of course, we are not safe there either. Some of the monsters pursue us in our homes, destroying them with their claws and teeth and pulling us out to be eaten. Oh, the horror of it all.

But I digress. I venture cautiously from my home because it is still. The night is so beautifully quiet and gloriously damp. Her scent, or is it his, wafts through the moist air. I taste it with my skin and the excitement grows. Suddenly we touch, barely at first, a taste, a growing sensuous feeling. We move slowly over each other, melding more and more of our skin together. Finally we are in full sexual embrace, my he and her she, her he and my she. Ecstacy oh ecstacy—mutual orgasm—exchange of bodily fluids—it goes on for hours. Finally and reluctantly we part and withdraw into our separate homes. The promise of a new family now rests within both of us. After several days a capsule containing our precious infants is cast off. In a few weeks our new family will set out to eat, grow, build a home, and love as we did to keep the race going.

Once again—in the beginning—or was it really the end? Maybe it was the beginning of the end, or was it the end of the beginning? Maybe it was the beginning of the end of the beginning—of the end?

Do you know who or what I am?

*   *   *

HINT 1 - Aw c'mon you guys. Keep fishing. You'll remember me.

HINT 2 - I am recognized as one of the most common and widespread critters on the globe. Among all creatures, I am exactly half way between the least complex and the most complex. One of the most successful and vital creatures on the earth, we populate all but the most severe ecologies. We range in size from less than a centimeter long to over four meters. We're an alien creature that lives right in your back yard—front too. We have more than 2,700 species worldwide and we come in red, white, pink, blue, grey and green.

Answer - You will know me if you have been fishermen. We are often hunted and used by small boys. That is the horror of being a fishworm, earthworm or angleworm, from the word, *angle*, for to fish. Among the annelids, we earthworms are special because we are super-streamlined, stripped-down, no-nonsense, highly evolved critters.

Check out these sites:

http://www.backyardnature.net/earthwrm.htm

http://www.kidcyber.com.au/topics/worms.htm

Each worm "eats" his/her way through the soil making burrows that are their "homes." These burrows can be several feet deep. Most, but not all species come to the surface at night to "mate" with the nearest member of the same species. They try to keep the rear segments of their body in their burrows so they can withdraw when danger threatens. Toads, frogs, salamanders, and small rodents are their main nighttime predators, while robins, flickers, and blue jays forage for them during the day. They are the main food source for moles, the only predator that digs them out.

Examine a section of lawn at night almost anywhere and you will see "night crawlers" stretched out of their burrows, their bodies "stuck" together mating or else alone, searching for a mate. Examine any earthworm and you will find a fleshy "ring" covering numerous segments. This egg sac is slightly larger than the worm's body and of a different color. Once the eggs are fertilized, the worm backs its body out of the sac leaving a small ball of flesh to protect the eggs and keep them moist until they hatch. A careful examination of soil in early summer will often produce several of these tiny balls of flesh.

They are fascinating animals most people know little about.

ARE EARTHWORMS IMPORTANT?

The actions and results of what earthworms do, certainly is not simple in ecological terms. The great naturalist Charles Darwin, after making a careful study of them, wrote this:

"...it may be doubted if there are any other animals which have played such an important part in the history of the world as these lowly organized creatures."

"History of the world," he said!

One important thing that earthworms do is to plow the soil by tunneling through it. Their tunnels provide the soil with passageways through which air and water can circulate, and that's important because soil microorganisms and plant roots need air and water like we do. Without some kind of plowing, soil becomes compacted, air and water can't circulate in it, and plant roots can't penetrate it.

One study showed that each year on an acre (0.4 hectare) of average cultivated land, 16,000 pounds (7200 kg) of soil pass through earthworm guts and are deposited atop the soil -- 30,000 pounds (13,500 kg) in really wormy soil! Charles Darwin himself calculated that if all the worm excreta resulting from ten years of worm work on one acre of soil were spread over that acre, it would be two inches thick (5.08 cm).

This is something we should appreciate because earthworm droppings -- called castings when deposited atop the ground -- are rich in nitrogen, calcium, magnesium, and phosphorus, and these are all-important nutrients for healthy, prospering ecosystems. In your own backyard you might be able to confirm that grass around earthworm burrows grows taller and greener than grass a few inches away.

❖   ❖   ❖

To reach another mind and heart and touch with loving care that most tender, guarded, secret being hiding within impenetrable protective walls is my most passionate desire.

To let that being know that there are others like it—frightened, lonely creatures—in the midst of the hostilities of the surface world of sham people;

To form that tenuous thread of understanding between these secret beings;

To share life, this brief flash between two black eternities, with others of my ken . . . would seem to give it meaning.

I would be, or strive to be, a poet perhaps . . . a reacher . . . a dreamer . . . an artist painting thought pictures . . . a thought sculpturer using words as chisels . . . an architect of phrases.

But whatever, I must reach out and try to touch others.

—Howard Johnson, 1974

Look to this day, for it is the life; the very life of life. In its brief course lie all the verities and realities of your existence; the bliss of growth, the glory of action, the splendor of beauty. For yesterday is but a dream and tomorrow is only a vision, but today, well lived, makes every yesterday a dream of happiness and every tomorrow a vision of hope.
Look well, therefore, to this day, such as is the salutation of the dawn.

—from the Sanscrit

My worst mistakes were when I doubted my own judgement.

—Unknown

Oh what a tangled web we weave when first we practice to deceive!

—Sir Walter Scott, 1771–1832

To which J. R. Pope replied:
But when we've practiced quite a while how vastly we improve our style.

—J. R. Pope, 1874–1937

Bigamy is having one wife/husband too many. Monogamy is the same.

—Oscar Wilde

Sea Cliff

He had a lifelong weakness for controlling things and people, but none of that mattered anymore. The vultures were outside, already fighting over the best morsels. He hadn't moved or spoken in weeks but, as she reached over to touch the artery pulsing in his hand, his eyes flashed open and he said, "What the hell are you doing with my hand. That hurt."

She jumped back. Lines of startled concern mapped her face. "Thank God. You've come to. I'm so sorry. I didn't know you were back with us."

"What do you mean by back with us? And who in hell are you?"

"Addy, I'm Lois, your wife. Don't you remember? You've been in a coma, haven't moved, spoken, or opened your eyes in several weeks. Ever since you took that terrible fall."

"I fell? How? Where? What happened?"

"We were hiking the trail by the sea cliff. You were right in front of me when you tripped over that tree root and plunged off the cliff."

"What cliff? I don't remember any trail or cliff. Not one by the sea. There are certainly no sea cliffs in Denver."

"I saw you lying on the rocks. I thought you were dead. Our friends, Sam and Georgia, were nearby on the trail. Georgia ran for help while Sam climbed down and shouted that you were alive."

"Sam? Georgia? I don't have any friends by that name. Who are they . . . And who are you? I don't know you. That's for certain."

"My God, Addy, that fall must have caused amnesia. I'm Lois, your wife. And what's that about Denver? We live in Capitola on the Pacific."

"Impossible. My Lois died in a plane crash that I survived years ago. That was right after we moved to Denver."

"Addy, you are out of your mind. I'm going to have the day nurse call the doctor and get him here. Something's wrong."

She called the nurse and told her, "Call doctor Kline. Tell him that Adam Cizneros is awake and talking. He is making no sense at all."

After hanging up the nurse said, "They're sending a psychiatric specialist right over, they say she is quite an expert for one so young."

"That's all I need, a nut doctor." He hears an especially loud vocal exchange from the back yard. "What's all that commotion outside? All that arguing and loud voices?"

"That's your three partners, arguing about who's in control of your company. I doubt those vultures will be happy to see you awake."

"I have no partners. I bought them out years ago. I hope this nut doctor is a sane, rational person who can tell me what's been going on?"

With that he tried to get up but discovered he had no strength and slumped back on the bed, nearly falling off.

"Damn it Adam, don't try to move without help. You're bound to be weak as a newborn after lying there for so long. The doctor should be here in less than fifteen minutes. Maybe she can make some sense out of what you are saying."

"I am beginning to understand something about the sea cliff. There was a trail Lois and I used to hike in the park not far from that little house in Capitola. It did pass along a high cliff, right on the shore."

"That's the one. That's where you fell. Right now you are in our bedroom in our little house in Capitola."

"That's nonsense. Impossible. That was years ago, when I still had those partners you say are out in the back yard arguing. Although that's like them, always arguing. That's why I bought them out and got rid of them. Things went much smoother from then on. But that's all ancient history, as is my Lois."

"Well, I'm your wife, Lois, and I am not ancient history. As you can see, I'm solid flesh and blood. Touch me if you doubt it. You're here, and barely out of a coma. That's absolutely for certain."

"I don't understand. You and all this . . . the house . . . everything, are from the distant past. Yes, I do remember more now. My head seems to be clearing. Things . . . memories . . . Sam and Georgia . . . the business . . . I'm beginning to remember. Still, something is badly out of whack. It's like I've gone back years into the past, a long-gone past. Yet it's impossible. I remember my Lois was killed in plane crash shortly after takeoff from Denver in April of 2000, yet here you are."

"Addy, it must be because of the fall. You must have hit your head and that brought on all these odd memories. You'll be all right in a day or so. I know it."

"Could all that have been a dream? No, that's impossible. . . . Yet . . . I'm remembering more about the house . . . and those three idiots out in the back yard. As soon as I can get about, I'm going to buy them out. I . . . quick, bring me a mirror."

"What for, dear? You look a mess."

"I want to check something out, please. It's important."

Lois stepped over to the dresser and came back with a large hand mirror. "Here, see for yourself."

Adam looked in the mirror and let out a gasp. "My God! That can't be me. My hair is black . . . and I look . . . so young."

"You're thirty-five, Addy, and we've been married for twelve years. You remember our wedding, don't you."

Adam dropped the mirror on the bed, his eyes glassy and staring straight at the ceiling. One of his partners walked into the room. After a quick glance at Adam he turned toward Lois.

"I wonder if he'll every come out of that coma?"

"Oh, he's already out of it, Ed. Now he's suffering some weird sort of amnesia. Thinks I died ten years ago, right after we moved to Denver of all places."

"Really? When did this happen? Have you called the doctor."

"They're sending a psychiatric specialist over. She should be able to give us some answers."

Ed swivelled toward the bed and leaned over toward Adam. "How you doin' ole buddy? What's this business about Lois being dead? She looks pretty lively to me."

When Adam continued staring at the ceiling, Ed waved his hand in front of his eyes. "Are you sure he came out of that coma? He isn't reacting at all now, look." Ed waved his hand again with no response. Lois ran to the bed.

"Addy! Don't mess with us."

Silence.

"Addy, please, this is not funny." Then to Ed, "I wonder if he slipped back into the coma. He was awake earlier. The nurse and I both spoke with him. Except for thinking I was killed in a plane crash ten years ago, he seemed fairly normal."

"Well, the jerk sure isn't normal. Look at that glassy eyed stare. It's scary."

Suddenly Adam turned his head, looked directly at Lois and said, "Get him out of here before I explode. . . Now!"

"Come on Adam, I was checking on you to see if you were coming around. What are you so pissed about, anyway?"

"Go! . . . Now!"

With that, Ed beat a hasty retreat out the door.

"Don't be so nasty, Addy. Ed was expressing his concern about you."

"Concern my ass. I remember. He and his brother have been stealing from me for years. I know, I know, I'm simply imagining all this . . . but I'm not. I can't be."

"I hope that doctor gets here soon. I'm beginning to worry about you. All this must have been a dream. A dream brought on and intensified while you were in that coma."

"You're probably right, Lois. Things are beginning to fall into place. It's hard to do, but I'm putting all those new memories, or whatever they are . . . I'm putting them in proper perspective, as dreams. Still, it all seemed so unbelievably real. I remember marrying a doctor that I met in the hospital where I was recovering from serious injuries I suffered in the crash that killed Lois. Her name was Sarah, Dr. Sarah Andros."

As he finished speaking, the doctor walked in the room. "I'm Dr. Sarah Andros. Did I here someone mention my name?"

It remained necessary to prove which side you were on, to show your loyalty to the black masses, to strike out and name names.

—*From Dreams of my Father - Barack Hussein Obama*

A stitch in time, saves. He who hesitates, is. He who laughs, lasts. Great minds think. A penny saved is a penny. Actions speak louder. Idle hands are the Devil's.

—*Howard Johnson, 1965*

Suagus and the Chetawk

Suagus snapped the laser lariat over the horns of the bull chetawk, pulled it tight, and dug his heels in for the expected battle. When the chetawk stood and looked at him, Suagus was flabbergasted. He yanked on the lariat several times finally pulling the chetawk over on its back. Still it made no effort to get free, or move. It simply lay there looking at him. At twice his weight and with three short blunt horns facing forward as its only weapons it seemed a poor match for the muscular Merlaner. It had no teeth to speak of because its food was the soft, leafy vegetation of Preator and its four feet were without claws or hooves. What were its weapons, and what could Manch have meant when he warned him to avoid these creatures? This one seemed harmless enough, docile and almost pathetic. He wondered how they had survived predation.

Merlaners are hunters who love battling and subduing strange new creatures on other worlds, and they are good at it. Since it seemed no threat, Suagus decided to release the chetawk and look for a more capable adversary, maybe a cassading or gerlew, both dangerous predators. He pressed the release on the handle of the lariat and it disappeared. As soon as the laser let go of the chetawk, it rolled to its feet. Then Suagus saw a furry blur and that was all he remembered.

He awoke in a medical facility, immobile and in considerable pain. The first thing he saw was Manch's face staring at him intently. When Manch realized he was awake he spat out, "You stupid fool! I warned you to stay away from chetawks. You are lucky one of the spotter craft pilots saw what happened or you'd be dead."

"What happened? All I remember was a blur after I released the damned lariat."

"Why didn't you do at least a little research before challenging a new creature. I warned you to stay away from them. "Those chetawks are unbelievably tough and deadly critters."

"Manch, you know it's the surprise of the unknown that drives our hunt. That's the whole point of hunting, overcoming the challenge of the unknown. Some people climb rocks, some people race rambots, we hunt."

"It's a stupid useless game as far as I'm concerned. Inconceivable danger with little or no valid compensation. Did you know you were dead when they brought you in here? The

263

impact of the chetawk striking your chest pulverized your ribs, collapsed your lungs and shut down your heart. It's a miracle you survived. You can't know it was six weeks ago you were brought in here."

"Six weeks? I've been out for six weeks?"

"Yes, and during that six weeks we returned to Earth with what was left of your body. You're in the Dixon Medical Center in Denver, hooked up to a Kessy life support system. You've been hooked up to the Kessy since we picked you up and you'll remain so for at least another eight weeks. You'll notice you are not breathing."

"Now that you mention it. How is it I can talk?"

"Don't you know anything about the Kessy?"

"Not much, just that it does some amazing things."

"Right now it's providing everything, full life support. It's connected to your neural net and does everything your body can't. Among other functions, it will provide your voice until your nervous system is rebuilt and you can again control your speech. Your entire chest has to be rebuilt, heart, lungs, nerves, circulation system, everything. It's in process. They're putting you back together one cell at a time."

"How come I still hurt?"

"Because you deserve to hurt! Maybe next time you'll heed at least some warnings. You were so badly damaged they must keep your pain network active to help find where all the damage is and how bad."

That's not very comforting, Manch."

"You don't need comforting. You need pain and discomfort to tell you not to do stupid things."

"You don't understand. It's what we do."

"You'd better find something else to do. It will be a long time before you heal and the pain goes away."

"OK! OK! Tell me, what happened? How did that chetawk do me so much damage?"

"Now you want to know about chetawks. It's a little late isn't it? You shut me up when I tried to warn you. Now you're all ears."

"C'mon Manch."

"All right. Long before you became a Merlaner you took marine biology didn't you?"

"So?"

"Do you recall a little ocean predator called the peacock mantis shrimp?"

"That's the little devil with the ultra high speed club that breaks shells, isn't it?"

"Right on. Do you remember how it developed so much power and was so fast?

"Yeah! It had a set of latches and tendons that were set by leveraged muscles storing a great deal of energy in a specialized spring. When the catch is released the tremendous energy stored in the tendons snaps their greatly enlarged and modified claw or hammer forward at unbelievable acceleration. This blow can smash the shell of most crustaceans and stun them into immobility. What's that got to do with the chetawk?"

"They have a similar mechanism in their hind legs that enables them to jump at incredible speed and over huge distances. They have been clocked at an unbelievable acceleration to over ten meters per second almost instantaneously. It's a defense mechanism as they are not predators. They aim themselves at a predator and let fly. When those three blunt horns hit any creature it is almost always killed by the blow. Most predators and many other creatures learn to avoid chetawks."

"Aren't the chetawks damaged when they attack? I would think that hitting anything with that much force with their heads would at least give them a serious headache."

"They have an unusual bone structure, much like that of a turtle. Their ribs are fused into a solid cylindrical shell beneath that loose skin. Before they jump, they lock their skulls into their shell. As soon as they jump they tuck all four legs into their shell and by the time they hit, they have become a rock solid single piece, a virtual missile. They walk away unscathed, at least the one that popped you did."

"Why couldn't you approach them from behind? They would be vulnerable from behind."

"I can read your mind. You're thinking about how to hunt and attack one when you get healthy, aren't you? Well forget it."

"You didn't answer my question. How about attacking them from behind?"

"Did you notice where their eyes were?"

"Yeah! Right on the top of their heads, on those little stalks."

"They have 360 degree vision so you can't sneak up on them unobserved. They turn to face any perceived threat."

"Yeah! I seem to remember he was always facing me. Then when I roped him and pulled him sideways, he didn't move. He finally fell over and lay there, docile as a lamb."

"That's another of their little tricks, playing possum. That usually causes their attacker to let down their guard. As soon as that happens they position themselves for a strike and POW!"

"That's about what happened. When I released the lariat he rolled onto his feet and turned toward me. That's the last I remember."

"Since I'm quite certain you are going to try again once you have recovered I suggest you do a thorough study of the chetawk before you attempt anything. There are two other little tricks they have in their repertoire that are effective and pose real dangers."

"Oh? And what would those be?"

"I assume the chetawk rolled on its back when you restrained him with your lariat?"

"Yeah! And those weird eyes on the stalks were watching me. They seemed to have moved so that when he was on his back they protruded on the side of his head."

"Did you notice his hind legs?"

"Not particularly."

"They would have been pulled up tight against his body. Should a cassading attack it would have to go directly over those hind legs to reach the chetawk's soft underbelly, then, POW! Those hind legs that sent him hurtling at you would hit the cassading and fling it hundreds of meters in the air. By the time it landed, the cassading would be dead from the blow of those hind feet."

"Aren't there any predators that can take down a chetawk?"

"Packs of gerlews have been known to attack and kill a chetawk on occasion, but then they usually lose at least one member of the pack. Then there's another thing. In addition to aiming themselves as a missile, they can cover lots of ground quickly."

"How?"

"The same muscle/tendon system they use to become a missile can also be used as an escape system. They lean back at about a forty five degree angle and let fly those hind legs. They fly 150 meters or more with their heads and legs locked in place. The hit the ground head first and then tumble until they regain their footing. They are back on their feet in about a minute. That's quite effective in getting away from any predator and especially a pack of gerlews. It also works to move them quickly in any direction."

"They sound like a real challenge. I wonder how I managed to get my lariat on him in the first place?"

"There are a number of other relatively harmless herbivores on the planet. A lot more of them than predators. He didn't see you as a threat until you lassoed him."

"Makes sense. I'll have to plan and execute my attack carefully."

"Well, you'll have plenty of time. Another eight weeks in the Kessy and after that, at least three, maybe four months in rehab. You'll need a lot of conditioning to get in shape for any strenuous activity. Remember, almost your entire insides are being rebuilt."

"That long? I'll go nuts."

"Hell, you're already nuts. That's what brought you in here in the first place."

✳   ✳   ✳

Manch had underestimated the recovery time. It was more than a year before Suagus was ready to face another challenge. It took several more months to arrange another Chetawk hunt on Preator. Once more his long time friend, Manch, organized and arranged for the hunt, "In spite of my misgivings." he commented.

Knowing what he knew about the chetawks took most of the challenge out of the hunt. Still, there remained some unknown danger stirring his adrenalin. Each hunt was nearly identical. He would walk up to a chetawk, lasso it with his laser lariat, pull it over on its back, and bind it with a laser loop. Then as it lay helpless on its back he would walk a safe distance away before releasing the loop. The Chetawk would get up, shake itself, look around, and then amble off as if nothing happened. Big deal. Several chetawks took that launch stance of forty-five degrees as if they were going to jump at him, but then, because he was so far away, they dropped down on all fours and ambled away, their eye stalks still focused on Saugus.

Challenge gone, he decided to return to the ship and consider another type of hunt. As he walked through the tall grass on his way he had a strange feeling. The hairs on the back of his neck stood at attention as his subliminal senses told him something was not right. He was being stalked. The hunter had become the hunted. From behind him and on both sides, he began hearing the swish of moving body against the tall grass. Almost imperceptible at first, he realized it was growing. When he stopped to listen, the sound stopped. As soon as he started walking the sound started again. Whatever it was, it was growing steadily louder and closing in.

Saugus took off at a run for a patch of forest ahead where he would be out of the tall grass and have better vision behind. He would be able to see his stalkers. Before reaching the forest he heard a new sound. It was a powerful thump, like something heavy hitting the ground. As he burst out of the grass, something crashed through the branches of the forest trees and hit the ground about fifteen meters in front of him with another loud thump. This was followed by several more egg-shaped grey shapes crashing to the ground around him. As he watched the objects unbundled and turned into a small herd of angry chetawks that surrounded him.

He did not want a repeat of his last experience in the Kessy, so before any of the chetawks could aim themselves, he threw his laser lariat around a branch some ten meters up one of the trees, and retracted it, pulling him up and out of harms way, or so he thought. As he watched, several of the herd folded themselves into attack posture. The first one to let fly struck the branch above his head with such force the branch shattered and pieces pummeled him as they fell. He would have to get higher up, much higher up.

He threw his lariat around a higher branch on the closest tree and pulled himself to a higher perch as two chetawks pulverized his previous foothold. Saugus knew he had to keep moving. Using his lariat, he pulled himself from tree to tree in the general direction of his pickup ship. He easily outdistanced the pursuing chetawks with their plodding walk. It wasn't long before the thump . . . crash of the chetawks told him they were using their jumps to catch him. He chose a zigzag path, making a sharp turn right, then left each time he heard a thump. This led him around the ship in a path he hoped the chetawks would not anticipate. He now had a much greater respect for intelligence of these creatures he once though stupid and a poor adversary.

As he approached the clearing where the ship waited he realized he had not heard any thumps for quite some time. Perhaps they had given up the chase. He pulled out his communicator and contacted Manch in the ship.

"Manch! Manch! Wake up my friend. I may need to get in the ship in a hurry."

"Saugus, you idiot, I don't know what you did, but there are about forty chetawks milling around the ship. They appear to be agitated, stomping their feet and shaking their heads. I'm assuming you are the cause of their anger."

"You didn't tell me they were pack animals. I thought they were solitary herbivores. One at a time I can handle, but a whole herd? That's a different matter."

"I told you what I knew. I also told you to study all you could find out about them. You obviously did not take my advice."

"You forget. The thrill of the hunt is the unknown. If you knew everything about your intended prey it wouldn't be a real hunt."

"How do you intend to get to the ship through our guests out there?"

"I've been thinking about that. How much vegetation is there around the ship? You know, that leafy stuff chetawks are supposed to live on?"

"Well, there's nothing near the ship. The landing blast fried all of that. Let me take a look with the viewer."

While Manch was looking around, Saugus began inspecting his position, looking for any cover where he might hide for an extended period. About a kilometer away was a large, flat rock, the only interruption to the dead flat grass-covered plain. The wooded area he had run through was level. The rock appeared to be fairly flat with steep sides. Perhaps he would be safe and hidden on top of the rock, at least from any creature on the ground. Saugus began working his way through the forest toward the rock, an idea growing in his mind. His communicator sprang to life.

"I found one area of those leafy plants. It's rather small. I guess a bit more than a hectare. It's over toward you and about a kilometer north of your position. Right by that big rock. I'm sure you can see the rock."

"Yes, damn it, I see the rock. That's where I was planning to hide 'til they got hungry and wondered off to feed. I'm working my way there as we speak."

"Well, I doubt they will be leaving soon. They don't appear to have any interest in food. There's a small patch about thirty meters away and not one of them is looking that direction."

"What's your fuel situation? Do you have enough to lift off and set down again, say right on top of that patch of vegetation over by the rock?"

"I don't know? That would be an expensive activity. We have plenty of fuel, but that would cut our safety factor for the trip home. Let me run a fuel analysis."

"If you have enough fuel, it would destroy their only nearby food supply and put the ship in a much better position for me to reach safely. They are herbivores, and my understanding is they will soon have to find something to eat"

"The fuel situation is this. If we make a perfect ascent and decent the first time, there's plenty of fuel. We could do that twice. But! And that's a big but . . . if anything goes wrong and we have to make several tries, we won't have much wiggle room for a landing back home."

"Manch, if anyone could do a perfect job, it would be you. Let's try it. That would solve a whole lot of problems. The lift off blast would do a lot of damage to those critters, and that too will help."

"How soon can you make it to that rock?"

"A half hour if I go carefully. About ten minutes if I throw caution to the wind and run."

"If I were you, I'd run. It will be dark in less than two hours and I'd like to lift off for home before it gets too dark to see clearly."

"I was afraid you'd say that. I'm off."

With that Saugus took off on a steady lope so as to cover the intervening distance and still have enough strength to climb the rock. Shortly after he took off the ship sent off a blast that lifted it off the ground. Saugus was not expecting that as he still had a long way to run.

Suddenly Manch called him on the com unit. "As soon as you started running, those critters all turned toward where they saw you moving and began hunching down in their take off position. I decided it was time to mess them up so I did a slow burn lift off and will head toward the rock. With luck, I'll get there about the same time as you. I watched and the blast took out about half of them. The rest all folded up into their missile configuration and are presently lying on the ground where they were standing. I'll let you know if they start moving."

Saugus was too busy running to respond. The forest was beginning to thin out and about 200 meters from the rock he would run into the patch of fleshy plants so favored by the chetawks as food. He had no idea how difficult it would be to pass through these head-tall plants so he changed direction to skirt them and stay within the thinning forest. He could see the ship and its tail of flames blasting along about forty meters above ground kicking up a huge cloud of dust. *That should make it difficult for those damned critters to see anything,* he thought as he raced through the trees. Then his com unit broke silence.

"Saugus! Be careful. I checked and all of those folded up chetawk missiles are gone, nowhere to be seen. I plan on setting down right next to that rock on the edge of that patch of vegetation. Head for there if you can. As soon as I set down, I'll limber up the photon cannon in case we need it."

Two loud crashes in the nearby trees told him the chetawks had caught up with him. He was close enough to feel the heat from the ship's engine blast. As he ran out into the open toward the rock, he noticed a big problem. There was a steep gully running through the patch of vegetation right in front of the rock. It was hidden by the vegetation. He had to warn Manch. He grabbed his com unit while continuing to run.

"Manch! Manch! Come in! Trouble!"

Manch answered, "What's up?"

"There's a steep gully running through that patch of vegetation right between you and the rock. You cannot set down there."

"If I don't set down soon, we definitely will not have enough fuel. I'm going to try for the rock. It looks flat and big enough."

As the ship lifted to get above the rock, Saugus ducked into a thick patch of the fleshy plants to shield himself from the blast. At the same time, a chetawk hit the ground near him and skidded to a stop no more than a meter away. Instinctively, Saugus threw a laser loop around him to bind him in the folded in condition. The chetawk remained helpless. He could not unfold, nor could he arm those powerful rear legs.

Suddenly the ship's blast stopped. Manch was atop the rock. Saugus released the lariat and ran the few meters to the rock wall and began looking for a place to climb up. He threw his lariat around a small tree growing precariously out of a crack in the rock and started pulling himself up. just as he reached the tree it pulled loose from the rock and he started to fall. Suddenly the tree stopped and then began pulling him up. March's lariat had found the tree as it pulled out. As soon as they reached the top of the rock they took off for the ship.

They ran for the ramp into the ship. Several loud thumps told them the chetawks were not finished with them. Fortunately it took at least a minute for the chetawks to unfold after they landed.

"Hit the dirt!" Manch shouted as he dove for the ground next to the ramp pulling Saugus with him. As they flattened on the rock, the roar of a photon cannon right above their heads was followed by a brilliant blue-white flash and an ear-shattering CRACK!

"Okay! I think we can get up and enter the ship." Manch said proudly, holding his remote firing device up and waving it. "I aimed the photon cannons at the ground about fifteen meters out from the ramp figuring we might need some protection."

Saugus looked. Everything including the chetawks and a fair portion of the rock surface had been vaporized. "Let's get inside and get out of here, NOW!"

"Aye Aye sir!" Manch said with a grin. "I hope you have learned to leave chetawks alone." As soon as Manch closed the ramp, two loud bangs put an exclamation point to his words.

"Manch? I am convinced. Let's lift off before they figure a way to punch a hole in this baby."

They never knew it, but their blast off knocked more than fifty chetawks off of the rock.

Manch repeated what he said after Saugus' first experience hunting them, "Those chetawks are unbelievably tough and deadly critters, clever too, especially when they are in herds."

❖ ❖ ❖

The superior man is modest in his speech, but exceeds in his actions.

—Confucius

Success comes in CANS. Failure comes in CAN'TS.

—Dartnell

There are times when admitting defeat is the greatest victory.

—Howard Johnson, 1973

The best feeling in the world is to do a kindness in secret and have it discovered by accident.

—Unknown

Deceit can destroy only deceivers. No liar can perceive the purpose of him whose heart is free from treason to himself. Guile is a form of wisdom that an honorable man may have and honorably use, persuading deceivers to employ their ill will ignorantly in the service of him whom they aim to destroy.

—Howard Johnson, 1968

[Karl] Marx never did a day of work in his life, and never took the trouble to find out how a worker really feels when on the job. He naturally assumed that workers were a lesser breed of intellectuals.

—Eric Hoffer

Most men, it seems to me, do not care for nature and would sell their share in all her beauty for a given sum. Thank God men have not yet learned to fly so they can lay waste the sky as well as the earth.

—Henry David Thoreau

I have no yesterdays. Time took them away. I may not have tomorrow, but I have today.

—Pearl Yeadon McGinnis

Reality is that which, when you stop believing in it, doesn't go away.

—Philip K. Dick

To a liberal Democrat, socialist, or almost any intellectual, reality simply does not exist.

—Howard Johnson, 1998

The Goo, the Ug, and the Badly

Ug shouted, "Remy, come, look," from his vantage point atop the rise in front of our camp. He was waving with two of his four hands, urging me to come up the slope to see what he had seen.

I jogged up the hill, trying to attach the harness of Stan, my BADLY, as I ran.

"Look! A human." Ug commanded as he pointed when I came near him.

"I can't see a damned thing."

"Look near base of big zuppa tree, there. Another human. Ug."

I mentally turned on my android robot, Stan's binoculars, and looked in the indicated direction. I saw the guy Ug was pointing to, only there were three of them, the range finder indicated almost four klicks. Handy gadgets, those head mounted binoculars. Their images were fed directly to my optic nerves by the optic implant. Stan did a quick search of the universal database of facial recognition and I soon knew who those three were, independent prospectors, like me. They were competitors, and not legal or friendly ones, either.

Josh Jay was their leader and the brains of the group. The other two were unknown to me until Stan gave me all the information he found in the BADLY database. Lumpy Lucas and Jethro Dylan were the others. They were both hardened criminals, real bad dudes. These three would not play by the rules.

Josh tried to kill me more than once back on Apodia 5, and I'm sure things had not changed in the years since we last mixed it up. Back then I was technically a little bit more legal than he was. Still, I didn't rob and murder my competitors at every opportunity. Josh had ten or twelve kills on his record. Who knows how many others went unrecorded? Fortunately I wasn't one of them, although he tried several times. Still, I lost a member of my crew, was missing two toes, and had a nick in my shoulder from our last battle. Josh was short a piece of his hip and lost his number one sidekick to my Galbo that time.

For obvious reasons they did not use a BADLY, in spite of the great advantages it offered. Every BADLY had a legal recorder that could be used in court to convict its user of criminal activity. Stan gave me a tremendous advantage, but he made certain everything I did was legal. Stan would prevent my illegal actions and record everything I did to catch any slip

273

up that crossed the legal line. That legal recorder was virtually indestructible, at least by the wearer.

BADLY is short for Body Android Defense Legal Yeoman. Some smart ass came up with that name for the robot/human interface system that enabled me to survive in some extremely hostile environments. It ties in wirelessly to the user's nervous system by use of six brain implants. Lots of guys don't like them because, well, each BADLY is an individual with an independent mind of its own. With its nerve connections, it can control your entire body, and override your own commands. This happens frequently in dangerous situations. I named my BADLY, Stan. I know he's a piece of high tech equipment, but he quickly became like a real person to me. Shortly after I acquired him, I started calling him Stan and it stuck.

"You know those?" Ug asked.

"Yep. Bad dudes. They'll try to kill us and steal our equipment. Let's head back to the crawler so I can get dressed." I had taken my weekly shower, courtesy of the crawler's water system, but hadn't yet put on my clothes and was busy hooking myself up to the BADLY, when Ug called me up the slope.

Ug, a local citizen of Argos 2, is my sidekick. Almost a pet, kinda like a dog on Earth, but quite a bit smarter, quicker, and with four hands with opposing thumbs. He has a limited vocabulary of English, and any conversation was peppered with *ugs,* thus his name.

"You want Ug hunt and kill?"

"No! I wouldn't want to take a chance to lose you. We'll avoid them. That shouldn't be hard with you and Stan on the lookout."

"They after gold too?"

"Gold, yes, but also platinum, palladium, and anything else of value they can steal."

"Why that stuff? Ug. Not good for anything."

"For some humans that **stuff** is worth a lot of money."

"What good money? Can't eat, can't drink, can't fight, can't make love, useless. Ug."

"Here, maybe, but in my world and in most developed worlds it can be exchanged for all of those things and much more."

"You my pooga. Have much power. Save Ug from goo worms with Galbo. I believe you, but not understand money."

"That's why I'm here on Argus 2. I heard there were a few, large deposits of precious metals to be found beneath certain types of rock outcrops. If I can find and recover one deposit, it will make me a wealthy man. I can live like a king."

"You like money, Ug like build grrruppa, big grrruppa."

"Boy, that's hard to pronounce. What the hell is a grrruppa anyway?"

"You call family or tribe. Ug call grrruppa."

"Whatever you say."

We folded up the shower platform and stowed it in the crawler, our lumbering transport. You could call it a king-sized RV/truck on tracks. Ug's four hands, each with three fingers and an opposing thumb made quick work of such tasks.

"Let's get the crawler headed away from those bastards. Hop aboard. We can head for that mountain range to the south. Prospecting should be good along those mountain streams."

As Ug hopped aboard he warned, "Lots of goo worms near those streams. Must be careful. Ug."

I ran the crawler in silent mode in hopes Josh and his men wouldn't hear us. Of course it's hard to keep a fifty-ton vehicle quiet while crawling through a dense forest, but at least the usually loud engine noise was greatly reduced. For insurance, I turned on Stan's scan mode and set it for any warm blooded life form near 100 kilos within three klicks. It showed nothing except Ug and me until I entered both of our ID scans. After that Stan would ignore our body scans. I thought my desired destination for Stan and then turned control of the crawler over to him. The heads up display showed up in my field of vision from inside my head. It took me quite a while to get used to those dancing symbols in my line of sight

I turned to Ug. "Remember how I found you?"

"Yep. Almost dead in worm goo."

"You were up to your belly with all four feet stuck in that stuff."

"I know Humans kill us. You take out Galbo. I think you kill me. Ug. Then blast goo. Kill many goo worms. Dry up goo. Save Ug. You now my pooga. I protect with my life. Hunt food for you. You big pooga."

I laughed. "I think our relationship is a mutually beneficial arrangement, Don't you?"

"What that mean?"

"It means you do good for me, and I do good for you. We both do good for each other."

"Ug knows. Get you food. Point out dangers. You protect. Give me Galbo. Teach how use. Ug one big pooga like Remy. We do for each other, good. Ug."

"No truer words were ever spoken. I couldn't ask for a better buddy."

Soon after I gave him a Galbo blaster and taught him how to use it, I found Ug could aim and fire a Galbo quicker than any creature I had ever seen. His aim was also amazing—he never missed. He could cut a zuppa fruit out of a tree at 50 meters without searing the tender skin of this delicious fruit, and then run and catch it before it hit the ground and smashed into mush. He always handed me the fruit since, like a dog, he is a pure predator. Ug had been my sidekick for almost a full year since I rescued him. He was a handy guy to have around on Argus 2 where life was somewhat reminiscent of the wild wild west of ancient American pioneer days, strange, lawless, dangerous, and full of surprises, especially for newcomers.

Suddenly Stan turned the crawler to the left. We went about fifty meters and then resumed our previous heading. On our right I saw we were passing a large reflective pool that had to be goo. It was at least a hectare. I'm fairly sure the crawler would have bogged down and we would have been digested. What a way to go. *Thanks, Stan*, flashed through my mind.

"Argos setting. Soon dark. Should stop. Camp now. Find high ground, away from goo worms."

"You're afraid of those critters. Aren't you?"

"Not afraid. Understand and avoid. Ug. Not stupid like before."

Stan had taken us a good deal east of where I suggested. The crawler was headed toward a little rounded rocky hill half a klick ahead.

"How's that hill for a place to camp?"

"Good. No goo worms there."

" I never thought to ask you, what were you doing when you got stuck in goo?"

"Chasing slartza not in pack. Try to mate. Ug."

"What's a slartza?"

"Slartza is female ready to mate."

"You were chasing a female you wanted to mate with when you got stuck in that goo? Passion makes males do stupid things," I said with a chuckle. "We're not so different, you and I."

"You chase slartza? Want to mate?"

"Like I said, we're not so different. You may have four arms and four legs, but you have only one . . . what do you call that thing you use to mate ?"

"Quorg."

I laughed. "We call ours a penis. It can get us into lots of trouble."

"Ug! Quorg make big trouble in pack. If pooga not approve, we no use. If we do, get hurt by pack. Not fun."

"Why didn't you rejoin your pack?"

"Mated with slartza belong pooga. He find out. Pooga drive out of pack. Kill if ever come back. Remy now my pooga. Lot better."

"Well, I will let you use your quorg whenever you want."

"That no good. Remy has no slartza."

I had a good laugh over that one. It took us about half an hour to set up camp which for me was a sleeping bag on the flat top of the crawler. Ug would climb a tree and settle on one of the lower branches. That got each of us above the ground away from a possible goo worm attack. Those little bastards were aggressive. They stayed underground most of the day, but came out at night to mate and hunt. They caught lots of tiny critters by exuding goo on parts of their bodies and digesting them with that goo. It was only when they worked together in considerable numbers that they could form those goo ponds and overpower and digest larger creatures. Ug headed out into the woods to hunt for his dinner while I fixed my own from rations I carried in the crawler. He would be back before dark.

It was almost dark and I had finished eating when Stan's warning went off. Something was moving, heading our way. I used the call of a local critter hoping to warn Ug. Then I turned on our recon scan to see what we had. I saw three figures coming up the hill toward us. Stan indicated they were human and about a klick and a half away. They were moving toward us at an easy jogging rate, following the path our crawler made earlier. If they kept up that pace they would be on us in about twenty minutes. Then they stopped and looked intently in our direction. It was obvious they were following the path our crawler made through the brush. When their images winked out I knew they had turned on a device that scattered Stan's signal making them invisible to the scanner.

I grabbed a light and began running downhill and to my left as fast as I could while trying to remove Stan from my body. I knew I could communicate with my robot from as far as thirty meters. When I reached the position I wanted, I left Stan draped over a small bush and then moved off to the right, looking for a large tree to hide behind. When I doused my light, I could see nothing in the pitch black. They would have to be using some kind of light to see their way without revealing themselves. IR was a possibility but that is easily detected using IR goggles. My guess would be far UV, not an easy light to see by, but the ionization of a powerful solar wind flooded the night sky of Argos 2 with it. I turned on Stan's UV detector and prayed the binoculars would work over the space between us. In the dim UV light I could see there was nothing moving in my extended field of vision. I turned up Stan's intensity. Still, nothing.

Then I spotted movement up in the trees. It was Ug, moving quietly through the branches at least twenty meters up. There was no mistaking those eight arms and legs. From the careful way he was moving. I could tell he was carrying something quite large and heavy. As my UV sight improved, I could clearly see he was also trying to position himself directly above the path our crawler had created when we came up the hill earlier.

I was startled when Josh's loud voice came from behind me, "Drop your weapons and don't move."

I turned and saw Josh about ten meters away with his Galbo aimed right at Stan. He didn't know where I was, but was homing in on Stan's UV signal. He was looking in Stan's direction through a huge set of bugeyed lenses strapped to his head. Crude, but effective. At the sound of Josh's voice the other two came lumbering up the path toward us. Ug dropped his heavy object which flattened one of them noisily. At the sound, Josh whirled around and fired at the loud thump sound cutting what remained of Jethro in two by mistake. Lumpy ran back the way he had come.

Josh returned his attention to Stan and began firing. Stan has one neat feature. It is an energy shield system that uses the energy of any weapon firing on him to redirect all energy away from his parts. The result of a Galbo blast is a huge shower of brilliant sparks, much of it heading back to the source in Josh's hand. While the Galbo is immune to its own energy, the hand holding it is not. Josh lost a portion of his hand and was writhing in pain on the ground. I raised my Galbo to put him out of his misery, but my finger wouldn't close on the trigger. Stan's voice in my head said, *It is illegal to kill an unarmed enemy unless there is imminent fatal danger.* At twenty meters and maybe beyond, Stan was still in control.

I turned to run over and disarm Josh. He still had one good hand and I wanted to remove his weapons while he was occupied with his wound. The instant I moved, a brilliant light flooded the scene. Lumpy had a search light. His following Galbo blast cut a big section out of the tree I had abandoned. Another sizzled above me as I flattened my body against the ground. A loud thump was followed by darkness and then silence with no further blasts from Lumpy's Galbo. Another of Ug's primitive, but effective weapons had found its mark.

All the activity or maybe my hitting the ground so hard, turned my UV off. I couldn't see a thing, but I could hear Josh moving. By the time my UV was back on line, he was not anywhere in my field of vision. I could hear him moving away through the brush off to my left. He was trying to get back to the path. *Keep going that direction Josh. You'll be in for a surprise.* I thought.

Ug lowered himself out of a nearby tree carrying another of those heavy objects. "Remy OK?" he asked.

"Yes, thanks to you. What the hell is that anyway?"

"Whamp nut. Grow big on trunk of whamp tree. Ug use to kill food. Work on enemy too."

I lifted the basketball size nut by its stem. It resembled a pumpkin and had to weigh at least twenty kilos. Its surface was hard and rough, like the surface of a concrete block.

"That makes some weapon: primitive, but effective. It worked this time, thank you."

"Ug learn about whamp nut long time ago. Use to kill food."

"Our friend is headed for that huge goo pond we skirted yesterday. Maybe he'll stumble into it."

"Smell much blood. He hurt bad. Goo worms get him."

"How the hell do you see? It's pitch black. I have to use Stan's UV to see anything."

"Ug see in night. Eyes get big. Light all around."

"You evolved far UV sight. I'll bet you see with your eyes what I see with Stan's UV only much better."

"Yep. Eyes get big. See at night. Use to hunt food. Ug."

I retrieved Stan from the bush and strapped him on as I headed back to the crawler. A quick look with the scanner indicated nothing. Josh's hiding device was still on.

Ug headed back down the crawler path. "Get weapons and other stuff," he said as he cantered off.

About fifteen minutes later the unmistakable fzzzzt, fzzzzt of repeated Galbo blasts came from the distance. I thought Ug had run into Josh. My fears eased as I saw Ug walking up the path dragging what appeared to be a large sack.

"Josh caught in goo." Ug said as he dumped the sack at my feet. The sack was one of their jackets tied up. In it were four Galbos, two sets of big-eyed goggles and several pieces of electronic equipment, all covered with blood. I was particularly interested in the two obvious data boxes.

"I wonder if he'll get away?" I said to Ug as the distant sounds of the Galbo continued.

"Big goo pond. Josh done for."

"I don't know about that. He's a tough hombre. I'll wager he had an a-pack kit he slapped on his hand to stop the pain and the bleeding. If anyone could cut themselves out of goo with a Galbo, Josh could. That's how I got you loose, remember?"

"Ug remembers. Goo pond much smaller than this one. Many goo worms here. They get Josh for sure."

"I wouldn't bet on it."

Suddenly the sounds of the Galbo ceased. "Goo worms got him." Ug said grinning.

A few fzzzzts farther away in the distance told us Josh was still fighting. A few more from still farther away and I was quite sure Josh had escaped the pond.

It was still the middle of the night and I was beat. "I'm gonna get in the sack and grab some more Zs. Can you stand watch in case our friend heads back this way?"

"Sure. Ug get in tree. Like the night. Sleep tomorrow. You sleep now."

I'll bet it was no more than two minutes after I climbed up on top of the crawler and into my sleeping bag that I fell asleep.

It was broad daylight when I awoke to the fzzzzt of a Galbo nearby. As I sat up and grabbed my own weapon, a flock of four winged kaloo birds flew by the crawler in a panic. Ug was hunting his favorite food. These critters are a local version of the Road Runners of the American southwest. Noisy, excitable, and unpredictable, they can administer a painful bite with their catlike mouth full of needle sharp teeth that sits atop a long neck. More than once I've seen them dig up goo worms with their long powerful legs, grab an end with their wicked mouth, and pull them out of their burrows like a robin does with an earthworm.

It wasn't long before Ug showed up, a headless, eviscerated kaloo in each of the lower two of his four hands. He held them up proudly, "Ug have food for several days. Good eating."

"Stick them in the cooler. They'll last longer."

"Ug stick **yours** in cooler. Like mine ripe."

For an instant, I visualized Ug consuming a ripe kaloo and it almost turned my stomach. "If your kaloo gets ripe, please eat away from me. I don't think I could handle it close up."

"Ug forget. You like . . . what you call food not ripe?"

"Fresh."

"You like fresh meat, OK?"

"Definitely. I think we had better get at what we came here for. These food discussions don't add any profits to pay for this little expedition."

"Look for useless metals again?"

"That's what I'm here for. I'll tell Stan to take us to those mountains to the south."

In about ten minutes we were packed up and on our way. Stan homed in on our destination so all we had to do was lean back and relax as the crawler ate its way through thick underbrush and around obstacles like trees and an occasional boulder.

"We should be able to get there before dark. Then we can fire up the metal detectors to run while we sleep. Maybe by morning we'll be rich."

"Rather hunt slartza."

"What if I said I could buy you a hundred slartza if we find enough of those metals?"

"You do that?"

"Sure, why not."

"How do?"

"I use the money we get for those useless metals to buy you many power stacks for your Galbos. You become really big pooga. Kill all . . . what do you call yourself? Slartza is female, what is male?"

"You say male, Ug say artza. Artza hunt for slartza ready to mate. Use quorg. Much pleasure."

"I get the picture. Anyhow you can kill all artza and take their slartzas. Then you will have many slartzas."

"Cannot do."

"Why not?"

"Other pooga hear. Hunt, kill Ug. Throw in goo worm pool."

"How can they do that? You have your Galbos, they don't."

"Last night. Lumpy have Galbos. Ug kill. Kill Jethro too. He have Galbos."

"I see your point. Tell me, how does a *pooga* get to be a *pooga* and form his pack?"

"Not easy. Find young gee want leave her pack, go with new pooga. Start you call family. We call grrruppa. Hard to say in English. Gee now slartza. Help find other gee to join new grrruppa. Take long time to get like old grrruppa."

"What's a gee?"

"Gee is female not mated. Not slartza"

"What do you call yourselves, all artza, slartza and pooga in all packs? Like we call ourselves people or humans."

"Grrrup."

"Boy, that is hard to pronounce. So you are all, grrrup. You told me why you were driven out of your grrruppa, but what about your life before that?"

Ug grow up in pack. Like all young artza, stay with mother in grrruppa. Many friends. Maybe thirty, maybe fifty in pack. Hunt together, sleep together. Life good. Then guorg grow big. Ug try mate with slartza. Pooga angry. Chase Ug. Try to kill. Ug get away. Pooga catch many artza like Ug. Hurt quorg bad. Quarg no more get big. Artza no mate. All pleasure gone. We call prud. Prud stay always with same grrruppa."

"I see. That's a lot like what we call a gelding among horses on Earth. They cannot mate but stay with the herd. Herd is another word for grrruppa or pack. What about the slartza? They don't mate with their fathers, or do they?"

"No. Pooga trade young gees with other poogas from other grrruppas. Pooga soon mate with gee. Gee then slartza. Slartza always belong some pooga."

"I think I get it, but how do you get slartza to join a new grrruppa? You have to steal them from someone, don't you?"

"Some pooga treat slartza bad. Slartza run away. Sometimes find new pooga. Must go far away. Ug meet young slartza that way. We mate many times. Move away, but not far enough. Angry pooga find slartza with baby while Ug hunting. Ug come home. Both dead. Ug wild angry. That pooga have big grrruppa. Too many for Ug."

"Sorry to hear that. What did you do then?"

"Chase other slartza from same grrruppa. Get caught in goo. Remy save."

"So you had just lost your mate and your baby when I found you. You must have been sad, and angry."

"Ug hurt inside. Slartza warm and helpful. Baby sweet. Now all gone."

"Wow. I never dreamed you could have such sadness. Why didn't you ever tell me?"

"Hurt too much to talk. Now plan revenge. Big revenge. Feel better inside."

"Oh? What are you planning to do?"

"Ug stay with Remy until Remy find metal and leave. Then hunt bad pooga. Bad pooga not afraid of Ug. Doesn't know about Galbos. Ug sneak up. Kill sentinels with Galbos. Pooga come see what happening. Cut off quorg with tight Galbo blast. Pooga now prud. Ug take over grrruppa. New pooga. Big pooga. Biggest pooga. Ug now top pooga all places. Treat all grrrup good. Use quorg with many slartza. Much pleasure. Many new babies. All thanks Remy."

"That's some plan. I sure hope you pull it off. When the time comes, maybe I can help."

"No. Ug must do alone. Maybe Remy come help celebrate after Ug new pooga."

"I understand. I do. I'll stay out of your way. I'll also stay around to help celebrate with you. I'd like to meet your new grrruppa."

"Remy one smart pooga. Good friend."

"That goes both ways, my friend."

It was late afternoon when we made camp a ways up in the mountains on a rocky outcropping. I set up the metals scanner to search for deposits then went about cooking part of the kaloo for my dinner. Ug took his up the only nearby tree to eat and scan the countryside.

When I saw the results of the metals scanner I let out a yell. "Holy shit, what a find."

Ug scampered down out of the tree and ran to me. "What Remy find?"

"Look at this scan. We are sitting almost on top of a large deposit of precious metals. I can't believe we found this much so quickly."

Stan's thoughts interrupted me. *Why do you think I took you up here? I noticed that metal from at least fifty klicks back. Did you forget I know what you are after?*

Stan, I forgot about you being in my head. I thought. *I'm sorry. I don't think I'll ever get used to you being there. You are quiet for a long time, then, click, you're right there, in my head.*

No need to be sorry. I have no emotional attachment to any of this, just instructions. I suggest you get busy with the mining equipment immediately. Josh could be back here and find you, or be lying in wait for you between here and the space port. I would project that he will have reinforcements with him.

Yes, but we plan on placing any metal we find in orbit directly from here. Once it's on its way Josh can do nothing about it. Our claim tags and ID information will be aboard the cargo capsule so no one can steal it.

And what will you do with your person between then and when you haul in the capsule from your ship? All that ID info will be useless to you if you are dead. I'm sure Josh has plans for just such a scenario. He may try here, but more likely while you are returning to the space port. That's when you will be most vulnerable.

If you hadn't kept me from killing him, we wouldn't need to worry now, would we?

There are rules and protocols I can't let you break or avoid. You know that.

Yeah, but I don't have to like them or agree with them.

After enduring my long silence, Ug looked at me strangely. "Why you no talk? How you get metal?"

"Sorry Ug, I was thinking about the best way to get to it. Since it is so concentrated we can simply drill a hole down to it, and pump it out."

"Ug no understand."

"Well old buddy, watch while we become miners. It's all mostly automated anyway. You can help me get the tools out and setup. OK?"

"You say. Ug do."

It took several hours for us to take out and set up the drilling rig. Before long the drill was grinding its way through the soft rock toward our prize under Stan's guidance. We pumped the drilling mud down and recycled it when it reached the surface. Our almost magic mud was separated from everything else in the shale shaker and then recycled. The several sieves in the shaker sorted everything that came up in the mud into various storage containers by particle size. We could see what we were drilling through by examining the material that came out of the shaker. I checked it every hour or so. Down to 160 meters it was unchanged. Then we hit hard rock. The time it took to drill a given distance doubled. Stan reminded me he had predicted this.

It took four days of constant drilling to hit pay dirt, literally. After we hit hard rock we had to change drill heads much more frequently and this added to the time. The sound of the drill changed suddenly. Stan informed me we were into the deposit, as if I didn't already know from the sound and the bright yellow pieces of metal that began showing up in the shaker outputs.

"Time to change the drill head." I told Ug, who then helped me pull the drill pipe.

It was a simple operation using slips to hold the pipe as each section was removed. Then we attached a grinder head to a long section of flexible drill pipe and put the whole shebang back down into the hole.

Down in the deposit, the grinder ate its way through the mother lode. All of the shaker outputs were almost pure metal, soft, yellow metal, mostly gold. The metal shavings were taken from the shaker, washed, and put in a press that turned them into almost solid blocks. As fast as one was made, Ug picked it up and placed it in the storage capsule of the orbiter.

It took us nine hours to pull up and pack all the metal we could pack in the capsule. The limit was the total weight that the capsule rocket could put in orbit, about ten tons. As soon as it reached the weight limit, we sealed everything up on the rocket, leveled the launch platform, and launched the rocket. I watched as my fortune lifted off, arced gracefully to the west, and out of sight. I hoped I would be seeing it in about a month.

We put almost a hundred more twelve kilogram blocks inside the crawler. That's more than a ton and Stan's advice said that would be the maximum the crawler could carry. I would have left all of the drilling equipment and taken more gold, but our agreement said we must leave nothing behind. We pulled all the drill pipe, lowered the tower, and stowed everything in the crawler. We plugged the hole, cleaned up the campsite, and headed for the space port. The port was at least four days to the north.

Late in the second day, Stan's voice came in my head. *We have company, at least forty of them, and they are not human, but about the same mass as humans. They are about five klicks west of us and moving on a parallel course about as fast as we are moving. I'm fairly certain they are a pack of creatures like Ug.*

"Ug, I think we've found some of your grrrup. Stan says there are about forty of them and they are nearby. Maybe we should stop and be quiet until they are gone."

"No. Keep going, Ug go take look. If my enemy's grrruppa, this my chance. If not, I catch up easy. Ug much faster than crawler."

"How will we know?"

"Ug come and tell. Maybe two, three hours."

"Good luck." I shouted as Ug took to the trees and headed west.

Stan kept track of Ug and the pack until an intervening hill blocked the scanner signal.

We lost contact, Stan's words came through clearly. *From the terrain ahead, we won't see them for at least two, maybe three hours. Then the hills flatten out all the way to the pass before the space port.*

With Stan controlling the crawler and picking his way through the dense forest of trees and underbrush, we were moving rather slowly. Sometimes he had to back up to get through tight spots. Fortunately the ground was fairly level. We were out on an open plain that sloped gently up towards a hill where Stan saw for an instant a blip that was a human.

Stan reported, *that grrruppa that has been pacing us, is crossing this same plain about six klicks to our west. They are moving rapidly in single file, heading for the same low place between the hill and the mountains beyond as we are. At the pace they are maintaining, they will reach the wooded pass at least an hour before us. The space port is the other side of those mountains, no more than half an hour away. There is no more sign of any humans on my scanner.*

OK, Stan. What's the best plan?

Do we not have an acoustical wave launcher aboard?

Hell I almost forgot we had that ancient weapon.

I suggest you raise the front turret and get it mounted facing forward. I can automate it to fire the instant we are fired upon by anything. That won't stop the first volley, but it will eliminate the source before the next shot, and do the same to any other weapon attack.

Even with the power assist, it took me almost fifteen minutes to muscle that heavy launcher out of its storage rack, onto the upper deck, and into the raised turret. Once the heavy power cable was attached and the acoustic lens control set up and tested, it was ready. We were less than a klick away from the pass.

OK, Stan. Do you need to test it?

Once I complete the diagnostics, I will know if it's operable. I don't want to test it. We'll save that so it will be a surprise to them. My guess is they have prepared a trap of some sort to stop the crawler, inside that forest ahead, right where the pass narrows. There is a high cliff to the right. That's where they will hit us. I'll take us well away from that cliff. They could blow it up and bury us under rock and dirt. Incidentally, that grrruppa? The individuals are all spread out through the forest in the pass. They are not moving. It looks like an ambush, but what for?

My mind raced. *What they hell are they doing there?*

The crawler rocked from a substantial blow.

Stan, what the hell was that?

They hit us with a blast from a large military Galbo cannon. No significant damage, but we won't be able to handle more than a few of those.

Did the launcher fire?

That was most of the jolt you felt. It fired at the same time the blast hit us. My guess is whoever operated that big Galbo is dead amongst the wreckage of his weapon. We will continue up through the pass. That grrruppa? They are all high up in the trees ahead and to our right. They certainly can move quickly through the trees.

If I had four hands and four feet, I could move quickly through the trees myself. I wonder what they are up to?

As we crested the pass and started down the other side, Stan stopped the crawler.

We have company, It's Ug. Stan said.

I opened the side hatch and there stood Ug with three more of his kind, obviously slartza.

"Remy. Ug now big pooga of big grrruppa. Revenge sweet. My slartza now. Big time happy."

"What about Josh and all his cohorts?"

"Many dead, include Josh. Others run away when big Galbo blow. Artza chase. Steal weapons, scare to hell. One stuck in goo pond. They never come back. Remy big pooga over all. Now every grrruppa know about Remy, and Ug."

"How'd you manage that against those guys and their Galbos."

"Ug have secret weapon. Look."

About fifty meters away in the woods was one of our attackers impaled by a wooden spear as thick as my arm and as long as I am tall.

"How in hell do you throw a spear that heavy?"

"Look up. Watch."

At least fifteen meters up in the trees, four artzas worked together lashing two fair sized trees together. They stuck a big spear in a cup like device between the two trees.

"Now look at man."

As soon as I looked at the dead man, Ug gave a signal. There was a loud twang from above and a second spear went through the man's body.

"That our secret weapon. Look up. See anything?"

All I could see up in the tree was tree branches and leaves.

"Now look up."

It couldn't have been two seconds later when I looked up again and saw the four with another spear ready to fire.

"Josh men never look up. Never see us. Great secret weapon."

We talked for about fifteen minutes. Ug told us his original plan worked like a charm.

"After pooga made prud, all grrruppa happy. Say he bad pooga. Hurt many in grrruppa. Ug new pooga. They say Ug good pooga. Make all happy. Ug quorg happy. Soon many babies."

After celebrating with the grrruppa for about two hours, I said goodby and headed through the pass down to the space port. Stan connected with security and my ship's crew of four were soon loading the metal bricks into our ship. After selling the crawler and contents to a local merchant, the total weight of the ship, including on board metal and the orbiting capsule with its stash, would be below max for our trip.

We lifted off smoothly and were soon in orbit tracking the capsule. Another four hours and the capsule was caught and stored in the cargo bay. After we were settled in stasis pods for the trip, Stan took over for the two month sleep on the trip back home.

✳   ✳   ✳

I was rudely awakened by Torba, one of my ship's crew. He and Farley stood there holding Galbos aimed at my head.

"What the hell?"

Farley said, "Slight change in plans. We redirected the ship's destination to the Altos Transfer Station. We're docking there now. If you cooperate, we'll drop you off along with the two in the other stasis pods. Then we plan to head off into the sunset with all that gold."

Stan's words reassured me. *They don't know that as soon as they changed the destination I redirected it back to the original. We are docking at Earth Orbit Station One where the police will be waiting. Their Galbos? Totally harmless, I drained their power stacks. Have fun.*

"Thanks Stan," I said aloud.

Torba looked startled. "Who's Stan? . . . I see. You're trying to distract me so you can jump me. Forget about it. I'm too sharp for you."

"Well asshole, go ahead down that exit ramp. The police are waiting to take you and Dumbo here off to the hoosgow. How do you like them apples?"

"I should blast you for being sassy with me. Shut up or I will."

"Go ahead, asshole. Shoot."

I laughed as they tried firing their Galbos and then ran down the open ramp into the waiting arms of half a dozen cops.

"You two have doubled the share of that gold for your two buddies here. I'm sure they will be forever grateful to you." I shouted as the cops cuffed them. It's great to see justice triumph.

I looked back at the bound stacks of precious metals and thought, *My gloriously sexy Jeannie will be immensely happy, even considering my almost three year absence.*

That made me laugh and think of Ug and his quorg. My quorg and I would soon be immensely happy too.

The Past Is That Portion of Time

The past is that portion of time we carry in the storehouse of our memory and all before it. It includes memories, pleasant and unpleasant according to our heart.

The future is that portion of time we carry in the dreams and wishes of our hearts. It is colored bright or dark according to our spirit.

The now—today—that brief demarcation between past and future is where we act, live, dream, remember, plan, think, cry, laugh, and love. It is good or bad according to our motivating desire at that instant.

—Howard Johnson, 1965

A Bag of Marbles

My name is Charley Woods. I am past eighty and need to share this bizarre story from my youth. Since my grandson, George, revels in my stories, I am telling it to him. I am also providing a great gift to go with my story and prove its veracity.

"George, I will tell you this story as I remember it. It happened more than seventy years ago when I was a boy about your age. Some of it is still difficult for me to believe, but I still have the strange things I will be telling you about. I have kept them hidden from the world until right at this moment. Many of the things I will describe may seem implausible, quite outlandish, even in today's highly technical world. So listen carefully"

"I will, Granddad. I love your stories."

It was July of 1937 and your granddad was ten when my parents took me to the Kosciusko County Fair in Warsaw, Indiana for the first time. It was about twelve miles from the tiny summer cottage on Lake Tippecanoe where we spent lazy summers when I was a boy.

The fair was an eye-opening experience for me. The midway and the rides were noisy and exciting with all the colors, booths, and barkers singing out the joys and excitement of their particular offering or ride. I was particularly drawn to the "mousey" game where a mouse, released from a can in the center of a table would run to and in one of the holes spaced around the edge of the table. The holes were in the edge of the table in little squares in six colors plus white. The object was to place a bet on which color hole the mouse would select to run into. There were six colored spots at each player's position at the counter that ran entirely around the booth. To play, you placed a coin on one or more of the colors. When the mouse ran down a hole, every spot that matched the color of the hole won five coins like the one bet. If the mouse ran down a white hole, no one won except the booth owner. I stayed to watch after I bet and lost the four pennies and the nickle I had in my pocket.

I hung around fascinated for at least an hour. I noticed an unusual looking boy standing near the corner of the booth. Something about his eyes and the way he moved was unusual. He looked to be about my age and was wearing a floppy hat that hid his face. When I looked his way and directly into his eyes I saw he was watching me. He turned his head, looked down, and shuffled through the dirt as he walked down the midway. About then my mom and dad walked up.

"Charley, Haven't you seen enough of that mouse game?" my dad asked with a grin. "Come along. We're going down to the church tent for some chicken and noodles."

"Aren't you about starved?" my mom asked. "You haven't had a thing since breakfast."

"OK, mom. I guess I am hungry now that you mention food." I felt terribly hungry, especially for the scrumptious chicken and noodles those church ladies made.

As we walked to the church tent at the end of the midway, I thought I saw that same boy who had been at the mousey booth. He may have fallen in behind us as we walked and I didn't see him again until after we ate.

"Charley! Now don't you go back to that mouse booth again." My mom warned. "That's gambling and you know what your father and I think of gambling. It's an evil that can take you over and ruin your life."

I didn't see how that little game could ruin anyone's life, but I respected my parents and usually followed their directions.

"Can I go down to the barns and watch the animals after we eat?" I was going to ask for a quarter to go on some rides but thought my chances were slim. It was during the depression when going to the fair was quite a costly treat. There would be little money for frivolous adventures, and I had already spent my entire allowance.

My dad looked at mom and said, "Mother, do you think that's all right? He's big enough to know what not to do. He's become quite a responsible young man." My father knew how to get me to want to behave well and please them. I didn't always do so. Still, it was a powerful, positive force in my young life.

Mom smiled and gave me a hug and kiss. "Yes, you are growing up. Before we know it, you'll be a teen, and I hope you won't forget all we've taught you the way some boys do."

"Aw, mom"

"Now you run along, but be sure to meet us at the quilt display in one hour. That's three o'clock. Can you remember that?"

"Sure mom." I threw over my shoulder as I started for the animal pens at a run. I loved watching them, especially the pigs.

As I trotted toward the pigs, I caught sight of that boy who was obviously following me. I wondered what he was up to, maybe a fight. Lots of boys our age looked for fights to prove—who knows what. I never looked for a fight, but when I had to fight, I fought to win.

After watching the pigs for at least fifteen minutes, I needed to head for an outhouse near the end of the midway. As I left the outhouse and walked around the back of the nearest booth, I almost ran into him. He was a bit smaller than I, so I figured if he wanted a fight he'd get the worst of it.

"What do you want? I've seen you following me. You lookin' for a fight?"

As I was looking at his face, I was struck with how it was, well, different, especially his eyes. They were dark blue and too big for his face, much too big.

In an almost musical voice he responded. "Certainly not. I wanted to give you something to read."

With that he pressed a small brown leather bag into my hand. It was tied at the top with what looked like leather shoelaces. It was soft and worn but sturdy. I took a long look at him. Not only was he different, but so were his clothes. He spoke haltingly in his musical voice, almost as if he was reading from a book in class. I stood there looking at him, spellbound.

"Please read these and keep them for me. Maybe some day I'll be able to come back to get them back from you, so take care of them. It may be a long time before I come for them, probably never, but who knows?" He spoke directly, looking into my eyes as he did. His presence seemed so friendly yet distant at the same time. Maybe it was the unusual sound of his voice and his strange appearance.

We stood gazing silently at each other for several minutes until he turned and walked away. I felt intensely curious as I watched him head down the midway and disappear into the crowd. I stood there almost frozen for several minutes before I decided to look at the contents of the bag. I sat on a packing case by the nearby booth and untied the plain bow knot that held the bag closed. When I poured the contents into my hand, seven marbles rolled out into my palm.

They were unusual as they were large and quite clear without the swirls of color in my other marbles. Though they were clear, each had a pale tint of the six basic colors. The larger one was perfectly clear, no tint of any kind. Each one seemed to glisten in the sunlight. Something on their surfaces almost sparkled. It was a gem-like, glistening quality, hard to describe. As I sat looking at them, I wondered what he meant when he said, "something to read." How can you *read* a marble? Maybe I had misunderstood what he meant.

As I started to replace the marbles, I realized there was something else in the bag. Reaching in, I pulled out what looked like the frame for a pair of eyeglasses, a frame with no lenses. They were of a black metal, quite strong in spite of how thin and fragile they looked. Then it dawned on me; the lens "holders" looked as if the marbles would fit them perfectly. The right side was enough bigger than the left to hold the clear marble while the left looked about right to hold the six colored ones. It was quite obvious this was what they were for so I plopped the clear one into the right frame and the blue one into the left. Then I hooked the ear pieces over my ears and looked through the marbles.

At first, the marbles merely distorted my vision so all I saw was the midway booths, greatly distorted and fuzzy. Then, as the view started to clear up, I noticed many tiny characters appearing. They were almost like oriental characters or pictograms, but at the same time quite different. They were in a gold color. I realized I was looking at what had caused the glistening I noticed when looking at the marbles. It was on the inside surface of the side of the marble away from my eyes. Then a view of the six colored marbles arranged in a row replaced the characters. They were in the same order as a rainbow with the red on the left and the violet on the right. They seemed to hang in midair, outside in a rolling green meadow. As I watched the red marble moved and headed straight for my left eye. When this repeated several times I realized I was to insert the red marble in the left eyepiece. When I removed the frame to insert the red marble, I noticed the tower clock above the midway and tents. I had barely enough time to meet my parents, so I dropped the marbles and frame into the bag, closed and tied it, stuck it into my pocket, and headed for the quilt barn at a run. I could investigate this strange bag of marbles later when we were home.

As we drove home, I decided I would keep the experience of the strange boy and the bag of marbles secret for the time being. At least until I figured out what the strange marbles were, how they worked, and what it all meant. Of course, a ten-year-old can only go so far before he must ask someone. In spite of this I knew these strange marbles and the glasses frame were not common objects. For this reason, I would deal with them unaided until I found valid answers.

Our little cottage had a small, stark attic reachable through a trap door in the ceiling of the only closet in the place. It was far too low for me to stand without striking the roof rafters. There were a few floorboards on top of the ceiling joists, so one wouldn't step through the ceiling of the room below. My father used the place as storage of things like curtain rods and poles and a few boxes of things wrapped in newspapers. There were small blocks of wood nailed to the inside of the closet door frame to provide purchase for the feet of those seeking entry to this forbidding domain. In spite of the difficulty of staying on those blocks, I could scamper up into this place and be hidden from everyone. It was a perfect secret hideout for a small boy. Recently my father had installed a light with a switch. I no longer needed to bring a flashlight to be able to see in the windowless space.

Monday after my father had returned to our home in Ohio, my sister came to pick up my mother and other sister to go to town shopping. When I said I didn't want to go with them mom said, "Promise me you won't take the boat out while we're gone. And stay on our landing. I don't want you going over to Black's or the pit."

"Sure mom, I'll stay here or maybe go down to play with Herc and Paul. I promise." I had serious, secretive plans for while they were away.

"There's sandwich makings in the ice box for your lunch, but don't drink your sister's pop. Make some Kool-aide if you want. There are grape and cherry packets there on the shelf." My mother gave me the pleasure of doing for myself. She was a terrific mother, not the hovering, over protecting mother some of my friends had to endure.

As soon as they left, I grabbed the bag of marbles and a cushion from the couch and headed up to my hideout. I turned on the light, positioned the cushion against one of the vertical two-by-fours, and took out the bag. I dropped the clear marble into its holder and the red one into the one on the left and slipped the earpieces over my ears. My first view was a fuzzy distortion of the roof rafters.

After a few moments the view cleared. Once more I was seeing the strange gold characters on the other side, inside of the marble. When I closed one eye and then the other, different sets of characters appeared in each eye. Yet when both were open there was but one image where the gold characters hung in space. It was a bit like looking at those double photos with a stereoscope. As I watched the characters began shrinking and my entire field of vision filled with tiny gold characters seeming to hang in space.

The characters finally grew so small they merged into a scene, a beautiful scene. It was outdoors in a field or yard. There were trees and bushes and mountains in the distance. Five people walked into my field of vision from behind me or so it seemed. They were so real I felt I could reach out and touch them. It felt like I was in the scene, not viewing it. After walking a short distance away, they turned to face me. The boy who had given me the marbles was one of them. They were obviously talking to each other, but I heard nothing. They were a couple and three children, two girls and the boy. They looked—well—different. Their clothes and movements seemed odd to me. They each had those same eyes I had noticed in the boy. Other than that they could have been a family like my own.

As I watched, the woman pointed off to the left, and the scene shifted, rather like a movie or video camera panning a scene. As far as I could see were green fields, trees, bushes, and then the forest. I watched, fascinated. Then I was moving above the forest at some height, as if in an airplane flying. As the scene moved I realized I had seen no buildings or roads of any kind, nothing man made or artificial. Everything looked wild, natural, and untouched. As the scene continued shifting I saw lakes and rivers and several mountain ranges, rugged and spectacular with snow capped summits. Beyond the mountains was an ocean or immense lake stretching to the horizon Next thing I knew, I was descending to the field where I started. The family was there, waiting.

They walked off toward a nearby clump of trees, motioning whatever or whoever was making the moving picture to follow. Among the trees, we came upon an enormous rock that split open as we approached. Inside were two large metal sliding doors that opened to a dark green room. As soon as we entered, the doors shut and we stood waiting for several minutes. I noticed a diagram on the wall showing the rock and a shaft that descended from it to far below ground level. I realized it was an elevator—a different kind of elevator from those at my Dad's office or the department stores where mom often took me. It was quiet, and there was no one operating it. The diagram indicated at least a dozen levels and the display showed our elevator stopped about halfway down the shaft.

The doors then opened into a lighted tunnel. At least I assumed it was a tunnel. We were obviously far below ground level. Everyone stepped into a small, open bus or cart that sat waiting by the elevator. The scene changed. We were emerging from the tunnel into a vast lighted cavern with light green walls. This cavern, the only appropriate name for this enormous underground space, had finished walls, not the rough rock I had seen in a cavern we once visited. It was so immense it contained several buildings at least ten or twelve stories tall. I was in a state of wonder that bordered on shock. I could not imagine a place like this existed, but there it was.

At the far end of the tunnel was a flat, vertical wall extending from the floor to the heights—the ceiling of the cavern. The little bus dropped us off right by the center of this wall. As I watched the wall split in the middle and began opening. The wall was a monstrous set of doors that could be opened and closed. The space beyond the doors was at least twice as large as the space on this side. In the center of the space stood a gigantic, square, white building, almost a cube, with no windows or openings of any kind in the walls. Then I saw that it was sitting on many blocks, cubes about twice as tall as a man. There were several stair cases leading from the ground up inside the building. There were also two wide ramps, one at each edge, that lead up into the building. As I watched, a truck of some sort carried a number of large unknown objects up one of the ramps into the building. None of this was in any way familiar.

I was in a state of total wonderment, mesmerized at what I saw. I realized we were going up one of the staircases and into the building. In a moment of temporary panic I reached behind me. The feel of the cushion and the two-by-four I was leaning against was reassuring. All this was a picture I was seeing. We moved through corridors and rooms with people working using things unimaginable on desks and tables and fastened to the walls.

The view changed. I was in another building walking down a grey hallway. Finally, we entered a large room, and faced a flat, barren wall of grey color. There were at least twenty people on seats in the room. The family joined them and were gazing at the grey wall.

At this point, the scene of the fields from my first viewing replaced the room. All but the red ball hung in mid air before my eyes. As I watched, the red ball moved from behind me and into its position with the others. Then the orange ball moved toward my left eye. The message was unmistakable. I took the frames off and replaced the red ball with the orange. I placed the frame in front of my eyes as before. It took a few minutes for the scene to move through the gold characters into the room with the grey wall and the seated people.

A picture of the cavern with the strange white cube replaced the grey wall. We seemed to be looking directly at the cube. It was terribly confusing. As I watched, the stairways and ramps retracted up into the building. All the people and machines that stood around and beneath the building were gone, and I could see that the towering doors were closed. All those in the room were leaning forward in their seats as if awaiting some astounding to happen. Then the building disappeared, vanished. A cloud of dust and a few small papers were flying about as if in a violent windstorm. As I watched they slowed and then gently fluttered to the ground. At this moment the wall turned grey and the people all got up and left the room. I was dumbfounded. What did this all mean? How did those pictures appear? It was like being there but without sounds or feeling of any kind.

The scene before me changed drastically. It was as if I was looking at the moon in the sky with stars all around it. I knew it couldn't be the moon as it was blue with white whisps, and blotches of brown and green in strange irregular shapes. A large white patch covered one area. I guessed it had to be a planet like our earth. Then I saw the buildings, hundreds of them stretched out in rows, hanging there in the sky with no support of any kind. As I watched, new buildings appeared, one at a time at the end of one row. It was incredible. Fortunately I was a young boy with an open mind, a Buck Rogers fan that knew all about different worlds and such. Even so, it was quite a frightening and extremely exhilarating experience. An older person who knew such things could not possibly exist might have gone right out of their mind. All I could think was, *what's next?*

As I watched, the scene moved rapidly toward the planet and past it into darkness. Then a tiny light appeared in the center of my view and began growing in size. When it was near enough I could see it looked like a big hunk of grey rock slowly rotating. It was roughly spherical but with an irregular surface of battered and chipped rock. Then the picture expanded to include both the planet and this rocky object which became extremely small, a point to be accurate. A dotted yellow line started from the object and moved in a barely noticeable arc until it touched the planet. When it touched, a huge, bright flash appeared on the surface of the planet, an explosion of unimaginable force that almost blinded me. Then it all made sense. I remembered the buck Rogers book, "The Doom Comet" where the earth was nearly destroyed when a large comet passed nearby. The only difference was that this was real. I was seeing a diagram of what had happened or was going to happen. I wondered, *why me?* My active imagination couldn't be playing tricks on me as there were several real things: the boy, the marbles, the glasses, and the bag. These were hard to deny. As I watched the six marbles once more appeared, the orange one changing places with the yellow. This time I removed both marbles and replaced them and the glasses in the pouch. I'd seen about all the marvels I could handle for the time being. The wonders in the yellow marble would have to wait. As I was climbing down I heard Herc calling from the back yard. I stashed the bag of marbles in the back of my sock drawer and went out to the real world to see what Herc wanted.

"What's up?" I asked through the screen door.

My friend and playmate from the end of our lane, Hercules Pronger, stood on the walk, one hand on his hip, the other holding a large potato chip can. "Can you come out?"

"Sure! What's with the can as if I didn't know?"

"I thought we could go cricket huntin'. Maybe earn some money sellin' 'em to fishermen."

"I like the money idea, but we've about caught all the crickets there are around here."

"I found a great new place for crickets, thousands of em. A place we never thought of before."

"Oh! Where's that?"

"You know that little cottage beside the channel to Little Tippy? The one sits all by itself?"

"Yeah, of course."

"Well, there's an old picnic ground right behind it. Runs north along that dug channel with that old, rickety bridge. Must be at least an acre of grass, soft fine grass, covering that picnic place. Walkin' through that grass I scared up tons of fat, black crickets. I found them when I went for a walk with my folks."

"I don't know, Herc. Mom's gone and she asked me not to take our boat out or go to the pit, and that's way the other side of the pit."

"Did she say you couldn't go out in our boat?"

"No—not really. She did tell me to stay on the landing though."

"I asked my mom. She said I could take our boat and that I could take you with me. She kinda likes you, you know."

"What about Paul? Isn't he around?"

"Na! He went home to Fort Wayne with his folks. Won't be back 'til Friday."

I wanted to go so badly. I could visualize those crickets jumping through the grass. Mom hadn't said I couldn't go out with Herc in their boat. That convinced me. "OK, I'll go. Let me write mom a note in case she comes back first."

I left a short note on the kitchen counter saying I was going out in the Pronger's boat, and that she could check with Herc's mom if she wanted.

The cricket hunt was a great success. We must have caught several hundred fat, black crickets that chirped merrily from their prison in the chip can. At five cents a dozen we had a small fortune if we could sell them to local fishermen. That would be our next chore.

By the end of the week we each had more than seventy cents in our pockets and enough crickets left for the weekend of fishing with my dad. He would be so proud. I was rich with three weeks allowance I had earned myself. I thought about the marbles many times but there came no opportunity for a secret viewing so them stayed in my sock drawer. Friday I could wait no longer. The draw of the magic marbles was overpowering so I thought of a plan. I gathered my treasure from my sock drawer and asked mom if I could go up to the road early to wait for my dad.

"Whatever do you want do go so early for. He won't get here for at least an hour and a half, maybe two?"

"I thought I might look for some of those *egg* stones in the gravel on the road while I waited." *Egg* stones as we called them, were dark-brown oval stones with bright yellow centers only visible if the stones were broken open. They were not common but all the kids collected them. I had three among my prized collection of odd things.

"Will you be careful of cars and stay out of the road? Keep to the sides of the road."

"Of course, Mom. They are never in the traveled part of the road anyhow. We always find them in the looser gravel on the shoulders." The little story I told my mom would mask my intent and protect my secret.

"All right then, but stay off the road. Some of those idiots go like sixty down that dirt road. If you hear one of those lamb-brained speeders commin' you get way outta their way, ya hear?"

"Sure Mom." I said reassuringly as I left the house making sure not to let the screen door slam to please her. As I ran down the lane I knew exactly where I would go. Where the lane met the road was an old abandoned building that once housed a tiny store. I could climb into the opening that once had a door, and sit in the far corner of the floor. No one could see me there unless they came and looked in the doorway and I would surely hear them comin' first.

My fingers trembled with excitement as I slipped the yellow marble in place and looked. A new scene appeared. I was looking down at a great height on what looked like the same spot I had seen in the first marble. There were the same mountains, forests and intensely green fields I remembered. A brilliant white flash appeared the other side of the mountains and was gone in an instant. Then a red glow almost like a sunset began expanding where the flash had been. As it grew upward, bright orange, then yellow, then white objects rose above the mountains. There were hundreds, then thousands of these bright objects moving higher into the sky above the mountains. It was spectacularly beautiful but somehow ominous at the same time. Then the scene shifted to what appeared to be a much higher view. The brilliant objects were raining down on the forest and fields which burst into flames and then disappeared in a brilliant orange-yellow cloud. A huge wall of turbulent water and foam flew over the mountains and engulfed the fiery cloud turning it white with moving spots of glowing red and orange.

Once more the scene shifted to a view of the entire planet. It was eerily similar to some of the drawings I remembered from that book, "The Doom Comet." The entire planet was covered with a grey haze punctuated here and there with small bright orange spots. An arc of jagged mountain tops rose above the haze halfway across the visible face. That was all that could be seen. Then the scene shifted far from the planet showing row upon row of those white buildings hanging in space. There must have been millions of them stretching as far as I could see in every direction away from the planet. Comprehension of this as a reality was difficult, maddeningly so. Of course I was wondering *what's next?* as the yellow ball exchanged places with the green one.

I placed the green marble in the eyepiece and looked. When the scene became clear I saw another blue, green, brown and white planet quite different from the first. From its resemblance to the globe that sat on the table in our living room at home I recognized our Earth. There were about twenty of those white cubes descending toward the dark, shadow side toward what I knew was the Pacific ocean. Then I saw one of the cubes up close as it settled into and beneath the dark water. What was happening was incredible yet unmistakable.

Suddenly the scene change to what I realized was the inside of the cube. There was a long open area with eight cylindrical objects about twice the size of a Pullman railroad car or city streetcar that filled the space except for an area on one end where an object that looked like it fit into the end of the cylinder sat on the floor. It was round, a lens shaped disk, at least a dozen feet high and about twice as wide as it was high. A large metal arm was attached to the top of the object and disappeared into the opening in the end of the cylinder. Quite obviously the disk fit into the end of the cylinder.

A number of people walked about on the floor near the object and through doorways in a wall that ran the length of the room flush with the openings to the cylinders beyond the wall. My imagination ran wild. Visions of Buck Rogers and Flash Gordon and their space ships filled my mind, but this was incredibly real, almost like I could reach out and touch them. The scene switched to the floor beside the disk where the boy and his family stood. As I watched, an opening appeared in the side of the cylinder and the entire family walked into the lighted interior. The opening then closed leaving no indication of where it had been. Then the arm began folding into the cylinder, lifting and turning the disk until it closed into the end of the cylinder.

There was a flurry of activity as doors closed over the disk and the end of the cylinder. People moved about watching brightly colored pictures in the wall, pictures that moved. After a short time the doors opened. The disk and cylinder were gone leaving only an empty metal tube that the cylinder had fit tightly into. I noticed a small stream of water pouring out of the tube and onto the floor. By this time the people had all moved down by the second cylinder and again the scene changed.

I was seeing the man and woman seated facing away from me toward what looked like a movie screen on a flat wall directly in front of them. There was enough light so I could see them and recognize them as the boy's parents. Stars appeared on the screen as rivulets of water ran or were blown off the outside of the picture. As I watched I realized the screen was getting lighter. Then a picture of North America appeared, quite obviously taken from far above. It was amazing and totally incomprehensible, but there it was, right in front of my eyes. The view was changing like it was being zoomed. First I could see most of the US, then the north central US. Lake Michigan got bigger and then disappeared as I was looking at Northern Indiana zooming up toward me. Then it stopped. I recognized what I was looking at. It was our lake, Lake Tippecanoe, and I was close enough to pick out things I knew, the Tippecanoe Country Club, the dance hall, Patona Bay, the submerged islands in the east end, all were clearly recognizable. It was winter as most of the surrounding lands were white with snow, my guess was early December as the lake was not frozen. It was amazing.

I was looking at our lake from close above and we were coming down. It was quite dark but there were lights here and there around the lake. All at once there were bubbles around the screen and I knew what had happened. We were submerging in the lake in its deepest part off Silver Point across from the Country Club. To a ten year old SciFi fan it was obvious what was happening. These people were hiding their small craft in the 100+ foot depths of the lake. I assumed they would use the craft to go and come, at night when no one would see them. I was beginning to understand what was being shown, but why to me? When the green ball exchanged places with the blue in the view, I did the same with the real marbles.

The scene that appeared next was absolutely impossible. First I saw the attic of our house, then Herc in our back yard asking me to go hunts for crickets. It was as if the camera or whatever device they used to take those photos was in my eyes. Everything I saw I had witnessed before: my friend Herc, hunting crickets, talking to mom, walking to the road and entering the little old store building. Everything in little snippits, from when I climbed into the attic right up to the present, everything.

I heard someone outside so I removed the marbles and placed them in the bag. I had no sooner done so when a knock on the door brought my eyes up. There in the doorway stood the boy with the big eyes.

"Come with me," he said and motioned. "We must hurry."

When I hesitated after jumping down from the doorway he grabbed my hand, pulled, and again urged me to hurry.

"Why?" I asked, almost frightened. Where are we going?" He led me to a delivery van parked on the side of the road near the old store. The door opened and the man I assumed was his father jumped out.

"Charley, you must come with us. It is important."

"But, my parents always told me not to go with strangers."

The boy pulled gently on my hand. "If we wished you any harm we could simply take you. You saw our capabilities. According to our laws you must come on your own, willingly. If you won't, we are forbidden to force you."

Something about his manner, his speech, seemed so honest, so sincere, I agreed and jumped into the truck. "My dad will be here in a while and I'll get in trouble if I'm not there." I said, pleading.

The man looked at me softly as he drove off. "It will be more than an hour until your dad arrives, by that time we will have you back here. I promise."

"How do you know that?" I asked.

"Right this moment he is buying gasoline in a town called Ridgeville Corners in Ohio. It will take him more than an hour and a half to get here."

I knew exactly where he was. He was at Ferd Bernfeld's station where we always stopped for gas. "How can you know that? That's impossible!" I blurted out.

"We have and use many things beyond your understanding, those glass books for example. Surely what you read or as you would say, *saw* in them, can convince you we could know where your father is."

My SciFi mind adapted and understood. "I guess I understand—sorta. But where are we going now?"

"We're going to our home as you would call it, not far from here. We'll be there in about ten minutes."

We headed up the dirt road past the Henwoods' house toward Syracuse. Before we started to cross through the deep ravine, we turned off on the right on a narrow road I had never noticed before. It was a narrow opening in a fence that led to what looked like a small, grassy yard. We stopped after we entered the yard and the boy got out and reconnected the wire fence behind us. Then we drove across the grassy spot and down a steep, craggy path into the ravine stopping finally at the base of a vertical cliff about thirty feet high. They checked a little grey box the man took from his pocket and the cliff opened up like a silent doorway. We stepped inside and the door closed.

"Come this way." they urged as they started down a long flight of stairs in a dimly lighted tunnel. We must have gone down more than a hundred steps before coming to another door. Once inside it was like a modern house all lighted up. Normal furniture, chairs, lamps, tables, it could have been the inside of any new house in town except there were no windows. In place of window glass were grey panels like the big one I had seen they had on their planet. I assumed they could show pictures.

"Please have a seat." the boy said motioning to a chair. "My family will be out in a minute to talk with you and explain who we are, what we are doing, and how important it is that you help us."

"I don't know what to say or do. This is all amazing to me and a bit frightening."

The rest of them came in and sat down near me. Then the woman spoke. Her smile and warm tone of voice were reassuring. "Please don't be afraid, Charley, I promise nothing bad or frightening is going to happen while you are here with us. You can call our son who gave you the books, John, as his name in our language would be hard for you to say. Please call me Mary, my husband Don, and our daughter's Lou and Jess."

I was still overcome with amazement. "How did all this get built? The tunnel, this place underground. Why are you hiding? Where did you all come from? You are all so—different."

Mary continued. "We will try to answer all of your questions, but first let me tell you about us and why we are here. By reading those books you have seen and know what happened to our home planet a few of your years ago."

I interrupted. "Yes but how did that work. I mean—you called those marbles books and said I should read them. They sure didn't look like any books I've ever seen."

John spoke up. "When you read your printed books don't you have mental pictures of what is going on? Don't you seem to see what you are reading?"

I thought for a moment. "I see what you mean, but that's a whole lot different from the way I saw those things in your books as you call them. To me they look like marbles, not books."

"Don't you remember seeing those gold colored characters before the pictures appeared?" John asked. "Those are our characters or maybe letters as you call them in your writing. When you see them, they bring those pictures or movies as you saw them up directly in your brain. It's complicated, similar to how your ears convert air compression variations to sounds you hear in your brain. Have you learned anything about that yet in school?"

"I don't think we've studied hearing much in school, but I read all about it in the book of knowledge. I sorta know what you mean. What you describe is like something straight out of Buck Rogers or Flash Gordon."

John laughed. "I have seen some of your "Big Little, Buck Rogers" books. They're quite far from reality, but I can see the similarity."

"But what I'd like to know is what I asked before. What's all this about and how did it happen?"

Don spoke up. "We don't have a lot of time so I'll explain briefly. After our planet was destroyed we needed a new home, a planet as much like our old home as possible. Yours was an excellent prospect until we discovered advanced life forms had developed. We're here for a short stay of up to about two of your years. We need time to replenish our depleted energy stores. As soon as that is completed we will be on our way to try to find a new home."

"Couldn't you stay here?" I asked.

Lou smiled and looked at her father. "My dad said there were already too many advanced life forms on this planet, that there wasn't room for us. Besides, do you remember how many of our ships as you call them you saw leaving our planet? Dad told me there were more than a thousand of our people in every one of those ships. In every one of those ships are several hundred small ships like the one our family used to come to your area. Our ship is hidden at the bottom of the lake where you live."

Don smiled at Lou as he told me, "When we left home we split up into thousands of small groups of twenty to fifty ships. There are twenty-three of our ships that came to your planet. We are searching this section of our galaxy to find a place suitable for all of us to live.

A place where we wouldn't have to displace creatures even those of limited intelligence. That's one of our most powerful laws. Also, we'd like all of our people to be able to live on one planet once more. We'd like to make it as much like the home we lost as possible. Should we find such a place we will call everyone so we can all head there and build a new home."

All kinds of questions flew through my mind. "I think I know what you mean—sorta. But energy? What kind of energy? Where can you get it and what do you do with it."

"Actually, it's electric energy. We store it in what you might call batteries, powerful batteries that can store tremendous amounts of power. Your best scientists would have a difficult time understanding our batteries and how they work. We go deep down in your Earth to where there is tremendous heat and convert that heat directly into electricity which we store in our batteries. All of that goes on in our ships as they sit on the ocean floor. That's where the internal heat of your planet is closest to the surface."

Something didn't seem to make sense to me. "That's interesting, but what does all this have to do with me? Why am I here? I'm just a kid. There's nothing I can do for you, nothing!"

Don and his wife looked at each other. Then Mary said, "There is something important we think only a boy like you can do for us, something simple and quite common that it would be dangerous for us to do ourselves. It involves flying a kite. You can fly a kite, can't you?"

"I'm one of the best kite builders and flyers around. I've even won a few kite flying contests, but why couldn't you do that yourselves? Couldn't John do that for you?"

Mary bent down and looked me straight in the eyes. "Look at me, Charley. What do you see?"

"A nice lady with a nice family."

"My face. Doesn't it look different to you? When you first saw John didn't you think he looked different from other boys, a lot different?"

"Well, yes. You all have those huge eyes, and you do look, well different than most people."

"That's why we don't go out among you except in extreme emergencies. When we do, we try to hide our faces under a floppy hat, especially our eyes. You noticed how different John looked at the fair, didn't you? We took a big chance that day with all those people around. What do you think would have happened if John had approached an adult like he did you?"

"I think I understand. An adult would have been surprised and maybe made a big fuss. I was curious."

"An adult would have called authorities, the police maybe. How would John have explained where his parents lived, or anything about himself. It could have become an extremely dangerous situation for us, a tragic exposure. Who knows where that could have led?"

"Why did you do it? And what about this kite thing?"

Don picked up a small brown box with buttons and a grey square on the front. "You see this, Charley? This is a communicator. It's a kind of 2-way radio with which we can speak to our other people all over your world. We have three of these. They are necessary so we can stay in contact with our friends. We all help each other and receive instructions and news on these boxes. These are our only link to any of our people."

"Those are neat looking things." I said. "How do they work?"

"The important thing is that now, none of them work. Two weeks ago there was a bad electrical storm, lots of lighting. We were transporting them from the ship to our home here when some downed power lines fell on our truck, right near your cottage."

"Yeah. I remember that storm. The thunder woke me up and I usually sleep through storms."

"Anyway, one of those broken power lines whipped around, broke out the rear window of the truck, and touched the stack of three communicators. That sent at least 7,200 volts of electricity through them and burned out the antenna system on all three units. We took one of them apart and tried to fix it, but the antenna signal amplifier was burned out and we have no replacement parts. We stretched out a wire and hooked it up to the fence to use as an antenna, but the signal was still too weak, We managed to get a weak incoming signal to our receiver, but got no response to any outgoing signal. It was too weak. I was trying to figure out how to get an antenna high enough to work when John suggested the kite. That's when we devised the plan to have John find and contact a young boy, and here you are."

"Wow. Flying an antenna using a kite. That's smart. Where are we going to do this and when?"

"We will contact you to let you know, but now I had better take you back to where you will be meeting your dad. He should be arriving in about twenty minutes. I trust you can keep this to yourself?"

"I sure will, and I understand why, really. It's like something out of those SciFi stories I read. You can count on me. Uh, how will you let me know?"

"You know about that big stand of Sumac in the field across the road?"

"Yeah. My friends and I go there sometimes to dig for fish worms in the little gully that runs through it."

"You said you flew kites. Do you have one here and do you ever fly it?"

"Yeah. I have a neat box kite I fly up at the top of the hill beyond that patch of sumac."

"I don't think that hill will get us high enough, but we can walk from there up to the much higher hill this road crosses. Will that be OK? Will your mom let you do that?"

"I know that will work."

"How about Monday at about ten"

"I'm sure that will be OK with Mom. I'll have to duck my sis though. She'll want to go with me if she sees my kite."

"Can you do that ?"

"Sure. I've done it before."

"Now we'd better get you back."

My dad arrived about ten minutes after I was returned. After our greeting he asked, "Why are you so excited. You seem like you are ready to jump out of your skin. Some big surprise you're dying to tell me, but promised your mom you wouldn't? Some fresh cherry pies, maybe?"

"Nope. I'm lookin' forward to goin' fishin'. I hear the bluegills are biting in the flats."

"Got any crickets?"

"A whole new bunch Herc and I caught. They're in the big can all ready to go."

"You get the boat ready while I kiss your mom, unpack, and get changed. Maybe she'll go with us."

It took almost an hour before we were all three out on the flats pullin' in bluegills. They were really bitin'. By the time we headed in before dark we had at least twenty of those hand sized beauties. They would be good eating over the weekend. All I could think about during the trip home was that space craft sitting on the bottom at the other end of the lake. We dumped our catch in the live box and headed in for dinner. Mom had fixed a plate of delicious fried bluegills I had caught a few days before. We ate them, corn on the cob and sliced tomatoes. Topped it all of with a piece of Mom's prize cherry pie with whipped cream on top.

After dinner Mom surprised me a bit. "Your son's had something bothering him since last week when we went to the fair. He says not, but I know my boy."

"Mom. I told you. Everything's fine. I caught us plenty of crickets and sold a bunch. I made almost three weeks' allowance."

Dad looked at me with that look. "Whatever is bothering you, son, spit it out. Get it off your chest. Your mother is never wrong about you."

"I don't have any idea what all the fuss is about. I'm fine."

"Did you have a falling out with Herc or Paul? Maybe that little girl staying at the Hackerd's cottage is the problem. You are a little sweet on her, aren't you? What's her name, Nancy. Isn't it?"

"Aw for gosh sakes. I'm gettin' on fine with all of them. Paul's at home in Fort Wayne, Herc's gonna take me for a ride in their boat and Nancy's comin' over to swim with me tomorrow afternoon. Everything's great."

Mom wouldn't quit. "Charley, I know something's bothering you. I can tell. If you won't tell us what it is we can't help."

That was the last straw. I didn't say a word, but left the table and ran out to the end of the pier where I could have some peace.

"Young man you get right back here to the table. You were not excused." my Dad called after me.

I knew I would be in big trouble if I stayed out there so I walked back and sat down at the table trying to think up something to say.

"All right young man, out with it." Dad demanded.

I was determined to convince them. "There is absolutely nothing wrong, nothing's bothering me except you saying over and over that something's bothering me. Do you want me to lie and tell you a made up story?"

They looked at each other then Mom came over and put her arms around me. "Charley, I'm sorry." She turned to my dad and said, "We've both been pushin' him quite hard. I must have been mistaken. Let's drop it."

"Thanks, Mom. I promise if anything's ever bothering me, I'll tell you about it." My fingers were crossed in my pocket for that one, but it worked.

It took forever for Monday morning to come around. I grabbed my kite and string and told Mom what I was going to do.

"All by yourself?" she asked.

"Yep. I've done that many times before."

"Ok, but tell me where you'll be in case I need you."

"I'll be where I always am, on the hill on Tom's farm. I may walk up to the top of the hill by Henwood's. It's higher there and there's usually more wind."

"Can you hear me blow my whistle from way up there?"

"Mom, I can hear your whistle all the way down at the Kalorama grocery. That's farther away and with lots of woods in between."

"OK, I'll whistle for you when lunch is ready. Come right away when I whistle. Don't you dawdle."

John was waiting in the sumac when I got there. We walked up the hill, across the ridge through Tom's farm and up the road to the top of the hill. John's dad was waiting partially hidden in some tall weeds by the road. We tied some fine reddish string to the bridle of my kite. The string was on the spool of an ordinary fishing reel.

"Using a reel is sure a great idea. Are you sure that's strong enough? It looks flimsy to me." I asked.

"It's plenty strong," John said. "It has a fine wire running through it that will serve as our antenna. The string that runs over to my dad's communicator will let us fly the kite out there in the open while he tries to make contact. Could we have the parts sent to your mailbox? They are small, about the size of one of your quarters. That's the easiest way we could think of to get them here quickly and easily."

John held the kite up as I ran and launched it. In the brisk wind it climbed high into the sky. As I let out more string I asked John, "How will I explain that to my Mom? I don't often get packages sent to me."

"Is there anything you could order and have sent to you?"

"I'll think of something." I said as I continued to let out string and the kite rose higher and higher.

John's dad called out, "I have a confirming signal. They have us.?"

I thought of something I had ordered before. "I can order a model airplane kit. It would get here in a week or ten days. How in the world would you be able to get those parts in that shipment?"

"Tell us what kind of model airplane kit you want. We'll do the rest." John's father told me.

"An Aeronca C-2 flying model. I've always wanted one of those. I have enough saved to buy one."

"We'll gladly pay for it and the shipping. Be careful when you open it because those three antenna parts will be inside the box."

"How will I get them to you?"

"John can meet you among the sumac trees like we did today. Once we have those parts we won't see you again."

I felt sad. My adventure with people from another world was about over, and I couldn't tell a soul about it. I left them on top of the hill and trudged sadly home.

Mom took one look at my face. "Son, I don't know what's wrong, but you look as if you lost your best friend."

I thought fast. "I spent two hours with a new friend I met while flying my kite. We were instantly great friends. Then I found out he was from a long way off and was leaving right then to go home. He's from clear across the ocean and won't ever be back. He's a neat kid and yes, I am sad. We had such a short time together."

"I'm so sorry. You say he has already left?"

"Yes. His mom and dad are nice, and his sisters too. It's so sad you couldn't have met them. They are nice people."

"That is too bad. I would have liked to have visited with them."

"You know what, Mom? That Aeronca flying model I've always wanted? I told them about it and they promised to send me one as a gift. Can you believe it?"

"That's wonderful. They must be nice. And only after a couple of hours playing together too. They must have realized what a fine young man you are."

"Yep. That's almost exactly what they said."

"It only goes to show you what nice people there are in our world."

I almost choked on that saying to myself, *and other worlds as well.*

Wednesday a week my model airplane arrived. I opened it, removed the small package with the parts and headed for the sumac patch.

"Where are you going, son?"

"Oh, I thought I might go up to the hill and fly my kite once more. You never know whom you might meet."

"You miss that new friend don't you. He must have been special."

"Yep." I said turning to hide the tears in my eyes. "See you later Mom."

I walked into the sumac patch and found the whole family waiting for me. They told me some of their group had discovered a planet almost perfect for them and less than a hundred light years away from earth. They would be leaving in their space ship in a few days. There were hugs and teary goodbys all around until they reluctantly headed up the hill. Before they left, John handed me another pouch like the first one with those marbles, telling me there was more about them I could experience in these new *books*. I trudged up the hill behind them, let my box kite fly in the breeze at the top while watching them disappear over the big hill. I was about cried out when I brought down my kite and headed for home, lunch, and back to reality.

"Wow, Granddad, that's some story. How come you are telling it to me, now?"

"Well, George, you remind me so much of myself when I was your age, especially your imagination. You can reel off yarn after yarn for anyone who will listen. That's why we enjoy swapping yarns, you and I. I'm going to leave this old planet before too long and I had to tell someone my story."

"Yeah, but those yarns are made up. You told me this one is true."

"True as the sun shines and the sky is blue, at least it's blue some of the time. Young man, I can prove all of this to you with what I have in these two pouches."

"Are those the pouches John gave you?"

"The very same. And now I'm passing them to you to do with what you want."

"What's in the books in that second bag? You never told me."

"Think of them as books you've not read. You'll have to see for yourself, won't you?"

"Not a hint?"

"Some wonders are best left for a personal discovery. You'll find out soon enough."

"Wow, Granddad. . . . Can I show them to my friends?"

"That's up to you, but I would think it through before sharing with those who might be disturbed or wouldn't understand what they mean. You watch them first and then decide what to do with them. I will trust your judgement, like I did my own those many years I carried them. Until right now I have never let another soul know about them, not one. Also, I never tired of watching them. I seemed to learn something new each time, something I must have missed before. They have been an unbelievable aid and comfort to me many times during my long life."

"Don't you wonder what happened to John and his family? Where this new world of theirs is and how they got along when they got there?"

"Many times. That's just the way life is. Sometimes you never learn the outcome of happenings you are part of. It makes one a bit humble. Now, go find a special secret place where you can read those *books*, where no one will ever find you. Young boys always have such a place. I did. I won't ask if you have one and don't you tell me."

"Thanks Granddad. You're the greatest"

"Maybe sixty or seventy years from now you'll be telling that story to your own grandson and passing on those *books*. Think of me when you do."

"I will, Granddad. I will."

❖   ❖   ❖

If you can't live your life the way you'd love to live it, then love your life the best way you can live it

—*Howard Johnson, 1973*

The poet lauds its achievement as follows. Countless are the virtues of a good company. The dullness of intellect one accrues due to bad company is undone in the first place and it is no mean achievement by itself. A habitual liar finds himself speaking the truth much to his own amazement. The total stature of a man reaches new heights as it adds greatness to it. It lessens the negativity of the personality as the amount of sin is drastically reduced. This leads to the clarity and happiness of the mind. The fame reaches far and wide. So there is an all-round improvement of one's personality. The question at the end is in fact, a positive assertion that every conceivable good thing can be achieved by a good company. It remembers me the saying "Tell me who your friends are and I will tell you what kind of person you are."

—*from the Sanscrit*

Decide right, or decide wrong, but decide! It is sometimes better to make a wrong decision than to avoid making any decision. You must always decide among two or more options. Anyone can provide good reasons for selecting a particular course of action under consideration. If you wait for 100-percent proof, you'll wait forever; so make up your mind, make the decision, and carry it through. Errors can usually be corrected, but the indecisive person stays in a constant storm of confusion, seldom achieving anything. The only thing that truly needs weighing is the cost of the course and its probable rewards. But remember, not to decide is a decision which may have dire consequences.

—Howard Johnson, 1985

A smile is a light in the window of a face that signifies the heart is at home.

—Howard Johnson, 1968

When the conduct of men is designed to be influenced, persuasion, kind unassuming persuasion, should be adopted.

It is an old and true maxim that a drop of honey catches more flies than a gallon of gall. So with men. If you would win a man to your cause, first convince him that you are his sincere friend.

Therein is the drop of honey that catches his heart, which, say what we will, when once gained you will find little trouble in convincing his judgement of the justice of your cause if indeed that cause really be a just one.

On the contrary, assume to dictate his judgement, or to command his action, and he will retreat within himself, close all avenues to his head and heart, and though your cause be naked truth itself, and though you throw it with Herculean force and precision, you will be no more able to pierce him than to penetrate the hard shell of a tortoise with a rye straw.

Such is man, and so must he be understood by those who would lead him, even to his own best interests.

—Abraham Lincoln

It seems the American Voter has come to admire failure, corruption, greed, and gross dishonesty in their politicians as long as they spew lies about and hate for the successful.

—Howard Johnson, 1993

Section III

Memoirs

True tales from the life experiences of

Howard Johnson

Family Outing, 1934

Excitement ran high as our family climbed into our large Nash touring car for the trip planned for such a warm spring day in 1934. We were all packed up and soon headed for the zoo in Cincinnati. My father, mother, uncle Chase, aunt Edna, cousin Rita, sisters Lois and Bobbie, and I all sat snugly in the huge sedan. As the smallest person, my place was sitting on a little stool on the floor in the backseat between baskets of food and other meal preparations for the picnic we planned to have at a park near the zoo.

During most of the trip, I stood and leaned against the back of the front seat to watch the view out the front window. I was fascinated by the constantly changing scene. When a gentle spring shower began to wet the road, I watched through the swishing tick-tick of the wipers. The tires of passing cars sprayed spirals of water into the air where the droplets lingered and formed a thin, gray mist above the road. When I asked if it wouldn't be too rainy for a picnic, my mother assured me this was merely a passing shower, and it would be sunny and dry by the time we reached Cincinnati. My mother was always an optimist—a soft-spoken, pleasant optimist whom I adored. That was the only conversation I remember from the entire trip.

I was watching the road ahead as we crested a gentle hill on the wet two-lane highway. A car came over the top of the hill well over the center on our side of the road. My father steered the Nash to the right to avoid a collision. The right wheels dropped off the concrete onto the muddy shoulder of the road. When he steered the car back onto the pavement, the rear wheel caught on the edge of the concrete and turned the car to the left. We were not going fast, so my father turned to the left to try to head in the other direction. This was not the best thing to do. The Nash skidded down the slippery road in a slow revolution to the left and headed for the other side of the roadway. By the time it reached the edge of the concrete, it had turned facing the opposite direction. When the two right wheels dropped off the concrete, they caught in the wet soil and the car rolled over on its side and came to a halt.

Of course, everyone fell to the right side; those in the back landed on top of me. The next thing I remember clearly was my father standing in the front seat, lifting us out the front door dutifully held open by my uncle Chase. Thankfully, no one was hurt in the accident. As I watched from beside the top of the car, I could see my father's head protruding through the

315

open door space. This picture is emblazoned on my brain as one by one the others were helped out of the car. After everyone was out, the raised door suddenly slipped from my uncle's grasp and slammed shut, breaking the window over my father's head. This was the only injury from the accident. I can still envision my father: blood streaming down his wet face as he climbed out of the car. A quick examination by my mother and aunt Edna found a few superficial cuts on top of my father's bald head. Thankfully, the injury was minor. Soon they were joking about the deviled eggs strewn all over the inside of the car, and whether my uncle Chase had deliberately dropped the door on his brother's head.

I remember nothing of the rest of the day. Those subsequent anticlimactic happenings have faded from my memory. I only have one other clear memory of the incident. Some time later, possibly the same day or on another day, I went with my father to where the car had been taken so he could examine it to determine what needed to be repaired. When we reached the car, there were several other wrecks parked near it. In one car, there was bright red blood all over the windows. I was impressed. The sight of those gory windows is firmly stored in my brain and will not be forgotten. I am not certain, but I believe the man told my father people had been killed in the wreck. This was a strong, sobering impression on a six-year-old's mind which remains vivid some seventy years later. I've never been to the Cincinnati zoo.

—Experienced by Howard Johnson, 1934, written in 1979

How hard it is to draw the line between a necessary act of justice, and mere malice; between savagery, and proper punishment intended solely to prevent recurrence of something wrong. No matter what the provocation I have found it wiser to abstain from vengeance, but to beware of those upon whom I might have had it. Not always, but not seldom they find it harder to forgive the magnanimity, than it is to forgive their enmity. And it is natural enough: nature is not trusted to direct our motives, only to beglamour them with false names.

—Talbot Mundy

All that glitters is not! All's well that ends! A rolling stone gathers! When it rains, it rains!
—Howard Johnson, 1965

Smile and the world smiles with you. Cry and your face is a mess.

—Howard Johnson, 1981

The Unusual Experience of a Seven-year-old

This is my story of what happened to me nearly seventy-six years ago. It is written as told to my thirteen-year-old grandson, Joseph, last year. This is the first time it has been written down as a true happening. I wrote it as a fictional story more than thirty years ago. There are enough miles on this old soul I no longer have concerns about being believed or appearing foolish. I told Joseph this story as accurately as a vivid memory permits. This story has remained etched in exact detail in my memory for all those years.

"I was heading east on Eleventh Street, a few blocks from home, walking there for lunch from PS 54 in Indianapolis. It was a cold, crisp, blue-skied January day in 1935. As I passed the Buhers' house, about halfway home, I happened to look up and notice a shiny object through the naked branches of a wintering tree. At first I though it to be a new kind of balloon caught in the branches, but as I took a few more steps, I realized it was above the tree and almost overhead, far from the winter sun hanging low in the southern sky. I was fascinated, for this was an exciting new wonder for an inquisitive second-grader still in the intoxicating time of life when new things were constantly being discovered. I leaned against the concrete wall separating the Buhers' yard from the alley, right where Mrs. Buher would place food for wandering beggars during the Great Depression.

"The object was round and appeared to be the same size as the full moon in the sky. It was incredibly shiny, almost like a mirror, yet I saw no reflection. Transfixed, I watched it for a long time, for a small boy, probably a minute or two. At this point, it began to move toward the west. It made no sound as it accelerated rapidly and disappeared over the western horizon a few moments after it started moving. As soon as it was gone, I ran home excitedly to tell my mother and find out from her what I had seen. This was at a time when people ran outside to look when an airplane flew over, and my favorite was the new Douglas DC-3. This object was most certainly not like any airplane I had ever seen before.

"When I arrived home for lunch, my mother was angry. 'Where have you been? I've been worried sick!' I was astonished and didn't even get a chance to ask her about the marvelous shiny object when she said, 'You'll go without lunch young man. Now hurry back, or you'll be late for school.' I couldn't understand how I could have been so late. I had a full hour at least for lunch, and school was five minutes away. I only stopped to watch the strange object for a few minutes. What had happened to the missing time?

"I remember nothing of the rest of the day until I came home and my mother again questioned me about why I was so late. When I asked her about the object I had seen, she thought it was another of the *stories* I used to invent to liven up my life and amaze others. I had discovered the price you pay and the pain of what happens when you become a *story* teller and people learn not to believe you. Every single person who heard my story laughed at me and ridiculed my tale except one, my grandfather. A storyteller himself, he listened attentively to my tale and wondered with me what the object was and what happened to the missing hour.

"I was so humiliated at every attempt made to find out about the object I gave up. Since then I told no one else about my experience. My grandfather and I discussed it a number of times over a number of years, often when we were fishing together on the lake. I was probably fifteen the last time we talked about it. It was our little secret.

"When I was in junior high school, I began to experience a strange phenomenon which has continued unabated to this day. In Mr. Armstrong's science class, I discovered I understood a great deal about things I had not read about or had not been explained to me by anyone. For me, this started a fascination for things of science that would last my entire life. I also found I knew the answers to many questions I should not have known. My classmate and buddy, Fred Hunziker, who sat next to me, was utterly amazed at my knowledge. Even Mr. Armstrong was flabbergasted to the point where he quit letting me answer questions in class. Another classmate, a bright girl, said she thought I knew more about science than our teacher. They began calling me *the brain* and not always in a complimentary fashion.

"My sis, your great aunt Bobbie, brought home her chemistry book and gave it to me. In my mind, I can still see the diagrams in that book of atoms with electrons in circular orbits about a solid, compact nucleus. I knew those diagrams were wrong, but of course, my sister thought I was nuts when I asked her about it. I was crushed, but what could I say? She was six years older and ahead of me in school. A high school senior, she was far wiser than I about everything. Her book piqued my interest in chemistry, which then led to my selection of chemistry for my college studies. It was many years later when I read a nearly exact description of the indefinite *cloud* structure of electrons about a tighter cloud of protons and neutrons, the nucleus, I had tried to explain to my sister. It seemed this was the latest concept of atomic structure, developed many years after I tried to explain the same concept to my sister as a boy of twelve.

"There are many other concepts of our physical world I *know* without any idea from where the knowledge came. I keep searching and reading to gain confirmation of many of these things. For example, I *know*, or at least can conceptualize, an understanding of our universe yet to be discovered or explained by anyone. I envision a roughly spherical shape for the universe. At the surface of this shape, all mass would be to one side of any point on the surface. The center of mass of this universe would be roughly at the center. Light, warped by this center of mass, does not escape from the universe. The true speed of light is a factor of its distance from this center of mass. Our measurement of light speed is a function of our own distance from this center of mass. Light passing near or through this center of mass is moving much faster than when it passes us. When light reaches the limit of the universe, it slows and *falls* back in the same way one celestial body orbits another, controlled by the force of gravity. The gravitationally limiting surface of the universe acts, in effect, as a kind of mirror, returning light and holding it within the gravitational grasp of the universe. For more detailed information, see the essay, *Relativity and a Completely New Concept of Our Universe* on page 37 of this book.

"The first time I heard a *flying saucer* or *UFO* story, I thought immediately of my childhood experience. The experience was so clearly described by so many of these tales. The ridicule heaped on those who saw a UFO caused me to rethink my experience and continue not to talk about it to anyone. The first time I heard an *abduction* story, I thought about that missing hour so long ago and of the things I *know* that I have no reason to know.

"I have made no conclusions, nor have I made any claims other than those I have described. The mystery to me is now greater than ever, and I surely will not have an answer in my lifetime. I search and ask in every way I know how, yet the mystery only continues to deepen. If there are others with similar experiences I would like to meet them, yet I hesitate even admitting what I experienced. I am still a bit apprehensive about any ridicule that might come and destroy my confidence in my own knowledge. I tell you this now because you know and understand me. Besides, I am nearing the end of my life and have so much less to fear than when I was younger. You can decide for yourself if the story is for real or the ravings of a crackpot. A discovery confirming the view of the universe I described would most certainly change the acceptance of my story now, wouldn't it?"

"Would it ever!" Joseph replied, then asked, "What do you think it was?" Joseph was utterly amazed, barely grasping the implications of the story, but never doubting a single word I spoke.

"I gave up speculating on that years ago," I replied. "I only know what I told you. No more than that but no less either. Many years later, when the first UFO stories began appearing, I thought I might find an answer, but soon realized all I would do would be to make a fool of myself if I came forward with my story. It's obvious to me, that whatever or whoever *they* are, *they* have a purpose to their activities. I've wondered for years what their purpose might be if indeed they have a purpose. I never came up with anything logical. I certainly have not been able to determine if it bodes good or bad for humanity. It's a real mystery.

Why would they imbue a small child with advanced knowledge? I'm convinced ~~believe~~ ~~maybe~~ that is what happened to me so long ago. There must have been others who received similar treatment. Though I've searched my whole life, I've never found another person who shared my experience. I've met and spoken with people who reported sightings and abductions, but not one of their experiences were anything like mine. In fact, I doubted the truth of their stories like everyone else."

"The truth is like the proverbial needle in the haystack, isn't it?" Joseph commented.

"That's almost an understatement. Over the years, I've had many incidents where the announcement of a discovery was something I already *knew*. Did I actually *know* it, or was it a trick of the mind, a déjà vu experience? I've pondered that question many times. Once, while talking with a group of engineers about a particular metallurgical problem, I posed a solution to them. It was during a meeting of the Cleveland Engineering Society. It was a solution I didn't have to think about. I *knew* it. Several months later, the specific problem was solved by the method I proposed. One of the engineers from the group contacted me and asked how I had come up with the exact solution. He knew I was not a metallurgist and wondered how I knew the particular answer to a problem no one else could solve. I was at a complete loss to explain it. Had my lack of knowledge in the field let me think outside the limits imposed by an expert understanding? Was my solution one of those serendipitous *aha* experiences we all have on occasion, or had I *known* the answer? I don't know where the answer came from and certainly would not claim to understand.

"There are only two concepts I feel certain were placed in my mind by an extraordinary process. The understanding of the true nature of the particles in the atom and the general makeup of the universe with the gravitational effect on light and other electromagnetic waves or particles. The first was only postulated by particle physicists many years after I knew and described it. The second seems only to be a theory in my mind."

"Why don't you publish your theory? Then maybe someone would listen."

"Joseph, your granddad is not a lettered member of the scientific community. With only a BS in chemical engineering after my name, I have no standing in the rarified atmosphere of the elite scientists whose theories gain acceptance. That's a simple reality of life. It does not bother me in the least. Though I have studied and read many books, papers, and reports in numerous fields of science, and can converse with experts in the field, I still have no standing. Without those letters and what they represent, I possess no confirmation of my understanding, or of the formal knowledge I possess. Still, I benefit greatly because my lack of formal credentials places me in a uniquely free position to say what I want without fear of grant committees, peer reviews, or other intimidations. Because of this, I have no fear of making honest comments."

Joseph and I discuss all of this frequently. He is interested in science, so I encourage him, of course. Until recently, no universe theory I have read about is even remotely similar to mine. Of course, I don't know if it's correct, and I certainly have no idea how to prove it. One thing it would do besides drastically change the *dark matter, dark energy hypothesis*, would be to revolutionize current theories of how our universe works.

Since our first conversation, several intriguing things have developed. I told this story to my sister, Bobbie, who said she remembered several of those incidents clearly, including our discussion about the diagrams in her chemistry book. A recent article in *Scientific American Magazine*, titled *Forget the Big Bang. Now it's the big bounce*, confirms at least a part of my hypothesis. What the article tried to explain was amazingly similar to at least some parts of my theory. The other developments include several fascinating new explanations from theorists working on the *dark energy, dark matter* hypothesis. It seems there are a number of new and competing theories in this field, a common happening in the scientific community. All of these new ideas tend to confirm parts of my own theory. Isn't that a kick?

Life is made up, not of great sacrifices or duties, but of little things, in which smiles and kindness and small obligations, given habitually, are what win and preserve the heart and secure comfort.

—*Sir Humphrey Davy, 1778–1829*

Life is a process. We are a process. The universe is a process.

—*Anne Wilson Schaef*

Ethics cannot be based upon our obligations toward [people], but they are complete and natural only when we feel this reverence for life and the desire to have compassion for and to help all creatures insofar as it is in our power. I think that this ethic will become more and more recognized because of its great naturalness and because it is the foundation of a true humanism toward which we must strive if our culture is to become truly ethical.

—*Albert Schweitzer*

He who has no God is his own God and will live accordingly.

—*Howard Johnson, 2007*

You shall not go about spreading slander among your kinsmen; nor shall you stand by idly when your neighbor's life is at stake. I am the LORD.

—*Leviticus Chapter 19:16 NIV*

Dear Lord,

I want to say today I have not done a single bad thing. I have not been angry or nasty to anyone. I have not said an unkind word, acted selfishly, bragged, or cursed.

But in a few minutes, Lord, I am going to get out of bed, and from then on, I'll need all the help I can get.

—Unknown

Seek for the patronage of some great man and, like a creeping vine on a tall tree, crawl upward where I cannot stand alone? No thank you!

Dedicate, as others do, poems to pawnbrokers? Be a buffoon in the vile hope of teasing out a smile on some cold face? No thank you!

Eat a toad for breakfast every morning? Make my knees callous and cultivate a supple spine; wear out my belly groveling in the dust? No thank you!

Scratch the back of any swine that roots up gold for me? Tickle the horns of Mammon with my left hand, while my right, too proud to know his partner's business, takes in the fee? No thank you!

Use the fire God gave me to burn incense all day long under the nose of wood or stone? No thank you!

Shall I . . . struggle to insinuate my name into the columns of the Mercury? No thank you!

Calculate, scheme, be afraid, love more to make a visit than a poem; seek introductions, favors, influences? No thank you. No thank you! And again, no thank you!

But . . . to sing, to laugh, to dream, to walk in my own way and be alone, free, with an eye to see things as they are, a voice that means manhood . . . to cock my hat where I choose, at a word, a yes, a no, to fight, or to write. To travel any road under the sun, under stars, nor doubt if fame or fortune lie beyond the bourne. Never to make a line I have not heard in my own heart; yet with all modesty to say: my soul, be satisfied with flowers, with fruit, with weeds even, but gather them in one garden you may call your own.

So when I win some triumph, by some chance, render no share to Caesar . . . in a word, I am too proud to be a parasite, and if my nature lacks the germ that grows towering to heaven like the mountain pine, or like the oak, sheltering multitudes. I stand not high it may be . . . but alone.

—Cyrano de Bergerac

The Bracelet

It was a warm, sunny day in July 1936 when I made a momentous discovery. My buddy, Paul Harruff, and I were playing out on the point of Walker's Park that jutted out into Lake Tippecanoe in Northern Indiana. It was a place where an old native American trail came down to the water's edge. I had been told Indians used to swim across the narrow place in the lake to a point on the opposite shore called Government Point. It was a narrow part of the lake, about a quarter of a mile from point to point. It was the only place for miles not blocked by wide parts of the lake or swamps that followed the courses of the streams feeding the lake and the river that flowed out at the opposite end.

The shore of the empty lot we played on was defined by a rock seawall. The lot had been filled to be level with the top of the seawall with several feet of fill dirt. Out near the point there was a sandy area that had not needed filling. It extended back across the lot for at least 150 feet. The sandy area stood out clearly from the filled land. Digging in that sandy area we found numerous arrow heads, spear points and other stone artifacts, mostly broken. We would dig in the sand and sift it with screens, the best way to separate anything buried in the sand.

Paul and I dug a deep hole near the point. It was so deep water began seeping into the bottom of the hole until it was even with the lake level. We would take a shovel full of sand from the near side of the hole, and sift it onto the other end of the hole, gradually moving the hole inland. After several hours, we had found a few broken arrow heads and one perfect spear point. Right after Paul found the point, his mom called him home for some reason. I continued digging. I was even digging down into the gray sand I found at the bottom of the hole beneath the water line. After two or three shovel loads of this sand had been screened I found an unbelievable prize. There in the screen were several perfect arrowheads with parts of arrow shafts still attached. Among these was a crumpled string of beads and small shells strung on metal wire. All of these were completely black, apparently from being buried in wet sand for so long. The wire too was black, but still soft and pliable. I carefully straightened it out to reveal a ring shape of several strands of wire holding the beads and shells. It fit nicely over my eight-year-old wrist. *A bracelet,* I thought, *how neat.*

I picked up my new treasures and ran all the way to our cottage at the other end of the landing to show my mother.

"Those arrowheads are perfect," she said. "They look like they were freshly made before being buried. I'll see if I can clean off the black stuff."

"How will you clean them, Mom? I tried to wash them off, but the black stuff won't wash off."

"I'll try vinegar first, and if that doesn't work I have several other cleaners I can try. I won't use bleach because it might destroy the string or whatever it is they used to bind the arrowheads."

She placed them in a large glass bowl, took out the vinegar, and poured it over them. She let them soak while we had lunch.

After lunch, we looked in the bowl and found most of the black was gone. Mom picked up each piece, carefully scrubbed it with a vegetable brush, and rinsed it off.

"Wow, Mom. Your vinegar sure worked. Those arrowheads are clean as a whistle. The wire is still black though. I wonder why?"

Mom picked up the bracelet. "The shells are all clean. Only the wire is still black. I have an idea what might clean it. Let me try it first."

She took out a small jar of white polishing cream and rubbed part of the wire with it. Soon the wire she rubbed with the cream was bright and shiny.

"I know it is a surprise," she said as she continued rubbing. "I think the wire is silver. It's so soft it must be pure silver. Silver turns black when it tarnishes. This is quite a find."

After lots of hard scrubbing, we finished cleaning the wire, rinsed off the entire bracelet and placed it and the now clean arrowheads in one of my dad's Antonio and Cleopatra cigar boxes. I would show this treasure to many people over the next few years. The box now held other arrowheads and a growing collection of other precious objects including my collection of pennies dating back to the nineteenth century. I kept the box on a shelf in the garage behind the cottage. Shortly after I graduated from college, the box came up missing. I have no idea how long it had been gone, or what happened to it. All I know is I tried to find my box and couldn't. It had last been seen on a shelf in the garage with several dozen other cigar boxes my dad kept a variety of fishing things, small tools, and fasteners. A number of other things I kept at the lake cottage disappeared during the same time. They were probably taken by someone who knew they were there. I have no idea at all who might have taken my things, or when it happened. I'm positive these things were not accidentally thrown out.

Over the years, I supposed many scenarios of the origin of those arrowheads and bracelet, and how they came to be buried so close to the old Indian trail. I plan to use the information in my memories of those artifacts and their placement in a story I am writing about native Americans in Indiana, particularly the Potawatomi and Miami. Oh, how I wish I still had the box with those items.

Ol' Bugeye

It was early in July of 1939 when I first spied him. He was swimming slowly past my startled eyes while I stood on the end of our pier. He was by far the biggest bass I had ever seen. His fat green body cast a large shadow on the sand in the four feet of clear water as he moved into the shadows of some surface weeds about thirty feet from where I stood frozen. My heart beat frantically as I watched him settle into a motionless shadow. Thus hidden, he would wait for a small fish to pass within striking distance. This was major excitement for an eleven-year-old boy.

Knowing any quick movement would send him immediately out into deep water, I slowly crouched down, crept off the pier and out of Ol' Bugeye's view into the yard. Once there, I rushed to get my casting rod and attach a large pikie minnow to try to catch him. The sharp pain of treble hooks puncturing my skin greeted my clumsy efforts as I frantically tied the plug onto the line with fingers shaking with excitement. It was a small price to pay if I caught him. I rushed across the yard, stopping and crouching down as I approached the seawall of boulders across the front of our yard. Peering carefully over the rocks, I soon spotted the telltale torpedo shape near the weeds. He hadn't moved. Now I had a new problem.

Only about twelve feet from shore in about eighteen inches of water and in a small clear area surrounded on three sides with thick weeds waited the giant bass. It would be almost impossible to cast the plug without frightening him away. I would have to cast while lying prone on the grass behind the rock seawall. If my arm moved much laterally when I cast, the fish would see the movement, and it would spook him. If the plug hit the water within about ten feet of the fish, it would spook him. If it landed anywhere in the weeds, the effort to free it would spook him. The only thing I could do, that would not spook him, would be to drop the plug at least ten feet beyond where he hid and retrieve it past the open space in the weeds where the huge bass might see it and attack.

I aimed carefully and flipped the rod tip, sending the plug in a high arc a good twelve feet past where the fish lay waiting. He didn't move when it splashed down even though the line hit the water about a foot from him. I moved to the left in order to retrieve the plug directly past his hiding place while keeping it away from the thick weeds. I tried a slow, twitching retrieve past the green monster, all the while expecting a sudden rush and strike. He completely ignored the tempting artificial bait. My heart still pounding with excitement, I tried another cast. This time it landed closer, and the bass turned in the direction of the noise. As he turned, I had a clear view of his head as he faced me directly for a moment. I noticed a prominent white spot over his right eye. Perch frequently had fatty white deposits on the top of their eyes, and I surmised this was the same although much larger. "Okay, Ol' Bugeye," I thought to myself. "Let's see if I can get you to hit now."

This time I retrieved a bit faster. As the plug wiggled closer, Ol' Bugeye started to approach, opening and clearing his huge mouth about a foot from the pikie minnow. He turned from the artificial bait at the last instant and swam slowly away toward deep water. Frantically, I reeled in and cast out again in the direction he went as he disappeared. Next I charged out on the pier to see if I could spot him. No luck. I ran back to the yard and over to the Waters's pier next door where I noticed many bass moved after leaving the clear, sandy area in front of our yard. I cast in all directions for some time, but apparently, Ol' Bugeye had moved on. I slumped my shoulders and headed slowly for the cottage where breakfast and my terrible disappointment would be shared with a consoling mother.

Several more times I spotted Ol' Bugeye cruising slowly through what was obviously his hunting grounds near the shore all along Walker's Park, the collection of fourteen cottages served by one dirt road and named after the original owner, Hannah Walker. I was six doors west of our place, swimming off the Mehl's pier. It was a warm July afternoon when I spotted him moving slowly east. I rushed home, grabbed my casting rod, and headed for the rocks at the far end of Waters's lots. It would be the first place he would appear if he continued swimming eastward. Sure enough, his large shape with the identifying white spot soon angled in directly toward where I stood. Distorted by the water, he seemed to swim with his body at an angle as he moved close to a small patch of water plants we called arrowheads from the shape of their leaves. There were about six stalks in about twenty inches of water some eight feet from the shore. The stalks stood almost in a line about three feet long running parallel to shore. Their large leaves, a foot or so above the water, cast shadows, making an ideal spot for bass to ambush small fish. Most sunny mornings, there would be a green torpedo lurking amidst the stalks.

Hugging the grass behind the rock seawall, I peered over to sight Ol' Bugeye parked among the stalks about ten feet from where I lay. I watched his great jutting jaw moving rhythmically in tune with his gills as he opened and closed them, his way of *breathing*. I admired this beautiful fish from his bronze-green back to the mottled black stripe along his side to the enormous tail curling slowly up and down as he hovered motionless in his hiding place. His pectoral fins waved slowly holding him in position. If a prey fish came in sight, all fin and gill movements would cease. Should the unsuspecting prey get close enough, one sudden twist of his muscular body and his tail would flick him instantly forward. His gigantic mouth would suddenly open, drawing a great gulp of water including the prey into the cavern which would close quickly on the hapless victim. A few gulps and the prey would join others in his fat belly.

In my mind, I could smell the sweet, slightly acrid fragrance of fresh-caught fish, feel his hard scales in my grasp and the slippery slime from his body covering my hands. Excitement grew as once more I flipped the rod and sent the artificial bait arcing some distance past where he hid. He must have seen my arm movement as one flick of his mighty tail and all that remained was the cloud of find sand kicked up by his tail during his hasty departure. Again, I tried and failed. Ol' Bugeye was definitely wise to stalking fishermen. To reach such a size, he must be old and extremely wily. He was proving to be a worthy adversary. I would have to try a new approach the next time he appeared on the scene, which would not be until the following summer.

The following summer I only saw Ol' Bugeye once when he swam by our pier amidst a school of several other smaller bass. This was a hunting pack, not unlike a pack of wolves, which cruised along the shore attacking any small fish they encountered. I had seen packs like this one chase and confuse small fish, gulping them up as they rushed away from one hungry mouth only to be grabbed by another from a different direction. Ol' Bugeye had joined about a dozen of his smaller cousins to work the shoreline. The smaller bass in the group were wary of Ol' Bugeye who could have one of them for lunch. Cannibalism is quite common in bass, and many of them become prey of their larger cousins. As summer drew to a close, I wondered what happened to my old nemesis.

During the summer, I caught several good-sized bass from the patch of arrowheads. One weighed more than four pounds. I switched from casting plugs to live bait, thanks to our neighbor, Dan Hackerd. He taught me to use four-to-six-inch-long perch for bait. They were plentiful and the natural prey of large predator fish. A few minutes in the shallow water about two hundred feet off shore with a worm on a cane pole and my minnow bucket held a dozen or so of the tiny green fish. I would stalk the favorite ambush spots for the green torpedo shapes, live bait at the ready on a bass hook on my casting rod. Spotting a bass, I would cast the perch as near to my prey as I could without spooking him. Often the bass would strike the instant the perch hit the water. I would watch as he repositioned the perch before swallowing it head first. As soon as he began swimming away, I would set the hook. I didn't miss many. Sometimes the fish would circle the perch, head angled down, until he was positioned so he could gulp the bait head first. After examining the hapless bait from several angles, the bass would often back slowly away and leave. This always happened when the perch did not move. The instant the bait moved or even twitched, the bass would strike. They wouldn't attack or eat dead bait and would only take those they knew were alive.

It was late June of 1941 before I ran into Ol' Bugeye again. One sunny morning, my heart jumped as I spotted him during my daily hunting trip of bass ambush spots. Right next to the patch of arrowheads lurked a familiar fat torpedo shape with the unmistakable white eye spot. This time I was prepared with a lively perch hooked on my trusty casting rod. A flick of the rod tip sent the perch into the water about five feet in front of my quarry. I held my breath as Ol' Bugeye slowly moved toward the perch which struggled to free itself from the hook. Suddenly the perch spotted the bass and swam frantically away, dragging the line behind. A single flick of his tail and Ol' Bugeye engulfed the tiny perch in his cavernous mouth. My heart pounded as I watched him spit out and then gulp my bait back several times. Finally, he started swimming away with the bait firmly in his mouth. I waited until he was moving steadily away then locked the reel and yanked back to set the hook. All I did was jerk the hook out of his mouth. I'm sure he never felt the hook as I missed him completely and reeled in a bare hook. Dejected and disappointed, I headed in for breakfast.

I spotted him briefly several times in July, but by the time I had bait and rod ready, he was nowhere to be found. It was mid-August, and I was swimming home from my friend Paul Harruff's place six doors to the west. As I passed the Matthews's pier, I saw a familiar dark shape lurking motionless in the shadow. Moving cautiously, I walked on in the three-foot deep water about twenty feet from where he waited until I could no longer see his sleek shape. Then I rushed out of the water and ran home. I had no small perch to use for bait; so I grabbed my casting rod, a single cane pole and can of worms, jumped into our rowboat, and rowed frantically out to where perch were easy to catch. Fingers shaking with excitement, I threaded a small worm on my hook and whipped the pole to throw the line with its red and white bobber to where I hoped perch lay waiting. The few minutes the bobber stood without moving seemed like forever. Finally, a slight bobble, a quick jerk of the pole, and a struggling perch came swinging toward my waiting grasp. Before reaching my outstretched hand, the perch flipped off the hook and dropped into the water. Frustrated, I grasped the empty hook and rethreaded the worm in a frenzy. Once again I threw my rig out. Almost immediately the bobber disappeared, and another perch swung in toward my waiting hand. This time he didn't drop off.

I hooked the perch onto the bass hook at the end of the line attached to my casting rod and headed for where I hoped Ol' Bugeye waited. I stood up in the boat to see better, then rowed cautiously toward where I last saw him. The dark shape still waited in the shadows. I dropped into the seat and nervously picked up my casting rod. A careful flip of the tip sent the bait toward Ol' Bugeye. "Too close," I murmured as the perch splashed into the water less than a foot from my old quarry. I stayed down, not wanting to add more alarming sights to his field of view. As I waited, my boat drifted slowly toward the pier, not a good thing. After several minutes, I slowly stood up to view the spot where he last waited. The fat torpedo shape was gone. *The splash of the bait so close probably scared him to deep water,* I thought as I quickly scanned the shallow sandy-bottomed water in both directions. "There he is," I said out loud when I spotted the slowly undulating green shape heading east toward the Waters's pier and the small patch of arrowheads. I decided to beach the boat at Hackerd's directly in front of me. Then I could pursue my nemesis on foot.

Holding my rod with the struggling bait high, I ran through Hackerd's lot well back from the shore and out of his field of vision. Once more I crept up to the rocks and peered at the patch of water plants. Nothing lurked there. Standing slowly, I scanned the shallows, catching sight of a huge tail as he slipped under Waters's pier. "He's headed for our weed patch," I muttered under my breath as I dropped back from the seawall and crept onto the pier. Again I scanned the shallows thoroughly, especially the shadows beneath the surface weeds. Then I saw it, a huge eye with a small white cap stood out in the shadow. The rest of him simply disappeared into the background. His camouflage was almost perfect, but his *bugeye* gave him away.

Taking great care not to scare him, I positioned my rod to cast. A quick flip of the tip sent my bait toward an open spot about four feet beyond Ol' Bugeye's nose. I watched that eye shift toward the spot. Soon his entire body slid out of the shadows toward the now-struggling perch and my sharp hook. My heart beat frantically in my chest as I stood there frozen, waiting for what seemed like forever. "Go ahead. Take it!" I kept urging under my breath as he inched slowly forward. My bait now lay on the bottom, motionless and belly up, not a good thing. Ol' Bugeye, now only inches from my bait, stopped moving forward then began slowly backing away. Suddenly the perch struggled to right himself in the water. In an instant, my quarry moved forward, opened his mouth and drew in perch, hook, and all with one swift, fluid motion. *Don't spit him out,* my mind commanded as he repositioned the bait with several quick jaw movements.

I watched my reel spin slowly then faster as he began swimming away, pulling the line out. The terrible question now became when and how to set the hook. I watched ten then twenty feet of line pay out before grasping the reel handle and lifting the pole suddenly to set the hook. This time my pole bent as the hook sunk home. At last, Ol' Bugeye was on my line. "I got him!" I shouted out loud. "I got him!" To my cries, Mom rushed outside. As was often the case, she had been watching my actions for some time. Wise in the ways of fishing, she picked up the landing net from where it leaned against a tree and rushed over to help. This was the biggest fish I had ever hooked, and a momentous struggle ensued.

I rushed to the end of the pier to try to keep him away from the thick weeds in the dense patch between Waters' pier and our own some eighty feet to the east. I knew he would probably break the line in those weeds. Suddenly he exploded into the air, opening his giant mouth and shaking his head furiously as he walked on his tail across the water, trying to dislodge the hook. Instantly, I dropped the rod tip and pulled the pole in to thwart his efforts.

Then he ran out toward deep water, taking at least thirty feet of line. I played him cautiously with the spring of the rod to keep the line from breaking. I soon regained those thirty feet of line against his steady pulling. He abruptly stopped pulling, turned and headed toward the pier. Once more he exploded into the air, shaking violently. I reeled furiously to take up the slack, drawing the line taut in time.

Another run of about twenty feet and another series of smaller jumps, and I knew he was tiring. As I reeled him closer to the pier, Mom, now at my side and ready with the landing net, instructed, "Don't let him near those pier posts." We both knew one turn of the line around any of those pipes, and he will break it and get away.

"Don't worry, Mom," I replied. "I'll keep him away from those posts. He's tiring, so be ready to net him soon." He was tiring, but I knew he wasn't done yet because he still swam upright.

Mom placed the net in the water, in front of him, as I pulled him straight in toward the net. When he spied the net, he exploded out of the water once more, sending a spray all over both of us as he dived and again took out twenty feet of line. "He's not giving up easily," Mom said as she wiped the water from her face and we both watched him struggle to pull away.

"I hope he doesn't throw the hook," I remarked. "The last time he jumped, I saw the hook was barely caught in the corner of his mouth. One strong pull and he'll be gone."

For at least the next ten minutes, I worked Ol' Bugeye back and forth in front of the pier to tire him out. I kept the line taut enough to work him with the rod tip. This prevented any more frantic runs or explosive tail walking. When he swam by on his side, I knew it was time to net him. One quick dip of the net and Mom had him. He flopped and struggled in the net as we headed to the yard to remove the hook. When we reached the yard, the hook had come out on its own. We netted him in time.

Before placing him in our live–box, we measured and weighed Ol' Bugeye. At twenty-six inches long and weighing seven and a quarter pounds, he was the biggest bass I had ever caught or would ever catch in Lake Tippecanoe. I don't know for certain if anyone ever caught one bigger or not. At the time, few catches were recorded. Our neighbor, Dan Hackerd, said it was the biggest bass he knew of, and Dan was our local fishing expert.

We kept Ol' Bugeye in the live-box until the weekend, so my dad could see him. I was so delighted and proud to show Dad my prize when he arrived on Friday. After the customary picture taking and additional confirming measurements, Ol' Bugeye was cleaned and prepared for cooking. Sunday afternoon, the entire family enjoyed Mom's special baked fish, while I enjoyed family celebrity status.

The author and
Ol' Bugeye after the
battle.

If

If you can keep your head when all about you
    Are losing theirs and blaming it on you;
If you can trust yourself when all men doubt you,
    And make allowance for their doubting too;
If you can wait and not be tired of waiting
    Or being lied about, don't deal in lies;
Or, being hated, don't give way to hating;
    And yet don't look too good, nor talk too wise;

If you can dream, and not make dreams your master;
    If you can think, and not make thoughts your aim;
If you can meet with triumph and disaster.
    And treat those two imposters just the same;
If you can bear to hear the truth you've spoken
    Twisted by knaves to make a trap for fools;
Or watch the things you gave your life to, broken,
    And stoop, and build them up with worn out tools;

If you can make one heap of all your winnings
    And risk it on one turn of pitch-and-toss,
And lose, and start again at your beginnings
    And never breathe a word about your loss;
If you can force your heart and nerve and sinew
    To serve your turn long after they are gone,
And so hold on when there is nothing in you
    Except the will which says to them: "Hold on!"

If you can walk with crowds and keep your virtues,
    Or walk with kings, nor lose the common touch;
If neither foes nor loving friends can hurt you,
    If all men count with you, but none too much;
If you can fill the unforgiving minute
    With sixty seconds worth of distance run,
Yours is the Earth and everything that's in it
    And, which is more, you'll be a man, my son!

—*Rudyard Kipling, 1865–1936*

My First Career: *Plain Dealer* Carrier

As a young boy, there were two things I remember I wanted and dreamed about for a long time. One was to become a Boy Scout, and the other was to have a Cleveland Plain Dealer paper route. Both of these exciting new experiences required me to be twelve years old, and I thought my twelfth birthday would never come. A major life-changing age, it was my first step toward the adult world and a milestone I longed to reach. I remember my father, a kind and patient man, telling me, "Don't be so eager to get older. Those years will pass quickly enough." How true was his comment, I would learn later.

Several of my friends carried the *Shopping News*, a free weekly paper which eventually became the *Sun Press*. A few carried the *Cleveland News* or *Cleveland Press*. I could have had one of those afternoon routes but chose to wait for my ultimate goal, a *Plain Dealer* route. *The Plain Dealer* was the morning paper, and the only Sunday paper in the city.

Soon after my birthday, I reached one of my goals when I joined the Boy Scouts. I remember how proud I was when I tried on my first uniform at the Halle Brothers store downtown. My father's business, Johnson-Stipher, Inc., was in the Hanna Building a few blocks from the store. He took time off to come with my mother and me to share in the thrill of my first uniform. I remember how proudly I stood before them.

Once my mother had sown on my patches and emblems with the troop 273 designation, I went to my first meeting at the Baptist Church out on Fairmont Boulevard at the end of the streetcar line. Our scoutmaster was Oliver Henderson, and his son Ollie was an Eagle Scout in the troop. I held Ollie in awe as one of my first heroes. My dreams were of the day I could lead the troop as he did.

I thought a *Plain Dealer* route would never open up in our neighborhood. The route where we lived was carried by a boy two years older than I. It would be years before it would be open. I heard the carrier with the route on Queenston Road was selling his route. Queenston was several streets away, but close enough. Tony wanted twenty dollars for the route, a vast amount of money for a boy of twelve in the spring of 1940. My father loaned me the money I needed on the promise I would pay it back out of my profits, a dollar a week.

Thursday and Friday I met Tony in front of McFetridge Drugs at the corner of Queenston and Fairmont where the papers were delivered each morning by the district manager. We carried the route together, so I would learn where to place the paper for each customer. While placement on the porch in front of the door was acceptable for most, some wanted the paper placed under the doormat, in the screen door, milk chute, or mail slot. All special locations were duly noted on my copy of the route list as we walked. Tony carried the first day, and I carried the second under his curt instructions. I learned proper placement of papers during wet or snowy weather the second day when it rained steadily on the entire route.

The following Saturday I walked with Tony as he collected payments and introduced me as his replacement. I can still feel the excitement of walking up and ringing the doorbell at each house on the route. I noticed we passed by many houses that didn't take the paper. When I asked Tony, he made some disparaging remark as we walked by. Sixteen-year-old Tony was not a particularly happy youth and had little good to say about anything. I learned from friends he liked to fight and was frequently in trouble at school. I was careful not to anger him as he was twice my size and treated me as a bothersome tagalong as we walked the route, constantly referring to me disdainfully as *kid*.

When we finished collecting, we were to settle up. He was to give me his paper bag, the ring with the record cards used for collecting, the route list, and the cutters used to cut the wire on the bundle of papers. I was to pay him the twenty dollars for the route. My father had prepared a receipt for me to give him to sign. At this point, an unexpected contention arose. Of the fifteen or so dollars for the weekly collections, about six dollars remained uncollected. This amount was supposed to be owed by people Tony missed collecting for some reason. Some were several weeks back. Tony told me to pay him those six dollars now!

At first he refused to hand over the bag and records. I was devastated. When I asked for my twenty dollars back, he changed his mind and handed me the items, all but the cutters.

"I'll keep these 'til you pay me all my money," he snarled, adding, "When will that be?"

I would not go to my dad for any more money. "I'll pay you as I collect from those who haven't paid," I replied bravely.

"You better pay me all six dollars by next Saturday, or else!" he replied. I knew the "or else" meant a few blows from Tony's fists. At least I had time to collect the money. I decided then and there to spend early evenings all week if necessary collecting all the back amounts owed, so he could be paid on Saturday. What I learned while collecting was a revelation.

Sunday morning I began the daunting task of delivering the Sunday papers. Arising at about five, I loaded the color sections of the paper in my wagon and headed for McFetridge Drugs where the white pages would be delivered. Fortunately, it was a dry, rather warm April day. I borrowed wire cutters from my dad to snip the wires binding the white pages. Hardly able to sleep the night, I was definitely on a high as I kneeled on the sidewalk in front of the drugstore, placing the color sections into the white and stacking them carefully in my wagon. Clutching my route list in my hand, I pulled the wagon full of papers down the street and into my first business venture. It was heady stuff for a twelve-year-old. I felt so proud as I placed each paper carefully under the doormat, inside the storm door, or in the milk chute.

It was getting light as I noticed a car coming slowly down the street from Fairmont Boulevard. My father was checking to see if everything was OK with me. "Having any trouble?" he sang out as he pulled to the curb near my wagon.

"Nope," I replied, a bit miffed at being checked up on.

"Your mother thought you might like this," he replied, smiling as he handed me a cup of hot chocolate and a roll. "I told her that you wouldn't want me to be bothering you, but she insisted. She said she didn't want you to have to work on an empty stomach."

"Thanks, Dad." The pain of being checked on eased a bit as I bit down on the roll and took a sip of the chocolate. My parents were respectful of my desire to be self-sufficient and were careful to let me do my own thing when it was warranted. I would learn the value of those lessons years later when I had children of my own.

The sun was well up in the sky by the time I finished my route and delivered the last two papers to the big houses on Meadowbrook Boulevard. I whistled as I pulled my wagon up the hill on Taylor Road on the long walk home. I knew breakfast would be ready when I got there, and I would have to dress quickly for Sunday school and church. The first day of my new venture was history, and nothing drastic happened.

The following week I learned more about my predecessor on the route. Several people refused to pay the entire bill, saying Tony had forgotten to punch their record cards when he collected. Some of them showed me pencil marks they made on their own card to indicate they paid. All told at least three dollars of the six he claimed were owed had been denied. I had records of each discrepancy and made a list to give to Tony along with the money I had collected. My thought was for him to try to collect the money for himself. My district manager told me Tony should collect all the back payments himself, but since I didn't want any confrontation with him, I didn't mind doing the collection. The contested amounts were another problem entirely. When I posed my problem to my dad, he suggested I ask Tony to come to our house to collect the money. This would keep me safe for the moment should Tony get angry.

Tony arrived at our house Saturday morning. He was not in a good mood. We sat down at the kitchen table with the money and the papers I had prepared. I handed him the list and the three dollars I had for him. He immediately exploded with a string of foul language.

My father stepped quickly into the kitchen and told Tony in no uncertain terms, "That kind of language is not tolerated in our house. You will immediately apologize to us."

There was a stony silence, but Tony soon backed down and apologized. My father continued. "I wanted to let Howard handle this by himself, but because of your threats, I had ~~slunk~~ step in. He's half your size, and I'll not have him threatened in our house. He has treated you more than fair. He collected some of the money the *Plain Dealer* man said you were supposed to collect. He gave you a list of all those who didn't pay and their reasons. He has done far more for you than he needed to do, and still, you threatened him. You will now leave our house and not bother my son again. Should you bother him in any way, you will have me to answer to along with the local authorities."

Tony slunk silently out of our house, the money and paper in his hands. I never did get the cutters from him, leaving a small victory for him to gloat over. I never saw or heard from him again. I thanked my father for backing me up and stepping in when he did. Independence goes only so far with a small boy. Sometimes it's nice to have parental protection.

I carried the Queenston route until mid-June when my mother and I left to spend the summer at our cottage on Tippecanoe Lake in Northern Indiana. We stayed at the lake the entire summer, returning home Labor Day in time to start back to school. Before leaving, I found a substitute carrier for my route in a friend who wanted it only for the summer. More than a few boys wouldn't carry papers in the winter.

By the time I had the route for a year, I had built it up from about forty-five dailies and sixty sundays to more than sixty dailies and nearly eighty sundays. Several incredible things happened during the year. One cold, rainy, spring Saturday I was collecting when who should come to the door to greet me but Bob Hope. My customer, Mrs. Hope, was Bob's aunt, and he was there for a visit. He insisted I step inside and get warm, even offered me a hot cocoa and took my coat, placing it on a chair to dry. He sat and talked with me for at least half an hour while I dried out and drank the cocoa. He was so kind and friendly. I could hardly believe my good fortune. What a story I had to tell friends and family. I could have stayed for a longer time, but excused myself, thanked him, and headed out into the rain with his autograph on the back of my collection card.

More than twenty years later, I would again run into him in the Chase Park Plaza Hotel, in St. Louis. He invited our small group into his room as we walked by his open door. During our visit, I recounted my earlier meeting in his aunt's home.

He said he remembered, "Let me think for a minute . . . Your name is Howard Johnson, right?" I couldn't believe he remembered. "How could anyone forget a kid with a name like that—twenty-eight flavors, right?" I can still see him clearly in both meetings—a kind, gracious, friendly man in spite of his fame.

Several years before in early summer of 1937, I met another famous Clevelander at a downtown Kiwanis father-and-son luncheon. It was a baseball lunch with several members of the Cleveland Indians. I sat with my father on my right and Bob Feller on my left. It was another thrill of a lifetime for a nine-year-old Indians fan. During the luncheon, they held a drawing for several baseballs signed by most of the Indians. When my name was called, I was speechless. Not only did I win one of the baseballs, but Bob Feller walked with me to get the ball and personally handed it to me. I was in seventh heaven. I can remember the game we attended in League Park. Jim Bagby Jr. pitched a five-hit game against the Philadelphia Athletics in a six-to-two win. During the game, Murray Howell came up to bat as a pinch hitter. I clearly remember the announcer saying, "Here comes Murray Howell to pinch-hit, waving his unbelievable one thousand batting average." As I remember, he hit safely in his first five or six at bats as a pinch hitter. I don't remember if his streak continued in this game or not.

When I returned from the lake in September 1941, my friend decided he wanted to keep the Queenston route. He knew the Taylor Road route had opened up, and my district manager wanted me to have it. It was a longer route and closer to home, so I grabbed the chance. In nice weather, I rode my sister's bicycle on the route. I would sit in front of the hardware store on Fairmont between Taylor and Queenston and fold all my daily papers into a square I could throw onto porches as I rode by. I became expert at dropping papers right by the doors. Occasionally a miss would have me off the bicycle and searching through the bushes for the errant paper. Wet days I folded papers in the shelter of the entrance to the Fisher Brothers grocery store next to the hardware. I wore a heavy poncho which completely covered me and my paper bag, protecting all from the rain. I always walked the route when it was raining.

When cold and snowy weather came, I carried my papers into the East Ohio Gas Company building on Taylor. It was wonderfully warm and dry, a perfect place to fold papers and pack them into my bag before venturing out in the wintry weather. I enjoyed even those cold, snowy mornings long before sunrise, walking or riding my route. I don't ever remember being cold while delivering papers. Collecting was another thing all together. While I enjoyed talking to my customers as I collected, this was the time when the cold weather numbed my hands and feet. No amount of socks, boots, gloves, and scarves could keep you warm. Gloves had to be removed to handle the change and punch the record cards. On extremely cold days, I might only work a quarter of my route before heading home to get warmed. Fortunately, some of my customers took pity on me, inviting me inside to get warmed, sometimes offering a warm drink. I took special care of those customers.

The best time to me was spring, especially when the dawn would break before my route was finished. I reveled in the smells of early spring mornings when the grass began to green and plants came alive. While there were not many dogs out during winter, spring seemed to bring out every dog on the street in the early morning. I carried some dog biscuits with me and soon made friends with every dog on the route except for one—a huge black Newfoundland. One Sunday as I turned to leave after placing the owner's Sunday paper in the milk chute by his side door, I was met with a low growl and a menacing set of teeth. The dog stood in the center of the driveway between me and the sidewalk. I was trapped. Placing my bag with about five papers over my arm between the dog and me, I reached into my pocket for the dog biscuits as I inched slowly down the drive, hugging the house.

I stopped moving and slowly took two biscuits out of my pocket. Speaking softly to the dog, I flipped the biscuits into the driveway and back toward the door away from my escape route. The dog didn't move but did stop growling. He looked at the biscuits then back at me. He then walked slowly over and began munching one of the biscuits. I moved toward the street as quickly as I dared while facing the dog. I wasn't going to turn my back on him. My pounding heart began to calm as I delivered the next few papers. After a discussion with my district manager, the dog's owner agreed not to let him out in the morning until after I had delivered the paper. That was the only dog problem I ever had. Most dogs would romp about and greet me happily. They treated me as their best friend.

Early in the spring I noticed an old bicycle lying in the tree lawn along my route. After it lay there for several days and through a snow storm, I decided to inquire about it. A 26" balloon-tired boy's bike, it appeared to be in reasonably useable shape with one flat tire, badly bent handlebars, and no drive chain. When no one seemed to know to whom it belonged or why it was there, I decided it must be abandoned. On the fourth day, when I passed it on my way home, I picked it up and rolled it back to my house. There was no required city license, so the only identification was the serial number stamped into the frame under the pedal case. I called the police with the number, and they told me no one had reported it stolen or lost. I decided to take a chance and repair it. With money I had saved, I went to the local bike shop and purchased a chain, tire, set of handlebars, and a new pedal to replace a damaged one. Borrowing some tools from my dad, I set about repairing the bike. I learned how by trial and error as I went and by asking at the bike shop when I ran into an unsolvable situation.

When I installed the chain, I discovered the New Departure coaster break was frozen, so the pedals wouldn't work. I slowly disassembled the coaster brake, carefully cleaned all the parts, greased them thoroughly, and reassembled them as I remembered from taking the brake apart. After tests demonstrated it worked perfectly, I had my first real bike. The bike I had ridden on my route, my sister's, was an old 36" high-wheeled girl's bike. I wouldn't ride a girl's bike around my friends, but now I had a real boy's bike. I kept thinking the owner would show up, and I would lose my investment of time and money, but it never happened. I even went down, nervously, and registered the bike using the serial number and got a brand-new city license. The clerk never questioned anything. A year later, I gave it a complete new purple paint job and new chrome fenders. I rode my rebuilt bike until I sold it for fifteen dollars before I went off to college two years later in 1945.

After my last full summer with Mom at the lake, I returned home in September of 1942 and discovered the route on Idlewood where we lived was available. I jumped at the chance. This was the best route around the area as most of the houses on Idlewood and several on Shaker, the next street, were duplexes. Due to a change in the routes, I kept part of Taylor Road from my old route. I now had a much shorter route ending on my own street with many more customers than my old one. With the war now raging in Europe and the Pacific, inflation began to be a problem. First, the Sunday paper went up to twelve cents; then the daily papers went up to four. I heard many complaints but lost few customers.

Sometime in January, I was talking to Barbara Ackhart across the street from her house on Idlewood when I slipped suddenly on the snow-covered sidewalk. There was a crack and then a sharp pain in my upper chest as my shoulder hit the ground. A trip to Dr. Brett's office at 105th Street and Carnegie and my broken collar bone was reset. My right arm was placed in a restrictive sling. I would have to find someone to carry my paper route, an impossible task in the dead of a bad winter, the worst in many years. My savior? My father. Under my directions, he would carry my papers until I found someone for the eight-to-twelve weeks my arm would be useless. He would carry all my papers until March. Sundays I helped by pulling the wagon with my good arm and instructing my father where the paper was to be placed. I managed to start collecting after two weeks; my sling held my record cards and punch as I went door to door. My customers were wonderful, helping me by punching the record cards and by having the exact change when I came.

One sad incident happened in the spring. The family across the street from our house had a female dachshund. Pipshin gave birth to a litter of the cutest little pups I had ever seen. I adored those pups and played with them every time I had the chance. The mother dog's name was Pipshin Von Hepchein Thomsplatz, a name I'll never forget. The dogs were kept in a fenced-in backyard where they romped noisily. I was shocked one day to learn all of the dogs died of poisoning. Scraps of poisoned ground meat were found by the fence between their yard and the house to the south. The grouchy old man and his wife who lived next door hated the barking of the dogs and silenced them with poisoned hamburger. I was filled with remorse for the tiny victims and anger for this customer of mine who harmed them.

First, I began giving them the worst paper I had, one from the outside of the bundle, cut and torn by the binding wire and often dirty from being dropped on the ground. Then I started missing their porch, leaving the paper in the bushes, even on rainy days. When the man complained on Saturday when I collected, I decided I no longer wanted to deliver to them. Sunday, my district manager called to tell me that I had missed a delivery. Normally, I would immediately deliver the missed paper, but this time I refused and told my manager why. When he said he would have to deliver the paper if I didn't, I relented and took the paper across the street. That paper was in the worst condition of any I had ever delivered. Of course, I received a complaint.

When the full story reached the powers at the *Plain Dealer*, they decided to side with me. No paper would be delivered. To get a paper, they would have to pick one up themselves at a drugstore or newsstand. This was a monumental victory for a small boy. The story of the puppy murderers spread among the neighborhood kids. Another related event happened the following Halloween when the puppy murderers were the victims of every mean trick in the book. This included some nasty things dumped on their porch and in their yard. Kids have their own way of avenging evil actions.

With war raging in Europe and the Pacific, my folks decided to forgo a full summer at the lake and go there for the last three weeks before Labor Day. My sister Bobbie got a job as a waitress at Howard Johnson's restaurant in Shaker Heights, and I got a job there as a busboy. The manager, a Mr. Milar, was pleased to have a Howard Johnson working there and frequently trotted me out to be introduced to customers. Saturday night was always a long night, and I frequently got home just in time to start my paper route. When the weather was good, I would ride my bike to work with Bobbie seated on the crossbar. We rode the side streets and through Shaker Lakes on our way because it was safer than riding on busy Lee Road. One Sunday morning, the sky was beginning to lighten as we headed for home. When we reached the dam and remains of the Shaker mill, we stopped, sat down on the earthen dam, and watched the sunrise. It was one of several magical, unforgettable moments I would share with my sister over many years. It was hard to pull away, but my *Plain Dealer* route beckoned, We left and hurried the quarter mile home. I don't know for sure, but Bobbie may have helped with my route since it would be near eight before I would be able to go to sleep.

That was the last summer I had my own *Plain Dealer* route. Soon after my sixteenth birthday in November, I got my driver's license and then a part-time job as a delivery boy/soda jerk at McFetridge Drugs at Queenston and Fairmont. I was sorry to give up those early morning walks in spring and summer and the many friends I had on my route. I wouldn't miss those cold, wintry days collecting at all. Surprisingly, this was not my last experience carrying the *Plain Dealer* on cold, wintry days.

In 1965, my son, Howard M. Johnson, became a *Plain Dealer* carrier when we lived on Lynn Park Drive in Cleveland Heights. Like my father before me, I often helped my son on his route. Our trek took us through the neighborhood and included houses on Yellowstone Road. Frequently I helped him on wintry or rainy sundays. No matter the weather, it was a marvelous experience helping my son and reliving those early morning walks during my youth. I noticed both the daily and Sunday papers were much larger and more expensive than they were twenty-five years before. My son, now a Delta Airlines captain and ex-U2 pilot, has three boys. The oldest is Howard Russell Johnson, the same full name as my father and the fourth HoJo in a row. Russ and the others, Daniel and Michael, live far from Cleveland and never were paperboys as far as I know.

Eleven years later I became a de facto *Plain Dealer* carrier once more when my stepson, Brent Evans, got a *Plain Dealer* route on Blackfoot and Glenn Russ in Euclid. Many a cold, wintry morning we would get up early, load the back of our van with *Plain Dealers* and wade through deep-snow delivering. We bought matching pairs of popular *moon boots* to keep our feet warm as we trudged through the snow. I know Brent wore his out, but I still have and use mine when it gets cold and snowy. They are constant reminders of my many memorable experiences carrying the *Plain Dealer* spanning nearly forty years.

It is not easy,
To apologize,
To begin over,
To be unselfish,
To take advice,
To admit error,
To face a sneer,
To be charitable,
To keep trying,
To be considerate,
To avoid mistakes,
To endure success,
To profit by mistakes,
To forgive and forget,
To think and then act,
To keep out of a rut,
To make the best of little,
To subdue an unruly temper,
To shoulder a deserved blame,
To recognize the silver lining,
But it always pays grand rewards!

—Ohio Educational Monthly, Author unknown

When things go wrong as they sometimes will,

When the road you're trudging seems all uphill,

When the funds are low and the debts are high

And you want to smile, but you have to sigh,

When care is pressing you down a bit,

Rest, if you must, but don't you quit.

Life is queer with its twists and turns,

As every one of us sometimes learns,

And many a failure turns about

When he might have won had he stuck it out;

Don't give up though the pace seems slow,

You may succeed with another blow.

Often the goal is nearer than,

It seems to a faint and faltering man,

Often the struggler has given up,

When he might have captured the victor's cup,

And he learned too late when the night slipped down,

How close he was to the golden crown.

Success is failure turned inside out,

The silver tint on the clouds of doubt,

And you never can tell how close you are,

It may be near when it seems so far;

So stick to the fight when you're hardest hit,

It's when things seem worst that you must not quit.

> —*This has been credited to several writers, but the true author remains Unknown*

Grown-ups love figures. When you tell them that you have made a new friend, they never ask you any questions about essential matters. They never say to you, "What does his voice sound like? What games does he love best? Does he collect butterflies?" Instead, they demand: "How old is he? How many brothers has he? How much does he weigh? How much money does his father make?" Only from these figures do they think they have learned anything about him.

> —*Antoine de Saint-Exupéry, The Little Prince, 1943*

Weird and Unexplainable Experiences

I know some things can be hard to explain, but I have had a number of what many would call paranormal experiences during my lifetime. Some are so bizarre I can hardly believe them myself, and I experienced them.

The first one I remember happened in the summer of 1940 or 41. My sister, Bobbie, my parents, and my maternal grandparents were all sitting with me around the table right after dinner. I don't remember what we were discussing, but suddenly I realized I knew what each person was going to say before they said it. I grabbed a pencil that was laying on the table and quickly scribbled some words on a piece of paper we were going to use to keep score in a card game. I turned the paper face down on the table and told them, "Remember what you say."

My sister looked at me and said, "What's this all about?"

My grandfather said, "Howard, what kind of game are you playing?"

My Dad said, "Are we going to play cards or talk?"

I asked my mom to pick up the paper and read what I had written. What she read was exactly what my sis, my grandfather, and my dad had just said, word for word, after I wrote them down and placed them on the table. I have never had a similar experience in my life.

Sometime during the 60s, I had lunch with my secretary, Mary Lou, after a business meeting. I ordered a special salad from the menu. Suddenly, after the waiter had disappeared into the kitchen, I turned to Mary Lou and said, "I know it sounds crazy, but I know our waiter is going to come back to our table and tell me the salad I ordered is not available because of an accident in the kitchen. He will also suggest a Chef's salad instead."

Mary Lou looked at me, smiled and said, "How could you possibly know that?"

I said, "I have no idea, but it's going to happen."

She laughed and said, "You're crazy."

No sooner were those words spoken when—well, it happened, precisely as I predicted.

"Did you set this up as a joke to fool me?" she asked as soon as the waiter left.

343

"You know that would have been impossible," I remarked. "We've never been to this place before and coming here was a last minute decision made after the meeting. I didn't know this place existed until you pointed it out."

"That's right," She commented while making a weird, quizzical facial expression.

We spoke about this several time during the ensuing weeks, between the two of us of course. It was too strange to forget quickly.

During the late 60s, I was going with a lovely lady named Caroline who was to become the mother of my youngest daughter, Kristen. We had several unusual experiences, each around my birthday involving her presents to me. One occurred at the Tack Room on Shaker Square in Shaker Heights Ohio, Caroline handed me a small, wrapped package saying, "You'll never guess what this is."

"No," I said. "I'm sure I can't guess, but I can draw what it looks like."

I proceeded to draw a rather unusual rectangular shape on my napkin and handed it to her.

This brought a look of startled amazement followed by, "that's impossible. Open the box."

Inside was a beautiful Hamilton watch with a silver case exactly the shape I had drawn on my napkin. This was doubly strange because I had correctly guessed what my gift was on my previous birthday, a hamburg hat.

The following year she gave me a birthday party at her apartment where several friends attended. With everyone watching, she handed me a light weight box about a foot long and two inches square. She made the comment, "You'll never guess what's in your package in a thousand years."

To which I replied, "It must be a two-liter Maserati." The thought and words came to me out of the blue. This was a rare and unusual car, and not one I can ever remember talking about. Her friend Barb screamed and Caroline looked incredulous. They had deliberately selected a model of a red, two-liter Maserati, knowing it would be impossible to guess. Even the idea of a model car was unusual. You figure it out. I can't.

The next year, Caroline's birthday would come while she was away at Atlantic City at a dental convention with her friend, Barb. At the last minute, in fact on her birthday, I found I could get away, and fly to Philadelphia on business. I didn't know where she was staying, or any idea of her schedule. None of her friends I called knew either so I decided to go to the convention and try to find her. I rented a car at the airport and headed for Atlantic City. It was about five in the evening when I left on a nearly two-hour drive.

On the way there I ~~was~~ drove past Zaberer's, a truly great restaurant Caroline and I had been to before. It had a great piano Bar Something prompted me to turn into their parking lot. *What the hell?* I thought, *If nothing else, I can have a drink at the bar.*

There was a group of women celebrating a birthday at a table I walked past on my way to the bar. As I plopped down in the empty chair next to Caroline, Barb stood up and said, "I said he'd be here, didn't I? That's why I saved that chair."

I know it's hard to believe, but it was purely a serendipitous accident. Caroline **knew** Barb and I had cooked up this meeting. In spite of my protestations and denials, I don't think she ever believed my story. Hell, I can hardly believe it myself.

Over the years, there were several other unusual instances, usually involving someone close to me. Except for several I have been written about in their own individual stories, none were as unusual or as unbelievable as the ones related. I have no explanations for them, and I certainly wouldn't try to create any. Some happenings are simply weird and inexplicable. They certainly do make for great story telling.

Come with Me to Macedonia and Fight

Commanders should be counseled chiefly by persons of known talents; by those who have made the art of war their particular study, and whose knowledge is derived from experience; by those present at the scene of action, who see the enemy, who see the advantages that occasions offer, and who, like people embarked on the same ship, are sharers of the danger. If, therefore, anyone thinks himself qualified to give advice respecting the war which I am to conduct, let him not refuse assistance to the state . . . but come with me to Macedonia.

He shall be furnished with a ship, a horse, a tent; even his traveling charges shall be defrayed. But if he thinks this too much trouble and prefers the repose of the city life to the toils of war, let him on land not assume the office of the pilot.

The city in itself furnishes an abundance of topics for conversation; let it confine its passion for talking to its own precincts and rest assured that we shall pay no attention to any counsel, but as shall be framed within our camp.

—*Lucius Aemilius Paullus, Roman Consul, 168 BC*

A poet in a mirror is - a wind walker - a reacher for stars.

The poet speaks - the mirror reflects - a few truly understand.

—Howard Johnson, 1997

The world has no room for cowards. We must all be ready somehow to toil, to suffer, to die. And yours is not less noble because no drum beats before you when you go out to your daily battlefields, and no crowds shout your coming when you return from your daily victory and defeat.

—Robert Louis Stevenson

The holiest of all holidays are those kept by ourselves in silence and apart; the secret anniversaries of the heart, when the full river of feeling overflows; the happy days unclouded to their close; the sudden joys that out of darkness start as flames from ashes; swift desires that dart like swallows, singing down each wind that blows!

—Henry Wadsworth Longfellow, 1807–1882

We have no government armed with power capable of contending with human passions unbridled by morality and religion. Avarice, ambition, revenge or gallantry would break the strongest cords of our Constitution as a whale goes through a net. Our Constitution is designed only for a moral and religious people. It is wholly inadequate for any other.

—John Adams

Without violence nothing is ever accomplished in history.

—Karl Marx

The quickest way of ending a war is to lose it.

—George Orwell

Where is all the knowledge we lost with information?

—T. S. Eliot

The function of socialism is to raise suffering to a higher level.

—Norman Mailer

The sick in soul insist that it is humanity that is sick, and they are the surgeons to operate on it. They want to turn the world into a sickroom. And once they get humanity strapped to the operating table, they operate on it with an ax.

—Eric Hoffer

The Pictures

It was a warm summer afternoon way back in 1940 when my buddies and I were playing softball on the playground at Fairfax grade school in Cleveland Heights. At the west end, immediately behind home plate, was a tall chain-link fence backstop to keep wild baseballs from hitting the nearby house. Between the backstop and the house, was a concrete walk and about twenty feet of grass. The trim little two-story house had a porch across the front and a well-kept flower garden in the fenced-in backyard.

Soon after we started playing, my friend, Bill, threw me a pitch. I swung with all my might—*whack!* A pop foul. The ball arced upward, missed the top of the backstop, lost speed, and headed for the yard beyond the fence. I cringed as it plopped down in the flower garden.

"You hit it. You get it," one of my friends hollered.

While climbing the fence would be easy, dropping into the flowers on the other side was sure to make the people in the house angry. With no easy way to get into the yard without climbing the fence, I would have to face the unknown occupants of the house. I walked hesitantly past the fence, around to the front of the house, along the walk, up the steps between large spirea bushes, and onto the porch. My buddies kept up the constant chatter of harassment boys typically do to each other as I headed for the door.

Imagining all manner of responses from whoever opened the door, I gathered my courage and summoned the sweetest, most innocent manner a clever thirteen-year-old could muster before knocking. Soon after my hesitant knock, an elderly round-faced woman answered the door. I was encouraged by her pleasant smile and friendly appearance.

"Yes?" she questioned.

"I . . . we . . . our ball accidentally fell in your backyard. Could I go get it please?" I asked in my most polite, humble voice.

She opened the door wide. "Of course. Come right through here and out the back door."

"Thanks!" I said gratefully as I started through the living room. A quick look around struck a familiar scene. "I've been in your house before," I commented matter-of-factly.

347

"I don't think so," she replied. "I would surely recognize you if you had."

I looked up as I walked past the staircase. Everything looked familiar, too familiar to have been similar to another house I might have visited. "I'm sure of it," I replied.

She seemed a bit annoyed at my persistence. "What's your name? Maybe I know your family."

"Howard Johnson. My dad is Howard also, and my mom's name is Ethel. We live a few blocks from here, up on Idlewood. Moved here about a year ago."

She shook her head and once again smiled pleasantly. "I don't know your folks at all."

"I used to go to Fairfax school right next door. Now I go to Roxboro Junior High," I said proudly.

"Yes, young man, but you must be mistaken. Maybe you've been in a house like mine and are confused. There are a number of similar houses in the neighborhood."

I stepped back to the bottom of the stairs and looked intently at the landing where the steps turned and went out of sight. I had a clear mental picture of the walls around the corner of the landing—walls I could not see from where I had been.

"I can prove it," I said turning back to her and pointing up the stairs. "There are two pictures on the wall around the corner. They are paintings of the heads of two horses, a brown one and a white one."

The woman's face blanched and froze. Her mouth opened, but no words came. She staggered slightly, reached for a nearby chair, and sat down firmly, staring directly at me with almost a look of horror. There was a long, awkward moment.

"Are you okay?" I asked.

"How did you know about those pictures?" Her lip quivered as she spoke.

"I'm sorry, ma'am. I didn't mean to upset you." She remained silent, staring. After several moments I asked, "What's wrong? You seem so scared."

Again there was a long silence. She calmed down, reached out her hand, and grasped mine. She spoke in a quiet voice on the verge of trembling. "Would you describe the pictures for me? Tell me what you remember of them."

"They are just the horses heads, looking straight at you. The brown horse has a long white spot from his muzzle to between his eyes. The white one has no marks at all, just pinkish lips. That's about all I remember. Oh, something else, you can see green fields and blue sky behind them."

When I finished, she pointed up the stairs. "Go look on the walls where you think those pictures are . . . Go ahead."

I walked up the stairs, turned on the landing, and looked at the walls. There were several pictures hanging there, pictures of her garden and several of groups of people, probably family. I was surprised. My memory was playing tricks on me. I returned down the stairs to where she waited. "I must have been wrong. It must have been another house I remembered."

"No, you weren't mistaken. You described two pictures that once hung on the stairway wall, described them exactly. That's what so frightened me."

"Then I must have been here before and don't remember when."

She pointed to the door to the backyard. "Go get your ball. When you return, I'll tell you about those pictures."

It took several minutes to find and retrieve the ball hidden under some flowering bushes. Shouts of encouragement came from my pals outside the fence. I checked to see if the ball had done any damage. Seeing none obvious, I yelled, "I got it," and went back into the house. The woman's kind smile was evident, but her look of fear was not completely gone. She pointed to a chair.

"Please sit down. I want to talk to you about those pictures you described so exactly. It's obvious you knew about them, but I can't imagine how."

I fidgeted as she waited for my answer. My buddies would be anxious for my return with the ball now clutched in my hands. "I have no idea. I just remember them."

She noticed my urgency to go. "You must be anxious to return to your ball game. Don't worry. I won't keep you long. How old are you?"

"Thirteen."

"Well, young man, those pictures you described belong to my daughter. They were painted by my brother who gave them to her as a birthday present. They hung on the stairway wall as you remember them until she left home and moved to California."

"Then I must have been here before she moved."

"That's what makes it so incredible. She took those pictures to California about fifteen years ago, long before you were born."

A chill went down my spine. Now, I was a bit frightened. "How can that be? I still remember seeing them right there on the stairway," I mumbled as I sat there wide-eyed.

"Young man, I don't know how it's possible. I don't feel like talking about it anymore. Let's call it a strange experience and leave it at that. Tell your parents I think they have a courteous son with some unusual memories. Now, you have your ball. I'm sure you're eager to get back to your game."

"Thanks!" I said as I got up to leave, still a bit unnerved.

As I walked down the steps, my head was still swimming. I don't remember if we played ball afterwards or not. I only remember many questions running through my mind as I walked the few blocks home. I never played ball on the girls playground again. Many times when I passed nearby and noticed the playground or house, my mind would take me back through the events of that day and I would get that uneasy feeling once more. I never visited the house or saw the woman again. It was too unsettling even to think about it.

Recently, while visiting Cleveland Heights with my wife, Barbara, I drove past the house and pointed it out to her. As I remembered those two pictures clearly, I had the same uneasy feeling, even after more than sixty years.

Finally, a word to describe 21ˢᵗ Century American government.

Ineptocracy (in-ep-toc'-ra-cy) - a system of government where the least capable to lead are elected by the least capable of producing; and, where the members of society least likely to sustain themselves or succeed, are rewarded with goods and services paid for by the confiscated wealth of a diminishing number of producers.

—unknown

Foreign aid might be defined as a transfer from poor people in rich countries to rich people in poor countries.

—Douglas Casey

A government policy to rob Peter to pay Paul can be assured of the support of Paul.
—George Bernard Shaw

Anything . . . that man can conceive of and believe in, can be accomplished.
—Unknown

The Jazz Drummer

When I started the seventh grade at Roxboro Junior High in Cleveland Heights, Ohio I was an active new teen. Alphabetical seating put me next to Earl Johns, the only black boy in our school. We soon found we were kindred souls, excited and immensely curious about everything. We were destined to become the best of friends. We were always getting into minor scrapes at school. Nothing serious, just general young teen mischief. When we started in instrumental music class, we each had the opportunity to try several instruments from which to choose. Earl quickly chose the drums. I tried several instruments before settling on the French horn, I think with some coaching from the orchestra leader who needed a French horn for the school orchestra.

Earl's father soon bought him a great Slingerland drum set with all the bells and whistles—literally. I was frequently at Earl's house where I was soon taking turns with him playing the drums. We played in the recreation room in their basement. I learned right along with Earl. His folks had a phonograph and lots of music on 78 rpm records. We would play along with the music. It was a different music from what I heard at home on our Victrola. Earl told me it was called jazz and that lots of black folks listened to it. I loved it, the rhythms, the freedom of expression, the ever-changing sounds, often taking off in wild musical gyrations.

✳   ✳   ✳

Shortly after my fourteenth birthday I was wailing away on the drums in Earls' basement while he was upstairs getting something. His father, who I had never seen in the basement, came downstairs.

"Howard? I see you are playing the drums?"

I thought a smart remark, but true to parental influence, did not say it. "Yes, Mr. Johns. Is it OK with you if I play on Earl's drums? He says it's OK."

"Of course it is Howard. I came down to listen. You are quite good."

"Thanks, Mr. Johns. Earl's really good too."

"I know you two are friends, but there's something special about the way you play. I can always tell when you are on the drums. You have a real feeling for the sound of Jazz."

"Thanks, Mr. Johns. I like playing with your records. You have some great ones. I don't know any of the people on those records except Count Basie and Duke Ellington, but I sure like the music. It gets to me."

Mr. Johns laughed. "You sure were gettin' down with my Apolo Jam Session record. You have a great sense of jazz timing. How would you like to play with a real band, sit in on a jazz jam session with a real professional band instead of those records?"

"Oh, I don't know, Mr. Johns, a real band? I'd be scared to death."

"I know a band who would love to have you sit in on their jammin'. Did you know I was part owner of a club down on 69th Street?"

"Down in the city?" I asked wide eyed.

"Yep. Right off of Cedar Road. It's in the darkest part of town if you get my drift," he said sort of secretly, and with an over emphasized wink.

"I'm afraid my folks wouldn't let me go there. Isn't it pretty dangerous?"

Again Mr. Johns laughed. "Well, there ain't many white folks come there, probably because they're afraid to, but we run a tight club. Keep the real troublemakers out. Once in a while we get a group of white folks who come slummin'. We treat then real nice. Mostly because we like their money. But you? You would be different. More like one of us. There are a few white musicians who drop by the club for some of our jams, several are pretty big names too. They are all treated with respect, not extra special, more like one of our regular crowd. We all get down together."

"I sure would like to do that. It would be exciting. I still don't think my folks would let me."

"I know Earl comes over to your house. He likes your mom. Says she makes him feel real good."

"My mom's a real sweetheart. I know she likes Earl. Says he is the most polite boy I've ever brought over."

"My Earl? That makes me feel real good. Tell you what. Why don't you ask your mom if you can go on a little trip with Earl this Saturday to watch him play drums with a band. It would be the truth. We have open jam sessions every Saturday starting at one."

"I don't know. She might want to know exactly where we would be going. I couldn't lie to her."

"Have her call me or Earl's mom. I'll bet she won't, but a call to us could ease her worries."

To shorten a long story, I did go, and she didn't call, so I went to Cleveland's *Little Apollo* club the following Saturday with Earl and his dad. It was an enlightening experience. The musicians acted standoffish at first. The tension made me feel uncomfortable. As the only white face there, I stood out like a light bulb in a dark room.

When it came my turn to sit in, the drummer said, "You be good to my drums now, y'hear? Don't go poppin' a skin." It was not said in a friendly manner. I didn't answer and sat down at the set.

It took some time for me to get used to the feel of a different set of drums, but I caught on when they started playing one of the arrangements from the Apollo, one I had followed in Earl's basement. I could see the members of the band warming up to me. After about a fifteen minute run, the leader pointed a finger at me and said, "Take it."

I *took it* and then some. Soon members of the band were shouting encouragement. By the time I finished my improvision and the rest of the band joined in, everyone in the place was standing.

The drummer who admonished me earlier came over, put his arm around me and said, "White boy, you can play on my drums whenever you want. You was really diggin'. What was your name agin?"

"Howard Johnson."

"Can't be. You didn't know, did ya? I'm Harold Johnson. We must be brothers." He said with a grin.

After that, I was treated differently—like one of them. Several band members made the comment, "You must be black under that white skin," always with a sincere, friendly laugh.

I heard Earl's dad several times telling customers, "I told you he was good, real good," all the time wearing his usual broad, friendly smile.

I became a regular at their Jam sessions, playing with them almost every Saturday afternoon. When my guilt made me tell my mom what I was doing. She was a bit shocked.

"Howard, I didn't even know you played the drums. You say Mr. Johns owns the club. Don't they serve liquor there? You shouldn't ever be in such a place. It's too dangerous. Besides, it's against the law."

"But Mom, You know I would never drink anything but water or soda. Those folks are nice to me. They wouldn't even let me have a drink if I asked. Earls' dad told us never to taste anything anyone offered us unless he personally Oked it. He's very protective."

"I don't care. I still don't want you ever to go there again. Promise?"

I was disappointed, hurt and furious. Without a word, I stomped off to my room and slammed the door. Soon mom was knocking at the door. I said a lot of hateful things, all an angry fourteen year old could muster up and say to his mom. I know I had never spoken to her like that before. Things were quiet for about half an hour. Then there came a knock on my door.

"Son? Come on down stairs and let's talk. Maybe I was a bit hasty with my comments. I didn't realize how important it was to you to play."

Not only was my mom a nice person, she was smart. She knew how to get to me without putting me down. It wasn't long before we were sitting at the kitchen table talking.

"You know, son, I didn't even know you were playing the drums. Then finding out you were playing in a place where they served liquor—well, it was a huge surprise to me, an unpleasant one. I'm over the initial shock now so how about telling me what you've been doing, and why you like it so much."

We must have talked for at least an hour. Mom told me how she had taken piano lessons and dreamt of being a concert pianist when she was my age, but had to give it up when times got bad, and my granddad lost his job.

"I was devastated," she told me. "I cried for hours. It took me a long time to get over it. I sort of blamed your granddad and was angry with him for quite some time. I was wrong to be so angry. I felt much better when I mustered up the courage to apologize to him, and tell him I was sorry."

"Gee Mom, it wasn't his fault."

"No, it wasn't, and I don't ever want you to feel that way about me or your dad. Anyway, I called up Mrs. Johns and talked with her for quite some time. She seems to be a nice person, also concerned about Earl. She didn't like him going to the club either, at least in the beginning. Anyway, she convinced me that you would be safe enough as long as her husband was there and running the club."

"Gee, Mom, that's great. You mean it's OK for me to keep goin' to those jam sessions?"

"For the time being, yes. But one more thing."

"What's that?"

"I don't think we should tell your father. At least not yet. He doesn't appreciate music or much care for musicians. Of course, he's proud of your sister, Lois and her piano playing, but that's different."

"It sure is." I commented thinking about her playing piano accompanying my attempts at a french horn solo to be played with the school orchestra. They had moved our baby grand piano to Cleveland when we moved from Indianapolis when I was eight.

She finished our little conversation with a warning. "For the time being let's keep this between you and me, a secret from the rest of the family. It will be our little secret until I can figure out how to break it to your father."

I was elated to have a coconspirator, especially my mom. "Thanks mom. I promise I'll be good."

One other thing happened several months later during spring vacation at school. Mr Johns asked if I would play drums at the club the last few days of the week so Harold could take a trip somewhere. The club was open Thursday, Friday, and Saturday evening, until the wee hours. When I asked mom, she became worried.

"I don't know about you staying up so late. You have your paper route the next mornings at five. When will you sleep?"

"Mr. Johns said he would have me home by 1:30am. I can grab a few winks, get up and do my paper route and go back to sleep soon as I get home. It'll be OK, Mom, honest."

To make a long story short, I did, we did, and it all worked out well for me. Saturday night the band leader made a special point of braggin' about, "Our little white drummer boy" to the customers. Then he had me do a drum solo, I got a standing ovation. I was so pleased and so proud. I could only tell Mom about it. Oh yes, my father was away on a business trip the week I played. That was plain lucky.

About eight years later, I found out Cleveland's Little Apollo club was a notorious hang out for Cedar Avenue pimps, prostitutes, bootleggers, and gangsters. I found this out from my new father-in-law who was the local head of the federal ATF. He had replaced Elliott Ness when Ness moved to Chicago in the thirties. He knew all about the Apollo and my friend Earl's dad. Sometimes it's comforting not to know some things. We never did tell my dad or my sisters.

We Americans engage each other in fierce inner political warfare, while our enemies stand on the sidelines urging on all sides, and gleefully watching our self destruction.

—*from the book,* **Energy, Convenient Solutions**, *Howard Johnson, 2010*

Passionate hatred can give meaning and purpose to an empty life. Thus people haunted by the purposelessness of their lives try to find a new content not only by dedicating themselves to a holy cause but also by nursing a fanatical grievance. A mass movement offers them unlimited opportunities for both.

—Eric Hoffer

You see things; and you say, 'Why?'
But I dream things that never were; and I say, 'Why not?'

—George Bernard Shaw

In its original form (above), the quotation was said by the serpent in George Bernard Shaw's play *Back to Methuselah*.

Paraphrased, it was used by President Kennedy in his Speech to the Irish Parliament on June 28, 1963 as follows:

"Some men see things as they are and say 'why' - I dream things that never were and say 'why not.'"

Robert Kennedy made this quotation famous during his 1968 Presidential campaign. He paraphrased it as follows:

"Speaking as an Irishman [Shaw] summed up an approach to life: 'Other people see things and say: why - but I dream things that never were and say: why not.'"

—from Internet research

Even in our pursuit of happiness, there is some sadness and even evil. This is because we are less than perfect beings, with less than perfect minds and hearts, communicating in less than perfect means, in a less than perfect universe. The best we can do in this less than perfect world is just that, the best we can do.

—Howard Johnson, 2010

I can swear there ain't no Heaven, but I pray there ain't no hell.
*—A bit of religious philosophy from **And When I Die**, written by Laura Nyro*
*and recorded by **Blood, Sweat and Tears** in 1969.*

It is easier to talk about money -- and much easier to talk about sex -- than it is to talk about power. People who have it deny it; people who want it do not want to appear to hunger for it; and people who engage in its machinations do so secretly.

—Smiley Blanton

The Rattlesnake Bite

It was the summer when I was fourteen probably in July. We were at the cottage on Tippecanoe Lake. It was early in the day and I was in my shorts, cutting the weeds out of our seawall made of small boulders. I was using my dad's sheath knife, and the sheath was attached to my belt. I wore his knife on my belt as a way of expressing my grown up manliness. As I hacked away at some particularly woody wild bush, I felt a sudden, sharp pain in my left wrist. Looking at the reason, I saw what I thought to be a stick stuck in my wrist and causing the pain. I dropped my knife, grabbed the *stick*, and yanked it out and away from my wrist. I was in instant severe pain and noticed my wrist was bleeding profusely.

I immediately realized this was no stick. I now held tightly in my right hand, a small rattlesnake. Fortunately I held it firmly, so close to his head he could not get his fangs into me again. I picked up the knife with my left hand and hacked off the snake's head. As soon as I dropped the snake's body it squirmed and thrashed on the ground. Remembering my boy scout training about snake bites, I started sucking my bite and ran into the house to tell Mom.

My dad was out fishing for at least the next three hours. My mother was upset and unnerved by the situation. She could hardly get the keys into the car to drive me to town and a doctor. I decided to drive and asked her to move over into the passenger seat. I was more worried about her state of mind than about my bite. I had only moved the car a few times and never driven on the highway. Still, I had driven the farm tractor in the fields and even on the highway for a short distance. I drove into town and the doctor with no problems.

The doctor examined my bite, which by now was painful and still bleeding. When he cleaned it, I could see I had ripped the flesh when I jerked the snake away. I pulled the fangs through my wrist leaving two gaping, bloody openings.

"The bleeding probably flushed all the venom out." Doc remarked. "I have some old anti venom here, so I think we should put it in the bite and give some as an injection. It's been some time since I've even heard of a rattlesnake bite around here although I know there was a case over near Silver Lake a few weeks ago. I'll fix those tears in your wrist and bandage it up. I won't suture them until tomorrow when we are sure all of the venom is gone. You come back tomorrow morning, and I'll take a look. If everything is OK, I'll sew you up."

By this time and with the doctor's assurances I would be OK, my mother was much relieved. She drove me home. At the doctor's instruction, I lay down on the couch and kept my left arm elevated. I was told to call him immediately should my arm begin to swell. We checked to make sure we could use our neighbor's phone. Mr. Deputy, our neighbor, said of course, particularly under the circumstances. The Deputies were kind, friendly neighbors.

When my dad came in from fishing, he was quite surprised to learn of my mishap. "Maybe we should take him to the Goshen Hospital. What do you think, mother?"

"I don't think we need to," Mom said. "Doc Miller has lots of experience. He told us he had treated rattlesnake bites before. He said from the looks of the bleeding, all of the venom was probably flushed away. I think we should do what he said and go see him tomorrow."

When I woke up in the morning, my wrist still hurt a lot, but it was not swollen, a good sign. Mom and Dad took me to see Doc early the next morning as he was closed Saturday afternoon. After examining the wounds, Doc announced, "I see a tiny bit of yellow flesh, right there at the bottom. See it?"

Mom and Dad both said they saw it. I couldn't see a thing but red, bloody flesh that hurt every time it was moved. Doc brought out a syringe with a long needle. I cringed because I knew what was coming. "This will hurt, young man, but I must inject some pain killer or it will hurt a whole lot more. Try not to move."

I gritted my teeth as the needle point touched my skin and then was plunged in deeply. It hurt, but not nearly as much as I feared. Soon my arm and hand were numb. I watched, fascinated as he used a tiny pair of scissors to cut out those few telltale yellow spots from deep inside of each tear. I could see the yellow clearly on the small pieces of flesh he wiped onto the towel under my wrist. He covered each wound with a purplish liquid and began sewing them up. My arm was so numb, I couldn't feel a thing.

Doc explained, "Please see to it these wounds are kept clean. Use a loose bandage, and call me if the flesh around the sutures swells or turns red. After three days, clean them with mild soap and water, and continue to do so each day for at least two weeks. Then I want to see you again. You were lucky, young man. You could have been sick or even died from that bite. I think the way you ripped the snake's fangs out of your arm saved you from lots of misery. Oh yes, don't go in the water until these heal completely. It will take at least three weeks."

My wounds healed cleanly, and I had no further problems the rest of the summer. My biggest problem was not being able to go swimming with my buddies. Those three weeks seemed like forever. Of course, I was always a bit wary around our rock seawall. We never saw or heard of another rattlesnake around there. To this day I have two white scars on my left wrist where those fangs did their dirty work.

My Dad: The Bluegill Fisherman

Daddy, Dad, and Pop were the names I used to address my father during childhood, as a young adult, and then as a business partner and father to some of his grandchildren. Each change signified an important shift in our relationship. No boy had a more caring, loving, proud, or concerned father. Quite strict by today's standards but far more fair, he was a thoroughly honorable and patient man—a wonderful role model. I loved him dearly throughout our relationship over the forty-five years we shared on this planet. There are times now when I miss him terribly.

The one activity we shared our entire life together was fishing. I can honestly say there is no time I can remember when fishing was not an important part of being at the lake. Though I hardly recollect it, I was told when I was three I caught my first *eating sized* fish, a bluegill. I caught it on a handline over the side of a rowboat in front of our rented cottage on Lake Tippecanoe.

Dad put up with a lot, teaching me to fish. He endured countless verbal assaults like "I want to go in. Let's go fish in Grassy Creek. I'm thirsty. Can't we move?"

He responded to many of my less-than-patient activities with requests like "Can't you sit still for a minute? Quit slashing your pole around. Don't jerk your line out of the water. Can't you bait your own hook by now? We'll go in when it's time to go in." Years later, my son, Howard, paid me back for what I did to my dad.

Eventually, the lessons sank in, and our fishing trips became enjoyable, cooperative joint adventures. Too countless to remember, some were quite memorable. We fished together in all kinds of weather, mostly in the east end of the lake in an area called the flats. My dad liked to catch bluegills and spent countless hours sitting with two cane poles jammed under his seat cushion patiently watching those red and white bobbers with long spike tops setting in the water. At the first sign of movement as the bobber jiggled or started to go under, he would grab the pole and wait until the tip disappeared. Then he would yank the pole smartly upward to set the hook. This action was always followed with *Got 'im!* as he pulled up the struggling fish or *Missed!* as he hauled in an empty line. Over the years, I witnessed those thrilling actions and the concurrent comments countless times.

359

**My Mom and
Dad in 1940 on a
Sunday at the lake.**

As a small boy, I spent the entire summer at the lake with my mother and two older sisters. Friday afternoons, Daddy, as I called him then, would drive up from Indianapolis to spend the weekends. I remember so may times, late Friday afternoons, waiting to meet him after walking our local dirt, one-lane drive up to the main road. I would trudge from our cottage, around the bend in the dirt road between the Deputy's and Holscher's cottages, through the woods, and up to where our mailbox stood. There I would wait, sometimes for an hour or more, for him to come down the highway. The *highway*, as we called it, was a two-lane dirt road leading to town and the paved highway. Sometimes my sister Bobbie, six years older, and even Lois, twelve years older, would go with me to experience the meeting ritual. With every cloud of dust in the distance, my excitement would mount until our car appeared.

When he stopped to pick me up, a bound through the open door, a hug, and I would be greeted with "How are you, young man?" and then "How's the fishin'?"

I will never forget those delicious, happy, oft-repeated moments of reward at the end of a sometimes long wait. Rain or shine, I walked up to meet my daddy and ride with him those few hundred feet to the little cottage where Mom always waited to greet him with a big hug and kisses. When he got there in the afternoon or early evening, Mom would fix a snack while Daddy changed into his fishing clothes. Then we would go fishing before dark. My sisters were not interested in fishing, having *girl* things to do.

I remember clearly one trip when Daddy arrived early. As he loaded the tackle in the boat, he said we were going all the way to the *flats* to fish. With Mom in the back of the boat, and me in the narrow seat in the prow, we headed out. Since we couldn't afford an outboard motor, Daddy rowed nearly a mile to his chosen fishing spot, between two stakes in the lake about five hundred feet apart. After lowering the main anchor, a three-foot piece of two-inch tile pipe filled with concrete, we got ready to fish. We unrolled the lines from our cane poles, attached bobbers for about seven feet of water, threaded wiggly red worms onto our hooks, and tossed them out.

Mom caught the first fish, as usual, a nice, hand-sized bluegill, and Daddy made his usual comment, "Well, Mother, you did it again. Why is it you always catch the first one?"

It was typical of the loving jabs that often passed between them. I can never remember them exchanging a harsh word. By the time the sun reached the horizon, we had a pretty good catch—about a dozen bluegills.

Daddy announced it was time to go as he began pulling in the anchor. "We'd better head for home now. It'll be almost dark by the time we get there."

Reluctantly, we started pulling in our lines as Daddy began rowing slowly to the west. Suddenly, Daddy was struggling with a fish, saying, "It must be a dandy. Get the landing net."

That was evident by the arc of his pole. Mom and I moved our poles out of the way and prepared to help. He pulled an enormous crappie to the surface and toward the boat. Deftly, Mom slipped the net under the fish and into the boat, saying, "That's about the biggest crappie I have ever seen!"

"It's a big one all right, and on a worm to boot," Daddy replied while I could only stare in wonder. Crappie were usually taken on small minnows, rarely on worms. "That's a fitting ending to a great fishing trip."

My father in the front yard of our cottage in 1944.

Soon all our fish were in a sack placed in a bucket on the bottom of the boat and covered with a damp cloth, so they would stay alive for the trip home. It was my job to keep the cloth damp during the trip.

The first thing we did when getting to our pier was dump the fish into the live-box. Hopefully, they would all live until tomorrow when they could be cleaned in daylight. After pulling our boat ashore, we stored the poles, leaning them against their usual tree. We dumped the worms into their storage bucket and placed it back under the house. The landing net and fish sack were hung out to dry, and the seat cushions were taken into the house.

While Mom was fixing something to eat, Daddy and I went out to check on the fish and measure the monster crappie. By the light of our flashlight, we could see several white bellies near the bottom of the live-box.

Daddy remarked, "Looks like several of them aren't going to make it. We'll let 'em go 'til after supper. Maybe they'll come to. They are all still moving, see?"

As I watched, one of the white bellies turned over and disappeared as the dark upper part of the fish blended into the dark water. "You're right, Daddy. One of them swam right side up. Where's the crappie? Should I get the net?"

"See the big one right at the top? That's him. I can catch him in my hand, measure him, and put him back before you could get the net. Here, hold the flashlight for me."

"Don't let him get away," I warned as Daddy lifted the crappie and placed the yardstick along his length.

"Sixteen inches! Wow, he sure is a big one—the biggest I ever caught. I'll bet he weighs more than two pounds." With that, he carefully placed the fish back in the live-box. "Now let's go get something to eat. I'm sure supper's ready."

After dinner, the girls did the dishes while we men went out to check on our fish. As Daddy lifted the lid, I shined the flashlight into the live-box.

"I don't see any bellies. They must all have come around," I reported.

"I'll lift the box up to be sure," Daddy said as he pulled the heavy wood-framed hardware-cloth box slowly out of the water. Water flew everywhere as frantic fish flailed the water, trying to get through the wire enclosure. "They're all pretty lively," he sputtered through the spray as he quickly lowered the box back into the lake. "I'm sure they'll be okay 'til tomorrow. We can clean them along with those we catch in the morning."

Tomorrow, the family could look forward to a scrumptious fish dinner. In addition to Mom's delicious fried fish, there would be corn on the cob, parsley buttered potatoes, and sliced cucumbers and tomatoes on fresh lettuce. There would be chicken or a hamburger patty for Bobbie, who didn't care much for fish. If we were lucky, dessert would be one of Mom's luscious cherry or blueberry pies. Daddy and I would argue for years which was the better, cherry or blueberry. He settled on cherry as his favorite while I chose blueberry. Probably, it was my first effort trying to be an individual, different from Daddy. After our early morning fishing trip and a family breakfast, it was time to clean our fish. I was an apprentice fish cleaner to Daddy and was expected to participate fully in the messy job. I reveled in the blood and guts and in the terror my bloody hands could bring to my two sisters. That was about the only way I could get ahead of them, and I used it as often as possible. Now it was time to go to work.

Daddy took the landing net and scooped the fish out of the live-box, placing them in a bucket or even a washtub if there were lots of them. After carrying them to the backyard, he placed the bucket on the ground next to the wooden fish cleaning table he had made with two-by-fours and the wooden side of a dental chair crate for a top. I was official scaler while Daddy handled the surgery. While I was proficient at removing the scales, my training in removing head, guts, fins, and tails was in the early stages. Daddy would always save a few small fish for me to butcher. At six years of age, it would be some time before I was a fully qualified *cleaner*. My next job was to bury the heads and guts in the woods, usually among the nettles. These common plants often stung me and left me *itching like crazy* when I brushed against them. When the messy job of cleaning and filleting was done, and the tender flesh lay in a pan in the sink, it was time for me to go swimming.

When I wasn't fishing, I was swimming. Many a warm summer day while Daddy was in Indianapolis, I spent morning 'til night in the water, only coming out to eat, and then reluctantly. Daddy began referring to me as his *water dog*. When he was at the lake, I usually sacrificed swimming to go fishing with him.

By the time I was thirteen, I had caught *bass fever* and spent most of my time fishing for those bronze warriors of the deep. About this time, Daddy became Dad. Along with this came a change in our fishing. Now, with a tiny Evinrude outboard on the back of the rowboat, we fished mostly in the flats and for both bass and bluegill. Early in the summer, we would find a spot near the weeds where the water was deep enough for bluegill yet close enough to the weeds for bass. In addition to our cane poles, we used casting rods and reels, bass hooks, and small perch for bait. We had an ongoing contest to see who could catch the most weight of fish, he with bluegills or me with my bass. As I remember, it was pretty much of a toss-up. Late in the summer when the bluegill moved into deep water, I would forgo the bass fishing, and we would fish together for deepwater bluegill.

My father and I made many fishing trips together over the years. I still remember those cherished times and all I learned from a wise father while we fished. One incident in particular I remember. It was late October in the mid-sixties when we were in the dental business together and I had long ago changed to calling him Pop. We were on his pontoon boat, fishing for crappies after going through the channel and into Little Tippy Lake. It was a cold, wet, misty day with pewter sky overhead. As we huddled against the damp cold, two duck hunters came by and proceeded to set their decoys near the shore a hundred yards or so from where we were anchored.

As they climbed into their blind, my father commented, "Can you imagine? Those crazy guys came out in this miserable weather to hunt ducks." We looked at each other and grinned broadly.

I can still clearly picture him bundled in warm clothing and a raincoat, pipe clenched between his teeth, pole tightly gripped in freezing hands, sitting in his chair on the aluminum pontoon in the damp, penetrating, bone-chilling cold.

One of my favorite photos of my father as a young baseball player about 1908, long before I was born.

One valuable thing I learned about children is they belong to themselves, not to their parents. As parents, we are charged with guiding the infant into childhood, the child into a young adult, and the young adult into a full person. It's an awesome responsibility for which so many are ill trained. Parents who try to *possess* their children make a serious mistake and frequently drive them away. God loans us children for a short while, charging us with guiding them to independent adulthood and completely equal status. They are for us to teach, enjoy, love, and set free.

—Howard Johnson, 2010

I have learned only too well to understand—in part at least—the reason why I have felt my heart break and had to watch my pride burn in outward composure. Time and again, beneath a mask of composure and admirable manners, I have inwardly laughed at another's downfall. Yes, I know it. But I know no law that binds me to betray my grief when destiny permits another's malice to inflict a penalty I owe.

—Talbot Munday, from the log of Lord Tros of Samothrace

As you think, you travel; as you love, you attract. You are today where your thoughts have brought you; you will be tomorrow where your thoughts take you. You cannot escape the result of your thoughts, but you can endure and learn, and accept and be glad. You will realize the vision (not the idle wish) of your heart—be it base or beautiful, or a mixture of both—for you will always gravitate toward that which you secretly must love. Into your thoughts you will receive that which you earn—no more, no less. Whatever your present environment must be, you will fall, remain, or rise with your thoughts, your vision, your ideal. You will become as small as your controlling desire, as great as your dominant aspiration.

—Howard Johnson, 1971

Twenty years from now you will be more disappointed by the things that you didn't do than by the ones you did do. So throw off the bowlines. Sail away from the safe harbor. Catch the trade winds in your sails. Explore. Dream. Discover.

—Mark Twain

Hitherto I have found my real goal unattainable. But I persist, since the attainable is no more than a rung on the ladder of life on which a man may climb to grander view, though it will break beneath him if he lingers too long.

—Talbot Munday, from the log of Lord Tros of Samothrace

The Wonderful Summer of '42

I was a year old when my parents first took the whole family to Lake Tippecanoe in Northern Indiana. We stayed for a few weeks in a rented cottage on Walker's Park. The following summer, and until 1933, we rented Holloway's cottage for the entire summer. It was on nearby Blacks landing. In 1931, my parents scraped up $300 and purchased a small lot on Walker's Park. I watched as the swampy lot was filled. In 1932, our tiny cottage was built on the site. I can remember walking on the floor joists when they were all that existed of the soon-to-be lake cottage. The window-filled front of the cottage was fifty feet from the stone sea wall lining the water's edge. There was a ten-acre tract of woods we called *the Swamp* on the other side of the dirt road behind the cottage. I loved all those endless summers in the water and in the nearby woods. I formed a firm attachment to the outdoors and the wild during those glorious years of growing up.

Many boy's first love is his mother, and I was no exception. When I was young, I had special terms for each member of my immediate family. My dad was my *buddy*; my sister Lois, twelve years my senior, was my *honey*; my sister Bobbie, six years older than I, was my *pal*; my mother was my *sweetheart*. Kind, soft-spoken, a committed Christian, and a gentle powerhouse, my mother was quite an exceptional person.

As a youth, I frequently accompanied my mother to church when she worked preparing many church dinners. I was often asked to help, which I did begrudgingly. My mother not only helped with the dinners, she ran them with an unusually gentle control. Her peers held her in great respect. In fact, any organization or group she joined soon elected her president. These included church groups, the Cleveland Heights High School Choir mother's club when my sister Bobbie sang in the choir and six years later when I did, and my college fraternity mother's club—to name a few. She never sought these offices. They just came to her.

During one church dinner preparation, the minister sat down and talked with me about my mother. A member of the Euclid Avenue Christian Church in Cleveland since I was eight, I held our minister, Dr. Jacob Goldner, almost in awe. I was fourteen at the time, and it was the spring of 1942.

"Your mother is a real wonder," Dr. Goldner told me. "She does an unbelievable job organizing these dinners and getting the women of the church to work together. She can persuade women who usually won't even speak to each other to work together harmoniously. I've never seen anyone like her. I only wish I had some of the same ability."

Dr. Goldner was not an ordinary minister but one who led the same large church for more than forty years and was nationally well known. I was quite impressed, and for the first time, I knew my mother was not just special to me because she was my mother. She was a special and unusually talented woman, and was known as such by many people. From then on, I had even more admiration and respect for her.

The summer of '42 was a truly memorable, summer-long experience with my mother. It will be one of my fondest memories for the rest of my life. In many of our conversations since then, Mom recalled it warmly as well. It was a spectacular, learning summer of growth and realization for me under my mother's tender, loving, watchful guidance. Probably because it seemed a bit more grown-up, I began calling my mother, Mom, during the last free summer of my childhood.

Because of the war and gas rationing, my father couldn't come to the lake every weekend as he usually did. My sister Lois had married a farmer and lived six miles from the cottage. My sister Bobbie worked in Indianapolis for the summer and rarely came to the lake. Most of the time it was Mom and I. Having no car or phone, we were quite isolated and dependent on Lois and her husband, Harold, for transport and communications. Many of my experiences on and in the lake were shared daily with Mom, my attentive listener. We talked a Great deal at meals we shared at the drop-leaf table in the tiny breakfast room.

An avid fisherman, I went fishing every day, usually by myself. Our boat was a small slab-sided rowboat about fourteen feet long with three seats. My father had purchased a tiny one-and-a-quarter-horsepower Evinrude outboard motor, so we wouldn't have to row the one and a half miles to the *flats* where we usually fished. While my father was a *bluegill* fisherman, I had caught *bass* fever and constantly sought these larger fish. I had learned to fish for bass with live bait from our neighbor, Dan Hackerd, who had a cottage two doors to the west of our place. I admired the huge fish he frequently placed in the live-box attached to his pier. Once he told me his secrets about catching these *green monsters,* I was hooked. Mom enthusiastically supported my fishing. She walked to the end of the pier to see me off, and to greet me when I returned.

One typical Friday morning, I headed the little rowboat in from fishing. As I rounded the point and our cottage came into view, I saw a familiar sight. From a quarter mile or so away, I watched a tiny figure come out the door, down the path of concrete blocks I helped my grandfather make, and to the end of the pier. Mom always came to congratulate her returning fishermen or encourage them when they came in empty-handed. After checking my morning catch, she headed for the cottage, remarking, "Get those fish taken care of. By the time you come in, lunch will be on the table."

"Okay, Mom!"

After taking care of the requested chores, I headed in for lunch.

"You wash your hands," Mom said as soon as I walked in. "Lunch is on the table."

Mom on the pier as she watched and greeted returning fishermen.

I stepped to the stainless steel sink in the small kitchen and turned on the cold well water that came directly from an artesian well outside, near the kitchen. There was no water heater, so the water was the temperature of the deep ground, about fifty-six degrees. It was great for drinking, but quite cold for washing. There was always a kettle on the stove to heat water when necessary. As soon as my hands were dry, I walked into the tiny breakfast room and sat down at the round drop-leaf table across from Mom.

I faced the front of the house and had a wide view of the lake through the windows all the way across the front. I could also see the lake and cottages along the eastern shore out the window to my left. To the south, the sun made twinkling diamonds on the gently rippling surface. Children's happy voices floated through the open windows as they played in the shallow sandy-bottomed water near to shore. Other sounds could have been an occasional motorboat, a dog barking, a rooster crowing, and the songs of many birds and insects. The sights, sounds, and smells were positively intoxicating.

**Four generations: Mom, Patte Ann, Grandma
Dickinson, Dotte Mae, and sister Lois (1941).**

After I sat down for lunch, Mom looked at me and smiled. "All right, tell me about it."

Soon I was deeply engrossed in tales of the mighty struggles with the new occupants of the live-box, with Mom an attentive listener.

When my father arrived Friday afternoon, Mom greeted him with a hug and a kiss then turned toward me. "Your son has something to show you in the live-box."

I proudly showed my catch to my father who commented, "You've become quite a fisherman."

Saturday, Harold, Lois, their twin baby daughters, and the senior Swensons arrived bringing our midday dinner from the farm. As Harold and his dad walked by, I proudly pointed to the two new fish heads nailed to the tree in the backyard. My prizes would be the main course of our meal. While the women prepared dinner, I again wove the story of my catch for an attentive Harold and his father. Harold, some sixteen years older than I, was like a second father to me. I spent many happy hours walking at his side on their farm as he worked or hunted.

The little cottage as it looked from the pier in 1942

Many times during the summer, I told Mom tales of stalking, catching, and sometimes losing the wily fish in our lake. She never seemed to tire of hearing my stories. In turn, she told me about her life as a girl growing up with three brothers and then, finally, a sister after she had married and left home. She also told me tales about my two older sisters, stories of delightful mischief about them and me when I was tiny. We formed a special bond that summer, one neither of us ever forgot. I can still see her sitting across that table by the window, the sunlit lake behind her, her soft, pleasant face attentive to my words. Or in the morning at breakfast, the morning sun streaming through the trees and sparkling on her face. We must have enjoyed each other in this fashion a hundred or more times that summer.

The fishing ritual could happen as many as three times a day, weather permitting. There was the early morning trip before breakfast, the midday trip I described in detail, and the evening trip starting anywhere from four to six and lasting until near dark. Once in a while Mom would go with me for the evening trip, particularly when the bluegills were biting. When she went along, we always fished for bluegills. In any event, I could always count on seeing that tiny figure walk out on the pier when I returned.

When I couldn't fish because of bad weather, we would sit and play Aggravation, Parcheesi, or cards. Rummy, Hearts, and Rook were our usual card games, and Mom was positively brutal. If I beat her, it was not because she took it easy and let me win—a lesson for success in life.

Mom was a soap opera addict, so I listened with her on rainy days to *Stella Dallas*, *Ma Perkins* and her sidekick Shuffle, *Helen Trent* who "found romance even after thirty-five," *Our gal Sunday*, the foundling who married England's richest and most handsome Lord, Henry Brinthrop, and several other soap operas.

Late afternoons my shows would come on. There was *Little Orphan Annie* and Daddy Warbucks, *Jack Armstrong* - Wheaties' All-American boy, *Captain Midnight*, *I Love a Mystery* with Jack, Doc, and Reggie, and a few others.

Later in the evenings it would be the comedy of *Fibber McGee and Molly*: *The Great Gilder sleeve*: *Jack Benny and Mary Livingston*: *George Burns and Gracie Allen*: *Bob Hope*: *Red Skelton*: and *Fred Allen and Portland Hoffa*. On Sunday nights, we listened to the trials of *One Man's Family* by Carlton E. Morse, the howls of the hounds at the start of the *Tales of Sherlock Holmes*, and the creaking door that introduced the *Inner Sanctum Mysteries*. With no picture box to view, our minds imagined the scene as each action was so vividly described.

Sometimes, Lois and Harold would come down bringing their two-year-old twin daughters. Then we would play Pinochle or Euchre, usually after the twins went to bed. Other times we would go up to the farm and play cards with them along with Harold's parents, Anna and John Swenson. Many a story was told during those card games, and I learned a lot more about past family happenings. I even got a few tales in myself. When it got late, Mom would frequently close her eyes and drift off. Sometimes the cards spilled from her relaxing hands as she went to sleep.

"I'm just resting my eyes," she would say as she returned to the game when one of us reminded her to play. When this happened, we knew it was time to quit.

The war raging in Europe and the Pacific was a constantly expressed concern, yet seemed so remote. The rationing of gas, tires, and food and some shortages were the only direct effects we felt. Somehow, we managed to have enough gas for the tiny outboard. With an endless supply of fish from the lake, eggs and chickens from the farm, and produce from the neighborhood, we ate quite well. With a childhood during the Great Depression, I grew up accustomed to few material things. I also learned to leave a clean plate. Not knowing any better, I was happy with what I had which was far more than many children my age. Those were marvelous, happy times, the closing summer of a treasured childhood in a close-knit, loving family.

It was the last full summer I spent at the lake for more than forty years. From the following summer on, I worked or went to school for most of the summer. In the years to follow, there would be many happy, memorable times together with family and friends. It was wonderful with everyone crammed into our tiny cottage for meals or out in the front yard when the weather was good. Sometimes cousins from Indianapolis would visit for a weekend, often bringing my maternal grandparents. There would be bodies all over the living room floor when I awoke in the morning. The chorus of snores from my dad and my grandparents kept the house rocking all night. There was always room for one more to sleep on a cot or sleeping bag on the floor. Those glorious times continued for years.

As the years went by, there were a few more times when Mom and I ate at the little table by the window with the morning sun streaming in through the trees. When we talked, we would often recall the wonderful summer of close companionship we shared in 1942. Little did we know then that she and I would spend another full summer together at the lake some forty-two years later.

Laughter and love are the handmaidens of joy. It's impossible to keep one without having the other. There can be no love without laughter and little joy without them both.
—*Chelton Chum in the novel, Blue Shift - Howard Johnson 2002*

I've learned in my lifetime so far that you can't help who you fall for and no matter how hard you try and how much it hurts you everyday that you just wanna be with them or just talk to them you never stop trying to make them happy by the little things you say or do because that's what makes your life worth going on for.
—— *Unknown*

Love me Sweet, with all thou art,
Feeling, thinking, seeing;
Love me in the lightest part,
Love me in full being.

Love me with thine open youth
In its frank surrender;
With the vowing of thy mouth,
With its silence tender.

Love me with thine azure eyes,
Made for earnest grantings;
Taking colour from the skies,
Can Heaven's truth be wanting?

Love me with their lids, that fall
Snow-like at first meeting;
Love me with thine heart, that all
Neighbours then see beating.

Love me with thine hand stretched out
Freely -- open-minded:
Love me with thy loitering foot, --
Hearing one behind it.

Love me with thy voice, that turns
Sudden faint above me; Love me with

thy lush that burns
When I murmur 'Love me!'

Love me with thy thinking soul,
Break it to love-sighing;
Love me with thy thoughts that roll
On through living -- dying.

Love me in thy gorgeous airs,
When the world has crowned thee;
Love me, kneeling at thy prayers,
With the angels round thee.

Love me pure, as muses do,
Up the woodlands shady:
Love me gaily, fast and true,
As a winsome lady.

Through all hopes that keep us brave,
Farther off or nigher,
Love me for the house and grave,
And for something higher.

Thus, if thou wilt prove me, Dear,
Woman's love no fable,
I will love thee -- half a year --
As a man is able.

— *Elizabeth Barrett Browning*

At last we two are free! I of you, and you of me.
Love lying in its grave. Silence is all I crave.

—*Unknown*

For the things we have to learn before we can do them, we learn by doing them.

—*Aristotle*

My First Love, Dolores

I don't remember when it started, but sometime during the eighth grade, when I was in Roxboro Junior High, I fell for a girl I thought was the most beautiful girl in the world, Dolores Osborn. I pursued her for some time in spite of her lack of interest. Oh, she was pleasant enough with me, and I did manage to take her to a movie or two, but I knew I was not on her list.

I remember walking her home along Fairmont Boulevard after we had a soda at the Dairy Bar, one of the kid's favorite hangouts. She told me about *pididdle*, a word one says when a car with one headlight appears, an invitation to a kiss, much like standing under mistletoe. If the other one of the couple repeats *pididdle*, it means they would also like to share a kiss. No sooner had she told me this when a car with one headlight appeared. I bravely said, *pididdle*, but there was no response. I struck out again.

The ninth grade dance was *the* social event of the year at Roxboro. To have a date was important, and needed to be arranged early or all the girls would be taken, at least all of the popular girls. At least two months before the dance I took Dolores to a couple of movies and worked up the nerve to ask her, "Would you go to the ninth grade dance with me?"

"Let me think about it for a day or two," she replied. "I'll have to ask my parents. They don't want me to *date* as yet."

Apparently, her parents did not consider going to the Dairy Bar or movies as *dating*. I didn't know if this was her way to say no or not, so I would have to wait for her answer on Monday. I thought the day would never come. I sat with Dee at lunch and soon had my answer, yes. I was ecstatic. I now had a date for the biggest event of the year with my dream girl.

Two weeks before the dance my dream was scuttled. Dolores came up to me and said, "I've been going out with Freddie (my buddy Fred Hunziker) for some time now, so I can't go to the dance with you. I'm going with Freddie."

How could any girl be so cruel? Easy, I guess. It was now far too late for me to get a date. All the available girls had long ago been asked. I did get even in a way. Dolores loved to dance and Freddie did not dance. During the evening, I danced every dance, many with close friends of Dolores. She sat at their table with Freddie and never did get to the dance floor.

After the dance, I went with a group of friends, mostly female, to the Dairy Bar for an after party soda. There was one couple and two girls who did not have dates because their parents did not permit them to date. (Shades of a bygone day) I was seated between the two girls who were being quite chummy when who should walk in but Dolores and Freddie. The two girls knew how I had been shafted (as did everyone at school) so they both became, shall we say, amorous—right then and there.

One of the girls, a close friend of Dolores's, put her arms around my neck and called out to Dolores, "We're sure glad you dumped Howie for the dance. We're having a great time with him." It was not long before She and Freddie left.

When I started at Cleveland Heights High, one of my dreams was to sing in the choir. Even though I played an instrument and was in band and orchestra at Roxboro, I joined the chorus, a prerequisite to being in the famous, *Heights High School A Capella Choir.* When I auditioned, I earned a spot in the second tenor section. I was thrilled, knowing I would be in the next Christmas concert. Another choir member was my old crush, Dolores Osborn. Though I had taken her to the movies a few times, she was not interested in me. I had given up asking her for dates and was reduced to admiring her from afar.

December 17, 1943—everything changed. The choir was to sing downtown at Higbee's department store, an annual event. Given permission to drive the family car, I asked several members of the choir including one girl who lived near Dolores. When she asked if I had room for one more and told me Dolores needed a ride, I, of course, agreed.

After the performance, the choir was treated to a short party with refreshments in a room at Higbee's before we headed home. I sat with Dolores during the party where she did not seem so standoffish as before. As I took everyone home, I happened to end with Dolores as my last passenger. We spent the next two hours in the car in front of her house. I told her how I had felt about her since we met in the eighth grade. Before long we shared our first kiss. After some amorous conversation and quite a few more kisses, I walked her to her door. My feet wouldn't touch the ground, my mind was whirling, uncontrolled, and I was deliriously in love.

At the choir party after the annual Christmas concert, I gave her a card with the words "A penny for your thoughts" and enclosed a bright new penny. Sometime later, she returned the favor and the penny. I still have that penny.

We were to share many Christmases together. We would also walk onto the stage together at Heights as choir alumni many times, sing *Emitte Spiritum Tuum,* and share memories with choir friends. We managed to sing *Emitte* one last time in 1996 at our fiftieth class reunion with about forty choir alums.

In September 1949, we were married. I got even with her for standing me up for the ninth grade dance. I never did let her forget that. In the years that followed, we had five children and then were divorced after she betrayed my trust by having an affair with one of my friends. Ironically, our divorce was final on our twenty-fifth anniversary. After our divorce, she married her long time boy friend. I eventually got over it, and we had a friendly relationship for many years, something of great benefit to our children. Fortunately, I am blessed to be one of those who remembers the good parts of life and can forget or minimize the bad. Her husband died of a sudden illness a few years later, after they moved to Detroit.

She passed away some twenty years after our divorce. In a testimony to how positively we had dealt with our children, they asked my wife, Barbara, a Methodist minister, to conduct the memorial service for their mother. Dee and Barb had grown to be such good friends they often went Christmas shopping together for gifts for our children. The service was a celebration of all of the positive things in Dee's life.

In later years I often thought our problems and subsequent parting were a bit of a blessing in disguise for me, once I recovered from the resulting emotional trauma. Had that not happened my life would have been very different. I would probably not have had the several spectacular relationships that I have experienced. There are several stories about these women elsewhere in this book.

Without doubt the most famous love poem in English literature is this famous work.

> How do I love thee? Let me count the ways.
> I love thee to the depth and breadth and height
> My soul can reach, when feeling out of sight
> For the ends of Being and ideal Grace.
> I love thee to the level of every day's
> Most quiet need, by sun and candlelight.
> I love thee freely, as men strive for Right;
> I love thee purely, as they turn from Praise.
> I love with a passion put to use
> In my old griefs, and with my childhood's faith.
> I love thee with a love I seemed to lose
> With my lost saints, I love thee with the breath,
> Smiles, tears, of all my life! and, if God choose,
> I shall but love thee better after death.

— Elizabeth Barrett Browning

So, fall asleep love, loved by me... for I know love, I am loved by thee. Take away love and our earth is a tomb.

— Robert Browning

Two souls with but a single thought, two hearts that beat as one.

— John Keats

Aye . . . like a snowflake in flight
    between sky and ocean are we.
        Beauty for an instant . . . Never to be again.

—Howard Johnson, 1973

People are lonely because they build walls . . .
        instead of bridges.
        —Joseph Fort Newton

Ice-blue eyes - secret tears - pain hidden in the heart - bright laughter - running, running - broken child's world - stifled fears - steeled, hard-shell - soft love-warmth - sun and bright sky to black midnight - the now trapped - foiled understanding - tenuous dream-wish - listen, hear - being freedom - broken chains - fly to - reach - have - love - borrow - miss you . . .

—Howard Johnson, 1965

Of all Life's difficulties I have found it hardest to compel myself to recognize and concede a woman's right to meet me on even terms. But it seems equally hard for a woman to understand my attitude. No more than all the priests, philosophers and poets do I know what love is. Unlike many of them, I am unwilling to pretend that I do know. Neither do I know what life is. But it seems to me that if love or life lack dignity, neither the one nor the other is worth the sacrifice of half a moment's thought.

—Talbot Mundy in Purple Pirate

True beauty is not in the eye but in the **heart** of the beholder!

—Howard Johnson, 1968

To be alone and without love is a waste of the heart.
To be with someone and without love is a waste of the soul.

—How Chee Loo, 1863

Granddad Dick

My grandfather, George A. Dickinson, Granddad Dick to all his grandchildren, was a marvelous man and mentor to me. From my earliest remembrances until he died in 1964, he held a special, almost magical place in my life and in my heart. Second only to my father in my male admiration, he also had a quite different, almost opposite personality. My father was a steady, dependable, hard working, conservative man who had to quit school after the eighth grade to go to work and help support his family.

My grandfather was a story teller. I would sit by the hour, fascinated as he wove his tales for me. He had a rather varied, up-and-down life with quite a few adventures. He was the fourth of nine children all born in Ohio and Kentucky around Cincinnati. He and all eight of his siblings lived into their mid to late nineties. I have no idea how much schooling he had, but he told me of having a medicine show with an ex slave. They started when both were in their teens. They rode all over Kentucky and southern Ohio from town to town in a medicine wagon pulled by a single horse. He told me numerous stories of adventures he and his black partner had with their show. After a number of years on the road, he gave up the medicine show and apprenticed to a plumber in Cincinnati.

When he was in his late twenties, he fell in love with Eva Mae Webber. Old man Webber was a drunk who beat his wife and children. Granddad told of taking a ladder and rescuing Grandma from her upstairs bedroom after breaking into their fenced in yard. They barely escaped over the fence with her father in pursuit. She was just seventeen. That's one story grandma confirmed. By the time they were married, he had worked for several railroads, traveling quite a bit. A journeyman plumber, he moved from job to job until his family grew to the point where he left Kentucky. After I was grown, my dad said Granddad told him they moved because, "all those sportin' houses made Dayton Kentucky a bad place to raise a family."

There were four in the family when they settled down in Indianapolis, My mother was four and my uncle Harry two when they moved. Granddad found steady work as a plumber for the railroad. Eventually they built a two story duplex on Oxford Street. They lived upstairs and rented the downstairs to pay for the house. During these years, he sometimes worked far from home. For some time, he was in Florida, working on the railroad. He was there when one devastating hurricane struck destroying the railroad work camp. The money was great, but he decided he had enough of Florida, so he went home. Gradually, his traveling gave way to steady work in Indianapolis as a plumber. He ended up working for the Claypool hotel.

By the time I was old enough to listen to his stories, some thirty-five years after the move, he was a part time plumber for the Claypool Hotel and worked irregular hours. It was during the depression and jobs were few and far between. Grandma supplemented their income by baking, mostly pecan rolls and pies. I can still taste those pecan rolls my folks bought from Grandma. They were fantastic. She sold all she could bake, even with competition from a small bakery nearby.

I can still hear her voice from the kitchen when Granddad was telling me one of his stories. "George! You quit filling that boy's head with your nonsense."

Granddad would quiet down, but keep right on tellin'.

Woven into each of his stories were bits of his home spun advice about life and people. As the years passed, I realized more and more the immense value of those simple lessons. How I wish that he had written down some of those stories of his youth and family times. That is why I write memoirs and urge others to do the same. Oh, what I wouldn't do for some records, some writings from any and all of my forebears. I remember a few of the stories I was told, but most are gone forever. One particular lesson Granddad taught me was about people who put down and denigrate others.

It was during one of our fishing trips when I was between twelve and fourteen. Granddad was recovering from abdominal surgery and was quite weak. He and Grandma were staying with us at our lake cottage while he recovered. The day was cloudy, but quite warm. We were anchored on the edge of the weeds in about ten feet of water, right in front of the cottage. I was on the outs with one of my friends and telling Granddad how I felt.

"I'm gonna tell that Paul a thing or two, cut him down to size. He doesn't know what he's talking about."

Granddad scowled. "Why would you do such a thing? Paul's one of your neighborhood chums, isn't he?"

"Aw, he's got some cock-eyed scheme for finding Indian arrowheads and sellin' them. Says we could make a lot of money."

"I suppose you think it couldn't work. Either you wouldn't find any, or if you did, you couldn't sell them. Is that right?"

"Yeah! It will never work. He doesn't know what he's talking about."

"Why do you think it wouldn't work?"

"He don't know nothin' about Indians, or where they might have walked. If we did find any, who would we sell them to? He wants to talk to me about it after we come in from fishin'. Why should I listen to him? I've got better things to do. I'll tell him he can just bug off."

"Without listening to what he has to say?"

"I said he didn't know what he was talkin' about, didn't I?"

"You know that for a fact, even without hearing what he has to say?"

"Well, no. But how would he know anything about Indians around here?"

"Why don't you ask him and find out? You'd rather shut him up and never hear what he has to say. Is that what you mean?"

"What could he know? Why should I listen? He don't know nothin' about nothin'."

"Maybe you just don't like the idea he might know something you don't. You act as though you think you know everything about everything, that you're far smarter than he is, and couldn't learn anything from him."

"Well, I am a lot smarter than he is. I'm a year ahead of him in school."

"That doesn't mean you're smarter. It just means you've been exposed to more formal education than he has. Besides, why would you want to put him down? He's one of your best friends, isn't he?"

"Well, yes."

"Then why would you want to put him down, insult him? I would think, that as a friend, you would want to build him up, not cut him down. Maybe he has a really great idea. You won't know if you never listen to him. How can you say it's no good when you know almost nothing about it? Howard, it has been my experience that a person can learn things from practically anyone else, regardless of what you think of them. All you have to do is listen."

"I don't know, Granddad. Some people are just stupid."

"Think about what you have said. You said you wouldn't listen to what he wants to tell you, and you don't even know what it is. When we go in from fishing, why don't you ask Paul to explain his idea? Then listen carefully to what he says and think about adding your own ideas. I'm sure you'll have some. Maybe between the two of you a good, practical idea will come up, something that will work. You'll feel a lot better than if you put him down. That I guarantee."

I sat in silence thinking about what granddad had said. Maybe Paul did have a good idea. I sure wouldn't know unless I let him tell me. I knew that if I put him down he certainly wouldn't tell me. The biggest thing was that it was Granddad telling me this, and he was my idol.

"OK, Granddad. I'll try your way. It makes sense the way you explain it."

"There's another thing. Your friend, Paul, will be impressed if you really listen to what he has to say. It will solidify your friendship and he will respect you all the more. People will usually listen to the words of those who listen to them. In the same way, people will usually respect those who show them respect. That's pretty much human nature."

Well, I listened to Paul's ideas and together, we put them to work. Over the summer, we dug and sifted tons of sand in an area at the end of our landing known to have been near an Indian trail. At least it seemed like tons. As a result, we ended up with a collection of arrowheads and hand axes, some in pristine condition. I have no idea what Paul did with his share, but I kept and displayed mine on a board. It was a source of pride and object of many tales to friends and family.

And Paul? He and I kept in touch for many years. He became an archeologist, earning his doctorate in Arizona. Last I knew of him, he was a professor at the University of Arizona as well as an active field archeologist, specializing in the Anasazi Indians. We lost track of each other about thirty years ago.

❖　　　❖　　　❖

As dead flies give perfume a bad smell, so a little folly outweighs wisdom and honor. The heart of the wise inclines to the right, but the heart of the fool to the left. Even as he walks along the road, the fool lacks sense and shows everyone how stupid he is.

—Ecclesiastes 10:1–3

Words from a wise man's mouth are gracious, but a fool is consumed by his own lips. At the beginning his words are folly; at the end they are wicked madness—and the fool multiplies words.

—Ecclesiastes 10:12–14

"Let me be a free man. Free to travel. Free to stop. Free to work. Free to choose my own teachers. Free to follow the religion of my fathers. Free to think and talk and act for myself."

—Chief Joseph, Nez Perce Indian Tribe

The full use of your powers along lines of excellence.

—definition of happiness by John F. Kennedy (1917-1963)

I'm living so far beyond my income that we may almost be said to be living apart.

—e e cummings (1894-1962)

A Fishing Trip with Granddad

My grandfather, George Dickinson, was a terrific pal for a small boy growing up. Granddad Dick, as we called him, usually shadowed and adored by his little fox terrier, Patches, always had an interesting tale to spin. More than once, when Granddad was telling me a story, I would hear my grandmother calling from the next room, "George! You quit filling that boy's head with your nonsense." He would put on a contrite face, wink at me, pause his story, and then continue when she was out of earshot. Granddad Dick was a master storyteller.

Sometime in his early seventies, he had surgery for stomach problems. He had a Great deal of pain and lost quite a lot of weight before his surgery. It was early summer of 1940 and Granddad was recovering from his latest surgery when he came to the lake to stay with us for a while. I was an excited twelve-year-old because it meant fishing trips laced with interesting and animated stories.

The first hint of daylight was showing one foggy July morning as we loaded our tackle into the tiny slab-sided rowboat. There were three eighteen-foot cane poles with lines securely twisted around the pole. The line had a couple feet of gut leader with a split-shot sinker and long-shank bluegill hooks. The hooks were secured in the butt end of the pole. Each of us brought our tackle boxes and favorite casting rods, mine with a South Bend reel filled with black, twenty-pound test, braided line. We loaded a net bag with its mounting loop and clamp to hold our catch of panfish: stringers for bass: a landing net to help us boat the hoped-for big ones: two minnow buckets, one with two dozen crappie minnows: a box of crickets: and two cans of the old reliable earthworms from the *worm bed* by the hedge. We were prepared for every kind of game fish or panfish in the lake as we pushed the boat away from the pier.

A couple of pulls on the starter rope atop the tiny one cylinder Evinrude outboard and we chugged away into the fog. The fog was so thick I had to follow the shoreline by sighting the trees barely visible in the growing daylight. I picked my way slowly past the ends of piers which loomed suddenly out of the fog right ahead. I knew exactly where I was when I spotted Jimmy Lindley's pier and brown speedboat with its monstrous Evinrude Speedifour outboard. At thirty-three and a third horsepower, it was the most powerful outboard on the lake. Next I spotted Jimmy Batchelder's Wolverine outboard. We were right at Pierce's point and would have to turn east and continue following the shoreline. I went slowly, almost idle speed, peering steadily ahead for the three long piers spaced nearly a city block apart along this stretch of shore before the next point. When I spotted the last pier with George Scott's big inboard tied alongside, I knew we were at the sandbar jutting out from the point. I watched as we went over the sandbar, and the bottom dropped from sight. From now on it would be dead reckoning due east.

I ran the motor at full speed, lining up on the wake to keep the boat headed east. When I thought I had gone far enough, I slowed to idle speed and began looking for the ghostly green fronds of seaweed deep beneath the clear water. I was beginning to think we might be going in circles when I spotted a weed, barely visible in the depths. I shut off the motor and put the oars in place. We wanted to fish near the weed bed in water about six feet deep, the ideal spot for crappies, green gills, red-eared sunfish, and big perch this time of year. I rowed across the now-thick weeds until they suddenly disappeared. Fifty feet past the inside edge of the weeds, we dropped anchor. I handled the back anchor and Granddad lowered the front one.

As far as we could tell, we were the only ones on the lake. There was hardly a sound in the wind-free fog. In the distance and faint, the occasional barking of a dog was answered by the crowing of a rooster. The fog was an effective muffler of most sounds. The silence was occasionally broken by the ripping splashes of small bass feeding on surface minnows nearby.

We each took a cane pole, unhooked the line from the butt, set the butt on the floor, and spun the pole to unwind the line. As soon as the line was free to the tip, we threaded a worm on the hook, attached a bobber about five feet above the worm, and threw the line out. If we didn't start getting bites pretty soon, Granddad would be unwinding the other poles. In the meantime, I unwound the handline I used to catch small perch for bass bait. It wasn't long before a couple of small perch were swimming around in the minnow bucket.

Before I could set a bass line, I hooked a nice bluegill on the cane pole. Almost immediately, Granddad caught a large red-eared sunfish, then another. I decided to try for crappies, so I placed one of the tiny crappie minnows on my hook and threw it out. Zip! The bobber disappeared almost immediately, and I yanked. A giant perch, nearly a foot long, was dropped into the bag. I glanced over the side to admire the four fish caught in the ten minutes or so since we arrived here. "Looks like this is a good spot," I crowed to Granddad.

"You sure know where to find 'em," he replied, raising his pole suddenly and pulling in another big bluegill. After landing another big perch, I caught three crappies in a row, and Granddad decided to switch to minnows. It wasn't long before he added several more crappies and a perch to the bag along with a couple more I caught. Suddenly they stopped biting.

When neither of our bobbers made the least bobble, Granddad remarked, "There must be something big swimmin' 'round down there. Skeered all our fish off."

"Great!" I said. "I'm gonna see if I can catch it. This fat perch on my casting rod might catch something," I remarked as I took a perch out of the minnow bucket and hooked him securely through the mouth. "Doggone it. I don't have any big bobbers for my bass line."

"Have you got a floating plug?" Granddad asked. "You could use it for a bobber."

"Great idea," I said brightly as I looked through my tackle box for a suitable plug and picked up a red-headed popper. "This baby will do fine." Using the screw eye on the front of the popper, I secured it to the line by about three feet above the bait. "Now, let's see if we can catch the big one that scared all our crappies and perch away," I remarked as I cast out the perch with its improvised bobber. We must have sat there for nearly half an hour with no action when suddenly my reel began to spin. "I got a run!" I exclaimed as I grabbed for the rod.

I locked the steadily spinning reel and yanked back to set the hook. There ensued a battle unlike any I've had before or since. It was a mighty battle, and the fish really fought. The line jerked back and forth unlike the steady, orderly tug of a fish diving for the bottom or walking on its tail. When it came close enough to the boat to be seen, I knew why it was so different. Granddad handled the net, and as I maneuvered my rod, he scooped up not one but two bass, about three pounds apiece. One was on the baited hook, the other was on the plug I used as a bobber.

"I've never seen anything like that in all my born days," Granddad remarked, shaking his head.

"I can't believe it either," I said as I struggled to get the first bass off the plug and onto the stringer. "Look at those beauties," I proclaimed a few minutes later when I lifted high the stringer with the two fat bass before hoisting them over the side of the boat and into the water.

Granddad shook his head in disbelief. "That's one fish story we won't have to dress up," he said with a grin. "I never heard of such a thing, catching one bass on live bait and the other on the bobber. Cast your rig right back out there. Let's see if you can do it again."

"I doubt it, but you never can tell," I replied while hooking another small perch on the big bass hook. I immediately cast out the rig with its plug/bobber above the struggling live bait. It went almost exactly where it had been before the miraculous catch. I immediately sat and resumed fishing with the cane pole and my trusty side line.

We soon added two more bass to our stringer. They were smaller than the first two and were caught one at a time on our live perch bait. There were no more miracle catches. When the panfish resumed biting, we knew the school of bass had moved on. Already our fishing trip was a success. There would be plenty of food for the table.

We moved each time the fishing slowed, adding to the now-heavy fish bag hanging over the side with each move. We added no more bass to the stringer, but we had an almost unbelievable catch of bluegills, crappies, green gills, red-eared sunfish, fat perch, red-eyes, a couple of muckle-johns, a pumpkin seed, and one rare large-mouthed sunfish. Of all the edible breeds in the lake, we were only missing pike, silver bass, and catfish. When the sun burned the fog away, we decided to head for home and one of Mom's great breakfasts.

Granddad unloading tackle from the boat as Aunt Lola and cousin Donna watched. Notice my Dad's live-box in the water between the boat and the pier.

After the anchors were pulled up and all our gear was secured, a single pull of the starter rope and the tiny outboard coughed into life. Granddad grinned as he lifted the heavy fish sack into the boat and laid it gently on the floor beside the stringer of bass. "That sure is one great catch. Your mother and grandmother will be amazed. Do you suppose they'll fit in your live-box?"

I don't know. Maybe we'll have to use the live-box stretcher," I joked back.

Not too long after we rounded the point and our cottage came into view, we saw two figures walk out the door, down the steps, along the walk, and onto the pier, our welcoming committee. My mother and her mother, Grandma Dick, watched and waited as we steadily crossed the half-mile or so of water and approached the pier.

"How'd you do?" Mom sang out as we approached. Our catch would not be visible until we were alongside the pier.

"Not too bad," I replied, trying to keep from grinning like a Cheshire cat. It was the same little game we played each time we came in.

"We got a few," Granddad said, knowing the game.

"Lord o' mercy!" Grandma exclaimed when she spied the fish. "Did you leave any for next time?"

"I do believe that's about the biggest catch I've ever seen," Mom added. "Let me get the camera! No! Wait! Put them in the live-box right away. We don't want to lose any. I can get pictures after breakfast. It's all ready, so let's go eat as soon as the fish are safely back in the water."

I jumped onto the pier, the stringer with the four bass tightly in hand. I turned and helped Granddad lift the sack of fish onto the pier. Then I rushed all the fish to the live-box on the other of side of the pier and in shallower water. I opened the top of the box and dropped the bass, stringer, and all into the box. Next, I lifted the fish sack, placed the top inside the live-box, and slid all those fish into the water. After removing the four bass from the stringer so they could swim freely within the half-inch mesh wire screen and wooden-framed enclosure, I looked for white bellies that would indicate weak or dying fish. The ones I saw seemed to be moving, so there were no dead ones. By the time we finished breakfast, most of the fish would be revived.

Meanwhile, the women gave Granddad a hand getting out of the boat, then helped him tie it to its mooring spot beside the pier. As we walked in toward the house, I started telling about catching two bass on a single line. I was so excited the words wouldn't come out in the right order.

"Calm down there, young man. Why don't you wait 'til we're having breakfast?" Granddad prompted, chuckling at my explosion of words. "Mother, the young man has quite a story to tell. Let's get him fed and calmed down so's he can tell it proper." Granddad always called Grandma, Mother and Mom, Daughter. I could hardly restrain myself from blurting out the story, but Granddad had an almost magical power over me, so I waited, but not patiently.

As soon as grace was said and our juice downed, Granddad turned me loose, saying, "I suppose you won't be able to eat 'til your story is told, so go ahead. Tell 'em what happened."

Granddad and the author with most of the *big catch*. I was a proud thirteen-year-old. The *two-at-one-time* bass are in the center of the stringer. Mom took this picture in the front yard of our cottage on Tippy.

I launched into a detailed accounting of our miraculous catch of two bass on one line. Next came a fish-by-fish account of our entire trip. Whatever I left out, Granddad filled in. It was some story. Breakfast was long gone before we finished our tale, and we didn't have to stretch a single thing. We had the proof of our tale right out there in the live-box.

Now, the unfortunate reality of all those fish to clean brought us back to Earth. It would probably take us as long to clean them as it had taken to catch them, and it certainly would not be as much fun. Before cleaning our catch, picture-taking time delayed the inevitable messy job. Knowing a bucket would hold only a small part of our catch, we borrowed one of Mom's galvanized washtubs about thirty inches in diameter and fifteen inches deep. I promised to wash it thoroughly in the lake when we were finished.

It took both Granddad and me to lift the live-box full of flopping fish out of the lake and carry it to the yard. There we dumped all those fish into the tub. A count revealed twenty-two fish, including the four bass. After taking several pictures of us holding a bass in each hand, we took some pictures of the fish in the washtub and the whole catch on a stringer.

Picture taking completed, we lifted the washtub, fish and all, and carried it to the backyard where our fish-cleaning table stood. Built by my father of two-by-fours and a top made from the wooden side of a dental chair shipping box and all painted white, the table was the ideal height for stand-up work. It stood at the edge of our yard near our flowing well which was soon pouring cold water into the tub with the fish. One by one the fish were taken from the tub to be cleaned, still wiggling. The panfish were scaled, beheaded, gutted, and fins were cut out. The four bass were filleted and skinned. Altogether the cleaned fish weighed nearly twenty pounds. We decided that was good for a morning's work. We both looked forward to Mom's scrumptious fried fish for dinner as we carried the cleaned fish into the kitchen and turned them over to the women. They carefully inspected and finished cleaning the product of our efforts in preparation for cooking that evening.

A soon as we put the fish in the sink, Granddad took me by the arm and led me back outside. "How about we clean up your mom's messy wash tub?" he said. "It should be washed out before all that blood and fish scales dry hard."

Granddad spent the rest of the afternoon sitting in the front yard watching me swim from the pier. While swimming, I returned Mom's fishy washtub to pristine condition by scrubbing it in the lake.

You can discover more about a person in an hour of play than in a year of conversation.

—*Plato*

I'll moider da bum.

—*Heavyweight boxer Tony Galento, when asked what he thought of William Shakespeare*

The only true wisdom is in knowing you know nothing.

—*Socrates*

I find that the harder I work, the more luck I seem to have.

—*Thomas Jefferson (1743-1826)*

Each problem that I solved became a rule which served afterwards to solve other problems.

—Rene Descartes (1596-1650), Discours de la Methode

In the End, we will remember not the words of our enemies, but the silence of our friends.

—Martin Luther King Jr. (1929-1968)

Whether you think that you can, or that you can't, you are usually right.

—Henry Ford (1863-1947)

Do, or do not. There is no 'try'.

—Yoda ('The Empire Strikes Back')

The only way to get rid of a temptation is to yield to it.

—Oscar Wilde (1854-1900)

An honest man is always a child.

—Socrates

A hero is born among a hundred, a wise man is found among a thousand, but an accomplished one might not be found even among a hundred thousand men.

—Plato

To Do Is to Live

To do what you know you must when the time comes;

To be who you truly are in the face of ridicule;

To stand true and tall against strong opposition;

To give without hope of reward or repayment;

To believe for the sake of truth only . . .

And to love for the sake of love only . . .

. . . is to live!

—Howard Johnson, 1963

Granddad Dick, Stepping Stones and a Tree

While I was growing up, Granddad Dick was my idol. That may have been triggered by Grandma Dick constantly telling me, "Howard, you are just like your granddad, almost a second edition."

One of the things I loved about my granddad, besides his stories, was that he was a builder. A plumber by trade, he could build or fix anything, or so it seemed to me. One of his projects was a pair of small rocking chairs he made and gave to our family about the time I was born. I still have one of those rocking chairs. Sturdily built, it still works perfectly, for a small child.

He let me help him make concrete stepping stones for the walk in the yard of our cottage. Some of those concrete steps are still in use to this day. We made forms out of wooden 1 by 4s, placing them on sandy ground in our back yard. We mixed the concrete by hand using a hoe and filled the forms about two inches deep. Doing four at a time, we let them set until the next day. Next morning, we would remove the steps from the forms and set them out to finish setting. Then we would mix another batch and repeat the process.

"Why don't we put them in the yard, build the walk, right away?" I asked after they sat for several days..

"See how dry and white they are?" Granddad asked. "Now we must wet them down, so they will finish setting. Water causes the cement/sand/gravel mix to set hard and get much stronger. It takes several weeks for a full water set."

"Weeks? It takes weeks before they set strong enough we can use them in the walk?"

"That's right. They would break if we used them too soon. We must wait until they are aged and turn gray. While this is going on, we must wet them every day. They use water to set hard."

"Gee Granddad, how did you learn all this stuff about concrete?"

"Just by watching and asking questions when I was doing plumbing work. Plumbers often work with men pouring concrete. That's how I learned. There's another thing about concrete most people don't know."

"What's that, granddad?"

"After the water set is finished, the concrete continues to get harder and stronger, for years. The chemicals in the cement combine with a gas from the atmosphere called carbon dioxide. This reaction turns the cement into rock, a rock called carbonate rock. It's the same rock chemically as the marble cut out of quarries and used to make buildings. Five years after concrete first sets with water, it will have grown twice as hard and strong as it was when the water set finished."

"Then these stepping stones will keep getting stronger from stuff in the air?"

"That's right. Now how would you like to tackle another job?"

"Sure!" was my eager answer.

Granddad took me by the arm and led me back behind the house, saying, "Let's do something special for your folks." He pointed to an open area behind the tiny back bedroom of the cottage. "Wouldn't a nice maple tree look good right there?"

"Sure, but how's that going to happen? Trees grow really slowly."

"You remember the nice tree in our front yard down home? The one about as big around as a telephone pole?"

"Yes, of course," I replied remembering times shinnying up that straight trunk.

"Well, that tree was no bigger around than my thumb when I planted it way back when your mother was a girl about your age and now look at it."

I glanced over at the three small trees growing in the woods at the edge of our *worm bed* about forty feet from where we stood. I could still remember when they were much smaller. Only recently I realized those trees had grown in size since I first dug around them for warms some eight years earlier. It was my first realization of the slow, imperceptible but steady growth of trees, growth taking years to be noticeable. I looked back at Granddad and said, "I see what you mean, but where are we going to find a maple tree we can use?"

"Since Dutch elm disease killed all the elms in your yard and in the whole woods back there, small maples and oaks should be growing all over to replace them. I'll bet we can find a good one the right size for transplanting with a few minutes of looking. Why don't we start over there on the high ground near the edge of the swamp?" he said, pointing to the woods next to our neighbors, the Deputys'. "Let's start over there and see what we can find."

To cut through the tall weeds on the edge of the dirt road, I took the *weed whip* along as we headed for the woods some thirty feet away from our lot. The weed whip was a foot-long blade set parallel to the ground on a handle much like that of a golf club. It was swung like a golf club to cut weeds off at the ground. The three-foot-tall weeds along the road were all nettles, so I was careful not to touch any while I cleared a path into the woods. It took Granddad about five minutes to find a suitable prospect. Growing right next to the fallen trunk of a dead tree stood a straight maple tree about eight feet tall with a trunk about an inch in diameter.

"This one will do nicely," he said as he kneeled to inspect its base. "First let's clear away all these leaves and see where the main roots go. We want to dig up all the roots we can without cutting any big ones."

We soon traced all the main roots and were ready to dig. I started for the storage shed. "I'll get the shovel."

"Hold on there. We'll have to prepare the ground where it's going before we touch a shovel to this tree. We don't want those roots to be out of the ground any more than they have to. If they get dry, the tree will die, and we'll have to start all over."

Soon we were back in our yard, digging a hole about four feet away from each wall in the corner where the bedroom and bathroom meet. "Why so close to the house?" I asked.

"So the branches will shade the house and keep it cool in hot weather. It'll be far enough away, so it won't disturb the foundation and you can get a mower all around it. By the time your kids are growing up, it will be a real tree."

This idea was incomprehensible to me at the time. It would be forever before I was married and had kids.

By lunch time, the maple had been transplanted and thoroughly watered. After lunch, the surprise was shown to my pleased mother. "I can hardly believe you did all that since morning," she remarked. "We wondered what you were doing, but never dreamed you were planting trees. The yard has looked so bare since all the elms died. That's such a perfect spot for a new tree."

Granddad beamed. Not all of his surprises were greeted with such genuine appreciation. I was so proud to be part of this project. Granddad spent the rest of the afternoon sitting in the front yard watching me swim from the pier. While swimming, I cleaned out the box we had used to mix the concrete for the stepping stones by scrubbing it in the lake. It would be ready for the next batch of concrete in the morning.

Photo taken in 1935 of my sister Bobbie and my cousin Donna. The trellis post behind Bobbie is near where Granddad and I planted the tree several years later.

The same view of the cottage after major renovations in 1993. My son, Howard, and I worked together to completely remodel the old part of the cottage. The same inside corner in both photos is behind the tree Granddad and I planted now going through the porch and roof.

More than seventy years after the preceding story, in the early summer of 2012 I visited my youngest daughter, Kristen and her family in Lakewood Colorado near Denver. I had a marvelous time with she, her husband, Vince, and their sweet, lively, precocious daughters, Cadence or Cady, nine, and Madeleine, or Mady, seven. Time caused a reversal of roles and this time, I was the grandfather telling tales.

On the bulletin board in their kitchen I found posted some wise words from another grandfather from another time. Here are those words and the attached photo.

An old Cherokee told his grandson, "My son, there is a battle between two wolves inside us all. One is evil: anger, jealousy, greed, resentment, inferiority, lies and ego. The other is good: joy, peace, love, hope, humility, kindness, empathy and truth."

The boy thought about it and asked: "Grandfather, which wolf wins?"

The old man quietly replied: "The one you feed."

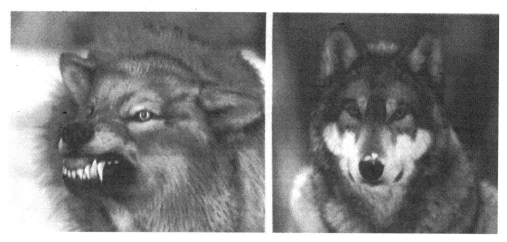

The story starting on the next page is of my own grandfather and I, is in a similar vein. It was one of the most powerful life lessons I ever learned, even so, it took me many years to realize its value and importance and strive to live up to such an important lesson.

When a person, child or adult, exhibits anger, resentment or any other negative emotion they are feeding the evil wolf within them. Once expressed in words like, "I hate . . ." the individual must act out the meaning of their words in order to demonstrate their commitment to what they have said. This is true even when they realize they are wrong. As in a child's temper tantrum, anger feeds on its self and grows often doing injury and pain to the one who is angry. The evil wolf is being fed and all those bad feelings grow. How much better it is for everyone to stem that anger, forgive and forget, and let the good wolf win.

Granddad Dick, Turnips and the Basement

My maternal grandfather was a terrific friend for a young boy. He was a master story teller who could keep me occupied listening for a long time, an eternity to a small boy. Many a time, as he would be telling me a particularly juicy tale, I would hear my grandmother call from the kitchen or other part of the house, "George! You quit filling that boy's head with your nonsense." Granddad would lower his voice and keep right on telling.

My maternal grandparents, the only grandparents I ever knew, lived in a two-story duplex on Oxford Street in Indianapolis. Until I was eight and we moved to Cleveland, we lived on LaSalle Street, a few blocks away, about a ten minute walk. Among his other talents, Granddad Dick (Dickinson) could build, fix or repair anything, or so it seemed to me as a small boy. Their house had a small basement under the kitchen. It was a dark, dank, foreboding place, at least to me.

About the time I was born, Granddad decided to expand the basement. He would first build a supporting system for the house, then break out the concrete block walls, dig out the dirt, and build a new foundation wall for the expanded basement. Granddad was about fifty when he started the project, right before the crash of '29 and the depression. Part of the same project was the construction of a round porch on the back of the house. The circular outside wall was about three feet high. The porch would be the depository for the broken blocks and other debris from the old basement wall. It would have a concrete floor level with the brick wall. The windows in the back (master) bedroom were to be replaced by a pair of doors opening onto the porch.

The depression hit while the project was still under construction. To make matters worse, Granddad began to have stomach problems. Several years after we moved to Cleveland, he went into the hospital for surgery. By this time he had lost a lot of weight, and many in the family were sure he had cancer and wouldn't make it through the surgery. True to form, he beat the odds, recovered from the surgery, and regained his lost weight Two years later he was back in the hospital with the same problems. In the interim, his surgeon had passed away.

Grandma told me the rest of the story. She was angry and frustrated and let me know it. She was angry at the original surgeon and frustrated because he died and she couldn't get back at him for what he had done to Granddad. She explained it to me this way:

"That evil doctor, the one who first operated on Granddad should have died before the first surgery according to the doctor who operated on his stomach for the second time. He told me that Granddad's stomach was a real mess, badly ulcerated. He had to remove half of his stomach to fix his problem. There was no cancer. This doctor surmised the previous surgeon had opened him up, took one look at is stomach, and sewed him back up figuring he had advanced cancer and was not going to recover from the surgery. There was no record of a biopsy. Granddad sure fooled him."

Granddad was still building the basement when I went off to college in 1945. He was seventy-seven at the time. As I remember, most of the digging and removal were done, and a block wall foundation was finally in place. The windows in the back bedroom were still in place, and some of the bricks on the perimeter of the porch had fallen out.

One evening, when I was probably seven, he and my grandma came to our house for dinner. I managed to sit right beside him of course. Among other dishes, my mother was serving stewed turnips, a dish I did not particularly care for.

When I was offered some turnips, I declined. I probably said, "I hate turnips," or something of that nature.

Granddad's face turned to a scowl as he looked at me, bent down, and whispered privately, "Howard, You should never say things like that. Never!"

"Why not? That's what I think."

He then gave me a lesson in human nature I will never forget. "Well, you think wrong, young man. You think wrong. Unless you change, you will deny yourself many enjoyable things."

I held my grandfather in awe and adoration, so I was extremely attentive as he continued.

"What you should say is, I love stewed turnips. . . . Say it now. I love stewed turnips."

It made no sense to me, but I said it anyway, "I love stewed turnips."

"Repeat it, several times. Announce it to everyone."

"I love stewed turnips. I love stewed turnips. I love stewed turnips," I dutifully remarked for all to hear.

"OK. Now, ask your mom for some turnips—right now. Say you love her stewed turnips when you ask. Go ahead, do it."

Though I was skeptical, I did what Granddad told me to do. "Mom. May I have some turnips? I love your stewed turnips."

Mom had a quizzical look on her face as she placed the serving of turnips on my plate, but said nothing,

Granddad leaned down and whispered, "Now keep saying, I love stewed turnips, as you eat them. You'll be amazed."

Well, I was amazed. Not only were they edible, but I even took a second helping. They were delicious. To this day, I love stewed turnips.

While he and I sat on our porch after dinner, Granddad tried to explain why this was so. "Howard, you ought to apply what you learned to everything. Consider all the things you say you don't like. Think about treating them the same way. Know why this works?"

I was in wide eyed attendance to his words. "No, I don't, so why did that work. I really did like those stewed turnips. Still, do. That's hard for me to believe."

Granddad smiled as he continued. "It's quite simple. It's the power of commitment. Once you announce something to your friends, to anyone, you then must back up your statement with actions. Let me ask you, do your friends like school?"

"Nah. They all hate school. Drather be doin' almost anything else."

"That's what they say, and once they've said that, it's a commitment. They then have to prove it to their friends. They must act as if what they said were true. Each time they say they hate school, they must act as if it were true, and that reinforces their feelings. That's human nature, and it is so with nearly everything in our lives."

"I don't know that I can do that with everything. Some things, like bad fish, taste terrible. They even smell bad."

"That's quite true, but you'll be surprised how many things, even people, are not nearly as bad as you say. Take school for instance. Do you like school?"

"I guess I do like learning about new things—even have fun solving word problems in math. School's not really so bad."

"Don't your pals say they hate school? How about you, do you ever say you hate school?"

"They always say they hate school. I do too, sometimes."

"Every time you say that, it's a commitment. Then you have to prove it—continually. The next time you talk with your friends, tell them that you love school and keep on telling them that. See what happens."

"I can't do that. They'll all laugh and make fun of me."

"You'll knuckle under and do what they want you to do. Is that it? You can't do anything they won't let you do?"

I bristled a bit at his words. "No, It's not that way, not at all. They don't tell me what to do."

"Sure sounds to me like they do."

Fortunately, I was fiercely independent. I had to prove I was not under the control of my friends, or anyone for that matter. "OK, Granddad. I'll do it. I'll tell all my friends I love school."

"Now you're on the right path. You'll understand better as you grow up."

Much later I would realize Granddad was teaching me two important life lessons, the power of commitment and the force of peer pressure. It would take years for the full effect of these lessons to become part of my nature, the person I am now. From that day on, I always said I loved school—and it worked. From the same lesson, I learned to ignore the taunts and deliberate insults of others. I realized they were the ones who owned the problem. I would make my own decisions about what I like or dislike, believe or don't believe. As a result, I have never been a joiner or member of a cause. Well, maybe some independent causes for what I believe to be the better good. Peer pressure? I can take it or leave it as it suits me. There is no doubt in my mind that those simple lessons my Granddad taught me have had a major positive effect on my entire life. To this day, I eat virtually any food that is well prepared with relish. I love everything edible.

A mathematician is a device for turning coffee into theorems.

—Paul Erdos (1913-1996)

Problems worthy of attack prove their worth by fighting back.

—Paul Erdos (1913-1996)

A Memorable Fishing Trip

When I was fourteen, I spent my last full summer at Lake Tippecanoe until World War II was over. After 1947, I would not spend another full summer there for fifty years. It was there I learned to love the outdoors and the wild during those glorious years of growing up. I also became an avid fisherman, going fishing every possible chance I had, usually by myself. We had but one tiny slab-sided rowboat with three seats and steel pipes for oarlocks. After years of rowing, a one-and-a-quarter-horsepower Evinrude outboard was added to the boat, so we wouldn't have to row the one and a half miles to the *flats*. A confirmed *bluegill* fisherman, my dad used long cane poles equipped with red and white bobbers to catch those tasty little fish. I learned to fish for bass with live bait from our neighbor, Dan Hackerd, who had a cottage two doors to the west of our place. I admired the fish he frequently placed in the live-box attached to his pier. Once he told me his secrets about catching these *green monsters*, I was hooked. I had caught *bass* fever.

The summer I was fourteen, there was one particular fishing trip I remember quite vividly. It started after we finished clearing the breakfast dishes. I first walked to the worm bed in the woods east of our yard to unearth some wiggly red, gray, and pink earthworms for bait. I used the worms to catch small perch for bait to tempt the giant bass. With a fresh supply of worms, I headed for the boat carrying my two trusty steel bait-casting rods with the South Bend reels loaded with black, twenty-pound test braided line. My fishing tackle included a small tackle box, a landing net, a stringer to hold the hoped-for catch, a bucket for the bait fish, and the can of worms. Other gear included a small can of gas/oil mix for the tiny outboard, a seat cushion for the wooden seats, a spare can for biological necessity, two anchors with their ropes carefully wound on cleats, a pair of oars, and a couple of cane poles for fishing in the shallows. As always, Mom came onto the pier before I left to wish me a cheery "Good luck and be careful" as I pulled away.

I headed southeast toward the *flats*, my most productive fishing spot. I looked back at the tiny figure on the pier as she waved a few times and then headed in for the cottage to clean or sew. We performed this ritual as each fishing trip started. To this day, whenever I pass through the part of the lake near where our old cottage and pier once stood, I can envision her standing there on our pier, watching. The new, much larger, remodeled cottage we sold several years ago fades as memory places the tiny white building with windows across the front behind Mom's image.

During the twenty minutes it took to reach my favorite fishing grounds, I decided where to start fishing. I usually anchored outside the visible weed bed in about ten feet of water. I knew the shape of the weed bed running north and south, parallel to and about five hundred feet out from the shore. There were several favorite locations from north near the channel from Little Tippy and south near the outlet of Grassy Creek. After making my choice, I approached the weeds slowly until I saw the first ghostly green fronds deep beneath the surface in the dark green water. After pulling away from the visible weeds about thirty feet, I dropped one anchor. I then pointed the boat toward the shore and dropped the other to hold the boat in position.

Once firmly in position, I took my *side line* out of the tackle box, baited it with a small worm, and dropped it over the side to fish for bait. My side line was a green, white, and red cylindrical *bobber* with twenty feet of line wrapped around it. The water was calm, so I merely dropped it into the water beside the boat. I watched as it spun in the water and the line unwound under the pull of the sinker near the baited hook. When the bobber stopped spinning, I knew the sinker was on the bottom, eight to ten feet down. Picking up the bobber, I wound up enough line to hold the bait a few inches above the bottom. Holding it over the side in my hand, I waited for the telltale tug of the tiny perch I sought. I didn't consider myself as really fishing until I caught some bait, usually within a few minutes.

I had to react quickly to those weak tugs, and many a time my quick jerk would miss. When I did connect, the tiny fish struggled all the way to the surface as I pulled in the line. It was a perch about six inches long, perfect for bait. I immediately took one of the casting rods, freed the large hook, and slipped the point carefully through the mouth of the perch. Hooked this way, it would stay alive and active. I cast the bait fish some thirty or forty feet toward deep water as thoughts of monster bass lurking in the depths ran through my imagination. Casting away from the weeds permitted the tiny perch to swim freely. Sometimes the reel would spin as the perch dove for the protection of the bottom. This time the line twitched several times as the perch struggled to get away and then sank slowly into the depths. It lay at a gentle slope from the tip of the rod away from the boat until it disappeared into the dark green water. I wound in the line until it no longer lay slack. This slight tension was necessary so a strike would be noticed as soon as possible.

On rare occasions, the bait would drop into the water so near to a hungry bass it was taken immediately. The spinning reel then indicated I had a *run*. This was not one of those occasions. Sometimes I would fish for hours and never have a run. As soon as I placed the baited rod on the seat, I returned to fishing with the side line. I needed more bait. It wasn't long before another lively perch was cast out on the other rod in another direction away from the first. I continued using the side line until six or seven perch were in the bait bucket. With enough bass bait in the bucket, I rolled up the side line, placed it in the tackle box, and unlimbered the cane poles. I set the bobbers to hold the baited hook about two feet above the bottom and threw the lines in the water near the weed bed. Maybe a hungry bluegill or crappie would come along. Now came the quiet waiting.

I looked around, surveying the scene. Here and there a gull flew above the lake, searching for food. Several mallards paddled toward the shore a few hundred feet away. Occasional ripping splashes here and there betrayed bass chasing schools of minnows near the surface. Now, in mid-July, they were mostly small ones six or eight inches long. The big ones were deeper down, seeking larger prey. Occasionally, a loud splash would give evidence of giant fish I knew inhabited our lake. I learned the jumpers were usually carp of twenty to thirty pounds and not real game fish. A few puffy white clouds stood silently here and there across the sky, and there was absolutely no wind. High overhead a large hawk circled over the shoreline. It looked like a redtail, but I couldn't be sure. It was too far away for positive identification. Redwing blackbirds croaked from among the cattails lining the wild shore where there were no buildings or other indications of human interference with nature. What a gorgeous day. I silently drank in all this beauty for at least half an hour.

Suddenly one of my bobbers disappeared, and I lunged for the cane pole on the right. "Missed it!" I muttered as the line came in freely. Lifting the pole straight up, I grabbed the line and checked the worm. The fish had pulled it loose from the hook. I rethreaded the hook into the worm and threw it back precisely where it had been. Maybe the fish would bite again. This time I would be ready. I sat tensely, hands on the cane poles with their butts under my seat cushion. It wasn't long until the bobber bounced a few times. I held the pole, ready to strike. The bobber disappeared, and I struck. Another small perch was added to the bait bucket. I baited the hook and threw it out on the other side of the boat. I didn't want any more perch, unless they were big ones.

Another ten minutes of quiet passed during which I changed the water in the bait bucket. I had noticed the perch gasping at the surface and knew their oxygen was depleted. With the fresh water, they once again settled to the bottom.

Zzzzzzzzzzzzzzzz, one of the reels sang out as the line was pulled rapidly from the spool. This was what I had been waiting for. I moved the cane poles out of the way and picked up the rod with the singing reel. As I picked it up, the reel stopped spinning. This was what drove me crazy. Was the fish still on, or had he spit out the bait? I knew to set the hook only while the line was going out and the fish was swimming with the bait. I waited a minute, another minute. I slowly and carefully wound the line in slightly to take out the slack. Nothing, no movement. After a long wait, I drew the line tautly and yanked the pole back to set the hook.

Our little slab-sided rowboat on the shore next to our lot. Look carefully and you can see the pipes used as oarlocks and the round mark on the transom left by the mounting clamp of the tiny outboard. Could that be a worm can on the ground next to the boat?

The line was free, and the fish was gone. "Drat!" I said out loud as I reeled in the line. Examination of the perch revealed a large section of scales scraped from each side. *Probably a bass,* I thought. Cuts in the skin would indicate a pike or gar had taken the bait. I took the now-dead bait off, threw it far back in the weed bed, and reached into the bucket for a fresh one. I had no sooner cast the new bait out and placed the rod on the backseat when the other reel spun. A new run began as the fish swam away with the bait in its mouth betrayed by the steadily spinning reel. As the line payed out, I picked up the rod and prepared to set the hook. After an appropriate wait for the fish to swallow the bait and with the reel still spinning, I locked the reel by grabbing the spinning handle and, when the line tightened, jerked back mightily on the pole to hook the fish.

Then followed the real thrill of fishing, reeling in a fish fighting on the end of the line. By the pull on the line through the sharply arced rod, I knew this was no small one. It would definitely require the use of the landing net to get it out of the water and into the boat. As I reeled in the line, I glanced about to make sure the net was within reach. Suddenly the line went slack. I was again dejected as I thought the fish had gotten away. I began reeling furiously when I saw a long loop in the line as the fish went straight under the boat. He was still on the line. Now I had a new problem, the two anchor ropes. I would either have to bring him back under the boat between the ropes or pass the rod, line and all under one rope to the other side of the boat.

I moved forward, stepping carefully around the equipment on the floor. Then the line tightened, and I was again fighting the fish with the tip half of the rod in the water. I knew if the rod happened to point the direction the line headed, during the pass under the front anchor rope, it would lose its spring action, and the line would probably break. Holding the rod down in the water, I carefully passed it beneath the prow of the boat and inside the taut anchor rope from my left to my right hand. As precarious as was my perch on the narrow front of the boat, I managed the pass without losing the line tension and arc of the rod. Water dripped from my hands and the rod as I lifted it out of the water and continued reeling in the mighty fish.

As the fish came close, the line pointed nearly straight down into the depths. I maneuvered the pole tip to hold the fish away from the two anchor ropes. Should he manage to get around an anchor rope, it would probably be all over, and he would get away. As I held the rod in my right hand, I used the other to grasp the landing net. Looking over the side as I lifted the rod high, I saw flashes of an enormous white belly from the dark depths. He was tiring. My heart pounded with excitement. He struggled to stay down, making smaller and smaller circles as he was slowly drawn to the surface. Suddenly, he jumped into the air and walked on his tail, shaking, furiously trying to dislodge the hook barely caught in the side of his mouth. Prepared for this move, I quickly slid the landing net under him while he was airborne. Instantly the hook came free and flew into the air. At the same time, the monster bass dropped into the net. *Wow! How lucky,* I thought to myself as I pulled in the net and placed the fish, net, and all on the floor of the boat.

The fish still struggled, banging around on the bottom of the boat. I dropped the rod and net to reach for my stringer with the metal hoops. Opening the bottom hoop, I checked to make sure the top end was firmly attached to its ring fastened to the side of the boat. I then reached inside the net, grasped the lower jaw of the bass, and slid him out of the net. I forced the sharp tip of the metal hoop through the soft, thin membrane between his tongue and the side of his jaw and snapped it shut. I lifted my prize proudly and surveyed him from his saucer-sized mouth to his wide tail. *He must be almost five pounds*, I thought, watching him give a few vigorous shakes as I lifted him over the side and into the water. He immediately righted himself and swam slowly against the restraining hoop in his mouth. I admired my catch for several minutes as he breathed steadily and regained strength. He would soon begin to fight the stringer.

Zzzzzzzzzzz, the other reel sang out. Another run interrupted my viewing. I quickly straightened out the equipment disarrayed by the previous struggle and reached for the rod. Soon, another struggle began. I could tell this was a smaller fish, but he still put up a fight by coming to the surface and jumping high in the air several times while shaking vigorously to dislodge the hook. It wasn't long before the new two-pounder joined the monster on the stringer. Two good fish was better than average for my fishing trips. Although I was seldom *skunked,* one good fish was the norm for most.

I quickly checked both rods, baited them from the bait bucket, and cast them out. As soon as the second bait hit the water, the line jerked furiously a few times, spinning the reel in short spurts. I soon realized this was not the bait, and when the line began a steady departure from the reel, I grabbed the rod and hooked my third fish. He must have been right there where I cast the bait and grabbed it as soon as it hit the water. Another good fish, a bit larger than the second, was added to my catch. This was indeed a great day already. I had only been fishing for about an hour and had three beauties on my stringer.

Things were quiet for the next hour and a half. I pulled up anchor and moved south a hundred feet, hoping to find another productive spot. After moving once more without any luck, I headed across the lake toward Pierce's point where a shallow sandbar bordered a deep hole stretching south from the point. It was on my way home, and I often stopped there when I was headed in empty-handed. After anchoring off the edge of the bar, I cast two perch out and waited. I decided to add to my dwindling supply of bait, so I would have plenty for the evening trip. Soon my side line was producing the bright green and white bait fish. I was thinking about pulling up and heading in when another tug on my side line resulted in a perch being drawn up. Suddenly the line was almost jerked from my hand. Something had grabbed the perch on its way to the surface. I was not prepared for the fight of a large fish on my side line. It was an old line with a tiny hook and with no pole to absorb the shock of sudden yanks, the line would probably break.

After the first violent tug, the line pulled steadily straight down. My mind rushed through many scenarios of how to handle the situation as I loosened the line from the bobber around which it was wrapped and looked around for something on which to fasten it. There was nothing appropriate. I wrapped the line once around the vertical pipe that served as an oarlock. This would provide some drag, so the line wouldn't cut my hand should the fish pull suddenly. Then I slowly pulled the line up with my left hand while pulling in the slack around the pipe with my right. Something heavy had been hooked and was hanging on the line. There was no struggle from below. As I continued to raise the line, the pull remained steady, but the line began angling out away from the boat. There was definitely something big on the other end. When it came close enough for me to see it, I gasped. It looked like a log under the water. I was drawing a monster pike slowly to the surface. It had to be at least ten or twelve pounds. I knew there was no way I could boat this monster with my side line, but I was going to try.

Pike are funny. They usually pull steadily on the line, without much fight until they are close to the boat. Then suddenly, when they see the boat or possibly movement on the boat, they lunge into headlong flight. Unprepared for this sudden rush, many a fisherman is left holding a rod with a broken line. I prepared for this by quickly wrapping the loose line back onto its bobber. I held the bobber loosely in my right hand with the line passing between the index and middle fingers. Extending my arm up with this bobber, I used it as a substitute for a rod, letting the line out by permitting the cylindrical bobber to spin in my hand. Next I unwound the line from around the pipe and spun the bobber to take up the slack. Thus prepared, I stood slowly and began pulling the fish to the surface.

True to form, the pike suddenly lunged away from the boat. I struggled to let the line out as slowly as possible while using my arm to counter the violent tugs. It worked. By the time all twenty feet of line unwound from the bobber, the fish stopped running. Carefully I began rewinding the line on the bobber and drawing the fish in. Again it was like a dead weight on the line until the fish came near the boat and again dove away. The cycle was repeated several times until I thought I might boat the pike. The next lunge didn't stop when the end of the line was reached, and with a quick snap, the line parted from the hook and the pike was gone. I sat down excited, dejected, and spent. I knew there was little chance to land such a large fish on a short, fragile handline, but I was still disappointed.

I could tell the sun had passed the zenith, so I knew it was about one and time for lunch. A light western breeze was now blowing, and the gentle ripples made a pleasant *slap-slap* against the flat side of the boat as I began pulling in my lines. First I raised the cane poles and spun them to wrap the lines tightly for storage. I removed the remaining bait from the hooks and threw it in the lake. When the hooks on both poles were secured, I stored the poles lengthwise, butts on the backseat and tips at the prow out over the water. Once more I freshened the water in the minnow bucket then began raising the rear anchor. As the boat swung freely in the breeze on the front anchor, I reeled in one casting rod then removed and freed the bait. I watched as the tired perch slowly swam down onto the depths, probably to be eaten by a hungry bass or pike.

As I started to reel in the other line, I noticed it had drawn taut. The reel hadn't moved. *Maybe there is a fish on the line*, I thought. I waited for a few moments holding the line taut, but still there was no movement. I decided I might as well assume there was a fish on the other end and so yanked back to set the hook. To my pleasant surprise, a fierce battle began. This was no small fish. When he came close enough to see, I was amazed at his size. It was an immense bass. It looked quite different as I fought him to the surface. Rather than jump, he chose to keep trying to dive to the bottom, an unusual tactic. When I scooped him up with the net, I realized why he looked different. It was a smallmouth bass, only the second I had caught in all my fishing and a real rarity on Tippy. He was a monster, twenty-six inches long but thin, unlike the fat largemouth bass already on my stringer. Thin as he was, he still weighed in at four and a half pounds. The largemouth on the stringer was only twenty inches long and weighed about the same.

I pulled up the other anchor, secured both anchor ropes on their cleats, and sat down in the backseat to start the motor. After opening the gas line cock on the outboard and the air bleed screw on the gas cap, I primed the carburetor, wrapped the rope around the flywheel, and yanked. When the tiny motor coughed into life, I pulled the stringer of bass into the boat and headed for home. I smiled, admiring my catch as the fish flopped for a while, then lay quiet. They would survive the trip home lying in the bottom of the boat. Soon after I rounded the point and our cottage came into view, I watched a tiny figure come out the door, down the path of concrete blocks I helped my grandfather make, and onto the pier. Mom always came out to congratulate her returning fisherman or encourage him when he came in empty-handed. She was going to be amazed at my unbelievable catch and at my story of the big one that got away. As I approached the pier, we went through the usual ritual greeting.

"How'd you do?" Mom asked while I was still too far away for her to see my catch in the bottom of the boat.

"Not too bad," I replied, wanting to surprise her when the two big bass were shown. We played this little game nearly every trip. She always knew when I was skunked though. No matter what I said, she knew.

As the boat came alongside the pier, she spied the fish and smiled, saying, "I see you did a lot better than not too bad. Look at the big one. It must be a five-pounder at least. The smaller ones are each at least a couple of pounds. Looks like we'll have plenty of fish when your dad comes."

Grinning from ear to ear, I replied, "Yeah! It was a pretty good trip." The instinctive pride of the hunter bringing home food for the table welled in my chest. Instantly, I became a Native American of several hundred years back, drawing my canoe up on the wooded shore with food for the tribe. I was sure scenes such as this had been carried out on the shores of the lake for centuries. The stone arrowheads and spear points I occasionally found gave testimony they had been there. I threw the stringer of fish over the side so they would revive, stepped onto the pier, and then secured the boat.

"Stay right there," Mom ordered. "I'm going to get the camera right now."

She scurried in for the camera while I unhooked the stringer and prepared to have my picture taken. When she returned, she walked past me on the pier, turned, and motioned for me to hold up the fish. I raised the stringer and held it outstretched between my hands, so all the fish could be seen clearly. "Now, let's get one with the big one by himself," she suggested. I took the smallmouth off the stringer and held him tightly by the lower jaw as she took the picture. "Your father will be so proud when he sees this catch. He'll be here Friday, you know."

"I hope they stay alive until then. I want him to see them as they are," I said proudly as I carefully placed the smallmouth into the live-box.

After taking the pictures, she headed for the cottage, remarking, "Get your fish taken care of. By the time you come in, lunch will be on the table."

"Okay, Mom!"

I lifted the minnow bucket out of the boat, made sure the lid was secure, sank it in the shallow water next to the pier, and tied the rope from the handle to a pier post. Next, I opened the live-box my father made and went to get the stringer of fish. They would probably be safe on the stringer, but large turtles, including snappers, prowled these shallows and had occasionally eaten fish left on stringers. The metal hooks of the stringer rattled as the fish struggled while I carried them to the live-box. I dropped stringer and all into the box, so I could unhook them without fear of their getting away. One by one I removed the clipped metal hooks from the lower jaw of each fish and freed them into the three-foot-by-three-foot wire mesh and wood frame live box, their prison until I cleaned them. I gazed in admiration at the four sleek greenish bronze shapes moving slowly around in the box. As soon as I lifted out the now-empty stringer, I closed and secured the lid of the live-box. Without the lid, the bass would soon jump to freedom.

"You wash your hands," Mom said as soon as I walked in. "Lunch is on the table."

I stepped to the stainless steel sink in the small kitchen and turned on the cold well water. The water came directly from an artesian well outside the kitchen. There was no water heater, so the water was always the temperature of the deep ground, about fifty-six degrees. It was perfect for drinking, but quite cold for washing. There was always a kettle on the stove to heat water when necessary. As soon as my hands were dry, I stepped into the tiny room between the kitchen and main room and sat at the round drop-leaf table across from Mom. I faced the front of the house and had a wide view of the lake through the windows all the way across the front. I could also see the morning sun sparkling on the rippling lake surface out the side window and a cloud of smoke drifting over the lake from a fire of some sort on the lakeshore. Beyond the smoke, summer cottages dotted the shoreline all the way to Pierce's Point. There the shoreline curved to the east and went out of sight. The myriad sounds of a summer morning drifted in through the window. I had experienced the same scene, sights, sounds, smells, and all the rest many times before, but they never ceased to please me. It was quite intoxicating, beautiful, almost primordial in its fascination. I loved every minute of it.

After saying grace, Mom asked, "All right, tell me all about it."

I immediately launched into a detailed and excited tale about all those fish I had caught and especially about the monster pike I almost caught on my side line.

When my father arrived early Friday afternoon, Mom greeted him with a hug and kiss and then said, "Your son has something to show you."

When I lifted the live-box and he saw six bass including two added since my most successful day of fishing, he was overjoyed. "Have you shown these to Mr. Hackerd? He'll know he has some stiff competition from the Johnsons when he sees those babies."

Dan Hackerd, our neighbor two doors down, was generally recognized, in our landing at least, as the best bass fisherman on the lake. My proud father now touted me as worthy competition. This pleased me. Nothing made me happier than seeing my father so proud of me.

My father also had a knack for the practical and brought me quickly back down to reality. "Isn't it about time you cleaned those fish?" he commented. "We sure don't want to lose any. My mouth's watering already."

Saturday and Sunday dinners were still in the live-box and needed to be cleaned and prepared for Mom's frying pan. "I'll clean them first thing in the morning, soon as we come in from fishing."

"How about we do it right now?" he responded with a smile. "I'll give you a hand, and we can be finished in time for dinner." Unlike me, my father didn't like to put things off.

My father spied Dan Hackerd walking toward his pier and motioned him to come over. I watched as he lifted the live-box to show him my catch.

"I think that's the biggest smallmouth I've ever seen," Dan said. "How'd you manage to catch him?"

I told Mr. Hackerd, as I always called him, all about my successful trip and about the monster pike that got away. I was a fourteen-year-old talking man-to-man with an adult, a heady and maturing experience, unusual for me. It was a treasured moment, for I knew my father was listening proudly.

❧   ❧   ❧

But at my back I always hear Time's winged chariot hurrying near.

—*Andrew Marvell (1621-1678)*

Everybody pities the weak; jealousy you have to earn.

—*Arnold Schwarzenegger (1947-)*

Whenever I climb I am followed by a dog called 'Ego'.

—*Friedrich Nietzsche (1844-1900)*

The Jam Session

Once in a while I was the drummer in the dance band at Tippy dance hall on Lake Tippecanoe. One sweltering Sunday afternoon in July, 1944 I was playing during our jam session. Tippy held a two hour jam session on Sunday afternoons starting at two o'clock. Almost all of the afternoon patrons came wearing bathing suits. Then in the evening we played the usual dance program. This would turn out to be no ordinary Sunday.

My cousins, Dale and Donna Dickinson, were also at the lake for the weekend. Dale was an accomplished trombone player, and Donna played the accordion, an instrument not normally found in a band. I invited them to the jam session and to sit in when the opportunity arose.

When we arrived for the setup, I saw there were an unusual number of other musicians assembled. "This is going to be some jam session," I said to my two cousins. "Look at all these musicians."

Dale grinned and replied, "This will definitely be a blast. There are several guys here from Indianapolis I've played with before."

The group of Indy guys came over. Dale introduced us, and we were soon comparing notes about our music.

I explained, "Pardon me guys. I have to get set up. I'm sure I'll be seeing you on the bandstand."

Our sessions started off with a few fairly standard numbers. Then our band leader, Lew, would holler, "Are you ready?" and point to our tenor sax guy who then took off improvising. This led to each member of the band taking the lead for five or six minutes. I was one of the last ones jamming before we took our first break. I knew today would be different. I counted at least fifteen musicians with instruments waiting near the band stand for their turn.

By the time for our usual break, several guys had already *sat in* and things were beginning to jump. Lew called out to the band, "Should we break now?"

411

Most of the audience were standing on the dance floor, clapping in time to our music. They shouted, "NO! Keep playing."

At four o'clock we were so into it, there was no sign of a break. A drummer I knew was from Indy and who played in a nearby dance hall, suddenly tapped me on the shoulder and asked to sit in. I was glad for the relief. I grabbed a coke and took a seat at one of the tables in the back. As I sat there, I realized new people were steadily arriving, both audience members and musicians. I stopped a guy with a cornet and asked him where he was from.

"I'm from Columbia City. One of our guys told me this was happening, and I wanted to join in. Five of us came down to play."

"You came all the way from Columbia City?"

"The guy who told us is from Fort Wayne. That's twice as far as our town. Who do we see about sittiin' in?"

"Well, I'm the local drummer. I've been relieved by a guy from Indy. I suggest you head over to the left of the bandstand. There are quite a few there already. You'll have to wait your turn."

"We had a hell of a time finding a place to park," he said. "When we showed the cop our instruments he said we could drive to the door, but our driver would then have to find a place to park and walk back. I haven't seen him as yet, but he'll have to go right past here."

"Go ahead over," I said. "I'll send everyone I see with an instrument the same way."

"Thanks. Jack will be carrying an alto sax. He's a tall guy in a white shirt," he said as he hurried off toward the other musicians. We never exchanged names. About fifteen minutes later I told Jack where his friends were.

Thinking to see what was happening outside, I walked out into the parking lot. It was filled with cars. People were walking down the part of the lot that sloped up to the right from the door all the way to the road coursing through many small summer cottages. I walked up the hill and found cars parked along that road as far as I could see in either direction. To the west, the road wound through at least a hundred small cottages before going past the Stoney Ridge Hotel. After the hotel, the road passed the Hob Nob restaurant. curved left through a swampy area and became southbound. It then went due south for half a mile to the highway into town. Everywhere there were people walking toward the entrance to Tippy.

As I walked back to the door, I could see many boats pulled up on the beach where the lake bordered the parking lot. Usually there were only a dozen or so boats pulled up on the beach next to mine. Now there were thirty or more nestled closely together all along the beach.

The entrance was so crowded I almost had to force my way back in. They made room when I shouted, "I'm the drummer." I found Lew sitting with two other members of the band at the same table I had left. All three were in their late forties, or early fifties. Because of the war, most bands were missing guys from eighteen to at least forty. Our band had five sixteen and seventeen year olds, like me, and eight or ten guys too old for military service. This made for some interesting experiences, especially with what would now be called *band groupies*.

"What's happening?" I asked. "Who's leading?"

"I think it's completely out of our hands now. We've created a monster," Lew said shaking his head. "I wonder what the people coming for our evening show will think?"

"They'll hear long before they will see," our trombone player, Dan, said. "And Howie, your cousin is one hell of a wailer. He's playing now, ya hear?"

"I knew he was good. That's why I invited him. What time is it anyway? I don't have a watch."

Lew again shook his head. "I hate to tell you, but it's past six. We're supposed to go on with our evening program in less than two hours."

"You think you should stop this?" Dan said to Lew.

"I don't think I could if I tried. I already decided to let it play out to the end. Hell, the dance hall people have to be happy as clams. The guy at the register told me that they have never taken in so much money. Incidentally, after four they doubled the price of admission."

Then I heard Donna's accordion take over. "What the hell is that?" Dan asked.

"That's my cousin Donna and her accordion. Sounds pretty great to me."

"I can't believe it. An accordion playing jazz," Lew said.

"You gotta admit, she sounds pretty good," Dan commented. "Great, in fact. I've never heard of a jazz accordionist. She takes runs a lot like a sax. Wow."

I headed for the band stand to watch and listen to Donna.

On any normal Sunday night we quit at midnight, but this was anything but a normal Sunday night. We'd been going steady for ten hours, People, mostly musicians were beginning to take naps, sleeping in cars, and even on the ground. I was fortunate as my cousins' family car was parked right outside the door. Dale, Donna and I took turns sleeping in the back seat, usually for an hour or at most two. None of us would have believed what was happening until it happened. People were showing up in an almost steady stream from as far away as Indianapolis and Chicago. Somehow the word was getting out something magical was happening at our Tippy dance hall. I heard there were cars ~~were~~ parked all along the road from the highway. The State Police were even on hand to help keep order and direct new arrivals. There were no fights or other disturbances that often occurred on Saturday nights. Maybe the presence of the State Police helped. It was orderly chaos.

By three in the morning, there were a dozen State Police at the dance hall. At their direction, the dance hall people were giving out different colored tickets which entitled the bearer one hour inside. There were then more people out in the parking lot than inside. Of course, everyone within several blocks could hear everything because it was an open air dance pavilion with only about a third under roof. The main bandstand and dance floor were open to the sky.

After some initial problems, musicians were given colorful armbands, so we could come and go as we wished.

At about five, the first light of dawn began showing in the east. Lew and I were outside talking.

"Howie, did you know there are more than a hundred musicians in and around the dance hall?"

"How can you know that?"

"They have passed out more than a hundred arm bands to musicians. The dance hall manager told me."

"I wonder how many non musician spectators there are? I'll bet there are a couple thousand."

"We may never know. I doubt the owners will tell us. They should have a good reading from their cash receipts."

"I never thought of that. . . Right now, I'd better head in. I think it's about time for me to do another gig."

As tired as we all were, no one even hinted at stopping. With musicians now playing for about half an hour at a time, we could rest as long as an hour or more. Who knew how long it would last? We were still going strong 'til shortly after noon on Monday when things began to slow down a bit.

Then suddenly, it was all over At about one I noticed the crowd was beginning to thin out. My cousins soon left for their rented cottage. Shortly after they left I noticed the hall was almost empty. By one thirty, the band and all those musicians had simply disappeared. I walked out to the beach, stepping over paper, bottles, and cans strewn all over the nearly empty parking lot. There were maybe a dozen cars left, mostly with sleepy or drunk occupants oblivious to everything.

When I reached the beach, my little wooden outboard was right where I had pulled it up, albeit with several empties tossed in the bottom. I chucked them into a nearby overflowing trash barrel, pushed my boat off the beach, and headed for home at the other end of the lake. It had been an exciting time, one I would never forget—or ever repeat.

Oh yeah. The Tippy Dance Hall? It's concrete block retaining walls around the open air dance floor, and wooden half roof over the inside dance floor, still sits unchanged at the same spot with the same parking lot and the same beach near the same bunch of small cottages. Inside there is still the indentation of Chick's knuckles in the aluminum door frame where he missed my head and struck the wall one raucous Saturday night. Ah, but that's another story. Tippy is virtually unchanged since that jam session almost sixty-six years ago.

See the next page for an aerial photo of the dance hall.

Pain and suffering . . . they are a secret. Kindness and love . . . they too are a secret . . . But I have learned that kindness and love can pay for pain and suffering.

—*Alan Paton* in *Cry the Beloved Country*

Those who are able to see beyond the shadows and lies of their culture will never be understood, let alone believed, by the masses.

—*Plato*

here are things that I believe that I will not say, but I will not say that which I do not believe.

—*Emmanuel Kant*

A prudent question is one-half of wisdom.

—*Dartnell*

Good people do not need laws to tell them to act responsibly, while bad people will find a way around the laws.

—*Plato (427-347 B.C.)*

Aerial photo of Tippy Dance
Hall taken in mid-April 2002

❶ Piers, leased to boaters in season - These were added many years after the jam session

❷ Where my little boat was pulled up on the beach

❸ The parking lot which extends up a slope past **❻** to the entrance at **❾**

The level of the road at the parking lot entrance at **❾** is at least ten feet above the level of the parking lot at **❸** and sixteen feet above lake level.

❹ The only entrance to Tippy is in this northeast corner of the roofed over section

❺ The inside dance floor and concession stand are under this roof

❻ About where my cousins' car was parked

❼ The open-air dance floor. The shadows at both ends show the wall height. Tables filled the area around where the number is, at the opposite end and under the roofed section. The top of the block wall is a good sixteen feet above the dance floor, but only six or seven feet above ground level outside by the road.

❽ The covered band stand on the outside dance floor

❾ Intersection of the road west through the cottages and the road south to Armstrong Rd.

❖　　　❖　　　❖

Never interrupt your enemy when he is making a mistake.

—*Napoleon Bonaparte (1769-1821)*

I have nothing to declare except my genius.

—*Oscar Wilde (1854-1900) upon arriving at U.S. customs 1882*

Human history becomes more and more a race between education and catastrophe.

—*H. G. Wells (1866-1946)*

The difference between 'involvement' and 'commitment' is like an eggs-and-ham breakfast: the chicken was 'involved' - the pig was 'committed'.

—*unknown*

Talent does what it can; genius does what it must.

—Edward George Bulwer-Lytton (1803-1873)

If you are going through hell, keep going.

—Sir Winston Churchill (1874-1965)

Give me a museum and I'll fill it.

—Pablo Picasso (1881-1973)

Don't stay in bed, unless you can make money in bed.

—George Burns (1896-1996)

I don't know why we are here, but I'm pretty sure that it is not in order to enjoy ourselves.

—Ludwig Wittgenstein (1889-1951)

There are no facts, only interpretations.

—Friedrich Nietzsche (1844-1900)

Nothing in the world is more dangerous than sincere ignorance and conscientious stupidity.

—Martin Luther King Jr. (1929-1968)

The use of COBOL cripples the mind; its teaching should, therefore, be regarded as a criminal offense.

—Edsgar Dijkstra (1930-2002)

C makes it easy to shoot yourself in the foot; C++ makes it harder, but when you do, it blows away your whole leg.

—Bjarne Stroustrup

Try to learn something about everything and everything about something.

—Thomas Henry Huxley (1825-1895)

There is Some Evil in Friendly Neighborhoods

It was early fall of 1943 after school started. I was back delivering the morning Plain Dealer as I had done for the past five years. During each early morning walk delivering the paper, I made friends with many dogs the owners let out early in the morning before there were many people astir. I carried a bag of dog biscuits and handed them out freely. Most of the dogs stayed in their yards, but a few followed me. At times there would be one or two happily wagging tails close behind me as I walked my route.

One of my favorite four legged friends belonged to the family across the street. She was a sweet and friendly female dachshund named Pipshin Von Hepchein Thomsplatz, a name I'll never forget. I played with her almost every day of my route for at least the last year. She had given birth to a litter of the cutest little pups I had ever seen. I adored those pups and played with them every time I had the chance.

The dogs were kept in a fenced-in backyard where they romped noisily. While the pups were kept in the backyard, Pipshin was out to greet me many mornings as I delivered the paper. I always sat on their steps with her for a few moments of greeting when I received lots of doggie kisses. Many late afternoons after school I would go over and play with the puppies in the back yard. The family who owned them allowed me to do so whenever I wanted.

One afternoon when I went to play with them the lady of the house greeted me with tears. I was shocked to learn all of the dogs died of deliberate poisoning. Scraps of poisoned ground meat were found by the fence between their yard and the house to the south. The grouchy old man and his wife who lived in the house hated the barking of the dogs and silenced them with poisoned hamburger. I was filled with remorse for the tiny victims and anger for this customer of mine who harmed them.

First, I began giving them the worst paper I had, one from the outside of the bundle, cut and torn by the binding wire and often dirty from being dropped on the ground. Then I started missing their porch, leaving the paper in the bushes, even on rainy days. When the man complained on Saturday when I collected, I decided I no longer wanted to deliver to them. Sunday, my district manager called to tell me that I had missed delivering their paper. Normally, I would immediately deliver the missed paper, but this time I refused and told my manager why. When he said he would have to deliver the paper if I didn't, I relented and took the paper across the street. That paper was in the worst condition of any I had ever delivered. Of course, I received a complaint.

When the full story reached the powers at the *Plain Dealer*, they decided to side with me. No paper would be delivered. To get a paper, they would have to pick one up themselves at a drugstore or newsstand. This was a major victory for a small boy. Because of my actions, I became somewhat of a hero in the neighborhood, for a short time anyway. The puppy murders were the talk of the neighborhood for quite some time. Needless to say, many of the neighbors pelted insults on the couple whenever they stepped outside or walked down the street.

The story of the puppy murderers spread among kids in nearby neighborhoods and even several streets beyond. All of this pent up anger exploded when Halloween came. They became the victims of every mean trick an angry bunch of kids could devise. This included some nasty things dumped on their porch and in their yard. Their fence gate disappeared and a small tree he had planted was broken off several feet above the ground.

During the next winter none of the local kids would shovel their walks or driveway. Several times after they had hired people to come clear the snow out of their driveway, piles of snow mysteriously appeared the next morning, even more than was deposited in the last snowfall. Kids have their own way of avenging evil actions.

By the following summer, things had simmered down a bit, except for the continuing general social mistreatment by most of the neighborhood. During the summer I enlisted in the Navy, was released from my enlistment when the war ended, and then started at Purdue University. I would only be a visitor to the house on Idlewood after that. I did hear from my parents the couple who murdered the dogs sold their house and moved. My guess is they couldn't take it any more. I almost felt sorry for them—almost.

He who has a 'why' to live, can bear with almost any 'how'.

—Friedrich Nietzsche (1844-1900)

Many wealthy people are little more than janitors of their possessions.

—Frank Lloyd Wright (1868-1959)

I'm all in favor of keeping dangerous weapons out of the hands of fools. Let's start with typewriters.

—Frank Lloyd Wright (1868-1959)

Some cause happiness wherever they go; others, whenever they go.

—Oscar Wilde (1854-1900)

Out of Nowhere

I became aware, at least, I seemed it was awareness, even though I didn't know the word. I wasn't thinking in words, only sensations. It's difficult describing awareness coming out of non awareness when there are no words. I hung there for a long time, or so it seemed—almost forever. There was no sense of my body. I was a disembodied thought, floating in blackness and silence, without senses or physical sensations of any kind. They were all turned off—didn't exist.

Slowly I was becoming aware of something new. It was frightening at first, then comforting. For the first time, I became aware of something outside of my thoughts—something not a part of me—something from away from my center of awareness. Still, I had no words to define this new sensation. Later I would learn there was a word for it, sound. I was hearing sounds for the first time. They were pure sounds with no memory or meaning, but I knew they were not in my awareness, but came from outside.

Then a new sensation, a new experience, another something from outside my being invaded my awareness. It was different from the first. It was a dull redness, sight. I would later learn its name. After time, it was followed by other sensations and awareness. I was beginning to sense my physical body. I first felt a pressure in back of me, a soft sensation from the opposite direction of the redness. Then I felt pain, slowly and progressively I was sensing my face, then my abdomen, then my arms and hands. All were sensed with differing amounts of pain. It was pleasurable at first, then unpleasant. My body hurt in several places, mostly about my face.

The sounds were no longer a dull fog, but were varying in tone and intensity. Some I could barely hear, some were loud. My awareness was growing. Then I opened my eyes. Intense brightness immediately and painfully shocked me. I closed my eyes quickly to shut out the pain. All of these new, unnamed sensations were frightening, yet I was also curious about them. A battle between curiosity and fear raged in my thoughts, all without words or understanding. Still, I was learning and becoming more aware of things outside my initial awareness. There was now more to my being than raw awareness.

The sounds growing in variety I recognized suddenly as voices. Still, they were sounds, almost musical sounds, yet they carried no meaning into my consciousness. Then I felt a touch to my face, another sensation from outside of my being—a pleasant sensation. I opened my eyes slowly. The intense brightness was gone. In its place was a foggy collection of indistinct shapes, some of them moving. I had a sense of people around me, others like myself. I was beginning slowly to remember, yet still there were no words in my consciousness. The sounds were still just sounds, but I was beginning to realize those sounds had meanings, meanings I could not understand or remember.

The fuzzy shapes moving about were clearer now. They were becoming faces, arms, hands and clothes. One of the faces got quite close and made sounds, unintelligible sounds yet vaguely familiar. Then I understood a word and knew it. "Howard," the voice said. It was my name and I knew who I was. I was remembering and it was a treasure.

"Howard, can you hear me?" the nearby face said. "It's your mother."

The face in front of me was more clear now, and I recognized it as my mother's. I tried to speak, but no words came into my head or from my mouth. I heard myself groan.

Someone said, "I think he's coming around. He's trying to speak. See, his eyes are open."

Things were now coming back into my mind in a flood. I was understanding words. Still, I had no idea where I was or when. Then I recognized someone else, Dee, my girl friend, who would years later become my wife. She was standing next to my mother and speaking to me. I recognized them both. Their words were sometimes jumbled and sometimes clear. There was another woman there, in a white dress, a nurse I began to realize. I tried speaking to them, but still my voice would not work. It was terribly frustrating not being able to ask where I was or the day or time. Why wouldn't they tell me? Then I got angry. The anger cleared my head a bit.

"Where am I?" I asked.

The nurse immediately announced, "He seems to be coming to. That's a good sign." Then she spoke to me, slowly and distinctly. "You are in Lakewood Hospital. You received a concussion playing football. You've been unconscious for three days."

I could not remember playing football. The last thing I could remember was Sunday dinner with my family after church. We were at Clark's restaurant on Carnegie.

When I explained this to the nurse, she smiled and turned to my mother. "He's lost more than a week. The dinner had to be on the Sunday before last. Is that the last time you had dinner at Clark's?" she asked my mother.

"We had dinner there each of the previous three Sundays. It's our usual place to eat Sunday dinner after church."

Then she turned to me. "Do you remember what you had or what you were wearing? It might help."

"No, but Mom was wearing her pretty new blue dress for the first time. I remember because we talked about it."

"It was the last Sunday we were at Clark's for sure," Mom said. "So he has only lost one week. Will it come back to him?"

"Concussions can cause temporary memory loss, recent memories mostly, and they come back over time. I think your son is going to be OK. He'll need bed rest for a while after he gets home. I'll wager the doctor will want to keep him hospitalized for several more days to make sure he's OK before we let him go home."

By the time they let me go home, most of my memories returned, all but those of the game. When coach came to visit, he described what had happened.

"You were running through what was supposed to be a hole over right guard. You dove through the hole and hit one of those big Lakewood guys coming the other way helmet to helmet. You both dropped like stones. How you managed to hang onto the ball I don't know because both of you were out cold. Incidentally, he's down the hall with a similar concussion. Both of you broke every one of the suspension straps inside your helmets. That's how hard you hit. I've never seen more than one or two straps broken before. I'll show you your helmet when you come back to school. I'm glad both of you will be OK. You had us worried for a while."

I thanked coach for coming to see me and for being so concerned. My doctor told coach he didn't want me to play for the rest of the season because of the severity of the injury.

"You'd better not take a chance, young man. You can't shake off an injury like that in a few weeks. By the time you're ready to play the season will be over," he explained.

A bit about my short high school football career. We were in full uniforms, pads, helmets and all for these tryouts. When I tried to make the team, coach didn't think I showed much promise. I was by far the slowest runner in the tryouts. Then they had us try to evade blocks. I got by several times without being touched.

Coach noticed, tossed me a football and said, "Johnson, let's see if you're clever at dodging tacklers. See how far you can run through the rest of the guys."

I evaded every tackler coming at me from in front and both sides. I twisted, spun, and wiggled my way through the entire group. I was caught and tackled from behind. I made the team as a halfback. That's what they called running backs in 1944. I did fairly well in the four games I played in, nothing spectacular, but I did make a number of running first downs in critical situations. The rest of the team began calling me *crazy legs* after the famous *Crazy Legs* Hirsch, because of the erratic way I ran. I had exceptional balance and often recovered and continued when other runners would have gone down. Except for being run down from behind, I was hard to tackle. Still, because I was so slow, I was never more than a mediocre runner.

I don't know if I ever fully recovered my memory of that play. It's hard to say. Possibly, so many people described what happened I may have gained my *memory* from their stories. Besides, that was a long time ago.

❖   ❖   ❖

God is a comedian playing to an audience too afraid to laugh.

—Voltaire (1694 1778)

He is one of those people who would be enormously improved by death.

—H. H. Munro (Saki) (1870-1916)

I am ready to meet my Maker. Whether my Maker is prepared for the great ordeal of meeting me is another matter.

—Sir Winston Churchill (1874-1965)

I shall not waste my days in trying to prolong them.

—Ian L. Fleming (1908-1964)

If you can count your money, you don't have a billion dollars.

—J. Paul Getty (1892-1976)

Facts are the enemy of truth.

—Don Quixote - Man of La Mancha

I begin by taking. I shall find scholars later to demonstrate my perfect right.

—Frederick (II) the Great

Maybe this world is another planet's Hell.

—Aldous Huxley (1894-1963)

Blessed is the man, who having nothing to say, abstains from giving wordy evidence of the fact.

—George Elliott (1819-1880)

Once you eliminate the impossible, whatever remains, no matter how improbable, must be the truth.

—Sherlock Holmes (by Sir Arthur Conan Doyle, 1859-1930)

Black holes are where God divided by zero.

—Steven Wright

Sunrises and Related Experiences

I don't remember my first magical sunrise experience as a single one, but rather a series of many over time that morphed into a mingled memory. My merged memory was certainly of fishing trips with my father and possibly my grandfather as well, over a number of different times. It was on Tippecanoe Lake in Indiana during repeated early morning fishing trips. These trips happened frequently during warm, lazy summer mornings starting when I was as young as five.

The scene before my wide expectant eyes is emblazoned on my memory like a motion picture or video. I see the pervasive grayness of a misty or foggy first light over the mirror smooth waters of the lake. The ghostly, barely visible black of forest trees at the water's edge is outlined by the pale first light of approaching dawn. The stillness, the cool dampness, the relative quiet save the voices of awakening birds, was intoxicating. I can see the gentle bow wave, hear the rhythmic splashes of the oars in time with the slow surging of our small rowboat as I feel my father's repeated strong pull and return of the oars. I can sense the boat reaching for a favorite fishing place as I peer from my perch on the front seat.

The unmistakable cry of a loon adds an almost mystical aura to the serene scene. The quiet magic is soon broken by the raucous crow of a rooster, then the barking of a far away dog. As we move past Pierce's Point toward the wider expanse of the east end of the lake, even the shadowy ghosts of the forest trees at the shore on our left fade into the grayness and disappear. We glide in silence over the shimmering surface, making a V of small waves. The oars create pairs of expanding circles of waves punctuated by sets of small whirlpools on each side. These regularly spaced disturbances reach off behind the boat in parallel rows, dissipating slowly as they fade away stretching out behind the boat and into the smooth surface.

A sudden splash gives evidence of hungry fish as they chase minnows at the surface. After nearly an hour rowing to the best fishing spot, my father peers deep into the clear water to watch for the sight of ghostly green water weeds barely visible in the depths. When he spots the first of the weeds, he reverses the oars and backs the boat a few lengths into slightly deeper water, there to drop the anchors. He lifts the length of concrete filled two inch pipe serving as the front anchor, checks to see if I have a secure hold on the rope and am prepared before he lets it go. I control the rope until it reaches the bottom, then snub the rope and wind it around the front davit. Meanwhile, he places and secures the rear anchor.

About this time, a bit of gold showing through the fog announces to the world the sun is winning its morning battle with the fog. Directly overhead the center of the dome of gray is slowly turning blue.

The repeated barking of dogs, the voices of children, and other morning sounds carry crisply across the still waters, announcing to us and the world that people ashore are starting their day. The gray dawn fog is no longer. A brilliant sun rises higher in a clear blue sky and takes control. A slight breeze begins ruffling the satin smooth surface of the lake. All the activities of a summer day on the lake are about to burst forth. As water activities liven the lake, I know our fishing trip will soon be over.

I can recall many similar dawns, stretching back to the time before memories. I hope to enjoy many more dawns of different kinds over many years to come. I love dawns and sunrises. The pastel, reserved and quiet colors of sunrise are so different from the bold displays of sunsets. In similar fashion, there are vast differences in lake activities between the quiet mornings and the busy afternoons and evenings with all types of water activities in full swing. Pale sunrise colors are in stark contrast to the bright pinks, brilliant oranges, and deep crimson colors of sunsets as are displayed on the cover of this book.

Another memorable sunrise, a specific single event, happened in the summer of 1943. My sister, Bobbie, and I worked at the same Howard Johnson restaurant in Shaker Heights, Ohio. At twenty-one, she was a waitress and could serve drinks from the bar. I started as a bus boy, then graduated to the ice cream, counter, and finally to the sandwich table where I learned all manner of sandwich building from hamburgers and BLTs to westerns and clubs.

Frequently I rode the four miles to work on my bicycle with my sister seated on the cross bar. We kept to the side streets rather than chance busy Lee Road with all its traffic. One late Saturday night—actually early Sunday morning after closing and after we had performed all of our clean up chores—we headed for home on my bike. First light was beginning to show in the east as we wound through deserted side streets. By the time we reached the dam on the upper Shaker Lake, the sun was beginning to peek over the horizon. Enthralled by such a gorgeous setting, we stopped, got off of the bike, sat on the grass, and watched the sunrise. We sat—and talked—and marveled—and drank in all the magical sights, sounds, smells, and other sensations of an awakening and thoroughly fascinating world. I clearly remember a robin perched on a high branch, giving voice to his morning song with great enthusiasm. We must have stayed there mesmerized for half an hour or more. Then reality snapped down on me. I had a hundred plus Sunday papers to deliver before breakfast and getting ready for church. I don't remember for sure, but I think Bobbie helped me with my Sunday paper route.

Another magical dawn, also at Lake Tippecanoe, happened in late October of 1945, before I entered Purdue University. I am fascinated by wildness and go to wild places whenever I can. The swamp and woods at the east end of the lake, near where my first recounting of sunrises occurred, was the wildest place I could get to easily. For more than a week I lived on my own, off of the land in the high ground woods and the nearly dry swamp in an area of several hundred acres. I packed an arctic sleeping bag, a boy scout cooking kit, a few utensils, matches, some crackers, salt, and butter. On my belt was a hand ax and a sheath knife. In my jacket I had my emergency rations, a large bag of peanuts and several Baby Ruth candy bars. I dumped everything into the same little boat used in the first sunrise story. It now sported a tiny outboard that powered it as I headed for my adventure in the wild.

Living off of the land is another story for another time, but I learned about finding and cooking edible water plants, mostly their roots. An accomplished fisherman, I had no trouble getting adequate protein from the fish I caught with my tiny pole and hand line. About the fourth night, a fierce fall storm blew in from the northwest. I was prepared, or thought so. I located my camp for the night on the eastern side of the high ground. Knowing a storm was coming, I picked a spot of dry sloping ground with good drainage. It was beside the downed trunk of a large tree. I spread my waterproof ground cloth and rolled my sleeping bag out, the head almost against the tree trunk. I rigged the rain guard over the head of the sleeping bag before fixing my supper.

The first blast of wind and rain hit suddenly, before I finished cleaning up from supper. I scurried inside my sleeping bag putting my boots under the rain guard near my head. I was wet, but not soaked. From beneath the rain guard I saw my view of the woods to the east disappear in a fury of wind, water, and leaves. I knew I was in for a rough night. Branches of different sizes fell all around me. The woods to the west gave some protection, but in return contributed hazardous missiles in the form of wind-blown branches and pieces of branches. After the first furious blast of wind, the storm slowly died down. By this time, my sleeping bag was covered with leaves and wood debris. Fortunately, though several large branches crashed down nearby, none of them hit me. After the violent storm front passed, it was followed by steady rain. The sound of the rain lulled me to sleep, probably around eight or nine.

I awoke suddenly, a bit wet and cold. My rain guard gone, I was looking at a clear, starry sky overhead. I found the rain guard draped over the log above my head. The pegs holding it had pulled out of the now soaked ground when a swirl of wind had flipped it over the log as I slept. It was cold, probably below freezing, but I was still warm inside my now damp sleeping bag. I checked on my spare clothes, mostly socks and underwear, wrapped in a towel inside the foot of the sleeping bag. I was happy to find my feet, socks, and all my clothes were quite dry. I decided to stay put until daylight.

First light of a clear, crisp October day crept slowly through the trees to the east. I watched the telltale pale blue sky turn pink, then orange, then yellow, all pastels. Then a sparkle of sunlight burst through the forest. The billions of insects silenced weeks before by the first frost left the woods deathly still. An occasional bird call gave some comfort. Then the raucous quacking of a nearby flock of ducks rent the silence. A scuffling sound startled me as a squirrel ran frantically past a few feet away. Hot on the squirrel's trail was a black fisher intent on making the squirrel its next meal. The world was awake to a new day.

One of my Dad's favorite quotes describes the scene, "Suns and skies and clouds of June, and flowers of June together, you cannot rival for one hour, October's bright blue weather." This would prove to be just such a day. I always thought this to be a James Whitcomb Riley quote. He was my dad's favorite poet. In looking it up for this book I found it was written by Helen Hunt Jackson. See page 61 for the entire poem.

It was December of 1955 when my wife, Dolores, and I together with our daughter, Deb, five, and son, Mike, two, were camping in the Florida Keys in our new Plymouth station wagon. We were in Bahia Honda State Park camped right at the shore. Deb was sleeping in the front seat while Mike and Dee were in the back of the wagon on a twin sized mattress. I was outside in a sleeping bag atop an air mattress on the ground near the car. It was quite cold, probably near forty, when I woke up to an early beginning dawn. Not yet prepared to get out of the warmth of the sleeping bag and face the day, I repositioned my body, so I could watch the sun rise over the Florida Straight. As it grew lighter, the scene appeared virtually colorless. The water and the fog above the water blended seamlessly into the gray of a haze-filled sky. Water, haze, and sky were all the same brightening gray with a slight blue cast. When the sun began to burn through the haze, it was colorless as well, virtually white against the gray.

The only sound early was of the waves lapping against the seawall. As the dawn light brightened, sea gulls began calling as they flew overhead. The world was waking up. The full sun was shining silvery through breaks in the clouds. Still the scene was almost colorless. The haze gave way to a bright, crisp, cool, colorful day with blue sky and white pillows of clouds drifting past green trees and above a blue and white ocean. I pulled my clothes on and ventured out of the sleeping bag. It wasn't long before I had a fire going to warm my hands and later cook our breakfast. By the time I had baited and cast our fishing lines out into the water hoping to catch mangrove snapper, my crew began tumbling out of the car. Today would be a new adventure for our little family from far away Ohio.

It was 1981, and I was on the beach of Grande Isle about ten miles out from Olongapo in the Philippines. I had sailed out to this R&R island on a Hobie 16 with Jingo, the Lieutenant Commander and XO of the US Carrier Coral Sea. We had with us the bathing suits and boat shoes we were wearing, two large beach towels, a waterproof bag containing our billfolds, socks, and a few pocket items, and a fifth of Chivas Regal provided by the XO. We spent the day exploring the island where we found the remains of the American fort built after the Spanish American war. We climbed over two long six inch guns still mounted in their emplacements and dated 1906. When finally we stumbled out of the forest into the beach area, we were tired, hungry and a bit woozy from the portion of scotch no longer in the bottle.

We stopped at the beach restaurant and had a hearty meal, the contents of which I have no memory. When we had finished our meal, and much of the remaining scotch, we were in no condition to sail the fifteen miles back to the mainland. It was either take the motor launch back and return in the morning for the Hobie, or sleep on the beach. There are no sleeping quarters on the island, and everyone including workers are supposed to leave for the naval base on the last launch at 11:30pm. With the only option being sleeping on the beach during a warm Philippine night, we said in unison, "The beach."

We had to sneak off and find a comfortable place to hide from the Shore Patrol who searched for and herded the last stragglers onto the launch. They paid no attention to the bright white sailboat pulled up on the sandy beach not far from the pier where the launch tied up. After the launch had pulled away, we found a smooth area of soft sand, spread our beach towels out, polished off the remaining scotch, and promptly fell asleep.

Very early the next morning, Jingo woke me and said, "Do you hear that noise? Something's on the beach, and I think it's coming toward us."

At first I couldn't hear a thing. Then I heard it, a soft, scraping sound as if something was being dragged across the sand in short spurts, *scrape - scrape - scrape.*

"Yea, I hear it, but you can't see a damned thing on this beach on a moonless night. What do you suppose it is?"

"Are there salt water crocodiles around here? I've heard they come out of the water at night and eat people."

"Thanks a lot, Jingo. That's a comforting thought. I've not heard anyone mention crocs. Lots about pythons, monkeys and wild pigs, but no mention of crocs."

The rhythmic sound continued and seemed to be coming closer. With no flashlight or matches, we were locked in place by the blackness. We couldn't even see to move our sleeping quarters. Unable to see our watches, we had no idea what time it was.

"I seem to remember crocs have excellent night vision."

"Damn, Jingo, That's encouraging. You really know how to cheer a fellow up."

"Let's be prepared to run. Both of us in the same direction toward the water. Don't head up the beach or we could run into those trees we passed on our way here."

In fun, I said, "I can see the headlines now, 'Two Americans disappear from Grande Island. Local authorities suspect crocs got them.' I wonder if they'll send out a rescue party?"

"Now who's telling the scary stuff?

"Shit! There are no crocs out here. They're all in Australia." As I finished my tease, I noticed the sky in the east was barely starting to show light. "Look! It's morning and will be light soon. You know how quickly the sun rises in the tropics."

It can't be too soon for me."

The sound changed as he spoke. Each scrape was now accompanied by a rustling sound like papers being ruffled about, and it was quite close by.

"What in the hell is making that noise?" I said sharply.

"I don't know, but it is definitely quite close. Damn I wish I could see."

In the pinking blue light of a rapidly expanding dawn, faint silhouettes of shapes were becoming visible. Before long we could see those indistinct shapes on the beach more clearly. Then we both saw something move a few feet away. It became obvious the rustling sound was caused by a creature moving through the dried palm fronds littering the sand near the tree line. Whatever it was, it was not a crocodile. It was oval shaped and about three feet long. In the rapidly increasing light, we saw and recognized a huge sea turtle moving away from the water.

"That damned critter must be mixed up. It shouldn't be heading away from the water at this time. It will dry up and die in the heat of the day," Jingo remarked. "Let's carry her back to the water. She must have gotten turned around in the dark on her way back from laying eggs."

It turned out to be a monumental task. I have no idea how much it weighed, but we couldn't lift it, especially with amazingly powerful legs and flippers flailing away. We managed to turn it around and head it toward the water. We watched for nearly an hour as the exhausted critter flopped its way down the beach and into the safety of the sea. During its journey, we watched the pinkish, sea-gray dawn morph into a bright, clear, sunny day. We stayed out of sight while the first launch arrived, then picked up our belongings and headed for the restaurant to fill our empty, growling stomachs. No, we were not hung over, the advantage of drinking high quality scotch over an entire day. The sail back to Subic was fast and fun now that we could see where we were going.

The Audition

A lot of unexpected things happened to me in the late summer of 1945. July 30, I enlisted in the Navy pilot training program. August 3, I graduated from Cleveland Heights High. I went to summer school two years, so I could graduate early and not be drafted out of school on my eighteenth birthday. The next day I headed for Great Lakes for my basic training. En route, I stopped at our summer place on Lake Tippecanoe for one last fling. I was sure I was going to die in the coming invasion of Japan. During the few days I was at the lake, they dropped the bomb on Hiroshima, then Nagasaki, the Japanese surrendered, and I had an emergency operation for appendicitis. It was an unusually busy few days.

My mother called the Navy and told them I was in the hospital and wouldn't be at Great Lakes when I was supposed to be. A few days later, two Navy guys showed up at the hospital.

"Your enlistment is still active," one of them told me. "We're here to do several things other than confirm what your mother told us."

"Oh? What else?"

"With the war being over, we can let you out of your commitment if you want. The next basic training program has been postponed for several months until we know what we will be doing."

"Do I have to get out of the Navy?"

"Not if you don't want to. We're making a number of changes in our training program. There may be some delay in starting basic."

"No, I want to stay in. I have dreamed for a long time of being a Navy pilot."

The two exchanged glances before the senior guy spoke. "If you stay in you will not be able to take flight training. We now have more pilots than we need so all of those programs have been shut down. If you choose to stay in the Navy, you will be going to radar school."

I was devastated. One dream shattered. I'm sure my disappointment showed. "If I can't go to pilot training, and since the war is over, I think I would rather get out."

After Mom and I had signed some papers, we said goodby to the two sailors. I was officially in the Navy for twenty-two days.

Mom said, "They were certainly fine young men."

I agreed while trying to hold back the tears. I was still in a hospital, my dream and planned future were shattered, and I had no idea where I might go or do. I was adrift on a sea of unknowns.

Near the end of September, after my surgery had healed, I headed back home to Cleveland. As a drummer, I thought I might find a job with a band. During my search, I found the Johnny Allen band was auditioning for a drummer two days later at the Palace Theater downtown. Late as it was, I called the phone number listed and applied for an audition spot. The next day I received a call and was told I could have the last spot. I was also informed someone there knew of my playing and put in a good word for me. I never found out who it was. I was also pleased when the gal said I would be playing on the Palace's studio drum set, Slingerlands like my own.

My audition was scheduled at four. Mine would be the last audition. I arrived at the theater about three dressed in my best drummin' outfit and carrying my own set of sticks and brushes. While settling in a back row seat, I saw there was no band. On stage were two sets of drums with a drummer at each set. The band was the Palace sound system playing records. I listened to the three o'clock audition and decided he wasn't very good. The drummer who was doing the audition sat at the other drum set with his back to me. He was obviously a pro. I was too far away to hear either of their voices, so I had no idea what was going on other than the playing. The man giving the audition cut it short at about three thirty. I knew that drummer wasn't going to get the job.

"Who's next?" the drummer called to a lady sitting in the front row. "I hope to Hell he ain't late."

"We've got a Howard Johnson scheduled at four."

"I'm here now," I shouted from way back in the theater.

"That's encouraging," she told the drummer, then called to me, "Come on down. Let's see what you can do."

I ran down the aisle and checked with the lady who had called me. "Go ahead up on the stage. Let's find out what you can do." Then she turned to the drummer. "Here's the last one, Buddy. Name's Howard Johnson."

The drummer turned, saw me and said. "You're just a kid, how old are you?"

I found myself looking into a face I immediately recognized, Buddy Rich. It took me some time to regain my voice. "I'll be eighteen in November. Been playin' in mostly jazz bands since I was fourteen."

"Well kid, I won't hold that against you. How about dance bands? Johnny's band is definitely a dance band. It's a new band I'm helpin' get started."

"Played with a dance band in a resort the last three summers."

"Ya read music?"

"Yep. Drum scores and regular. I can read 'em both."

"Sit down and get used to those drums. We can start as soon as you feel comfortable."

I sat down and took a look. "I got a set almost identical. I'm comfortable already."

"I see you brought your own sticks. Smart kid."

"Thank you mister Rich."

"Hey. You want me to start callin' you mister Johnson? It's Buddy, and you're . . . how about HoJo? I love your ice cream."

His easy way of talking and friendly manner immediately put me at ease. I liked this gruff character right off. "HoJo's fine with me, Buddy."

"There you go. How about we get at it. Let's see what you can do. We'll start with the score right there in front of you. Know it?"

"Of course. We often played Stardust. It's one of the most popular dance numbers in the country."

"Hit it Irene," Buddy called out. Immediately the intro to Stardust began.

I caught it immediately after the three bar intro. The score I was reading was a bit different from what I was used to, but not difficult. Buddy played with me, watching me carefully.

Mid way through the song Buddy stopped playing and shouted. "Cut." After the music stopped, he instructed, "Why don't you play the score you played with the dance band. Forget the score in front of you, OK?"

"Sure, Buddy."

He was following my score by listening to me play. We tried several other numbers both with and without a printed score. As I remember they were Frenisi, Tuxedo Junction, and I've Got a Gal in Kalamazoo.

"Kid, you're pretty good. Now that we've done the simple stuff, how about something a little tougher. You say you're a jazz drummer, can you follow?"

"I've played in a number of great jam sessions. I'll sure give it a try."

"O like you, kid. You're smart . . . and not a smart mouth. I'll play a riff, and you repeat it. Then we'll play it together. Got it?"

"You bet."

He started with a simple series and then made them more difficult. Then he hit me with a long and complicated improv piece. I tried, but it was too long for me to remember. When we played it together, I stayed with him, pretty much.

When we finished, he turned to me and said, "Kid, you are damned good. I threw that last one at you for fun. I never expected you to be able to follow, but you damned well did almost all of it. Never had that happen before in an audition, never. I don't need to hear any more. The job is yours if you want it."

I was floating on clouds."Really? You like my playin'?"

"Sure do. You remind me of me at your age. You're good, and you know it. You're confident, but not a braggart. You're also respectful. I like that. Let's go sit in the office. There are a couple of comfortable chairs there, and I want to talk to you for a bit."

"OK, Buddy." I was so thrilled I could hardly take a regular breath. All I could think of was I had played with the famous Buddy Rich, and he thinks I'm good. It was heaven on earth, unbelievable. I was overwhelmed with joy and pride.

"We're done, Irene. Wrap it up. I want to have a private conversation with HoJo here, so we're going in the office. See you later."

He pulled a couple of long necks out of the frig, took off the caps, handed me one, and motioned me to sit. "I know you're too young to drink, but I never knew a drummer who didn't relish a cold one now and then. If you don't want it, leave it, and I'll drink it."

"On those hot summer days in the dance hall, we kept a cooler hidden among our instrument cases. Nothin' like a cold beer to quench your thirst on a hot day."

"OK kid, here's what concerns me. Bein' in a road band means lots of travelin'. Most of it cramped in one of those charter busses. A few days in Cleveland, a few days in Omaha, a few days in Denver. You get my drift? It's a rough life. Are you sure you want that?"

"No, not really sure. I was suddenly cast adrift after having my future all planned out. I like playin' drums, so I thought I'd try it out."

"Tell me what happened. In fact, tell me about your life so far."

We must have talked for at least half an hour. I told him about my life and he told me I should consider another line of work, that drummin' in a traveling band was a rough life for a young kid like me.

He summed it all up. "Let me tell it to you straight. The girl you're so in love with? Unless she follows the band, you'll see her just a few times all year, when the band plays near Cleveland. Your family the same. The pay is generous, $200 a week for starters. If you don't spend it all on booze and women, you could save up quite a bundle in four or five years. There will be lots of temptations, women and drugs especially. A lot of musicians fall in with bad people, really bad people. You're so young and might not be able to resist those influences. I've seen it happen many times. Think about it. Johnny won't be able to start you for at least a month, so you've got time. I like you, kid. Lookin' at your background and knowin' what you'd be lettin' yourself in for, I hope you turn it down. Find some local band to play in. Hell, I'll even write you a recommendation. Don't hit the road for at least the next five or six years 'til you get your feet on the ground."

After much soul searching, I took his advice. A month later I enrolled at Purdue University to pursue a career in engineering. Playing drums requires such dedication I decided to make a total break. It had been too essential a part of me requiring many hours of practice to stay on top. I sold my drums and gave away everything else save my favorite set of sticks, which I still have. I have never even sat in or played drums since. I have no regrets although I do get an occasional twinge when I hear certain 40's songs or see a good drummer in operation.

In the spring of 1949 I was at Ohio State, visiting my high school sweetheart. (She later became the mother of five of my children.) While there, we attended a popular dance. The band? Buddy Rich and his brand new band were playing. They were fresh from Jazz at the Philharmonic. Our group had a table right near the band.

Buddy walked right near our table at the first intermission. My frantic waving caught his attention, and he came to our table..

"I came over because you look vaguely familiar," Buddy commented. "Do I know you?"

"Cleveland 1945, the Palace Theater audition."

"HoJo. Sure, now I remember."

"I took your advice. I'm a senior in engineering at Purdue. Had to give up drums completely, I was that much into them. Oh, and my high school girl friend? This is Dee. We're engaged. Plan to be married next September."

"Congratulations and good for you both"

He stood there talking with us and recounting my audition for about five minutes.

"Now I have to get back to work. The demands of being a drummer." He grinned and winked before heading back to the band stand.

Dee was impressed as were our friends, the four couples with us at the table.

I never saw Buddy Rich again. Oh, and the Johnny Allen band? Soon after they started, a new band leader took over . . . fellow by the name of Ray Anthony.

The optimist proclaims that we live in the best of all possible worlds, and the pessimist fears this is true.

—*James Branch Cabell*

A friendship founded on business is better than a business founded on friendship.

—*John D. Rockefeller (1874-1960)*

All are lunatics, but he who can analyze his delusion is called a philosopher.

—*Ambrose Bierce (1842-1914)*

You can only find truth with logic if you have already found truth without it.

—*Gilbert Keith Chesterton (1874-1936)*

An inconvenience is only an adventure wrongly considered; an adventure is an inconvenience rightly considered.

—*Gilbert Keith Chesterton (1874-1936)*

I have come to believe that the whole world is an enigma, a harmless enigma that is made terrible by our own mad attempt to interpret it as though it had an underlying truth.

—*Umberto Eco*

Be nice to people on your way up because you meet them on your way down.

—*Jimmy Durante*

Three o'clock is always too late or too early for anything you want to do.

—*Jean-Paul Sartre (1905-1980)*

The Purdue Attitude

I was a green seventeen year old when I started at Purdue in the fall of 1945. The ratio of males to females stood at about 5.5 to one, and most of those males were vets returning from WWII on the GI bill. This put a cash poor, non vet at a distinct disadvantage in the competition for dates. Most freshman females developed what was disparagingly called the *Purdue attitude* by the time they were there a few months. It was a haughty, *aren't you lucky to be in my company,* attitude with many ramifications. Dates when they could be arranged were usually six to eight weeks into the future.

My sophomore year I was pledged to Pi Kappa Phi Fraternity and living in the frat house. Our fraternity had an arrangement with the Pi Beta Phi sorority with whom we had frequent *trade* dances where their available members came to our house for a get acquainted dance. Several of my brothers were dating Pi Phis and even married them later. At our first trade dance, one of these girls put me together with her room mate, a tall, lovely freshman with long straight blonde hair. We hit it off immediately. We danced, walked outside, talked and genuinely enjoyed each other. I thought, *Wow. She likes me. Maybe I have a chance for a steady girl friend.*

We had a long, lingering good bye including a soft, warm kiss before she left. When the dance was over, we set a date for the movies in, well, she was busy for the next four weeks, so the date was set for Saturday, five weeks ahead. After several afternoon soda dates at the local Sweet Shoppe during the next two weeks, things seemed to be going well. I should have been suspicious when her study schedule began interfering with those soda dates, but what did I know?

On the Saturday of our date, my folks were visiting family in Indianapolis. As a result, I had the family car for our date. Two of my frat brothers and their Pi Phi dates would be going with us. We planned a movie and then a visit to the Five O'Clock, one of the students' favorite watering holes in down town Lafayette. This was a date certain to impress—the new car, a great movie, and fun at the Five O'Clock—all positive things.

The three of us were waiting in the main living room of the Pi Phi house when our dates came down stairs. Ellen, my date, was even lovelier than I remembered her. I was impressed. When we got in the car, Ellen sat with me up front, and both other couples squeezed into the back seat together. There was a great deal of snickering and soft chatter from the back seat, but the front was quiet. I realized something was not copacetic when Ellen sat far from me looking straight ahead. When I tried conversation she remained distant and almost cold, quite a change from our earlier meetings. During the movie, she hardly spoke in spite of efforts at conversation on my part. By the time the movie was over, I was pissed. This was not a date, it was an endurance contest.

When the movie was over about ten, we all got in my car and headed for the Five O'Clock less than a mile away. Our plan was the usual fun and companionship. We were going to stay there partying, then leave in time to get the girls back home by one. I pulled up to the front door of our partying place. The four in the back were disgorged and headed for the door. Ellen would have gone with them, but I reached across and held the door shut so she could not leave. The following conversation ensued as I pulled away from the curb.

"What are you doing? Where are you going?" Ellen asked in more words than she had uttered in the entire evening thus far.

"Well, It's obvious you are not enjoying the evening, and I know I am not, so I decided we would both be much happier if I took you home."

"You . . . " was all that came out of her mouth. Her eyes flamed with anger.

Silence and a dead ahead steady stare was all I got until I arrived at the Pi Phi house. I got out, opened the car door for her, and watched as she stomped up the walk to the door, each step declaring her anger and defiance. I did not go with her, the usual and courteous thing to do. I honestly believe it was the only time ever I didn't walk a date to her door. As soon as she entered the house and closed the door, I headed back for the Five O'Clock to rejoin my two brothers.

When I got there, I had to explain my actions. The two girls were sympathetic, but warned me there might be consequences.

"No one treats a Pi Phi that way without some negative action by our sorority," Sue said. "You may be sorry for your actions."

"My actions?" I asked incredulously,. "I was treated like crap, and still treated her like a gentleman should. At least until I took her back to the sorority house. She was obviously unhappy, and I took steps to end a miserable time for both of us. How come I am at fault?" I never learned whatever problem Ellen had with me.

Ann, the other gal who was also in several of my classes, added, "Well, nobody treats a Pi Phi like that without consequences. You should have stayed with us here."

By the time I dropped them off back at the Pi Phi house, both girls were much on my side. Unfortunately, they did not prevail in the special meeting the Pi Phis held to determine how to punish me. The verdict? No Pi Phi was ever to date me under any circumstances. All Pi Phis were to shun me on campus—yea—like the Amish. I was to be treated as if I didn't exist. I was called into a meeting with our Archon (president) and the Pi Phi president. The verdict was to be read by our Archon, of course, since no Pi Phi could speak to me.

When he finished, I laughed. "Is that all they are going to do? I thought I was going to be hung, drawn, quartered and boiled in oil." The Pi Phi looked daggers at me. "Ask her what about the two Pi Phis who usually ride home with me to Cleveland. Will they be forced to find another ride, will the sorority insist a spy ride with us to make sure they don't talk to me, or will they rely on their promise not to speak? This whole thing is ridiculous, and I'm the one who is laughing. I can't believe college women would sink to such immature posturing. Ellen is the only Pi Phi I have ever gone out with, and I seriously doubt my life will be affected in the least if I never speak to another Pi Phi in all of my days."

Phil, our Archon, cautioned me against saying such things. Then he ushered the Pi Phi president to the door. When he returned, Phil said, "I can't believe those women. You were right on, calling them immature."

"Why in the hell didn't you say so when she was here?"

"Come on Howie, I was trying to keep this a one person battle and not get the whole fraternity involved. There are seven brothers dating Pi Phis, and they don't need the flak an all out battle would bring on."

"You're right, of course," I said. "It's no big deal as far as I am concerned."

I hardly noticed the shunning. Every Pi Phi with whom I had classes acted as if nothing was amiss. One of my brother's Pi Phi girlfriend refused to speak to me, but they broke up soon after anyway. The others treated me quite normally. I never had the slightest inclination to ask another Pi Phi for a date. There were lots of other attractive females available.

Fate sometimes makes for strange pay backs. Two years later at our senior prom during the introductions of significant guests, the same Ellen had the job of making the following intro: "This is Dolores Osborn, Homecoming queen candidate at Ohio State University and her date, Howard Johnson." No, she wouldn't look at me.

Oh yes, years later my daughter became, what else, a Pi Phi at Wittenburg. Her sorority treated me royally, even after I told them this story.

A doctor can bury his mistakes but an architect can only advise his clients to plant vines.

—*Frank Lloyd Wright (1868-1959)*

It is dangerous to be sincere unless you are also stupid.

—*George Bernard Shaw (1856-1950)*

If you haven't got anything nice to say about anybody, come sit next to me.

—*Alice Roosevelt Longworth (1884-1980)*

A man can't be too careful in the choice of his enemies.

—*Oscar Wilde (1854-1900)*

True Hell is when you realize you've made a stupid mistake and that someone else has to pay for your error.

—*Howard Johnson, 1966*

Forgive your enemies, but never forget their names.

—*John F. Kennedy (1917-1963)*

Lord, protect me from my family and friends, my enemies I can handle myself.

—*unknown*

Logic is in the eye of the logician.

—*Gloria Steinem*

No one can earn a million dollars honestly.

—*William Jennings Bryan (1860-1925)*

Everything has been figured out, except how to live.

—*Jean-Paul Sartre (1905-1980)*

Well-timed silence hath more eloquence than speech.

—*Martin Fraquhar Tupper*

Thank you for sending me a copy of your book - I'll waste no time reading it.

—*Moses Hadas (1900-1966)*

From the moment I picked your book up until I laid it down I was convulsed with laughter. Some day I intend reading it.

—*Groucho Marx (1895-1977)*

It is better to have a permanent income than to be fascinating.

—*Oscar Wilde (1854-1900)*

The Yellow Buick Convertible

In November of 1945, I began attending Purdue University, a seventeen-year-old fresh out of high school. With the ratio of guys to gals at Purdue standing at more than five to one and the campus swarming with returning WWII vets on the GI Bill, a seventeen-year-old hardly had a chance. Dates were hard to come by unless you had a steady or a car and lots of money. Most girls had dates four to six weeks ahead, and after a few months at Purdue usually developed what we disparagingly referred to as the *Purdue attitude*. This common condition was a blend of haughty superiority and an "aren't you lucky to be with me?" attitude. Even the local high school girls added little to improve the mix.

For this reason, I frequently hitchhiked from Lafayette to Indianapolis for weekends visiting my cousin Dona. She introduced me to her high school girlfriends, so I seldom had a problem getting a date, even on short notice. These girls were friendly and even somewhat impressed with a *college man*. I quit looking for dates on campus and even brought several *Naptown* girls to West Lafayette for Purdue social events: dances and parties. *Naptown* was Purdue slang for Indianapolis.

With great anticipation, I rode the Lafayette city bus to the end of the line. That was as close to the US 52 bypass as I could get on public transportation. From this point, I thumbed cars as I walked the mile or so to the bypass near the Alcoa Aluminum plant. Sometimes I would get a ride as I walked, but usually I ended up walking all the way to US 52. There were always a number of other guys lined up along the road, so as a new arrival, I would take my place at the end of the group. Even with six or seven others there before me, I usually got a ride in less than a half hour. Hopefully it would be with someone going all the way to *Naptown*.

One warm Friday afternoon in May, I hadn't been waiting long before a yellow Buick convertible stopped to pick me up. The driver, a middle-aged rather paunchy man, smiled pleasantly as he pulled onto the road headed south.

"You a Purdue student?" he asked as the Buick picked up speed. When I nodded in the affirmative, he continued, "I hope you don't mind the wind. This is the first warm day, and I want to leave the top down and enjoy the warm spring air."

"Fine with me," I replied. "I like top-down driving." I was, however, beginning to be apprehensive as he continued accelerating to a high speed.

I noticed he was going more than eighty, quite a bit too fast even for this dual, two-lane modern concrete highway. As we sped on, he explained that he was a salesman on his way home and drove this route every Friday afternoon. He explained his rush to get there was because of a dinner date with a new lady friend. He went on talking about his date in an animated fashion with exaggerated gestures. I was suspicious he had been drinking, and this heightened my fears. He was weaving over both lanes as he talked and gestured. Fortunately there were few cars on the road with us. When he did come up on a slower car, he seemed to be reasonably careful as he passed, but as soon as a car was passed, the gestures and weaving resumed.

We had gone about twenty miles when it happened. During one of his weaves across the center line, the rear wheel dropped into a sizeable winter pothole. The car twisted quickly to the left and then right as he overcorrected. The back of the car seemed to bounce higher each time he tried to correct. Suddenly we were airborne, and I saw the driver's feet fly over my head as he flew out of the car. At the same time, I dove for the floor and wrapped my arms around the pedals and steering column to hold on and stay with the car. It was an instinctive reaction which probably saved my life. All I felt then was the car bouncing and grinding along the ground upside down. Suddenly I was being pressed harder and harder against the floor. There was no pain, but there was tremendous pressure, and it was dark, very dark. Mercifully, the car ground to a halt.

In the silence, I took stock of my situation. I didn't seem to be badly hurt. In fact, I felt no pain, just a constant pressure on my body, a wet, cold pressure. My head was in a small air space under the dash beside the steering column. My right shoulder was jammed against the floor, and my right arm was pinned under the pedals. My left shoulder was cold, wet and held firmly in place by constant pressure. My left arm was immobilized against the front of the seat. With my torso and legs jammed against the floor and seat, they too were compressed, cold, and wet. My chest was compressed, but I could still take shallow breaths. There was absolutely no sound.

After what seemed like a long time, I began to hear voices, muffled and distant at first, but clear enough to be understood. I tried to cry out, but the pressure barely permitted shallow breaths, and I had a hard time getting enough air. I couldn't make any sound as all parts of my body were immobilized. It was about fifteen minutes before people were close enough to the car for me to hear them. I caught parts of conversations, enough to know the driver had been killed, the police had been called, and everyone there was certain the driver had been alone. No one had been close enough to witness the accident, and they were all wondering how it happened. It was apparent I was in for an ordeal, so I concentrated on staying calm and clear headed. Gradually the voices died down as the people moved away and back to their cars, leaving the upside-down Buick where it lay. When it became dead silent, I knew I could be in for a rough time.

It was a long time later when again I heard voices. The police were on the scene along with a few spectators. I guessed it was at least half an hour after the accident. I was relieved when someone, probably a policeman, explained they needed to turn the car over to see if there was anyone else in the car. They talked briefly about trying to turn the car over manually, but decided there were not enough people, considering the poor footing. It was then I realized the car had come to rest in a freshly plowed field. The wet dirt had filled the car and packed around me, holding me in place firmly. Again, I heard someone, probably the policeman, explaining they had called for a wrecker able to negotiate the muddy field. Once more, the voices died away as the people left the wreck where it lay.

My fingers, toes, arms, and legs were not completely immobilized. I found I could move my toes in both feet and even wiggle my right foot. Working my foot carefully, I tried to open the space around it. Part of my right hand also worked loose. It was close to my face, so I used it to clear the mud oozing up around my nose. I knew I had to keep the space clear to be able to breathe. When I realized the car was slowly settling in the mud, I almost panicked. It was forcing the wet soil up into every airspace. It crept slowly, almost imperceptibly up around my right arm toward my face. Fearing it would soon block my nose and cause me to suffocate, I worked frantically to free my right hand and push the oozing mud back. Finally, Either I began holding my own or the car stopped settling. In any event, the mud no longer crept up toward my nose. My concentration on the struggle with the mud took so much of my attention I seemed to lose track of time.

Finally, I heard a new sound. It was the motor of a large truck slowly approaching. *Was this the wrecker?* I wondered. Before long, there were again voices coming to where I lay imprisoned. After many shouted instructions and the loud sound of a nearby engine laboring, a loud clank of metal on metal told me the wrecker was about ready to lift the car. Suddenly the car shuddered and began to move. The wrecker was lifting the car from the right side. When I felt my feet drop free, a woman screamed.

A man's booming voice announced suddenly, "There's a body in there."

My immediate frantic kicks were greeted with "He's alive! His feet are moving."

A shout, "Hold it!" stopped the lift of the wrecker. I felt hands tugging on my feet, trying to pull me free, but the mud held me solidly.

"We'll have to lift some more, slowly," a man shouted. "Careful now. Care . . . ful!" the man instructed. As the car was slowly lifted from the right side and the mud fell away, I felt the pressure subsiding. Finally, the mud pulled away from my left shoulder, removing the crushing pressure. I could once again breathe fully. It was a marvelous feeling. I freed my right arm from around the pedals, drew my head out from under the dashboard, rolled out from beneath the lifted car, and stood up. A cheer went up from the small crowd watching. I was wet, muddy, stiff, and a bit sore, but otherwise, unhurt.

When I asked for the time, someone answered, "Seven-thirty." I had been in my muddy prison for nearly four hours.

Once out of the wet mud, I warmed quickly in the eighty-degree air. Surveying the scene, I could envision what happened. The car had left the road, dug its front into the soil across the ditch, flipped end for end, and landed upside down in a freshly plowed field, sliding at least two hundred feet. The windshield and dashboard acted like a scoop, packing mud tightly around me and filling the entire interior of the open car. Its path through the fresh soil told the whole story. The policemen were insistent they take me to a hospital and I was equally determined not to go. I was unhurt and able to move freely without pain.

Soon I found myself in a small hospital in Lebanon, a few miles away from the accident. The first thing I did was take a shower and clean off the remaining mud. After a short examination, the doctors were convinced I was unhurt, other than a few small scrapes and bruises. I doffed the hospital gown and changed into fresh clothes from my suitcase. The police had retrieved it from the field, relatively undamaged.

With the assistance of the Indiana State Police, I arrived at my cousin's party at about ten. Armed with a most intriguing story for the young lady who thought I stood her up, I was forgiven. In fact, the young lady was quite impressed. I was the only one who came to the party in a State Police cruiser.

The significant problems we face cannot be solved at the same level of thinking we were at when we created them.

—Albert Einstein (1879-1955)

Basically, I no longer work for anything but the sensation I have while working.

—Albert Giacometti (sculptor)

There is more stupidity than hydrogen in the universe, and it has a longer shelf life.

—Frank Zappa

There's a limit to how many times you can read how great you are and what an inspiration you are, but I'm not there yet.

—Randy Pausch (1960-2008)

An Unwanted Fight

It was a Friday evening in early July 1948, and I was enjoying an evening at the Tippy dance hall on Tippecanoe Lake in Indiana. I was pursuing a drop dead gorgeous young lady I met that morning on the diving float near our summer home at the other end of the lake. She was from the little town of Huntington about forty miles south of Tippy and was staying in a rented cottage a few doors east of our place with six other fine young ladies. All seven of them were at the dance, but I was only interested in Barbara.

Being a summer resident and having played in the band for several years, I had many friends there. (and a few not so friendly) With me were three of my fraternity brothers from Purdue. They were staying with me for the weekend. They included Pete Brewster, Bill Adamson, and Clarke Thornton, each one over six-three. Pete was a first string tight end on the Purdue football team who later became an all-pro tight end for the Cleveland Browns.

The girls were seated at a table near the dance floor, and we were at another table behind theirs. Actually, we were soon a mixed group of guys and gals around both tables. When the band took a break, my three buddies and three of the ladies headed outside for a little liquid refreshment stashed in our car. We couldn't bring any into the dance hall. I was sitting next to Barbara, relaxing with my feet sticking out on the dance floor when four of the local hoodlums from Elkhart strolled by looking for a fight as always.

One of them, Max, kicked my foot as he walked by, turned and said angrily, "I heard what you said about me."

I stood up, and Max moved around behind me. When we stopped moving, I was facing Max. "I didn't say anything about you, Max. Why would I? I hardly know you." I was not in a good situation. Max stood in front of me between the girl's table and the dance floor. His three buddies stood behind me on the dance floor. That is typically how they "fight," always at least two against one. While one is covering up to protect himself from any blows, the other hits from behind. Should their victim turn to fight the guy hitting from behind, the two reverse roles. I did not want any part of that kind of fight.

"I'm going to teach you not to talk about me like that," Max said, itching for a fight.

"I've never said anything about you," I said, trying to keep the situation calm and buy time. Then my three buddies walked in past our table. They walked up quietly and stood behind Max. He was concentrating on me and didn't notice them. Out of the corner of my eye I could see his buddy on my left frantically waving his arms to warn Max he had company. He still didn't notice.

I took the offensive. "OK, jackass, let's have at it," I shouted.

Max immediately drew back to throw a punch only to be pinned by Pete and lifted off the floor. Max and his buddies were all short, so my guys towered over them. The other three scattered leaving Max to his fate. Max struggled, but Pete held him in a crushing bear hug. He wasn't going anywhere.

"What should I do with this jerk?" Pete asked.

Clarke joked, "Throw him back. He's way too little to keep."

Pete unceremoniously dropped Max on the dance floor who climbed to his feet and then scampered for the door. The considerable audience the ruckus had drawn broke out in applause. Most knew Max and his buddies and how they operate. In all the years since then, I never again saw Max.

When the dance was almost over, the girls decided to continue partying at their place. Two of them had paired up with Pete and Clarke. They would go in Clarke's car while Barb, Bill, his gal, and I would go in my car. On our way out, the rest of our group went ahead when I stopped to talk to one of the guys from the band. He was an old buddy who had played in the band I was in several years before. We were standing by one of the inner doors when a general brawl broke out. Seeing a giant fist aimed at my head, I ducked. The fist, which was attached to a rather large, muscular young blonde guy, whistled past my ear and slammed into an aluminum panel covering a concrete block wall. The blonde guy doubled over in pain holding his arm. Oh yes, the imprint of his knuckles in the panel can be seen there to this day.

I ducked quickly out the door to avoid the growing brawl and headed for my car. Suddenly I was grabbed from behind and spun around. The blonde guy was in pain, seeking revenge, and had made me the object of his anger. He was also drunk. I had no reason or desire to get into a fight with him, but I had no choice. As he staggered after me, I realized how drunk he was. I easily ducked out of his clumsy attempts to grab me, and the roundhouse lefts he threw at me. He wasn't about to use his injured right. I tried running away, but drunk as he was, he ran faster than I could. He caught me and knocked me down with the only blow he landed. It was to the back of my left shoulder, and it hurt. He was very strong.

"I don't want to fight you," I yelled at him several times, but he kept boring in. Every time he swung and missed, I would flick a half closed fist at his face when he was off balance. This went on for quite some time. I could see Barb, Bill and his gal following with the group of watchers gathered in the parking lot. I waved Bill off when I thought he might come in to help. I knew that as long as I danced away from his wild left hand I would be OK. His face was getting quite bloody. The sharp flicks from my half open fist were cutting him up. Still he came at me, and I danced away backwards. Twice a couple of his friends tried to get him to quit, but he pushed them away.

We had to have been at this for a long time, maybe half an hour, when suddenly he simply collapsed—passed out on the ground. The combination of too much alcohol and all the exertion did him in. I'm sure my blows were a minor factor. When his two friends came over I was thinking now they're going to fight. I was surprised when they apologized for his behavior and for his drinking. They seemed genuinely concerned when I told them about his right hand and how he hit a concrete wall with it. They had only seen him chase me out in the parking lot.

The older one asked, "Do you know who this is?"

I replied, "Haven't the slightest idea, but I'm sure glad he wasn't sober, or he'd have beat the hell out of me. He's one big, strong hombre."

"His name's Chick Jagade. He's the varsity fullback on the Indiana University football team. Actually, he's a pretty nice guy. This is the first time I've ever seen him get in a fight. He drank far too much tonight, another thing he doesn't usually do."

"You must be a good friend of his."

"Actually, I'm his older brother, Tad, and this is our younger brother, Grafton," he added indicating the other young man.

We talked for a while when Bill and the gals came over. We also helped them put him in their car. He was sure heavy.

"How much does he weigh?" I asked.

"Oh, about 240 now. He'll get down to 225 when he's playing," Grafton answered.

I had been fighting a guy who out weighed me by almost seventy pounds. I was sure glad he was drunk.

As they were about to leave, Tad said, "Why don't you come over here tomorrow evening and join us."

I begged off saying, "I'm afraid he'll want to kill me as soon as he sobers up."

Tad smiled. "No, Chick is not like that. We'll get him patched up and keep him sober. He'll want to apologize to you, I'm sure."

When the four of us appeared at the girl's place, we had quite a tale to tell.

It was with mixed emotions and a lot of urging by my friends that we went to Tippy Dance hall the next evening.

When Chick and his brothers arrived I was nervous. Chick came over and immediately stuck out his left hand for a friendly shake. His right hand was bandaged, and there were several small bandages on his face. He smiled and apologized profusely. Before long we were like old friends. When he found out Pete was on the Purdue team, they were soon huddled and talking football. It was certainly a far more enjoyable evening than Friday. The other three girls paired up with the three brothers. We all danced and talked the whole evening What a difference a day makes.

Oh yes, Barb and I dated for the rest of the summer. I even invited her to Purdue for the junior prom and a few other occasions, but we drifted apart. Fifty-five years later we met once more at the funeral of a mutual friend we didn't know we shared. We both recognized each other immediately. We quickly recounted some shared memories including the nights after we met. She was still an attractive and gracious women at seventy-four.

❖   ❖   ❖

It is far better to grasp the Universe as it really is than to persist in delusion, however satisfying and reassuring.

—Carl Sagan (1934-1996)

All truth passes through three stages. First, it is ridiculed. Second, it is violently opposed. Third, it is accepted as being self-evident.

—Arthur Schopenhauer (1788-1860)

Many a man's reputation would not know his character if they met on the street.

—Elbert Hubbard (1856-1915)

Perfection is achieved, not when there is nothing more to add, but when there is nothing left to take away.

—Antoine de Saint Exupery

Life is pleasant. Death is peaceful. It's the transition that's troublesome.

—Isaac Asimov

If you want to make an apple pie from scratch, you must first create the universe.

—Carl Sagan (1934-1996)

A Christmas Tree for Carol

My first year in college at seventeen was right after the war ended while Purdue was still on a wartime-accelerated schedule of three semesters per year. I was placed in a special English composition class as part of a new program to "broaden the horizons" of engineers by adding a number of liberal arts courses. In this course, we wrote at least one essay or short story each week. With Christmas coming up, I decided to write a Christmas story I named *A Christmas Tree for Carol.* The story follows within this memoir, of which it is an important part.

Carol Sims was a bright, excited ten year old looking forward to Christmas. It was early December of 1932, and for more reasons than the depression, Christmas looked bleak for the Sims family. Carol was blind and unable to walk. When she was three, she slipped out the front door and fell down the stone steps, hitting her head savagely on the corner of one step. Rushed to the hospital, she lay in a coma for almost a week while her family and friends prayed for her recovery. When she awakened, the Sims family's joy was short lived. They found she was unable to feel or move her legs and couldn't see. Carol was paralyzed and blind.

Many trips from her small town home to specialists in nearby Cincinnati led to the conclusion she had permanent brain damage. She would never walk or see again. The specialists who examined her said they couldn't find a reason for her blindness or loss of use of her legs as all their tests indicated she had recovered fully. As far as they could see, she should have been back to normal. They referred the Sims to a neurological specialist in Philadelphia, a top man in the field, who could help her. The Sims family needed a lot more money than they had for the cost of the trip and necessary treatment.

Carol's father, Carl, owned and operated a small grocery store near their modest home. The family struggled to save the most money they could from store profits to build up enough to fund Carol's trip and treatment. Before they had saved enough, the depression hit. Then they had to struggle even harder to feed themselves and keep the store open. Their savings began to shrink along with all hope for treatment by the specialist.

Late in 1932, the owner of a grocery chain offered to buy their grocery for cash. It would give them more than enough for Carol's treatment but would leave the family without their business. The chain owner offered Carl a job running his old store, but the low pay would make it difficult for them to pay their home mortgage. After much back and forth discussion, Carl and his wife decided selling the store was the only option they had if Carol was to have a chance to recover. They were to meet with the buyer to sign the sales contract on the third of January. Carol's trip to Philadelphia would be arranged as soon after as practical.

With Christmas coming, Carol begged for a Christmas tree. She wanted it placed in her room next to her bed. "But you can't see it," her mother said. "That's a lot of money for a tree you will be unable to see, honey."

"But Mom, I will be able to smell it, and if you put it next to my bed, I can reach out and touch it. . . Please."

Carol got her tree. It was a small one, decorated with lights, ornaments, tinsel, and foil icicles on every branch. She reached out and fondled the branches and the icicles. She breathed in deeply the pine fragrance.

"Thanks Mom and Dad. This is the best Christmas ever," she said as they bade her a loving goodnight on Christmas eve.

As her father pulled the covers up around her, he warned, "It's a miserable night outside with lots of wet snow falling. Looks as though we'll have a white Christmas after this storm goes through. Stay warm, sweety."

Carol lay listening to the wind whistling through the branches of the tree outside her window. She smiled as she drank in the smell of the Christmas tree and reached out to touch gently the icicles on the nearby branches. *This is indeed a wonderful Christmas.* She thought to herself.

Suddenly there was a blinding flash and fierce crash of thunder as a lightning bolt struck the tree outside her window. The bolt jumped to the roof of the house, went through the roof and the ceiling of Carol's room, transferred to the metal icicles on the tree, turning them into a shower of sparks that filled the room. It then coursed through Carol's metal bed before going to ground through the floor.

Carol's mom and dad rushed into her room and found her sitting upright on the floor amidst and covered with the burned out remains of thousands of icicles. She was dazed but apparently unhurt by the lightning. They turned on the light and saw their daughter moving her hands strangely in front of her face.

"I can see. I can see. I can see," she said with increasing excitement as she looked up at her parents with eyes no longer dead. As her father picked her up gently and placed her back in her bed, she kept repeating those precious words over and over again.

"It's a miracle, a Christmas miracle," her mother repeated joyously several times.

Then another miracle, "I can feel my legs," she shouted as her father lowered her gently back into her bed. "I can feel my legs."

It was a long Christmas eve before the Sims family settled down to rest and sleep.

The next morning the hole in the roof and the dripping of melting snow into a room covered with burned icicles was of little consequence and certainly did not dampen their joy. Amazingly, other than the burned icicles, the tree was undamaged. Carol kept repeating, "The tree is beautiful. The snow is beautiful. Everything is beautiful," as she continued getting used to her regained sight.

Over the next year, Carol had to relearn to walk. Her wasted leg muscles gradually gained strength, and by the following Christmas she could even run a little. Eventually she would grow strong and become a dancer on those once withered limbs. The Cincinnati Inquirer did an article on her miraculous experience, and she became known as *The Lightning Girl*, the name they gave her in the article. With all the publicity, people flocked to the Sims' tiny grocery store for a chance to see the miraculous *Lightning Girl*. With the new and unexpected wealth this generated, the Sims family made many donations through their church and directly to the needy in their small town. Each Christmas following there was a small tree placed in Carol's room and decorated, especially with icicles.

I took the story home with me for Christmas vacation where I gave it proudly to my sister, Bobbie, to read. Starting at an early age, Bobbie and I had many verbal battles. At six years older than I, she knew a lot more than I did about many things and did not hesitate to tell me so. She ridiculed the part about the lightning saying, "Lightning never strikes in a snowstorm. Everyone knows that."

I was devastated. She was my big sister and even though we had many verbal battles, I still loved her, respected her, and constantly sought her approval. Her words hurt, but of course, I was not about to let her know that. I don't remember, but I probably went off to lick those verbal wounds all by myself.

It was another Christmas several years later, our first as man and wife, when Dee and I were visiting Dee's parents when another wet snowstorm blew in. I remember so clearly, standing at the back door, wet snow swirling and hitting my face, when a bright flash of lightning lit the night sky. I immediately remembered, with ample satisfaction, what my sis had said about the story. As the clap of thunder followed, I shouted to my self, *Vindication! I was right, and she was wrong. Lightning can strike during a snowstorm.* It brought me such satisfaction I stayed outside in the storm and watched several more lightning flashes followed by thunder. I was enjoying and savoring each flash. I don't remember if I ever told my sis about it or not. I know I wanted to phone her immediately, but whether or not I did, I don't remember. I wonder if she will read this memoir and remember after nearly sixty years.

❖   ❖   ❖

A witty saying proves nothing.

—Voltaire (1694-1778)

Sleep is an excellent way of listening to an opera.

—James Stephens (1882-1950)

The nice thing about being a celebrity is that if you bore people they think it's their fault.

—Henry Kissinger (1923-)

Education is a progressive discovery of our own ignorance.

—Will Durant

I have often regretted my speech, never my silence.

—Xenocrates (396-314 B.C.)

It was the experience of mystery -- even if mixed with fear -- that engendered religion.

—Albert Einstein (1879-1955)

If everything seems under control, you're just not going fast enough.

—Mario Andretti

I do not consider it an insult, but rather a compliment to be called an agnostic. I do not pretend to know where many ignorant men are sure -- that is all that agnosticism means.

—Clarence Darrow, Scopes trial, 1925.

There are people in the world so hungry, that God cannot appear to them except in the form of bread.

—Mahatma Gandhi (1869-1948)

It is much more comfortable to be mad and know it, than to be sane and have one's doubts.

—G. B. Burgin

Earthquake

In March of 1952, My wife, Dolores, our year old daughter, Debbie, and I moved from Cleveland, Ohio to Long Beach, California. I had taken a job as plant supervising engineer for MacMillan Petroleum in Signal Hill. We rented a one bedroom second floor apartment on Carson Street near Long Beach Boulevard. It was a short, ten minute drive down nearby Cherry Avenue to the refinery.

Four months after I started working at the Signal Hill refinery, we were running overnight tests on the fractional distillation column. There were several outlet pipes distributed in a line on one side of the column. These outlet pipes were for the various "cuts" of the petroleum separated within the bubble cap column from, petroleum gas at the top to gasoline, kerosene, jet fuel, heating oil, lubricating oil and heavy residue oils and asphalt at the very bottom. The raw petroleum was heated in a boiler and fed into the column on the other side from the outlets and near the bottom of the column. The column was about four feet in diameter and sixty feet tall.

It was before five in the morning on July 21, 1952 when I climbed up the steel ladder of the column to the 40 foot level to take the five a.m. samples and readings. At this level, there was a steel platform about 15 inches wide surrounding the column. There were three steel gussets welded to the column at the same level as the platform. Waist high there was a steel pipe railing providing a hand hold for anyone standing on the platform. One inch braided steel cables attached to the gussets were anchored by concrete tie downs. These tie downs were attached to the ground with long steel posts driven deep. These three cables alone held the column upright.

I had finished noting the readings and taking the samples when ground waves from the Tehachapi earthquake reached the refinery. The column began swaying. The first I knew something was happening was when I was slammed backwards (fortunately) into the insulation surrounding the column. When the column pitched back, I grabbed the pipe rail with my hands and was folded over this pipe at my waist. It was all I could do to hang on. As the ground moved in waves, the column swung back and forth, alternately smashing me back against the insulation and then forward against the pipe railing. This went on for what seemed to me to be forever until the swaying slowed enough, so I could descend to the ground.

453

While I was being slammed back and forth I noticed two things: there was a lot of heat from flames down below, and the steel cables seemed to be getting fatter. Beneath me, hot petroleum spewing from broken pipes burst into flames and almost got to me. Several pipes including the feed pipe from the boiler had broken spewing out petroleum and distillate which immediately caught fire. I watched the bravest man I have ever known or even seen, running through those flames wearing only foul weather gear and boots, shutting off valves and pumps. But for his efforts, I would have been cooked, literally. It took him three months to recover from his burns. Needless to say, I visited him often at the hospital and got him whatever he wanted. He was the chief operator of the column and everyone held him in awe when he came back to work. The company gave him a brand new Cadillac and some money, I'm sure. He was modest and unassuming about what he had done.

The second thing I noticed, the fattening of the steel cables connected to the gussets was caused by individual wires in the cable breaking from the stress. The cables were becoming fuzzy with broken wires. Luckily, they didn't weaken enough to break. It was at least half an hour before the fires below went down and the column swaying slowed enough, so I could chance scampering down the steel ladder. Someone threw me a pair of heavy work gloves, so my hands would not be burned by the still hot bottom section of the steel ladder. As it was, my shirt and pants were burned in several places, but I only received a mild burn on my arm. I had been incredibly fortunate. Oh yes, there was a complete imprint, several inches deep of my body in the insulation where I had been slammed against the column. The insulation had cushioned my head and body when I was repeatedly flung against the column. This was my one dangerous experience with a serious earthquake, 7.9 on the Richter scale, the third most powerful quake in California history.

Dolores told me how Deb's crib rolled back and forth across our bedroom while she held on to it for dear life. She looked out the window and saw palm trees and street lights waving back and forth as the ground moved in waves as if it were water. The brick and board book case in our living room was our only casualty at home. She also said our apartment creaked like the sounds of old wooden ships at sea in the movies.

We also experienced the aftershock of 5.8 that struck near Bakersfield on August 22 at 3:42 in the afternoon. I was at home in our apartment at the time and heard the creaks and groans Dee described in July and saw the light poles and palm trees sway like small trees in a wind storm. Our book case withstood this quake without collapsing. All I can say is it is a strange and unnerving experience to feel and see the solid earth move in waves like the surface of a body of water.

Discovered - Something Physical I am Good at

When I was young, I loved sports but was not particularly good at all but one. The one game in which I excelled was dodge-ball. In junior high, I was almost always the last player, the winner, in this elimination game. I was among the best at dodging, so of course, I loved the game. High schools do not have dodge-ball. It was a junior high game only.

When I went out for football and baseball in high school, I was ridiculed and humiliated, mostly because I was so small. I couldn't cut it among the bigger boys. Until my junior year in high school I was small for my age. I was a full head shorter than most of my peers. Then, in the summer between halves of my junior year I grew six inches and put on quite a bit of muscle. I gained thirty-five pounds, probably because I did a lot of swimming and running at the lake.

Late in the summer I went out for football. They no longer kidded me about being small, but they did about my running. For whatever reason, I was the slowest one on the team by quite a bit. The coach soon discovered that though slow, I was extremely hard to tackle, an art learned from my dodge ball days. Not heavy enough for a full back, I played half back. I was fairly successful for the first few games of the season until our opponents realized they could easily catch me from behind. After that, I had a fairly successful season running inside and dodging tacklers in front of me. I was good at getting five yards through the line.

I tried pitching in base ball with little success and was a mediocre hitter and 2nd baseman. I tried my hand at softball in both church and industrial leagues with limited success. I was a fair hitter and third or first baseman, but certainly not as good as others on the team. Still, I loved the competition. When we moved to Long Beach California, I played third base in an industrial league at about the same level as before.

Then, In June 1952 We sold our old Dodge convertible and bought a sports car, a new MG TD. Gene Sommers, the salesman who sold us the MG, was a racer who raced midgets on Southern California courses with some success. He invited us to attend a sports car time trials the following weekend in Costs Mesa, a few miles south of Long Beach. He even offered to let me drive his car in the time trials as my car was far too new and hadn't been broken in.

The Costa Mesa time trials turned out to be a fairly big deal. A number of Hollywood celebrities were there, many of whom would be competing in their expensive Ferraris, Mercedes, and Alpha Romeos. There were 44 MG TDs and TCs entered, enough to have a separate class for MGs. We would be competing against the same cars with identical engines. It would be a fair test of purely driving ability.

We first walked the entire 1.5 mile course to familiarize ourselves with the two lane city streets in this housing development with no houses as yet. We would be racing on asphalt roads with no curbs. Cars would be waved off one at a time with the next car starting when the earlier car passed the half way point of the course. Starting times were drawn by lot, and I was right in the middle. There would be three runs for each car with the first run a practice run. The time counted would be the best time a driver posted of the next two runs.

My practice run went quite well. I calculated braking points and only over ran two corners. I would have to do better on those two corners. My first timed run went perfectly. I didn't miss a single braking point or corner. The time they posted for my first run looked quite good to me. It was right in the middle of the MGs, identical to the times of the next two cars. My next run was I thought much better. I waited, but there was no time posted for my run. It was still not posted when the last car finished. Something was amiss.

Then there was an announcement over the PA system. "Ladies and gentlemen. We discovered our timer has improperly entered many of your times, placing them a line below your number on the record sheet. Most of the times posted on the board are in error. We will be making the necessary changes in a few minutes, so please be patient. There is no doubt the new times will be correct."

There were lots of both cheers and moaning from the crowd when the new times were posted. All were listed in the order of their finish. I noticed my friend, Gene, had finished in seventh place, a good showing against tough competition. I kept looking down the list and couldn't find my posting anywhere. Then I happened to glance at the top and there it was. I had taken second place. I couldn't believe it. My friend came over to congratulate me. He refused to believe I had not competed before.

"You're pretty sneaky," he said smiling. "If you haven't driven in competition before, you should think about it now. You must be a natural. Both of your times are much better than mine and in **my car**. That's hard to swallow. If I were you, I'd see about entering the next few races. Golden Gate, Pebble Beach, and Torrey Pines races are all coming up in the next few months. I'll even help you get registered, tested, and entered in any of those races you can get to."

A twenty-year sports car racing career was kicked off in those three races. I finished well up in the standings at Golden Gate Park where I finished sixth in class and Pebble Beach where I finished fourth. My finishing position was excellent for a novice driver as there were almost forty entries in my class at both of those races. Torrey Pines was another story.

On the starting grid, I was far back in the right lane of the two lanes, at least fiftieth. Starting positions were chosen by lottery and I didn't do so well. It was to be a standing start. The starting grid was on a straight stretch of two lane asphalt with sloping grassy shoulders on either side. When the starter dropped the green flag, the first car in the left lane stalled. My lane moved out but for some reason was moving slowly, much slower than I wanted to be going. As soon as I passed the starting line I dropped off the pavement onto the grassy shoulder, accelerated and passed most of the cars in my lane.

At the end of the quarter mile starting straightaway, was a tight left-hand corner. I negotiated the corner in seventh position. I had gone from fiftieth to seventh in a quarter mile. I held my position through several corners, both right and left. When I pulled onto the back straightaway, I quickly went from seventh to twenty-fifth when the bigger and faster cars went roaring by. I managed to hold onto third in my class through the next few laps, fighting through the corners with a Jowett Jupiter and a Singer, both of which had larger engines than my 1250cc MG.

Then I made a mistake that cost me many positions but gained me some short-lived fame and a nickname. One LA TV station caught the entire thing on tape and showed it many times during the next few days. Basically, I over cooked the corner, slid out on the gravel shoulder sideways, over corrected and headed back across the road right in front of the Singer. I went off the other side and slid on the slippery *ice plant*, covering the roadside all the way uphill to the snow fence erected to keep spectators away from the course. My rear bumper hooked the snow fence, and I started back toward the course amid screams of spectators whose cameras were hanging from the snow fence being dragged down the hill toward the road. It took some time for those spectators to unhook the fence and send me on my way.

I ended up nineteenth in class but gained valuable experience. I had a much less serious mishap than the driver of another MG. I had been trying to overtake and pass a yellow MG for several laps. Together we approached a place where the road came perilously close to a cliff that dropped down to the beach several hundred feet below. The driver of the yellow MG slid off the road and disappeared over the edge right in front of me. I was horrified but kept going as corner workers were rushing to the scene. The next time past that point, I was greatly relieved to see the driver standing near the road, helmet off, waving to let the rest of us know he was OK. His car had landed and stayed on a ledge several feet below the road and out of sight from the road. The driver of the yellow MG was Dan Gurney, a now famous name in racing.

I drove in a number of other races in California and Arizona including one memorable one in Phoenix in a different race car. It was called the Cracker Box and was an old VW chassis with mechanical brakes, an aluminum body, and was powered by a Porsche 1500 cc engine, It was light and fast. Practice times had been used to set the starting grid. I found myself sitting in the front row between John Crean in a Porsche roadster, and John Von

Neuman in his famous Porsche. I was in fast and beautiful company with a funny looking, home-built car. Some of the spectators laughed, but there was no denying we had posted some fast times in practice..

When the starting flag dropped for our standing start, I shot immediately into the lead which I held for the first four laps. Von Neuman was on my tail but couldn't pass me. Then I goofed. I took a corner too wide, got a bit off the road, and Von Neuman shot right past me. I chased him for quite a number of laps trying to figure out where, when and if I could pass him. Then I got some help. We were coming up to lap a beautiful gold Jaguar on a short straightaway with a sharp right corner at the end. I decided to take him on the right, and John chose the left. When I pulled up to pass, he was so surprised to see this weird car passing him that he lost it. I was going a bit faster and knew he would have to brake long before I would.

The Jag driver must have been so startled by my appearance he forgot his brake point for the next corner, locked his brakes up much too late and slid straight ahead down the escape road taking the unfortunate Von Neuman with him. John never caught up with me, and I coasted over the finish line, out of fuel. We had to push the car to the impound area, an easy chore. Soon after that race, we moved to Cleveland ending my California racing. In the next race for the Cracker Box, the driver flipped it and was seriously injured. The car was completely demolished and was never rebuilt.

After moving to Cleveland I continued racing—an MGA at Put-In-Bay, Nelson's ledges and Akron, a Corvette at Elkhart Lake, A Porsche Carerra, several specials, two Saab GTs and an MG TF, a Vanguard 5 with a Saab engine, and of course, the ill fated Lotus 30 in the story, ***Flying High, A Racer's Tale*** in this booklet. I even managed to run a Saab GT in the twelve hour race in Sebring, Florida two years running. I partnered with my buddy, Don Wolf at Sebring and several other races.

Hopefully I will add several other interesting racing stories to this booklet including one where I hit a deer at 145 mph in a race, and another where one of my competitors ended up in a tree. Maybe even the story of watching an Allard wrap itself around a tree in Golden Gate Park.

Obstacles are those frightful things you see when you take your eyes off your goal.
—*Henry Ford (1863-1947)*

When you gaze long into the abyss, the abyss also gazes into you.
—*Friedrich Nietzsche (1844-1900)*

Knowledge speaks, but wisdom listens.

—*Jimi Hendrix*

The Palm Reader

It was June 1955 when my wife and I attended the American Dental Trade Association annual meeting at the Homestead, an elegant, old, southern luxury hotel up in the mountains near Hot Springs, Virginia. Our daughter, Debbie, five, was with us. My father and mother were also in attendance. I had been working with my father for more than a year in our dental supply business, Johnson-Stipher, Inc. in Cleveland.

After my father's partner in the business, Henry Stipher, died suddenly, I helped my dad buy out the Stipher family's interest in the company. In May 1954, my wife Dolores and I moved from Long Beach, California, to Cleveland, bringing our daughter, Debbie then four, and son, Howard, eight months old, with us. I left a job as a pilot plant engineer to go into the dental business with my father, also Howard Johnson, but with a different middle name.

After a year and a half in the business, I had developed enough to have a part in the program at the meeting. It had been quite a change from my engineering position. The meeting at the Homestead was my first time making a presentation, and I was quite nervous. This annual national meeting, was where all of the officers, owners, and principal players in the industry met to get to know each other personally. It was five days of morning socializing, golf, and tennis followed by four hours of programs about many management phases of the dental industry. Evenings there were formal dinners with entertainment.

One evening they had a palm reader set up in a black tent in the lobby. She was a part of the entertainment. My wife, Dee, went and had her palm read and came out so pleased and excited she urged me to have mine read. I laughed and declined. As a trained engineer and scientist, I have no interest in fortune tellers or anything remotely associated with astrology or the occult. The group we were with razzed me unmercifully so I caved to the pressure and joined the line waiting to have their palms read.

I had mixed emotions as the lady pulled the black curtain back and ushered me into her darkened palm reading alcove. We sat on opposite sides of a small table. She was younger than I expected, about my age, and did not have any kind of exotic look like the fortune tellers in the movies. She was a rather attractive woman in a long black, sleeveless dress, almost a girl-next-door type. She took both of my hands and placed them, palms up, on the table. She asked me a number of questions to put me at ease and then began examining my hands. She traced her fingers down several of the lines, turning my hands to see where the

459

lines went around the sides. She made a number of notations on a pad of letter-sized paper on her side of the table. I'm guessing it was at least ten minutes before she spoke.

"You have extremely interesting hands," she said. "I can see you are going to have a long and interesting, even exciting life. Let me show you."

She began pointing out features of the lines on my hands and what they foretold. She told me a lot of things that made me chuckle, things that could apply to almost any young man my age. Then she told me a few things so ridiculous I laughed to myself about them. Here's what she said that I remembered because they were so far from what I could even imagine for myself.

"You will be married three times and have five, maybe six children, mostly girls."

I couldn't imagine my marriage to Dee not lasting my entire lifetime or that we would have so many children. We had two and were considering one more. But six? Impossible. Then she continued.

"You will travel a great deal and live in a number of different places, some quite far away, even in a different country. You will change how you make your living numerous times and in several different fields."

Again, I remember almost laughing, thinking how wrong she was. Then she finished with the following comment, emphasized by pointing at the lines on my hands and explaining why.

"You will become a writer and then become wealthy, late in life."

I laughed at these impossible predictions, thanked her, and rose to leave. As I did so, she handed me the paper she had been writing. It listed all of the predictions she had made—in detail.

"Keep this. Save it for a long time. You may be surprised. The palms do not lie."

I took the paper, and walked out through the black curtains to join Dee and some friends. At dinner that evening I shared some of her predictions with Dee and those at our table commenting to the effect fortune tellers were ridiculous. We all laughed at my experience. I placed her paper in a small briefcase I had with me and much later transferred it to a folder containing some things I had written about the meeting.

The last night of the meeting I sat at the head table because of my part in the program. Dee was seated on my left and a young woman a bit younger than I was seated on my right. The man on her other side was her father. I noticed she was quiet, and expensively dressed in high fashion. She spoke with her father a few times but hardly responded when I tried to engage her in table conversation. She was polite but reserved. In contrast, her father and I had several animated conversations during dinner. He was friendly and quite charming. He carried on several conversations with others at the table, while his daughter said little. By the end of dinner, he appeared to be a bit drunk.

After dinner, one of my friends asked, "How did you get along with Black Jack?"

"Who's he?" I asked.

"The guy sitting next to his daughter, Caroline, at your table. She sat right next to you. You don't know who they are?"

"No. Should I?"

"Well, she's Jackie Kennedy's little sister, and he's their father, a wealthy stock broker who knows a lot of people. Didn't you hear his talk about investments?"

"Wow, really? Jackie's sister. Now that I think of it, she does look a bit like her. And no, I didn't hear his talk. I attended another session when he was speaking."

"They call him Black Jack Bouvier, probably because he's so dark, but maybe because he's a bit shady. His wife divorced him years ago because he likes booze and womanizing. I'm surprised her mother let Lee—they call her by her middle name, Caroline Lee Bouvier—I'm surprised she let her go out with him."

By the time we headed to our room for the night I had forgotten the palm reader. In fact, I never again thought about the reading until almost thirty years later.

It was 1982, and I had just returned after living in the Philippines for a year. My second marriage was over, and I was moving lock stock and barrel from Euclid, Ohio, to LaGrange, Illinois. I was packing my things in my cargo trailer when I dropped an old box which fell open. Inside was the folder into which I had placed the paper from the palm reader those many years ago. I had never seen it since I first placed it in the folder. When the paper fell out, I picked it up and looked at it. In fact, I stared in disbelief. So many things she had foretold had come true. My second marriage was over, I had five daughters and a son, was now embarking on my fourth career and had lived in a foreign country. It was scary reading what she had predicted, damned scary.

Fast forward to today. Incredibly, nearly all of her predictions have come true. She missed on a few of the smaller predictions, but many of the biggest have come to pass including my being married three times. Her prediction about my becoming a writer now blows me away. Uncanny! Of course, I'm still hopefully and patiently awaiting the fulfillment of her last prediction, the one about great wealth.

❖   ❖   ❖

Everywhere I go I'm asked if I think the university stifles writers. My opinion is that they don't stifle enough of them.

—Flannery O'Connor (1925-1964)

Too many pieces of music finish too long after the end.

—Igor Stravinsky (1882-1971)

Anything that is too stupid to be spoken is sung.

—Voltaire (1694-1778)

When choosing between two evils, I always like to try the one I've never tried before.

—Mae West (1892-1980)

I don't know anything about music. In my line you don't have to.

—Elvis Presley (1935-1977)

Dancing is silent poetry.

—Simonides (556-468bc)

If you can't get rid of the skeleton in your closet, you'd best teach it to dance.

—George Bernard Shaw (1856-1950)

No Sane man will dance.

—Cicero (106-43 B.C.)

Music speaks what cannot be expressed, soothes the mind and gives it rest, heals the heart and makes it whole, flows from heaven to the soul.

—Angela Monet

Hell is a half-filled auditorium.

—Robert Frost (1874-1963)

Show me a sane man and I will cure him for you.

—Carl Gustav Jung (1875-1961)

If I were two-faced, would I be wearing this one?

—Abraham Lincoln (1809-1865)

The time is near at hand which must determine if Americans are to be free men or slaves.

—George Washington

An Unexpected Betrayal

<u>An important note:1.</u>

The following is an exercise of a memoir in pure dialog. I want people to understand this is close to the actual conversation we had right after I discovered my wife's infidelity. I was desperately hurt, angry, disillusioned and violently exploding. We were all fortunate I never carried through on any of the immediate actions I contemplated. One of those actions is vividly described in another memoir titled, *The Drop of Rainwater*, which follows this one. My apologies to anyone hurt or offended by these words. Consider them the ravings of a newly tormented soul whose world had collapsed. Actually, after the next memoir, which followed this one by a few days, things returned slowly from the depths back to as normal as they could under the circumstances. I'll have to admit, with a few changes of direction and the right opportunity, our roles could have been reversed. As it was, we both struggled with our personal demons for the next fifteen years until we divorced. What held us together was our family, our children and parents. But for them, and the cohesive influence they provided under the direst circumstances, who knows what might have occurred.

Yes, things could have been different, and I went through some difficult and trying times, but so did Dolores. Fortunately, neither of us used our children as weapons against the other as happens in so many similar family situations. As far as I can recall, neither of us ever said a bad word about the other to any of our children. As a result, we were both blessed with the love and affection of all of our offspring. That, I believe, is a rarity in today's world. I, in fact, got over my anger several years later, and Dee and I had a friendly relationship for the rest of her life.

I can not speak for her, but our break up opened a whole new life for me once our children were grown and I put away my anger and self-destructive life style. I had several great relationships with some wonderful women, and one fifteen year marriage that ended tragically when Barb passed away at much too early an age. Each of these ladies still living are dear and loyal friends. I have come to like the person I have become and bear no malice toward anyone, for any reason. I am the only one of the principals in this memoir still alive, so it should bring no pain. I've come a long way since the time of this dialog. Luckily, I have become a happy and much more tolerant individual.

A Memoir in Pure Dialog

"You miserable God damned bitch."

"What! Why! Put me down. Stop. You're hurting me."

"I'm gonna dump you out in the street where you belong."

"No, put me down. Why are you doing this?"

"I'll put you down as soon as I get you out the damned door. Then you can think about it, outside."

"Ow! That hurt."

"You'd be hurtin' a hell of a lot more if I was into wife beatin'. In spite of it all, I can't hit you. I never hit a woman in my life, and I am not gonna start now."

"Help me up. Please. Let's talk."

"Talk? Talk? What about? I put a little recorder on your damned phone and heard you talkin' to your lover boy. You sure talked enough then. I couldn't believe you would cheat on me, ever. Hell, I'd have bet my life on it . . . and I would have lost."

"Wait, let's talk this over, please. Let me explain. Let's go inside."

"You think goin' inside and talkin' will solve anything? What's to talk over? You've been screwin' my best friend behind my back. How long has this been goin' on, anyhow?"

"I'm so terribly sorry. It just happened, out of the blue."

"Out of the blue my ass. I can't believe it. I'll tell you one thing. I'll never believe another God damned woman again. And what about the kids? Did you ever think about them?"

"Don't bring them into this. They're not any part of it."

"They damned well better not be any part of it. Hell! How do I know they're mine? How long you been sleepin' around anyway. You're nothin' but a cheatin' whore. I been married and faithful all these years to a cheatin' whore. I'm gonna kill that bastard."

"Don't you ever talk like that. It wouldn't solve anything. All it would do is deprive our kids of a father."

"Are you tellin' me that miserable bastard is their father?"

"No, of course not. It's that you would be in jail if you did that, or worse. That's what I meant. Damn. This gets complicated."

"Complicated? . . . Complicated you say? I find out my dear sweet wife of ten years has been sleepin' with my once best friend and you say it's complicated? Never once was I unfaithful to you in all those years and you say it's complicated?"

"Come on now, you expect me to believe you went to all those conventions with all those women and you never once cheated?"

"Never once . . . **never**. Oh, I was tempted enough times, but all I had to do was remember my dear faithful wife at home with our kids and I couldn't. What were you doing while I was away on those trips? Were you screwin' Chuck in our bed? And with our kids in the house?"

"No. No, never."

"Was he tellin' you how I was probably sleepin' with all sorts of women? Was that how he got to you? I'll kill that bastard."

"Don't you talk like that. You're scaring me. It wasn't that way at all. We never did it in our bed, or with the kids around. I couldn't."

"What are you afraid of, that you'll lose your lover boy? It's a fine mess you made of our lives, a fine mess. I can't even think I'm so upset. My whole life has been turned upside down. You never answered me. How long has this been going on?"

"It started about a year ago, while you were in Chicago for a convention. I was lonely . . . and . . . well, it happened."

"I'll bet. I'll bet he told you I was chasin' all those females at the convention. That's it, isn't it?"

"Chuck never said anything like that. I was the one who thought it, all on my own. I figured you were like all the other men at those conventions. I've heard the stories."

"Then you thought wrong. You don't know a damned thing about who I really am, do you? All these years together and you don't know me."

"I really thought you were sleeping with those women. You mean you never did?"

"No, I never did . . . I'll say one thing though."

"What's that?"

"Next convention I'm gonna try to make up for lost time."

"Damn you. You don't mean it."

"Oh, don't I? I'll even relate to you all the gory details."

"You wouldn't."

"Why the hell not? You've got your lover boy. No reason now why I should stay pure. Hell, you know the blonde lady down at the end of the street, Shirley Thomas?"

"What about her?"

"She's been comin' on to me for years. Next time I won't laugh it off. You can count on it."

"She's a terrible person. Everybody knows she sleeps with almost any man that comes along."

"And what makes her different from you? Besides, she's not married."

"All the neighbors would talk. Think what it might do to our kids if they find out."

"And what do you think the story of your escapades will do to our kids when they find out about you and Chuck. Hell, maybe they'll cheer. I don't know."

"You wouldn't tell them."

"Of course I wouldn't, but not for you, for them. I'll do everything in my power to keep them from knowing. They're great kids and don't need to know their mother's a lyin' cheatin' bitch. You damned well better keep it from them. If they ever find out from you, I'll . . . I'll . . . Shit, I sure don't know what I'd do, but it will be bad. I guarantee."

"I have no idea what I can do. I'm so ashamed. I got caught up in something I couldn't handle, that's all. What's going to happen to us?"

"Well, for one thing, my number one priority now is my kids, our kids. I don't give a damn what happens to you except what might hurt those kids. I don't want them to find out, and I certainly don't want either of our folks to find out. Your mom would be devastated. And your dad? You're still his little Dolly. He would be deeply hurt if he ever knew. I sure as hell won't tell anyone."

"I know, they would be disappointed. Won't you be asking for a divorce?"

"Not unless you want one. I'll put up with all this crap for the kids sake. We'll probably both want a divorce when they're grown and on their own. Of course, if you insist on one now, I'll deal with it."

"I certainly don't want a divorce. It would kill my folks."

"Well then, we'll have to put up with each other as is. Remember, as far as I am concerned, we are no longer married. We'll be acting it out for the kid's sake. And Chuck? I won't do anything to him because it would devastate our kids. However, I will work hard to find ways to make life miserable for him—for the both of you. That I promise."

The Drop of Rainwater

My eyes followed the drop of water as it zigzaged down my windshield—a tear slipping down a face. The summer shower had ended, depriving the drop of enough new liquid to run so it crept downward in agonizingly slow starts and stops. A tiny image of the street lamp overhead shone clearly inside the drop. In the surrounding blackness, glistening drops on bushes shone each with its own image of the street lamp. Around me lurked vague, shadowy shapes—bushes, a tree, a driveway framed by houses, a side door, and a fence—seen through the glistening glass.

The revolver between my knees felt warm to my hands, a nearly living thing with a mission driven by fierce, all consuming wrath. I carefully caressed the demonic power in my hands.

My passion kept saying, *Go! Do it! Get it over with!* Another barely audible voice kept echoing a single word, *Think*. My mind, indeed my whole body, buzzed with confusion keeping me frozen, waiting, almost as if stopped in time. Waiting for what? An explosion? An angry act of restitution? A bellowing of carnal blood lust? Or a retreat into less-troubled contemplation. Primal juices flooded through my arteries, blinding reason and logic. A million years of driven instincts screamed for action, revenge. A few thousand years of civilization tried desperately to stem the tide. The battle raged within me, each force gaining momentary advantage and then losing it.

I was about to step out into the damp night, walk up the driveway to the door, open it, mount the stairs, and then empty the missiles from the warm, almost living thing in my hand into my best friend. A vivid scenario swirled through my head as I waited—the dull sound of my feet on the carpeted steps—the soft swish of the inner door as I open it over the carpet—his look, first of recognition, then surprise, then realization, then terror—the thunder and flash as the shells explode—the soft thump as he folds and falls to the floor—my sense of primordial satisfaction at repayment of the ultimate betrayal by a friend. I then turn, exit the house, enter my car, and drive away into the dark, damp night. This act of vengeance was so appropriate, so fitting, so demanded. Would it be so satisfying?

The forces that lead me to this brink of action were inexorable, omnipotent, irresistible. My mind reeled with passionate hatred—a constant, anarchic seething through every part of my being—a fire-breathing dragon burning in my head. Yet still, the voice of reason kept at me, a small child tugging at my coat. *Think! Think!* It whispered, but how could it possibly stand against the mob action of instincts and emotions screaming throughout my entire being. Then came the time, the point of no return, the instant of final decision, the bitter sweetness of the moment of retribution—or?

I watched as the drop reached the end of its tortuous path and suddenly was no more.

The instinct of nearly all societies is to lock up anybody who is truly free. First, society begins by trying to beat you up. If this fails, they try to poison you. If this fails too, they finish by loading honors on your head.

—Jean Cocteau (1889-1963)

Everyone is a genius at least once a year; a real genius has his original ideas closer together.

—Georg Lichtenberg (1742-1799)

Success usually comes to those who are too busy to be looking for it

—Henry David Thoreau (1817-1862)

While we are postponing, life speeds by.

—Seneca (3BC - 65AD)

Where are we going, and why am I in this handbasket?

—Bumper Sticker

God, please save me from your followers!

—Bumper Sticker

First they ignore you, then they laugh at you, then they fight you, then you win.

—Mahatma Gandhi (1869-1948)

If thou of fortune be bereft, and in thy store there be but left,
Two loaves, sell one, and with the dole, buy Hyacinths to feed thy soul

—Muslih-uddin Sadi

A Hidden Reality for My Family, Ancient History Revealed, I Think Necessary for Understanding

I believe it was in January of 1958 when I discovered my wife, Dolores, was having an affair with one of my closest friends, Chuck. You cannot imagine what that discovery did to me. My whole world collapsed, and my life was over, or at least so I felt. The day I confronted her, I flew into a rage, physically picked her up, and threw her out of our house. When I cooled down a bit, I went out, apologized for my anger, and brought her back into the house. In that short time, I made the decision to try to work things out, so we would stay together.

Unfortunately, my decision was not rational. It was not many months before the demons of anger and retaliation took control of me. I came so close to committing murder it scares me to think of it to this day. I took Dolores's father's revolver and went to extract revenge. What happened was described vividly in my short memoir, *The Drop of Rainwater*. Their affair continued unabated over my expressed anger and loud objections.

For as long as two years I was physically affected to where my body *buzzed* for days on end. My doctor said it was a nervous reaction to intense stress. I didn't tell him what the stress was. In fact, I couldn't tell anyone. I did everything I could to hide what was happening from everyone. I developed a calm, controlled attitude to hide from everyone the hatred and passionate anger seething within me. It took a serious, gnawing toll by absorbing almost all of my energy.

Within a few months I had embarked on a course of *wine, women, and song* to take revenge on my errant wife. She was furious when she found out, but so what? She and Chuck were continuing their affair. Still in this angry state of mind I was well on my way to a life of destructive debauchery. Convinced I could never trust or believe another woman, I embarked on a path of retaliation you can probably surmise.

After two furious years, my life was forever changed by, of course, another woman. At the Ohio dental meeting in the fall of 1960, I had lined up a lunch date with an attractive hygienist I had known for at least a year. She was to be my next conquest. I took Caroline out for lunch in Cincinnati and we walked through downtown for several hours, talking. There was something different about Caroline, something soft, kind, and emotional. Something that disarmed my considerable defenses against feelings. As we sat on a wooden bench in a small park, I blurted out the painful story of my unfaithful wife. I was by now fully disarmed, vulnerable, and developing those telltale, warm, fuzzy feelings I wanted so desperately not to have. I didn't want any good feelings toward any woman.

I decided not to ask her out to dinner for fear of becoming involved, the last thing I wanted to happen. My inclination was to get as far away from this lady as possible. So guess what? We sort of gravitated to the point where I don't remember if I asked, or if it was a mutual decision, but we went out to dinner. The magic began then and there. We enjoyed each other during a lovely dinner at La Maisonette with its wine, soft music, and romantic ambiance. During the evening she explained that she recently had an unpleasant sexual experience with a man. I don't remember the details even though I remember her telling me that she was quite apprehensive about men. For whatever reason she felt safe and secure with me. (She was) By the end of a romantic dinner that lasted at least three hours, I was hopelessly in love. After dinner, we went up to my hotel room for a glass of wine and more conversation. We had a little kissy face, but I could not bring my self to go any farther.

We dated off and on for some months, quite platonically. One evening after I returned with her to her apartment she asked me why I never tried to romance her. Was there something wrong with her? Didn't I like her? This led to some serious smooching and eventually to her bedroom. I ended up staying the night. I discovered she felt about me as I felt about her.

We were soon hopelessly in love and involved in a torrid affair. We became deeply committed to each other, spending as much time together as we could under the circumstances. This caused my moral turn around and the beginnings of the reversal of the downward course I was on. (Moral turn around seems strange terminology under the circumstances, but it did happen.) Make no mistake, this reversal was brought about by the love Caroline brought into my life. Of course, my kids, all four of their grandparents, and our greater family were also positive factors. However, I don't think it would have happened, if it had not been for Caroline. And remember, this was fully two years after Dolores's treachery, so Caroline had nothing to do with those related difficulties. In fact, she probably made things better between Dolores and me.

Dee and I tried one last time to salvage what we once had. We took a long trip west in our new Porsche in 1964. It was an enjoyable, memorable experience. Those three weeks were almost as if nothing had ever happened. Then, in LaPort, Indiana, during the last motel stop before reaching home, I suddenly knew it was all over. We both did. Everything changed back to the way it was before we left. She and I should have divorced right then and there but didn't because of our children.

I'll say one thing for Dee. She was a terrific mother. She would sacrifice virtually anything for our kids even to depriving herself. Everything except her lover. She was a lot better mother than she was a wife, at least for me. I'm quite certain there were things about me that she did not like. Perhaps this brought her to turn to another man. She never shared any information about this with me, so all my related thoughts are pure conjecture. I have often thought the reason she took up with Chuck was because he needed mothering. Alcoholics need serious mothering and I did not. That's one major need of Dee's I couldn't fill.

For me, Caroline, Iola, Barb, and now Daphne were each better wife material than Dee. I'm not berating Dee. She was a nice person for a friend or mother or even daughter. For some reason, we no longer fit together well as husband and wife. Obviously, she found her soul mate in Chuck.

When Dee became pregnant, I knew her baby could not be mine for obvious reasons. She was in complete denial when I told her that. I decided I was going to raise *our* new baby with the same love and caring as all the rest of our children. When Mindy was born and to this day that is the way it has been. I have a fantastic relationship with Mindy and her family after a few rough spots early on when we were each having our own private struggles with life. My relationship with Mindy is one of the true treasures of my life. It is an accomplishment both of us worked on diligently. I consider my part in creating this treasure as one of the crowning achievements of my life.

Why Caroline put up with me during the years leading up to 1966 I have no idea. I had been unfaithful to her several times, a hold over from my years of debauchery. On January 1, 1967, I made a vow to be faithful to Caroline. I never broke my vow. In fact, that vow was so important to me that I never cheated on anyone again through several committed relationships and two marriages.

On January 27, 1968, Caroline gave birth to a beautiful little girl we named Kristen Leslie. Her first name was Caroline's choice, and Leslie was my choice. Things went along as smoothly as could be expected until Kristen was nearly three in late 1970. By then Caroline was through with me. (My words and opinion) It must have happened because I was still married and showed no signs of changing things. Finally, in desperation, she asked me to stay away from her and Kristen in no uncertain terms. Not angrily, but definitely.

It took me until 1974 to divorce Dee. Was I stupid not to have done so ten years earlier? Right or wrong it was purely for our children, all of them. Was it a mistake? Impossible to know! By the time we were divorced, Caroline had married. When I found this out, I beat up on myself, got crazy and married Roberta. I was a real mess. That marriage should never have been. It was over the evening of the day we were married, but I hung on for nearly five years with one of the most evil human beings I have ever known. I knew some of her evil acts, but how truly evil she was came to light recently when Caroline told me what Roberta had said to her and Kristen. But I get ahead of my story.

Roberta cheated on me several times with an old lover, a Euclid judge, married and with several children. The vow I had made in 1967 was so strong, I did not retaliate. Oh, I thought about it, but my vow was about being true to myself, not Roberta. It was a different and powerful thing, that vow. I have kept it until this day and I will never break it.

When Kristen was twelve or thirteen, Caroline called me one day, out of the blue and said, "Your daughter would like it if you came to watch her skate." It was early in 1980, and she was a competitive figure skater. I went with her mom and watched her skate several times over the next year. Caroline was now divorced and, what else, I was married. Caroline was pleasant, polite, and yet quite cool to me. I didn't tell her, but my divorce from Roberta was imminent.

After I separated from Roberta, I spent a year in the Philippines. When I returned I packed up all of my things from our house in Euclid into my cargo trailer and headed for Chicago. My self esteem was as low as it had ever been. I may have appeared calm and unruffled on the surface, but inside seethed those old demons of self deprecation. In spite of this, I stopped to see Caroline and Kristen in their apartment in Rocky River. I thought maybe we could try to put our lives together once more. That was the last time I saw Kristen until the summer of 2009. She was fourteen. I then walked out of their lives because I thought they would be better off without me, nothing more, nothing less. Once more I chose a foolish and hurtful path. Had I known what I learned in Colorado, our paths might have been quite different. Since the past cannot be changed, it is all missed opportunities and reminders of poor decisions. Why are we so soon old and so late smart?

For more about Kristen and what happened in 2009, read, *A Life Changing, Wonderful Experience - Reunion with My Daughter, Kristen* on page 625 in this book.

I moved my business and personal belongings to Chicago, got divorced from Roberta, (That's quite a story in itself) and proceeded to try to put my life back together. With determination and the considerable help from the lovely Iola and her two teenage daughters, I began regaining my self respect and self esteem. By the time we had been together for almost five years, I reached the point where I liked myself. I hadn't felt this way for more than twenty years. For several reasons we never married. Then, in 1987 we decided mutually to go our separate ways. It was an amiable separation. We remain friends and keep in touch to this day. Her oldest daughter, Kim, and Kim's husband, Paul, bought a place on Tippy. They stop by for a pleasant visit once or twice each summer.

❧   ❧   ❧

A Big Gun

It was November 1958, and I was attending a convention at the Edgewater Beach Hotel in Chicago. One evening, Several of my young male friends and I took the opportunity to visit a strip bar a few blocks from the hotel for some guy carousing. The main bar was on both sides of a long, wooden runway where the ladies danced, pranced, and otherwise displayed their wares.

The front door was near to the midpoint of the runway, and the bar ran all the way around the runway. The seats at the bar were packed as close as possible. Directly across the runway/bar from the front door was a short hallway leading to the rest rooms. When we came in, we checked our overcoats and took four empty seats close to the entrance.

We had enjoyed the show and a few bottles of beer for at least an hour when nature called, or maybe the rent came due on the beer, and I headed for the gentlemen's room to—well—you know. All of the urinals were occupied so I headed for the room with the flush toilet. When I pushed open the unlatched door, I ran into a man with a bundle of bills in his hand and another with a small briefcase. Later, I figured out I had interrupted a serious, high level drug buy. The man with the briefcase pulled a revolver and aimed it directly between my eyes. My forward motion carried me to where the gun barrel was against my forehead, I can still see the ends of the hollow point bullets in the cylinder. While the gun was hard against my forehead, I heard a loud click. This entire action happened in at most a second.

In what was an unbelievably quick response, I did what amounted to a half gainer backwards and took off straight out of the john. I hit the men's room door full speed, slamming it to the floor. I continued fleeing up and over onto the bar. Then the cannon behind me worked and the powerful bang spurred me on even more. No, he didn't hit me. I charged past the startled dancer on the runway, jumped over the bar and my buddies on the other side , and hit the massive front door at full panic speed. Now, I was running for my life down a Chicago street, clad only in a sport coat and slacks in zero weather. I turned into the first alley that showed up and ran like hell until I ran out of breath.

I stopped, leaned against the brick wall of a building hoping I wasn't breathing my last breath. No one was chasing me, at least as far back the alley as I could see. Then I heard sirens and reflections of flashing red lights lit up the end of the alley where I had entered.

473

After I had regained my breath, I began retracing my steps. When I emerged from the alley, I saw two Chicago Police cars angled in at the door of the bar, lights flashing. Two uniformed cops were talking to my friends. The other two were inside.

"There he is," one of my friends said, pointing in my direction.

As I walked up and joined them, one of the cops said, "This is the guy who took down that big door? I find that hard to believe."

It was then I realized the solid wood door to the bar was laying flat on the pavement. It had been torn out of its stone opening. The frame was still attached to the door, and the fasteners were still sticking out of the frame. I could hardly believe I had done so much damage. I didn't even remember hitting the door. All I could remember was my panicked flight away from that cannon. Then I realized my forearms hurt, a lot. When I looked at them, I saw that both sleeves of my jacket were ripped and bloody. My sport coat had been destroyed when I hit the door. The outside of my forearms were abraded and bleeding. They looked like raw meat, oozing blood. Immediately they began hurting big time. When the cops saw this, they called for an ambulance. They sat me down in one of the cars and immediately removed my jacket and started cleaning my injuries and applying what I took to be antibiotics, and then some loose bandages. I was freezing, so they wrapped a couple of blankets around me.

One of the cops said, "We would like to ask you a few questions. Can you tell us what happened?"

"Yeah. Some guy shot at me."

"We're pretty well aware of that, but what led up to the shooting?"

I had calmed down enough to be thinking more rationally. I explained the entire thing as I saw it, from when I first got up and started for the john. I finished my explanation with, "Man, when I saw that cannon in my face I thought I was dead. Everything afterwards is a blur until I ran out of breath. Then I walked back and saw the door flat on the sidewalk and found my arms were torn up."

One of my friends said, "When we saw him coming right at us over the bar, we couldn't believe it. He flew over the runway and the bars on both sides. I was amazed at how high he jumped. Then he hit the door, and it disappeared. It happened so fast. Did you catch the guy with the gun?"

"No. We have a pretty good idea who it was so we'll round him up." He looked at me. "Do you think you can identify him?"

I thought for a minute. "I don't know. I only saw his face for an instant, then all I could see was that revolver in my face."

"You sure it was a revolver?"

"I could clearly see the hollow pointed bullets in the cylinder when he pressed the gun between my eyes."

"It was definitely a revolver. Do you know what size?"

"All I know is it looked like a cannon. I have no idea what size it was."

"Was it a snub nose or did it have a longer barrel?

"I would guess it was a snub nose. It had a short barrel. The cylinder was close to my face. I was lucky it misfired."

"Why do you think it misfired?"

"I heard a loud click when it was against my face."

"Are you sure the click wasn't the hammer being cocked?"

"I have no idea. All I know is it gave a loud click I assumed was the trigger being pulled. It did not go off until I was going over the runway."

Then the ambulance pulled up. "Let's get those arms fixed up." They walked me to the ambulance.

"What do we have here?" the medic said as the cop put me into the rear of the ambulance.

"He's got some nasty abrasions on his arms. Not serious, but nasty and painful."

"You can say that again," I said as the medic pulled the loose bandages off my arms.

"How in the hell did you do that?" he asked as he was washing the blood off of my arms. "That's a strange looking wound. Did you fall on concrete?"

The cop laughed. "You wouldn't believe it. This guy took down the main door to Silby's. Hit it so hard it ripped right out of the stone work. It's layin' right over there on the ground."

The medic shook his head. "That's why it's so clean. Usually a wound like this is full of gravel, small rocks, or dirt. Yours is quite clean. I can fix this up, and you can be on your way in short order. You won't have to go to the hospital. Unless you want to, that is."

"Not if I don't have to."

"I'll give you some antibiotic cream to use. Wash your arms every two days and apply the cream. Then wrap it gently with gauze. You should heal fairly quickly. Your skin is not in nearly as bad shape as I thought it was when I first saw you."

Treatment over, I hurried into Silby's to retrieve my overcoat and hat. A heavy drape had been hung in the doorway to keep out the cold, and the girls were back at their bumps and grinds. I doubted I could salvage my sport coat, but I took it anyway. One of the guys flagged a cab and soon we were back at the Edgewater with quite a story to tell our friends at the convention.

Wit is educated insolence.

—*Aristotle (384-322 B.C.)*

My advice to you is get married: if you find a good wife you'll be happy; if not, you'll become a philosopher.

—*Socrates (470-399 B.C.)*

Advice is what we ask for when we already know the answer but wish we didn't

—*Erica Jong (1942)*

Show me a woman who doesn't feel guilty and I'll show you a man.

—*Erica Jong (1942-)*

I've learned that people will forget what you said, people will forget what you did, but people will never forget how you made them feel.

—*Maya Angelou (1928-)*

Egotist: a person more interested in himself than in me.

—*Ambrose Bierce (1842-1914)*

A narcissist is someone better looking than you are.

—*Gore Vidal*

Wise men make proverbs, but fools repeat them.

—*Samuel Palmer (1805-80)*

Many a great proverb is read, shared with friends, and then forgotten. What a pity all that wisdom goes to waste.

—*Howard Johnson, 1998*

It has become appallingly obvious that our technology has exceeded our humanity.

—*Albert Einstein (1879-1955)*

The secret of success is to know something nobody else knows.

—*Aristotle Onassis (1906-1975)*

The power of accurate observation is frequently called cynicism by those who don't have it.

—*George Bernard Shaw (1856-1950)*

My Middle Daughter, Diana

There's no getting around it, my daughter, Diana—also known as Dee Dee, Dee, Na, and maybe a few others—is a hoot. She has been a live wire as far back as I can remember, and that's quite a while. Oh, the nickname, Na? She gained that one when kids would call her Diane which she always corrected by adding, "Na." Soon friends were calling her simply, Na.

A creative, energetic, friendly, determined, sometimes in-your-face, lady, she is the *Auntie Mame* of our family of five ladies and one gentleman. Of course, there are also six granddaughters, six grandsons and a great of each sex. They are quite a crew, but of course what would you expect to hear from a proud patriarch.

I have many memories of her as a little girl. Dee Dee never had a bottle in her life. Mom was all she ever needed. That made for a close bond between her and her mother.

Once I was mowing the front lawn when I saw her fall on the sidewalk next door. That little blond two-year-old picked herself up and headed my way on the sidewalk. I went to check if she was hurt, but she walked quickly by without a whimper or tear, a determined look on her little face. I followed her up the front steps, into the house, through the dining room and into the kitchen to her mother. The instant she was within sight of her mother, a torrent of tears accompanied by a loud wail proclaimed her injured status. Her mom picked her up to offer loving solace for her injuries. The crying soon gave way to those pitiful sobs little girls use so effectively. Dad helped clean the scraped knee, then applied antibiotic ointment and a band aid while she clung tightly to Mom. Similar scenes were repeated several times following minor injuries. No, that never bothered me a bit.

Dee Dee loved cookies. Hand her one and she would quickly stick out her other hand while depositing the first cookie in her mouth. Place a second cookie in her outstretched hand and it would instantly be replaced by the first hand, then empty. Off she would go carrying her three cookies. Some system.

Dee had stored some candy in a dish on the topmost shelf of our corner kitchen cabinet. She had to use a step stool to put it there where it would be safe from little hands. Oh yeah? I had been reading in the living room when I realized it was quiet, too quiet. All parents know what I mean. I got up and softly padded into the kitchen. There was Dee Dee, standing on the second shelf of the cabinet, her head almost against the ceiling, stuffing pieces of candy into her pockets. I didn't want to startle her for fear she would fall so I stood quietly by the doorway watching. She finally climbed down the cabinet to the counter using the shelves as a ladder. She then climbed from the counter onto a kitchen chair she had pushed against the lower cabinet and onto the floor. As she struggled to move the chair back to where it belonged, she looked up and saw me watching. It was almost as if she had received a physical blow. Her whole body jerked. She hesitated for a moment and then took of at a run for her bedroom. I scooped her up easily.

"And what were you doing up in the cabinet, young lady?"

Caught red handed she immediately pulled an old reliable female response—she cried. I, of course, couldn't hold it anymore and burst out with uncontrollable laughter. This caused her to cry even louder. Dee rushed in to find out what all the noise was about.

Dee didn't have any idea about the reason for the confusing signals, but took her cue from my wild laughter. By this time, my overwhelming mirth had me rolling on the floor. She caught the contagion and began laughing as well. She picked up our wailing little girl and tried to comfort her, but her own growing mirth didn't help the situation. Dee Dee cried even louder.

Dee paused enough to ask, "What on earth are you two going on about?"

At this point, Dee Dee blurted out between wails, "I . . . took . . . some . . . candy," and proceeded to cry even more.

I was now on my back, doubled up, and in some pain from the prolonged, consuming bout of laughter. I held out my hands in the universal palms up sign of helplessness. Almost immediately, Dee was on the floor joining me in debilitating bouts of contagious laughter. Then there were three. All three of us were sitting on the floor, paralyzed with laughter in which effort Dee Dee had now joined in. Her crying quickly reversed into a smile and then into roaring laughter. She was not going to be left out of all this fun and hilarity.

It took a while, but I calmed down enough to try to explain to Dee what I had seen. I got out about four words before once again exploding in . . . well, everyone knows that once your funny bone gets control, all you need do is look at the other person, and it takes over again. Tears were running down our cheeks, we could hardly get our breath, and those abdominal pains were bending us into balls. Can you imagine what someone walking in on this scene might have thought? The men in white coats would have to be called.

I have no idea how long this lasted, but we gained control, and rational behavior once more prevailed. I told Dee I would explain what brought this all on at a much later time when I would not be so likely to return to the state of consuming mirth. It was probably a couple of weeks before I bit the bullet and told Dee the story. Even then we experienced another small bout of hilarity.

I also retrieved the candy, I let Dee Dee keep two pieces. Of course, there was no way we could have punished her. We did try to explain how dangerous climbing those cabinets was and told her there would be serious consequences if she ever did that again. Even then, we both had a terrible time keeping from laughing as we talked with her.

We had a summer cottage on a little lake in Northern Indiana. (I still spend time there during the summer where we have annual family reunions.) The summer Dee Dee turned two, we put her in a bright red life jacket whenever she might go out on the pier. I was sitting on the pier, dangling my feet in the water, watching her older siblings jump in the water when there was splash right next to where she had been standing. Instantly her blond head popped out of the water, buoyed by the jacket. She blinked her eyes and wiped the water off of her face as I jumped in to rescue her from her fall. I picked her up and stood her back on the pier. As I was climbing back onto the pier, I watched that little imp squat at the edge of the pier, then jump in feet first with a loud squeal.

This time I watched as she paddled over to the ladder unaided, and climbed back onto the pier. She had seen her older sisters do it, so why shouldn't she? After that episode, we watched her carefully, so she wouldn't jump in at the wrong time or with no one around. Dee Dee would always need a little extra attention as she continued being a dare devil. Luckily she never had any major incidents.

During her early teens, before she could drive, I ended up as the chauffeur for she and her dozen or more friends. Her whole group would pile into the back of the maxivan I used in my business. I would pick up and deliver them to this party or that show, and then retrieve them and reverse the process, delivering them to their homes around the neighborhood. Many an evening I would be carrying a squealing, laughing, cavorting, and always joyful batch of fourteen-year-old girls and loving every minute. During the many trips I made hauling those girls, not once did they cause me any grief or make problems. They were a delight, and I often told them so, usually when they thanked me as they left the van for their homes.

About this same time, Dee Dee began going regularly to a *teen canteen* held Sunday evenings at a nearby church. I heard from friends that while the canteen was well chaperoned, there were sometimes problems in the parking lot outside. These problems were booze, drugs, and fights. When I heard this, her mother and I became concerned. We had issued some carefully considered restrictions and warned all of our kids about the dangers of drugs, alcohol and sex, but you know teens. A few of her friends were less than desirable in our opinion, but we never forbade a friendship. We talked about the dangers of getting involved with the wrong kids and hoped our words were at least heard. Fortunately all of our children were fiercely independent and not usually followers. We worked hard to teach them this from an early age.

One evening I picked up the phone and carried on the following conversation:

"Mr. Johnson?"

"Yes."

"This is Anette Brown. You don't know me, but your Diana is my daughter, Sheri's dear friend."

I immediately imagined a problem. "Is there something wrong?"

"No, absolutely not. Very much to the contrary. I called to tell you what a wonderful person Diana is. I want you to know what she has done for Sheri, and how grateful we are."

"Oh? That's nice to hear. It makes me so proud of Diana."

"Well, you should be. We've had a problem with Sheri because of some of her friends. Thanks to them, she's been on drugs for quite some time. We tried everything from consulting with social services, to drug rehab and nothing worked. Then Sheri met Diana at the church teen meeting, oh, about four months ago. I don't know how she did it, but she got Sheri away from drugs, completely."

"That's amazing. I had no idea. Maybe all the effort we put in with our kids about booze and drugs paid off."

"I'm sure it did, certainly it did for Sheri and our family. I wanted you to know how much we appreciate your daughter. We can't thank Diana enough. And one more thing, Sheri knows I'm calling you about Diana. She asked me to ask you not to mention my call to Diana. You know how teens are. They don't want a big fuss made over anything they do, especially by parents."

"I understand. Mine too are that way."

"I could go on and on, but I've had my say. So I'll say goodby. Maybe we'll meet some day."

"Thanks again Mrs. Brown and goodby."

I sat there truly amazed. Then I joined Dee in the living room and shared the news about our daugter.

"You never know, do you?" she said. "The one you are concerned the most about turns out like that. What a great feeling."

It was probably near the same time a different situation erupted. Dee called me at the office one afternoon. She was obviously upset. Through tears and a bit of anger, she informed me that Dee Dee had been arrested at Higbee's department store in the Severence Shopping Center near our home.

"Can you go quickly to the city police station on Mayfield? I have several ladies here from church, and it will take me some time to get away, so you may get there before I do. Please don't do anything drastic."

"Of course I won't. You know that. And calm down, please. Try to get your emotions under control, What we need now are some steady, careful hands."

"But . . . I'm her mother. What do you expect at a time like this?"

"I expect we should both be calm supportive parents. We should learn all the details and not jump to any conclusions. I'll see you there. Bye."

Fifteen minutes later I walked into the police station. After I had identified myself, I was ushered into a room with a table and three chairs in the center. Dee Dee was seated on a bench at the end of the room. She avoided looking at me. I walked over, sat down on the bench and put my arm around her.

In the calmest voice I could muster, I said, "talk to me, sweetheart. Tell me what happened."

She looked up at me through teary eyes. "Aren't you mad at me?"

"Of course not. Concerned, but certainly not angry. I'm here to help you get through whatever is going on."

She reached up and wrapped her arms around my neck. "I thought you would be furious."

"What you need now, what we all need now, is love and understanding. I'm sure whatever happened can be straightened out. All I know is you were arrested with two other girls for taking costume jewelry, for shoplifting. Tell me what you did. Please remember, I am on your side. I am not going to punish you. Whatever punishment you receive will be up to the police and the court."

Dee Dee clung tightly to me, sobbing. She kept saying over and over again, "I'm sorry. I'm so sorry."

The arresting officer came in and motioned me to sit in one of the chairs at the table. He spoke softly to Dee Dee and told her to remain seated on the bench. He sat with his back to Dee Dee facing me across the table. He smiled at me and gave me a wink. I got the message.

"I'm officer Dudek, Cleveland Heights police. I'll tell you straight out, your daughter was caught shoplifting. It's a petty theft, but still a crime, a felony. She and two other girls took several dozen pieces of costume jewelry from Higbee's and left without paying. We stopped them outside the door, searched them and found the stolen jewelry. We brought them here and let them call home."

A ruckus erupted right outside the room. A large woman was waving her arms and screaming at the matron standing in the hall.

"That's the mother of one of the other girls. She's been here before. So has her daughter, Same thing, shoplifting. Excuse me for a moment. They may need help to calm her down."

I watched through the glass as Officer Dudek and two female officers restrained the still screaming woman. They got her to sit down and shut up by waving a pair of handcuffs in front of her. Whether he would have used them or not, they worked to quiet her and get her to sit. I saw my wife enter the hall in the midst of this. She stayed as far from the ruckus as she could while moving toward the desk. The officer at the desk pointed to the room we were in, and Dee headed our way. She stayed as far from the woman as possible. She reached the room as the woman sat down.

"What was that all about?" she asked as she closed the door behind her.

"That's Vickie's mom," Dee Dee said as she ran to her mother and hugged her.

Officer Dudek reentered the room. "Mr. Johnson will you please come with me. Your wife and daughter can wait here."

I went over and kissed Dee and said I would be back. Officer Dudek led me to another small room across the hall. There was a small desk and two chairs against one wall.

"Please sit here," Dudek said pointing to one of the chairs while sitting down in the other across the desk from me. "I'm sure you understood my silent message in there. Your daughter had a single ring in her pocket when we searched her. The other two had more than two dozen each. Also, they have both been arrested before for doing the same thing. Stupid, I call it."

"I would say so," I replied. "What happens now?"

"Honestly . . nothing. I've been working as a juvenile officer for fifteen years. I can almost always tell the good ones from those who will probably become criminals later on. Your daughter seems to me to be a good teenage kid. We'll release her to you with a warning. Higbee's will not press charges. One look and I was quite sure you and your wife are going to do the right thing with your daughter. My guess is she learned her lesson, especially after hearing what I said deliberately so she could hear. When we found the one ring, she started crying and saying she was sorry. I'm sure she has never done this before and probably will never again. We get quite good at reading these kids, and their parents. Your daughter got caught up in a typical dare situation with girls I don't think she should run with. That's something only parents can control, and even then it's difficult."

I thanked officer Dudek and walked over to where Dee and Dee Dee were waiting. They were both pleasantly surprised when I said she was free to go. Later, at home, we sat down for a little talk. It was a positive, clear-the-air sort of talk with a lot of positives from both of us to a contrite young lady. We never again would worry about her straying. All on her own and with no prompting, she promised not to go anywhere with Vickie and the other girl. True to form, she did some wild things even after she grew up. She was an excellent water skier and did some fantastic maneuvers behind our boat, things few of the others could do. She also went sky-diving several times. She took a white water rafting trip on the New River in West Virginia when the river was high and violent. I understand that is one of the most dangerous of its type of rivers. Then she managed to bungee jump off the New River Bridge, after her raft trip.

One other little incident comes to mind. It happened after her mother and I separated and I was living in a small apartment a few miles from the old homestead. Our oldest daughter, Deb was married and gone and our son was away at college leaving the three youngest girls at home. I stopped by the house one evening to do a few repairs and finished around eleven in the evening. It suddenly dawned on me. I had not seen Dee Dee.

When I asked where she was, Dee said, "She's staying over at a friend's place."

"You let her go out on a school night to overnight at a friend's?" I asked.

"Well, to be honest, she moved out to stay with her friend three weeks ago."

I did not like the sound of that at all. "Who is this friend she's staying with?"

Dee looked like she was about to choke, got up and walked into the kitchen from the living room. I followed. "What's going on here you don't want me to know?"

"Well, you know Dee Dee. She's pretty damned independent."

"Yes? . . . What else?"

"It's Jimmy Seivers, she moved in with Jimmy Sievers. He has an apartment on Warrensville Center Road about a mile from us."

I did everything I could to keep from going ballistic. "You let her go live with a guy who is twenty-one years old? How could you?"

"Well, she didn't ask. She moved her things out one day while I was shopping. I found a note on the table when I got home. It was a surprise to me."

"You know he could be arrested, don't you. She's a minor for God's sake."

"Please don't do anything drastic. Jimmy is a nice guy, has a good job and is responsible. He says he wants to marry her. She could certainly do a lot worse."

"At fifteen? When did he tell you that? Why did you keep this from me? Damn."

"After she moved, they came to talk to me, together. He's a real gentleman."

"In my book, any grown man that lets a fifteen-year-old girl move in with him is no gentleman. So what do we do about it?"

We probably did the best thing we could do under the circumstances, nothing. After a few months, she moved back home. They continued dating, and every once in awhile she stayed overnight. We disapproved but stayed away from any serious restrictions. Knowing Dee Dee, I'm sure clamping down on her would have driven her away from home. We decided to live with it.

Actually, the decision not to interfere may have been one of the best things we could have done. Jimmy got her to work hard at school, something we were never able to accomplish. She went from a student getting by to a great student. She graduated in the top third of her class, something we never dreamed would happen. I was sure they would get married soon after she graduated, but no, she entered college studying photography. She even won some State of Ohio awards and had several of her photos displayed in the State House in Columbus.

After five years together, Dee Dee broke it off with Jimmy. All he wanted to do was get married, and she wasn't ready. She had too much to do on her own. At least that's what she told me. Over the next twenty years, she developed a career as a travel agent, then worked her way up to become a manager for AAA insurance. During those years, she went with several young men I rather liked, but she wasn't interested in marriage. Then, right after her thirty-ninth birthday, she married a terrific guy. They've been together without a bobble now for eleven years.

Oh yes, several years before she met her husband to be, Dave, she gave up her management position and returned to sales. "I can make lots more money, and it is sure less pressure. I'm pretty much on my own, and I love dealing with my customers," she told me. That's Dee Dee.

Soon after the change she bought a neat little house in Ferndale, not far from Royal Oak where she worked. Since their marriage, Dave, a talented craftsman and finish carpenter, has remodeled much of her house. This includes a new kitchen, bathroom and a great bar and entertainment room where there once was a bare basement. He built a new garage/shop, and added a porch and hot tub. They are bright and fun people to be around.

Oh, for children they have two bull mastiffs, huge dogs for their little house but they all seem happy, two dogs and two people. That's what counts, doesn't it?

♣　　　　♣　　　　♣

Sometimes when reading Goethe I have the paralyzing suspicion that he is trying to be funny.
　　　　　　　　　　　　　　　　　　　　　　　　　—Guy Davenport

Ignorance can be cured with education and experience. Stupidity furiously resists all efforts at correction or cure.
　　　　　　　　　　　　　　　　　　　　　　　　　—Howard Johnson, 1977

Against stupidity, the Gods themselves contend in vain.
　　　　　　　　　　　　　　　　　　　　　　　　　—Friedrich von Schiller

We can easily forgive a child who is afraid of the dark; the real tragedy of life is when men are afraid of the light.
　　　　　　　　　　　　　　　　　　　　　　　　　—Plato

Most people would sooner die than think; in fact, they do so.
　　　　　　　　　　　　　　　　　　　　　　　　　—Bertrand Russell (1872-1970)

Hell on Wheels

It was May 1960, and I was entered in a road-race in the Waterford Sportsmen's Club in Waterford Hills, Michigan. I would be driving my Saab GT sedan in the touring car class. As we were setting up our paddock area, I noticed there were two other Saab GTs in the paddock. Others in our class included three Alpha Romeos, two Volvos, two chevy Corvairs, several other sport sedans, and a single, much-modified VW. There were enough GT cars to hold single class races on both days.

During practice, I had chances to dice with several GT cars and did quite well. I looked forward to doing well both in the preliminary 20 lap race on Saturday and the 40 lap main event on Sunday. Starting grid position would be determined by lap speed during practice. I never did well in practice as I spent practice laps finding the best way to handle corners by trying different lines. I almost always did much better in a race than during practice. As a result, I was usually back in the middle of the starting grid. In front of me were two Alphas then the one Volvo beside a Corvair, then one of the Saabs and the VW, I was in the fourth row on the inside of the third Alpha. The rest were lined up behind me. The preliminary race on Saturday was important because the starting grid for Sunday's main event would be the finishing order of the preliminary race.

I always move up quickly and aggressively at the start of a race when many drivers hold back because of the traffic. When the green flag dropped for our standing start, I jumped into the space between the Saab and VW in front of me, beating them into the first corner. After negotiating the first eight corners, I stared down the main straightway in fourth place behind two Alphas and a Corvair. It took me twelve laps to work my way into first. The Alphas were my main competition. They were faster on the main straightaway, but I could take them in several of the corners. There was a high speed half circle curve to the left, almost a quarter of a mile long called the Carrousel where I found I could pass any of the competition.

On lap thirteen, I came up on one of the other Saabs, a grey one. The driver must have spun out and was a lap behind, dead last. I set up to pass him in the next corner, but he moved over and blocked me. This blocking action continued all the way around the course. I couldn't pass him on the straightaway, so I planned to do it on the Carrousel curve. I set up to go inside, but when he moved over to block I quickly switched to the outside and moved up beside him to pass. I could hardly believe it when he moved over toward me. In the process of avoiding hitting him, I dropped two wheels off the pavement, hit the dirt, and threw up a billowing cloud of dust, right at the pits and start-finish line. I was now past him and still in first place. I looked forward to winning the race and being on the pole for Sunday's main event.

The next time I passed the finish line the flagman waved the black flag. I couldn't believe it. The black flag meant stop racing and go into the paddock, usually because of some obvious problem with the car, or a serious infraction. I knew neither of these applied to me, so I continued racing. The next time I passed the flag station he not only waved the flag, but pointed to my car. Throwing the black flag at a car leading a race is rarely done, but I pulled into the paddock. I was so angry I raced from my parked car, throwing my helmet down I charged the flagman. I wanted to kill him. He was on the other side of the race course, so I couldn't get to him.

Several of my guys got hold of me and dragged me away. It was probably a good thing since the flagman was a giant of a man at least 6-6 and outweighing me by probably a hundred pounds. I was pissed and let everyone within earshot know it. Now I would be dead last on tomorrow's starting grid for the main event. Who knew, maybe they wouldn't even let me race.

Next morning, after I had cooled down, I filed a written complaint. Within fifteen minutes, I was talking to the flagman who black flagged me. When he said I was black flagged for dangerous driving by getting all four wheels off the pavement, I almost went ballistic. Fortunately I held my temper and said that was totally untrue, I had only dropped my right wheels off of the pavement, a permissible event. He then admitted he hadn't seen exactly what happened, only the dust cloud. When I explained what had happened and that the other Saab was at fault for blocking, he backed down a bit. Several other nearby people had seen what happened. When they heard me talking to the flagman, they spoke up and confirmed what I told him. The flagman, who was also the President of the Waterford Sportsmen's Club, apologized and wished me well in the main event. His comment made me feel better, but I was still starting at the back of the pack.

When we lined up on the grid, the guy in the gray Saab who had blocked me was right in front of me. I walked over to him and said politely, "When I come up behind you and try to pass, do not move over and block. It is against the rules, and you should have been the one black flagged."

"I am faster in some places, and you are faster in others. Competitive racing is not blocking. All I was doing was racing aggressively," was all he would say

I held my cool. "If you block me, I will honk my horn. If you continue to block, or try to keep me from passing when I am obviously faster, I will put you in the bushes. That's a promise."

When the green starting flag dropped, I was so hyped I stalled and had to restart my engine. As a result, I was several car lengths back when I went into the first corner. It took me several corners to catch up with the gray Saab. At the end of the main straightaway, I was right up on his back bumper and trying to pass. Again he continually blocked, even as we went past the main flagman. The flagman waved the move over flag at the gray Saab who paid no attention. He continued blocking through all of the corners up to the main straightaway. I had enough and put in motion the plan I had devised to cover this eventuality. Earlier, I tried to get an angle on him by moving from behind him as we went together down the straightaway. Each time I moved he moved over to block. This time it would be different.

I am in the dark Saab leading the race The gray Saab is a lap behind. This is several laps before I was black flagged for passing him while he was illegally blocking.

Starting down the straightaway, I took a position on the left of the road. He was directly in front of me, blocking as before. I stayed wide open, closing to no more than a foot from his bumper. We were both lined up for the best wide line around the right hand corner at the end of the straightaway. Normally, I would have braked hard at one particular point approaching the corner to slow down to a speed where I could make the corner. I knew he would be doing the same thing. About fifty feet before my braking point, I backed off the gas and slowed a bit, dropping at least a car length behind him. When he hit his brakes to slow for the corner, I stepped on the gas. My front bumper hit his rear bumper directly, transferring the momentum of his car to mine and mine to his—simple physics. He was now going much too fast to make the corner, and I had slowed enough to make it. The last I saw of his car was the undercarriage as he rolled over.

Yes, I deliberately took him out, something I had never done before nor have I done anything like it since. I only hoped he wasn't hurt and was pleased on the next lap when I saw him standing beside his wrecked car, holding his helmet. Incidentally, I raced like a madman, a real battle charging through the cars. I passed the last car on the last lap to win the race. I fully expected to have been black flagged at any moment for my flagrant disregard of the rules. The flagman gave me an enthusiastic thumbs up when he handed me the checkered flag for my victory lap. A wildly cheering crowd greeted all a round the course when I passed with the flag.

At the awards banquet in the evening, the club president allowed what happened was poetic justice, and he saw no need to censure anyone. Then I was treated to a standing ovation by the racing crowd. I still wished I could have handled it differently, but not too much.

I ran into the same driver at three subsequent events. Each time, as soon as he saw me, he packed up and left. About three years later I saw he was entered in a different class of all out race cars in what was called D sport racer class, the same class in which I was now running a Vanguard SR5 powered with a Saab engine. I noticed his car too was powered by a Saab engine. I don't think he recognized me until after the race. I discovered from the program he was a Saab dealer somewhere in Michigan.

I started fourth on the grid and never saw him until I came up and lapped him near the middle of the race. Several laps before the finish I passed him again, putting him two laps down. As I pulled away from him, I raised my hand and flipped him the bird. After the race, he came over to our pit stall and introduced himself. He then proceeded to apologize for his actions of blocking at Waterford Hills. He said I had taught him a valuable lesson and he never blocked again. I told him how glad I was he wasn't hurt, and that he was no longer blocking. I then apologized for flipping him the bird. His comment? "I probably had it coming." I never saw him again in ten more years of racing.

When you have to kill a man, it costs nothing to be polite.

—Sir Winston Churchill (1874-1965)

Any man who is under 30, and is not a liberal, has no heart; and any man who is over 30, and is not a conservative, has no brains.

—Sir Winston Churchill (1874-1965)

The opposite of a correct statement is a false statement. The opposite of a profound truth may well be another profound truth.

—Niels Bohr (1885-1962)

Flying High, A Racer's Tale

For more than ten years, I had been racing sports cars of various kinds on many of the SCCA road courses around the country. During my racing days, I had numerous interesting and often exhilarating experiences, but none like the one from a Saturday in June 1964.

I was at Mid Ohio near Mansfield, sitting in the paddock in my 1954 MG TF-1500. I had pulled in after taking my first few shakedown laps of practice in my completely rebuilt car. The engine was still running when Roy Gelder walked up.

"You looked pretty damned good out there."

"Yeh, the old inchworm is runnin' better than ever after the rebuild." I had a big, green inchworm decal on the spare tire cover on the back. It was a joking reference to my running in H production class, the slowest performance class in SCCA racing.

I take a flag lap in the Inchworm at Mid-Ohio after winning a race where I battled an old adversary and won. We both broke the course lap record for class H more than once during the race—in the rain. One of my daughters is in the car holding the checkered flag.

"Well you know you were already up to your best time ever in your car, even with just a few laps of practice. I was timing you."

"Hell, my pit crew hasn't even started timing my laps yet. They're bound to get even better."

"You'll be giving the Pittsburgh Press gal in her pink frog a run for her money this year."

"I don't know, Donna May's got top notch equipment. My rig with me in it outweighs hers by at least 400 pounds, and she can drive."

"I came over to ask you a question." Roy looked serious as he spoke.

"Oh?"

"Do you think you can handle a really hot car?"

"Hell, Roy, I've driven everything from a TD to a Porsche spyder and last year I raced a '60 Vette at Watkins Glenn and Elkhart Lake. Anyone who can handle that horse can drive anything wih wheels."

"Ever tried any of those all-out racing machines, Lotus, Ferrari or Maserati?"

"I drove a 1.6 liter Ferrari around the course at Santa Barbara, but never raced one. What are you gettin' at?"

"Well, I happen to have a brand new, BMW powered, 2 liter Lotus over there in the next aisle. After I drove it around the course, I decided it was a lot more car than I wanted to try to race. How would you like to give it a ride?"

"Me? You want me to drive your rocket?"

"I know you can do it."

Immediately, I was so excited there was nothing but cotton in my mouth, no spit. All I could think of was this was a dream. "Hell yes, I'd love to give it a ride. Who do I have to kill to get in it?"

"Come on, How. Shut your damned engine off and come take a look. And bring your helmet."

He didn't have to ask twice. Within half an hour I was checked out, approved by the officials, and on the course in a drop-dead gorgeous, flaming red, rear-engine rocket. This was a dream car. It was more responsive in every way than anything I had ever driven before. I took a few laps quite slowly getting the "feel" of the car. I road-tested the steering, brakes and handling. All were superb. I stopped in the pit for a last minute parlé about pit signals before roaring out for some serious laps.

It took several turns around the track for me to find brake points and plan corner entries before I felt confident enough to begin pushing. During practice, there were several much slower cars out at the same time, so I had an entirely new (to me) dimension to deal with. The new challenge was the fast approach to a corner with a slow car right in my way. This required learning a lot of new cornering strategies—quickly. By the time I had gone twenty laps I was getting into it, so I decided to start lapping at speed, pushing it as fast as I could.

The back straightaway at Mid-Ohio is actually two half-mile straights with a slight bend or kink between them. Up to about a hundred and twenty, it is driven as a straightaway. At a faster speed, it has to be negotiated as a turn. At a hundred and seventy, it is a tough turn where you must use all the roadway available unless you brake.

On about the twenty third lap I came out of the keyhole at speed and used max throttle going through the gears before setting up for the bend. Ahead of me, approaching the turn was a black Austin Healey going no more than ninety. My closing speed was in excess of 80 miles per hour, nearly twice his speed. Suddenly there was a loud crack and the smooth ride became a tooth jarring sensation. Quicker than I can tell it, the fiberglass front of the car shattered and disappeared. The right front wheel and tire tore loose from the frame and hurtled toward my head. It bounced off my helmet, cracking the visor in the process, and was gone.

I had no time to think as things were moving so fast. Without a right front wheel, I had no control of the car, and I was rapidly approaching the black Healey. The brief thought, *I bought it*, flashed through my mind. Suddenly the car sort of pole vaulted into the air. All we could think of later was that the frame had dropped into a crack or hole in the road and dug in. This made the car rotate forward and fly airborne backwards and upside down. The flat rear fiberglass body acted like an airfoil, causing the car to float on the air and float slowly to the pavement. It must have gone fifteen feet or more above the roadway and directly over the Healey. For an instant, I found myself looking directly into the wide eyes of the Healey driver, upside down. The car landed, in front of the Healey and upside down. At this point, my head, actually my helmet, was caught between the headrest and the road as the car slid through the kink and off the road into the dirt.

I was completely disoriented from the rapid changes of direction, and the road pounding my head. Fortunately my head was encased in a tough, life-saving helmet. The remainder of the car ground to a halt upside down in a cloud of dust that almost suffocated me. I believe my visor, still in place, kept the dirt from my eyes, nose, and mouth. I was quite conscious all during this time.

At first, the safety people rushing to the scene couldn't find me as I was inside the aluminum box that held the seat with me in a six-point suspension harness. There were several pieces of the car laying around. Several looked much bigger than the box. Someone spotted my legs and shouted, "Here he is."

I was relieved when they turned the box over and began extracting me from the six-point harness. By the time they had me free, the ambulance was right there. Quickly, but with care and precision, they strapped me to a back board and placed me into the ambulance. Siren blazing, they raced backwards across the course, turned, and drove out the gate to the

Mansfield hospital. I took stock of my condition. I didn't seem to be hurting much in any single place, just an all-over ache. I knew I wasn't seriously hurt.

The accident happened about ten thirty in the morning. After several x-rays, lots of feeling around by several ER doctors, and a long argument about keeping me there for observation, I won out and was back at the track in time to drive the Inchworm to second place in H production during the Saturday race. I couldn't stay in front of Donna May. We won't talk about Sunday.

My friend, Roy, walked up and showed me a sheared right spindle from the suspension. The cause of the accident was definitely mechanical failure, not driver error. After totaling about $60,000 worth of race car that didn't belong to me, I had a few pangs of guilt. I'm not sure, but I think Roy sued Lotus and won. It was definitely their fault. As for me, two days after the accident I could hardly move. It took nearly two weeks for the aches and pains to go away.

I have as souvenirs, a bright red Snell helmet with a three-inch diameter hole abraded through the fiberglass and into the supporting foam, and a small, ragged piece of red fiberglass from the hood. I also have one hell of a story to tell.

The car in front is the same version as the one I drove with a 2-liter BMW engine.

Mid-Ohio Sports Car Course
2.4 mi (2.2 mi w/o chicane)

❖   ❖   ❖

We all agree that your theory is crazy, but is it crazy enough?

—*Niels Bohr (1885-1962)*

When I am working on a problem I never think about beauty. I only think about how to solve the problem. But when I have finished, if the solution is not beautiful, I know it is wrong.

—*Buckminster Fuller (1895-1983)*

In any contest between power and patience, bet on patience.

—*W.B. Prescott*

In science one tries to tell people, in such a way as to be understood by everyone, something that no one ever knew before. But in poetry, it's the exact opposite.

—*Paul Dirac (1902-1984)*

Anyone who considers arithmetical methods of producing random digits is, of course, in a state of sin.

—John von Neumann (1903-1957)

It is unbecoming for young men to utter maxims.

—Aristotle (384-322 B.C.)

Grove giveth and Gates taketh away   *—Bob Metcalfe (inventor of Ethernet)*
         on the trend of hardware speedups not being able to keep up with software demands

Reality is merely an illusion, albeit a very persistent one.

—Albert Einstein (1879-1955)

One of the symptoms of an approaching nervous breakdown is the belief that one's work is terribly important.

—Bertrand Russell (1872-1970)

A little inaccuracy sometimes saves a ton of explanation.

—H. H. Munro (Saki) (1870-1916)

There are two ways of constructing a software design; one way is to make it so simple that there are obviously no deficiencies, and the other way is to make it so complicated that there are no obvious deficiencies. The first method is far more difficult.

—C. A. R. Hoare

What do you take me for, an idiot?
         —General Charles de Gaulle (1890-1970), when a journalist asked him if he was happy

I heard someone tried the monkeys-on-typewriters bit trying for the plays of William Shakespeare, but all they got was the collected works of Francis Bacon.

—Bill Hirst

I have voted hundreds of times, but only five times have I voted **for** a candidate. Every other time I had to choose the lesser of the evils presented on the ballot.

—Howard Johnson, 1992

Tragedy is when I cut my finger. Comedy is when you walk into an open sewer and die.

—Mel Brooks

Life may not be the party we hoped for, but while we are here, we might as well dance.

—Rosie Giesie

A Valuable Lesson from Road Rage

It was probably May of 1965—memory from forty and more years ago is not terribly accurate, but the experience was emblazoned on my memory clearly, precise in every detail. It is a perfect example of learning by realizing how stupid some emotional outbreaks are, and the damage they can do.

I was driving to work from our home on Lynn Park, in Cleveland Heights, Ohio. It was a beautiful, clear spring day. I was driving west quite pleasantly on Mayfield, headed for my office in downtown Cleveland about twelve miles away. The morning rush hour meant Mayfield was jammed with traffic. Each driver heading for work was probably anxious and in a hurry. The drive through the center of Cleveland Heights, past Taylor, Lee, and then Coventry Roads was uneventful. Past Coventry, the bottleneck where the road narrowed to a single lane in each direction was jammed with bumper to bumper traffic as usual. Down the hill through Little Italy, Mayfield was stop and go as always.

Right after making the left turn onto Euclid Avenue, I passed through the Case Western Reserve University campus. Then some jerk whipped in front of me trying to gain the space I left between myself and the guy in front of me, in case of a sudden stop. I jammed on my brakes hoping not to be struck by the car behind, or hit the jerk who almost hit me. It was infuriating. This brought on an attack of what we now call *road rage*. I screamed curses at him as traffic began moving. Determined to get to where he could see me, I tried desperately to get in front of him and cut him off, or flip him the finger to at least partially assuage my anger. I changed lanes tying to get beside him while charging down Euclid towards the right turn onto Chester at the Art Museum.

My anger turned into real rage as I headed down Chester, changing lanes frequently in an effort to catch him. I probably cut off other drivers in the process. The harder I tried, the farther and farther behind I found myself. I seemed always to be changing into the slower lane of traffic. By the time I reached downtown, he was nowhere in sight. I was literally screaming in anger and frustration, The traffic gods were definitely working against me.

When I pulled into my parking space in the Bulkley building garage, my frustration was exploding. I sat in my car shaking furiously and panting with my heart beating frantically in my chest. Great gobs of adrenalin were surging through my blood vessels and into my tissues. I was unreasonably frustrated, my entire being in an explosion of disconcerting anger. I pounded on the steering wheel with my fists and forearms. Anyone watching would have thought me a madman.

Some small spark of sanity began returning. Looking at my face in the mirror, I saw a wild thing stared back at me. I was starting to regain sanity. I called myself an idiot for getting so upset and acting like one. I realized the damage I could be doing to myself while he who triggered my anger went blithely on his way oblivious to me and my agitated state. He probably didn't even realize he had cut me off. I was in the process of creating a major, possibly health-damaging event all on my own. It was quite silly. I vowed then and there to change my ways and deal with such bursts of responsive anger in a calming, self-deprecating manner. It took at least a year, but I have since had a rather laid-back approach to such things. Oh, I blow up now and then, but major, useless, self-destructive attacks are a thing of the past.

I am so thankful for the realization of my foolishness, and for the resulting efforts. They have stood me in good stead during the many traumatic or anger inducing experiences we each have from time to time. I realized among those types of experiences, some are real, some are self-inflicted and unreal, some are quite imaginary, and most are quite insignificant. I credit that one experience and my reaction to it with changing me to a much more pleasant and tolerant individual, less hyper than I was before, yet still with intense emotions. Several friends and members of my family remarked about the change, pleasant change, that had come over me. Sometimes a person is lucky, sometimes a person *gets it*, and sometimes a person listens to the soft voice of reason inside their head. It takes all three.

❖   ❖   ❖

Make everything as simple as possible, but not simpler.

—Albert Einstein (1879-1955)

In the end, everything is a gag.

—Charlie Chaplin (1889-1977)

The nice thing about egotists is that they don't talk about other people.

—Lucille S. Harper

You got to be careful if you don't know where you're going, because you might not get there.

—Yogi Berra

The only difference between me and a madman is that I'm not mad.

—Salvador Dali (1904-1989)

Fishing the St. Regis with Angus Thomas

Early one June during the sixties, three friends and I drove all night to get to the St. Regis Indian reservation north of Messina, New York, for some serious smallmouth fishing. We had arranged for two Mohawk guides to take us fishing at this special time. Our connection to the bass run was Angus Thomas, an iron worker who dropped everything else when the smallmouth were running. Angus worked for the construction company owned by one in our group and had called to tell us to get there pronto. The morning after dropping our clothes at the motel and stopping for a quick breakfast, we went early to his house. He and another Mohawk friend would be our guides, providing the boats and fishing gear needed for our trip up the St. Regis River from the St. Lawrence.

Angus's house stood on a grassy yard sloping all the way down to the river's edge some three hundred feet from the house. There were several large shallow-depth aluminum fishing boats with small outboards pulled up on the grass next to a broad, short, wooden pier that stuck out about twenty feet into the river. As we walked down the slope, Angus turned to greet us and introduce the other guide. They had our boats and gear ready, so we only needed to decide who went with whom and we would be off. As usual, I was paired with Bud, the father of my friend Chuck. He was a somewhat cantankerous man of about sixty-five who practiced one-upmanship skillfully. Angus walked Bud and me to his boat and directed the others toward the one with the other guide.

As we pulled away from the pier, I took stock of the sparse, but efficient equipment in the boat. As Angus suggested, Bud and I brought our own spinning tackle and artificial bait. There were two casting rods with sturdy reels and fresh line ending in a 3.0 snelled bass hook. Two large minnow buckets, filled with golden shiners for bait, sat on either side of the center seat which was a built-in live-box to hold our catch. A single anchor with about ten feet of rope sat in the prow on a short shelf. A single pair of battered oars completed our gear. There was no landing net.

It took about fifteen minutes for the small outboard to move us from the pier down the St. Lawrence to the mouth of the St. Regis River. As we entered the river, Angus slowed the motor to idle speed and suggested, "Use your artificial bait here. There is little current, and the water is quite clear and deep. We won't need to get to where we are going to anchor for at least half an hour, so you might as well try your luck while we're on our way. You might pick up a walleye or even a musky along this part of the river."

Bud and I soon were casting with our spinning rods. The river either was deep or had a black bottom where we were. As we moved slowly up the deep part of the river, we kept dropping our Repalas right near the water's edge ahead of the boat. No one had a strike all the way up to the shallow water where the rocky bottom suddenly appeared. Angus shut down the motor and started rowing.

"I don't want to bust a prop on these rocks," he commented as he rowed upstream. "There's a little current now that it's shallower, but we'll have to row the rest of the way."

By now the river narrowed from the calm, three-hundred-foot width to a mere fifty feet or so. The whitish rocky bottom looked to be no more than four feet deep. As we moved farther upstream, there were numerous rocks jutting through the surface here and there. The sound of the water rushing over the rocks—the fresh spring smell of wet woodland in the air—the cool and invigorating spring breeze—the thrill of expectation of battling those dark green bullets—all were eagerly shared by friends as we slowly maneuvered upstream.

Finally, Angus worked the oars vigorously, maneuvering the front of our boat onto a large flat rock near the center of the stream. He tossed the anchor into the water on the other side of the rock to hold us in position. The other boat was in a similar position about twenty feet away from us. By now the current of the crystal clear water was moving rapidly past the boat. The jumbled white rocks of the bottom could be seen clearly through about three feet of slightly turbulent water.

"We're a bit early, but you might as well bait up," Angus instructed as he scanned the water beside our boat. "I'll tell you when they're here. It shouldn't be more than ten minutes or so."

"How far should I cast out?" I asked Angus.

"You won't need to cast out at all," he replied. "Drop your bait in the water on either side of the boat up front and let it flow with the current. Don't let out more than a few feet of line. Once your bait starts dragging, lift it out of the water and move it upstream again."

It seemed a little strange to me, but he was the guide who promised us lots of fish, so I was going to do what he said. I was watching the clear water beside the boat when a dark shape suddenly flashed by headed upstream. Angus cried out, "They're here! Get your baits in the water."

Carefully, I dropped the shiner by the front of the boat and followed it with my rod tip as the current moved it past the boat. Bud was doing the same thing on his side of the boat. Soon the current pulled my line taut, and the shiner rose to the surface at the end of my short eight feet of line.

"Pull your bait out and drop it in front again," instructed Angus as another dark shape sped by in the clear water.

Suddenly, Chuck, in the other boat, cried out, "I got one!" as his rod arced sharply, pulled by a feisty smallmouth heading upstream. No sooner had the words reached our boat than a mighty tug on my pole let me know there was a fish on my line as well. I was amazed at the ferocity of my fish as he headed upstream, bending my rod sharply.

"Lift him out of the water and swing him to me," Angus instructed as I battled my quarry. "As soon as I get him off your hook, bait up again and go after another."

Finally, I had the green and white scrapper out of the water and swinging back to Angus as instructed. Deftly, he flipped out the hook and dropped the smallmouth into the live-box. I would guess his weight at about two and a half pounds, a nice smallmouth. I immediately fished another shiner out of the bucket and hooked him through the lips before dropping him into the water as before. Looking over the side, I now saw many dark shapes speeding past our boat, heading upstream. My bait never made it past the middle of the boat. *Bam!* Another feisty smallmouth, like the first one, was on my line. Soon he joined the other one in the live-box. Before I could bait up, Bud had one on.

Left to right: the author, Chuck, Bud, John, and our fish.
Mohawk guide, Angus Thomas, took the photo.

Fishing soon became a continuing series of frantic motions. There were so many dark shapes scurrying past the boat we could hardly see the white bottom. It took more time to bait your hook and drop it in the river than to catch a fish. No sooner would a shiner hit the water than a frenzied smallmouth would take it, and the battle would be on. Finally, there were so many fish they would try to grab the shiner before it even got in the water. Several times, a smallmouth would jump completely out of the water and grab the shiner in midair, usually getting hooked in the process. It was a madhouse. Suddenly, after about forty minutes, they were gone. Through the crystal clear water, the river bottom appeared rocky white once more.

"You caught 111 fish," Angus reported. "Not bad for an hour's work. They won't be back again until the same time tomorrow."

"We each caught more than fifty bass," Bud commented. "I think I caught a few more than Howard here."

"No," Angus replied emphatically. "I marked each catch right here on the seat. You caught fifty-two, and Howard caught fifty-nine. I don't know for sure how the other boat did, but my guess is Howard caught the most."

Bud was a fairly decent guy, except when a case of one-upmanship kicked in like it did then. "You must have mixed up the score," he countered. "I'm sure I caught more than he did."

Angus grinned and pointed to the marks on the seat. "Nope! I always place the marks on the side the fish come from. You caught all yours from this side with the fifty-two marks, and he caught all the ones on the side with fifty-nine marks."

Angus would not back off a bit. As we started back, Angus compared notes with the other guide. The other two each caught forty-eight bass for a total of ninety-six. We had a pool into which each man had given twenty-five dollars. The pot was split between who caught the first fish, Chuck, the most fish, me, the biggest fish, John, and the last fish, also me. Bud was quite unhappy about the outcome.

I chuckled inwardly at Angus's matter-of-fact comments. Bud, usually a congenial guy, had to one-up everyone about nearly everything. Like those kinds of characters we all have known, whatever he had or did was always better, bigger, newer, or more expensive than what you or anyone else had. Fortunately, he couldn't argue with Angus's firm comments, so he gave up.

It was about noon when we arrived back at the pier. The prospect of the messy job of cleaning and filleting two hundred seven bass loomed as we unloaded. Bud grinned as he piped up, "Howard has to clean fifty-nine fish, and I only have to clean fifty-two. You two only have to clean forty-eight each."

Bud, the author, and John with the author's walleye.

"Does that mean I get to take fifty-nine fillets home with me?" I asked, grinning. "I thought we originally agreed to split our catch evenly no matter who caught what. If you guys want to change now, that's okay with me."

"I was only joking," Bud commented as we set to work cleaning fish. Nothing more was said on the subject during the constant, friendly male banter accompanied our cleaning. I made certain I filleted more than my share without saying a word about it. It was my private way of satisfying my own needs even if no one else knew.

The rest of the week, we spent fishing in the St. Lawrence for walleye and musky with limited success, a few walleye. We went home at the end of the week with about 250 pounds of fillets packed in ice chests. It was a lot of scrumptious eating for several families over the next few weeks. Many fillets were frozen to be enjoyed later. As much as I wanted to return for the fantastic smallmouth fishing in the St. Regis River, I never again made it back there.

A note of interest: I was so impressed by Angus Thomas the man, his Mohawk heritage, and by his name, I used him as the model for the lead character in my novel, *Blue Shift*. One funny incident occurred during this fishing trip. I happened to mention to Angus that I had some Algonkian ancestry back some ways. His simple reply, said with a wide grin, "Algonkians, blood enemy to the Mohawk."

A related incident happened many years later while Barb and I were attending a wedding reception of the daughter of one of Barb's friends in Elkhart. I was telling someone the story of our fishing in the St. Regis and mentioned Angus Thomas. On hearing this name, a young, nearby couple turned, a bit startled.

"Are you talking about a Mohawk Indian Chief from St. Regis?" the woman asked.

"Yes, I am," I replied.

"He's my uncle," she said with a big smile.

They introduced themselves to Barb and me. They both had black hair, dark eyes, and solid, muscular bodies like Angus and many other Mohawks I met. They had been born on the reservation, and then married and moved to Atlanta. They were close friends of the newlyweds. We spoke with them, comparing notes for quite some time. What a fascinating coincidence it was.

❖   ❖   ❖

I love Mickey Mouse more than any woman I have ever known.

—Walt Disney (1901-1966)

He who hesitates is a damned fool.

—Mae West (1892-1980)

Good teaching is one-fourth preparation and three-fourths theater.

—Gail Godwin

University politics are vicious precisely because the stakes are so small.

—Henry Kissinger (1923-)

A clever man commits no minor blunders.

—Goethe (1749-1832)

Argue for your limitations, and sure enough they're yours.

—Richard Bach

To love oneself is the beginning of a lifelong romance

—Oscar Wilde (1854-1900)

Surprise! Surprise!

After twenty-two years of marriage and five children, my wife, Dolores, and I mutually decided to part company, it was during a terribly depressing time in my life. We parted, but remained on friendly terms after struggling with a difficult situation. Soon after we separated, I attended the annual meeting of the Chicago Dental Society, the same dental convention I had attended for many previous years. During the convention, a long-time friend of mine who lived in Chicago invited me to attend a musical show at the Schubert Theater near our hotel. It was a performance of what was to be the next TV special for the female star. My friend was a member of a support organization for the theater in Chicago. After the performance, he took me to a reception in the theater where we could meet and talk to the cast. I was quite excited to have the opportunity to speak with some famous performers who were in the show we had just seen.

While my friend had wandered away, I ended up talking with Donna Lynn, (her name has been changed to protect the guilty) the beautiful lady who was the star in the show, a bubbly singer and dancer who was and still is quite famous. For some reason, we hit it off immediately, almost like soul mates. Donna led me to one of the couches near the bar where we sat and talked. While there, many people came over to congratulate her on her performance and have a few words. After several people had engaged her in conversation, I decided it was time to get up to move on.

When I started to go, she grabbed my hand and said, "Don't leave. I'd like to talk with you some more." Of course, I stayed right there. She then began introducing me to the people who came up to her as "an old friend," using my first name only. Soon she got up, took my hand again, and steered me through the crowd of admirers and introduced me to other members of the cast.

After at least an hour, things began to quiet down. During a moment when no one was nearby, she leaned over and whispered in my ear, "Are you hungry? Would you like to skip out of this zoo for a while?"

"Of course," was my surprised answer. "Where would you like to go?"

"Get your coat and meet me out front," she said with a wink and a smile. "We'll catch a cab to one of my favorite places."

503

"Sounds like a winning plan to me," I answered as I tried to hide my surprise and amazement.

She walked away swiftly, greeting several people as she sped across the room and out the door. I couldn't find my friend to let him know I was leaving. I could explain the next day, so I walked to the checkroom, retrieved my coat and headed out into a near zero February night in Chicago. I didn't have to wait long before Donna came bouncing out of the theater.

When we grabbed the first cab, she told the Cabbie, "Chez Paul, please." Then she turned to me and asked, "You don't mind, do you? I mean, I've sort of run over you. . . . No, I can tell. It doesn't bother you, does it?"

"You surprised me, pleasantly, but no, it doesn't bother me."

"I like that about you. I had you pegged from the first few bits of conversation we had when we first met. You, Mr. . . . My God, I don't even know your last name."

"Johnson, Howard Johnson."

"Not **the** Howard Johnson."

"As far as I am concerned I am," I replied with the customary response to that frequently asked question. . . . "No, I'm in the dental business, not hotels and restaurants." I couldn't help but laugh.

"Well, Mr. Howard Johnson, you are a gentleman, and I am pleased to make your acquaintance." The twinkle in her eyes and the light punch to my upper arm told me a great deal about this little bundle of energy. If nothing else, this evening was going to be a memorable one. She was a real person and devilishly animated, much like her stage persona.

During our ride, she was suddenly quiet and pensive. Then she began telling me how she and her husband had separated a few months before. When I told her about my failed marriage, she replied dramatically, "We are kindred souls, destined to meet and have a wild, passionate love affair."

I must have looked a bit shocked. "Don't take me too seriously," she said quietly. "I can be a little crazy, at times, especially when in untraveled territory, rather like the boy whistling while passing a graveyard."

"I get your drift, but I don't know. Right now a wild, passionate affair sounds pretty damned good to me," I said to lighten things up.

"I like you, Howard Johnson. You're a bit crazy yourself, good crazy!"

I will not detail what happened later in the evening since that is not the purpose of this story. Suffice to say we spent numerous enjoyable times together during the next few months in several locations when we could get together. She even spent most of two summer weeks at my Indiana lake home with me, incognito for her of course.

After seeing us together in my sailboat, one of my close friends told me, "Your new friend sure looks a lot like Donna Lynn, the actress." Several other friends made the same comment under similar circumstances.

"Now that you mention it, she certainly does," was my usual reply, while trying desperately not to grin like a Cheshire cat.

For an ordinary guy with an ordinary income to become involved with a famous actress for a short fling is one thing, but by Thanksgiving, things were beginning to get a bit serious. That development unnerved me. I was quite certain Donna felt much the same. We were talking about my kids and she coming to meet them at our Thanksgiving dinner when reality came down on us. We had a serious and long overdue talk about the future in a downtown Cleveland hotel room. The details are irrelevant, but in the end, we decided it was best we part right then. It was a tough and sad parting for both of us. For the first time, our usual laughs were replaced by tears. We both knew it wouldn't work for us, regardless how much we cared for each other. Our worlds were too far apart. We decided that a clean break with no contact would be the best way and so it was.

Some time later I read she and her husband had reunited. As far as I know, they are still married and together. I wish her nothing but great happiness. For a few months, she was a marvelous, joy-filled part of my life. I will always treasure that ecstatic experience. It was a bright, exciting, light and carefree happening during a low and otherwise dark period in my life.

There was a single exception to our separation. It is the reason and purpose for this story. It was one of the funniest and most personally satisfying thing ever to happen to me. Several years after we parted I made one of the worst mistakes, stupid mistakes, of my life. With my personal life rather a shambles, I met and then married a divorced woman on the rebound. I knew it was a serious mistake the evening of the day we married, but I stuck it out for several years. WRONG!

This too is only a reference point to the story. About three years after our marriage we became Hannah theater supporters in Cleveland. The Hannah was the only major downtown theater to hold stage plays, and in particular, musicals. One show that came to town was—you guessed it, another of Donna's performances in preparation for her next TV special. I had never told Roberta about my relationship with Donna, or even that I knew her as she would surely have gone ballistic if I had. I did so to avoid one of those common, scenes of angry recriminations. Of course, my wife insisted we attend Donna's show and the reception afterward in the Hannah lounge.

I experienced some trepidation and excited tension when I walked into the lounge with my wife after the show. A social climber who was always looking to impress, she headed for the most prominent person she could find to engage in conversation. She abandoned me near the bar, so I could get our drinks. As I waited in line about twenty feet away, Donna walked into the room. It took her about five minutes to spot me.

When she did she almost screamed, "Howard Johnson!" ran across the lounge and threw herself bodily onto me, engulfing me with arms and legs. (It was her exuberant style) All of this took place as my wife watched in awestruck silence. In fact, nearly the entire room was silent for a moment. By the time Donna unwound herself from me and had begun a nonstop series of questions, my wife reached us. It was such a delicious moment for me; I had a hard time suppressing an ear to ear grin. My mind reeled as I wondered what she was going to say and do.

I hadn't a clue what Roberta was thinking, but I was certain there was a terrible conflict going on in her mind. I knew she wouldn't blow up—right then. That's because she would love to be in the limelight with a celebrity like Donna. I knew she wanted to kill me for not telling her that I knew Donna personally. I also knew she was totally baffled with the entire situation. I had never seen her so utterly speechless. There she was, standing with mouth agape and no words coming out. I was in deep trouble.

Then, I did the only humane thing I could do. I took Donna's hand, turned toward my wife and said, "Donna, I would like you to meet my wife."

Roberta stammered something like "Honored to meet you," while I'm sure she was thinking something like, *I'd like to see you both dead.*

I'll have to admit, she held her poise fairly well. In contrast, I was enjoying extreme gratification, loving every minute of my wife's confusion and discomfort.

Then Donna said, "Howard and I are old friends, dear friends. You must be a wonderful person to have such a husband, a lucky lady."

Her face a mask of total confusion, she replied, "Yes, he is a great husband."

At those words, it was all I could do to control my reactions. I'm sure Donna sensed my wife was extremely tense and irritated, and that I was enjoying things immensely.

"Nice meeting you, but excuse me, please. I have a necessary obligation to visit with more of these people. Stick around if you can. I'd like to talk to you later," Donna said clearly before she wandered off.

At this point, I knew my only safe sanctuary was there in the lounge, so when my wife suggested we leave I replied, "I am enjoying myself immensely and would much prefer to stay."

She scowled and, quietly hissed at me, "How is it that you know that woman, and why didn't you tell me?"

"It was a long time ago, and I didn't think it important."

"Not important that my husband knows a major stage star personally and didn't tell me? What kind of relationship did you two have, and when was all this going on?"

I didn't think she would start a battle right here, so I answered. "I never mentioned it because I knew it would piss you off. We had a wonderful close relationship for nearly a year, long before you and I met. We didn't think it would work out, so we cut it off. She's a wonderful human being, in addition to being a talented singer and dancer. She's a real, down-to-earth person, and both of us will always treasure our shared memories. End of story."

Without a word, she stormed out the door as she had done in previous temper tantrums. I couldn't have cared less where she went and what she did. Since I had the car keys in my pocket, I knew she would have to find another way home unless she waited for me. It was useless for me to chase after her and, of course, I wouldn't because I knew it was what she wanted me to do. By this time in our marriage, I was making it a point never to do what she wanted me to do. I was afraid she was so infuriated that she might come back and make a scene. Then I realized it was the last thing she would do in a place with famous people. If anything, she would come back with syrupy conversation laced with not-so-subtle sarcasm aimed at yours truly. When, after half an hour she didn't show, I was sure she was on her way home preparing a particularly virulent purgatory for me when I returned. I decided, what the hell, I might as well stay and enjoy myself 'til the last dog was hanged.

When I spoke with Donna after things quieted down, she made a classic remark. "Howard, I can't believe you married the wicked witch of the west."

"See what happened when you dumped me?"

"That's **not** funny."

"I'm sorry. My sense of humor is kinda screwed up."

"I can see why. Where is she now? I hope she's not lying in wait for me with a club. Boy, did I ever get some nasty vibes from that lady."

"Don't worry, she'll save it all up for me when I get home."

"You poor man. You deserve much better. Other than that I won't say another word."

"Come on, Donna, it will be a cold day in Hell when you don't have something to say about almost anything," I said with a chuckle.

"My Gawd, you do know me, don't you? Let's say I'll try," she replied with a grin.

"Seriously, how are things going with you?"

"I'm sure you know Ray and I are back together. You couldn't know, but you're one of the main reasons we came back together."

"Oh?"

"After you and I parted company, I began thinking Ray's a lot like you, a real person. He was still my manager, so we were often together, even though we were no longer living together. Within a month or two of when you and I parted, he asked me to have dinner with him. One thing led to another, and we've been back together now for nearly three years. We're sure it's going to work this time."

"Wonderful! I'm happy for you. Uh, . . . you didn't tell him about us did you?"

"Of course not. It would have served no purpose. There was certainly no point in rockin' the boat."

"Good."

"Anyway, I'm fairly sure that if it weren't for you, he and I would not have gotten back together. See what you did for us? I am so grateful."

"You're quite a lady, Donna. I'm so glad we had some good times together."

"Yes we certainly did, didn't we? I have some beautiful and exciting memories."

"Me too. Of course, you knew that, didn't you?"

"We both knew it. . . . I don't give advice, but there are lots of great ladies out there. Be sure you are careful when you look."

"I'm a bit gun shy after this one."

"I can see why. Next time, and I'll bet a year's pay there will be a next time. Next time don't move so quickly. Take all the time you need to make sure she's right for you."

"OK mom!"

We talked until the staff wanted to leave, had a goodby hug. That was the last time we ever had contact. I still retain those precious memories, and I'll wager she does as well.

A kiss of love once tasted, lingers in the heart, leaving the sharer wanting more. Sometimes it is not to be.

❖   ❖   ❖

Where your pleasure is, there is your treasure: where your treasure, there your heart; where your heart, there your happiness."

—*Saint Augustine*

Luck is the residue of design.

—*Branch Rickey - former owner of the Brooklyn Dodger Baseball Team*

Fill what's empty, empty what's full, and scratch where it itches.

—*the Duchess of Windsor, when asked what is the secret of a long and happy life*

The Joys of Selling a Family Business

In 1974, I sold my Dental supply company, Johnson Stipher, Inc. to Codesco, a division of IU International, a large, conglomerate. It was a time of change in many types of family owned small businesses, not in the dental supply business alone. For a number of reasons, many of them political, family owned businesses were being forced to sell to larger firms, which were sometimes gobbled up by still larger ones. It was the business version of the popular illustration of a small fish about to be swallowed by a larger one about to be swallowed by a still larger one about to be swallowed by yet another, larger fish.

In the resulting acquisition frenzy, many families were forced to sell or lose everything. It was a brutal time for many small businesses, few of which survived. One long time friend, a member of a family who had owned the same business for more than a hundred years, went from owning a profitable business to owing well over $100,000 more than the business was worth within two years. When his bankrupt business was sold, he was lucky not to lose his house. He was able to keep it only because he realized what was happening soon enough and transferred his home to his children.

I was a bit luckier than most, and hung on profitably for almost a year. I sold the business before going negative with our net worth. After a final inventory, conducted by IU's accounting firm, the company's net worth was reduced and went into negative numbers. According to their figures, I ended up owing nearly $35,000 to close the deal. A settlement meeting was scheduled for the following week. It was to be held in a small meeting room at a nearby hotel. Present would be the following from Codesco: Gene, the district manager, Larry, the general manager of retail operations, two of their accountants, and two of their lawyers. My attorney, Don, and myself comprised our entire contingent. Larry presided over the meeting. This is the Larry who had boasted he would, "cut that smart ass Howard Johnson down to size."

The two hours of discussions we had resulted in a stalemate. We couldn't agree on a final settlement price. Don, my attorney, had been going over their last offer while we were talking. He reached over, touched my arm and whispered, "Let's go out of the room. I found something I want to show you."

I turned to speak to Larry. "If you will please excuse us for a few moments. My attorney needs to speak to me in private."

"As soon as we were out in the hall Don said, "I think you had better agree before they realize they have made a serious error in our favor.""

"What error is that? Are you absolutely sure?"

Outside in the hall Don smiled and stopped me. "Did you see what I saw? They presented their first offer of settlement and whoever wrote up the agreement either put the decimal point in the wrong place, or wrote in the wrong amount. They put the amount of their settlement at $5,000 rather than the $50,000 in the original suit."

"How could those high powered attorneys make such a mistake? What should I do?"

"I'd sign the agreement as quickly as possible and hope they don't notice. You'll also have to pay their Attorney fees which are listed at about $3,000. Still, that's $7,000 less than you actually owe them and a lot less than what they were demanding. I'll bet they told their attorneys to add $35,000 to the $15,000 you owe them, for a total of $50,000, and they simply misunderstood or miscalculated and entered $5,000. Not one of them ever checked the amount or caught the error."

"Are we missing anything?"

"I read through the document twice carefully. It covers everything. I particularly like the disclaimer about this as a final settlement of all debts."

"Do how does this work? I mean my insurance, what I agree to pay them and everything else."

"After the agreement is signed, you will issue them a sight draft for $8,035. They will then release your insurance policy documents, worth in excess of $15,000, and that's it."

"That's it? What about the sight draft? I won't have that kind of money until I cash my insurance policy."

"They agreed to hold the draft until you have deposited your insurance money."

"Then I had better get in there and sign the deal before they catch the error, right?"

"Yes, but appear reluctant. Tell them I said that's probably the best you could do under the circumstances. Then pray until we've all signed the agreement and had it witnessed. Pray they don't notice their little mistake. They will issue you a certified check for $25,000 on the spot. They should also mark the note as paid and return it to you. I hope they don't put two and two together when they mark the note as paid. As soon as we break up, get to your bank and cash or deposit the check. Once you have done so, the deal is finished and cannot be reversed, even when they discover their mistake. They will discover it eventually. When they do, don't say anything, but have them contact my office, I'll handle it."

"You know they are holding about $20,000 worth of my personal property, some of the furniture, tools, and inventory of my manufacturing company, Jotec. Those items were and still are in the building. I moved most of my stuff into a building I have rented half way across town. They wouldn't let me take the items they are still holding until the deal was finished. They locked then in a storage room in the warehouse."

"That's all detailed in the sales agreement. Once those papers are signed they will have to let you take them. Also, they will have to honor your two year employment contract as manager. That is as long as you meet the obligations and performance listed in the contract."

"That's no problem as far as I'm concerned."

"OK. Let's get the contact signed and the check banked. I'll breathe much easier when that's completed."

Everything went smooth as silk. After the papers had been signed and notarized, the check cut, and my note returned to me marked paid, we all shook hands and left. No more than half an hour later the check was deposited in my bank. The deal was now completely closed.

It was two weeks later I received the call. They had found their little mistake. I was asked to provide them a check for $30,000, immediately. I could tell they were distressed when I referred them to my attorney. I never heard another word about the check.

About a month later, I was planning to move my Jotec remaining property to my new space. I rented a trailer and brought it down early Saturday morning. My assistant manager was to meet me with the key to the padlocked section of our warehouse where my stuff was being stored. When I asked him to remove the padlock he refused. He informed me that he had been ordered not to unlock the warehouse or release the contents to me. When I reminded him permission for me to remove those things was spelled out in the final agreement, he said the Codesco attorneys were working on the agreement, and I couldn't have my property until they had finished.

When I called Don, he told me. "They know it would cost you a few turns in court and a bundle of money to get your stuff released by court order. They're being difficult, trying to cause you grief. What do you think your inventory would bring if sold at auction?"

"The brass fittings and solenoids might bring close to my book value, the original purchase price. The finished pieces they might be able to sell to their customers, but they wouldn't do well in any kind of distressed sale. All of that stuff is worth a whole lot more to me as property of my company than to anyone else. They have no idea what it's worth or how to dispose of it profitably. Each item was specifically removed from the inventory listed in the sales documents. I would estimate the best they could get would be $8,000 to as much as $10,000."

"You might have to lie low for awhile and say nothing. See what they do if you simply ignore them. Do you need your stuff right now?"

"No, I have enough inventory in my new location to cover any sales. Besides, I doubt Jotec will have much business for the present. I plan to concentrate on my job managing the store. I would like to earn those promised bonuses."

"In the meantime I'll write them a couple of nasty letters indicating dire consequences if they don't release your property. I doubt it will move them, but it might. It depends on how angry someone there is at you."

"OK Don. Thanks for the help."

Things went without further incident through the summer until October when I was to speak at the ADTA national convention at the Peach tree Center in Atlanta. See the next memoir in the book titled, **Sparky**, for what happened there that further ingratiated me to the powers at Codesco.

❖　　❖　　❖

The graveyards are full of indispensable men.

—Charles de Gaulle (1890-1970)

You can pretend to be serious; you can't pretend to be witty.

—Sacha Guitry (1885-1957)

Behind every great fortune there is a crime.

—Honore de Balzac (1799-1850)

If women didn't exist, all the money in the world would have no meaning.

—Aristotle Onassis (1906-1975)

I am not young enough to know everything.

—Oscar Wilde (1854-1900)

The object of war is not to die for your country but to make the other bastard die for his.

—General George Patton (1885-1945)

Once is happenstance. Twice is coincidence. Three times is enemy action.

—Auric Goldfinger, in Goldfinger by Ian L. Fleming (1908-1964)

SPARKY

In November of 1974, I was on the program of the annual meeting of the American Dental Trade Association at Peach Tree Center in Atlanta, Georgia. I had recently sold my company to IU International, a major conglomerate, and I was staying on as branch manager. Two months previously I had finalized my divorce from my wife of 25 years. These were major changes in my life with far reaching effects. The new owners would not have permitted me to go to this meeting had I not been scheduled to make a presentation on computerization to more than a thousand industry participants. My presentation was on the second day of the four-day meeting, so I would have two free days after I spoke. My immediate boss, Gene Owen, the district manager and his boss, Larry Valant, the general manager of retail operations, were there. I was advised to stay with them and not go off on my own. It was apparent they were going to ride herd on me during the entire meeting. I had other ideas.

Things went smoothly at first, including my presentation, which was well received by the conference participants. My two bosses admitted grudgingly it was a good presentation. As a reward, or so it was intimated, they decided to take me out on the town. We ended up at a dance hall/restaurant/bar somewhere in Atlanta. Shortly after we sat down at a table near the dance floor, a group of five ladies came in and sat at the table next to ours. We started talking with them and discovered they were celebrating the divorce of one of them. I was attracted to one of the ladies, a tiny woman they called Sparky, and immediately asked her to dance. We had been dancing no more than two minutes when Larry cut in. I returned to the table and struck up a conversation with another of the ladies whose name I have forgotten. I did remember she said she was married.

When Larry and Sparky returned to the table, he steered her to a seat on the other side of the table away from me. It was obvious he planed to monopolize Sparky for the rest of the time we were there. The gal I was talking to said, "Your friend is some Lothario. He'll get nowhere with Sparky the way he is acting. He sure thinks he's God's gift to women."

I thought no more about it for the next hour or so while we told stories, joked, laughed, and had a few drinks. I played it cool so as not to upset my bosses who made it plain to everyone what our relationship was. I never did get another chance to dance with Sparky—there.

Out of the blue and a startling surprise to me, Sparky stood up, reached across the table for my hand, lead me around the end of the table and said, "How'd you like to go somewhere quiet? Let's get out of here, now." As we walked toward the door, she added, "I drove my car so I could do as I please in case I met someone I liked."

I was amazed, but at least had the presence of mind to agree and walk out with her. I can still see the incredulous look on Larry's face as we left. As we drove, she said, "I know he's your boss. He is so in love with himself that he made my flesh creep. I hope I didn't get you in trouble."

"No problem," I replied. "I'm happy as can be to be away from those two anyway. They're not my kind of people. They're a bit too much into corporate slavery for me."

In about half an hour, we were at a bar in a quiet little restaurant on the Interstate loop around Atlanta. After a few drinks, some dancing, and several hours of friendly conversation, we ended up at her apartment in Decatur. Needless to say, we hit it off very well—kindred souls you might say. I learned she was divorced from a member of the Sparkman family, a politically active southern family that included John Sparkman. John was a conservative southern Democrat and US Senator who ran as Adlai Stevenson's vice-presidential candidate. Her nick name, Sparky, was from her last name, Sparkman. Her actual name was Imogene, but everyone had called her Sparky since she was married. She had a twenty-something daughter who sometimes stayed in her apartment and whom I would meet several months later.

The next morning she drove me to Peach Tree Center in time for the meeting. Who should be standing near the entrance when we drove up but, you guessed it, Gene and Larry? Avoiding them, I hurried inside to get showered and ready for the day's meetings. I had plenty of time. I was wondering what my bosses would say when I joined them for breakfast as was planned. I soon found out. I received a pointed lecture about proper behavior for IU employees at meetings (underlings that is) and was reprimanded several times until I had enough.

I looked Larry straight in the eye and said, "I am single, footloose and fancy free. I see nothing illegal or immoral in my behavior. I suppose if Sparky had invited you to spend the night with her you would have refused." (Larry was married, and I knew he spent the first night with a hooker because he bragged about it.)

Gene looked at me with an air of superiority. "You could have been a bit less blatant. Having her drop you off right at the front door was a bad idea. It was obvious to everyone what you had been doing. IU men don't do things like that."

"You mean it's OK as long as we sneak around and hide what we do?"

From then on the conversation went downhill. I was not making points with my new bosses. Later, a friend said Larry told the story of my insolence to several people, carefully leaving out his part, of course. It was obvious Larry did not like being *one-upped* and especially by one below him on the corporate food chain. He and Gene would do what they could to mess up my new relationship. The same friend informed me Larry announced to a large group of people he was going to, "cut that smart ass Howard Johnson down to size." How he tried and how badly it backfired on him is another fun story for another time.

By the time breakfast was over, I had decided, to hell with it. I was not going to be cowed by my two bosses. I didn't care what they did. During the rest of the meeting, Sparky and I spent as much time together as we could. I did not miss any of the scheduled meetings and other activities. I spent the nights in Decatur and had taxi service wherever I wanted to go.

Thursday evening there was a formal dinner dance. I made arrangements for Sparky to attend with me and moved our seats from the IU table to one with several of my old friends at the opposite end of the hall. Looking drop-dead gorgeous, Sparky showed up in a stunning blue evening dress. She was clearly one of the classiest ladies at the dinner. We had a terrific time dancing and talking with my friends who asked me later how I had met such a lovely lady. Of course, we shared both of our stories with them.

The meeting finished on Friday morning, and my flight home was scheduled for Friday afternoon. Against Gene's specific orders, I changed my flight to Sunday evening and spent a lovely weekend getting to know Sparky better. We had a fantastic time touring Atlanta including a visit to Stone Mountain. We spent hours enjoying each other.

I had some consulting work in Atlanta scheduled for December and January, so we made plans to be together then. As part of my employment agreement, I was permitted to complete all of the consulting contracts I had signed before I sold the business. There were eight scheduled over a period of a year including three in Atlanta with AT&T Longlines. Via the corporate grapevine, I heard Larry was livid about those consulting jobs and wanted me to cancel them. He never told me, and I wouldn't have cancelled them, even if he had.

Unfortunately, Sparky's daughter didn't like her mother being involved with a "Yankee" and began causing Sparky lots of grief after my December visit. She was especially nasty with me during my January visit when she was around. This lead to a friendly parting of the ways for Sparky and me. It was a sad time for both of us.

Almost a year later Sparky called me. She wanted to see me again, and her daughter had relented. Unfortunately, I had to tell her that I was in a serious relationship. I was living with a woman I would later marry. Maybe I should have dumped the lady and gotten back together with Sparky. The marriage would prove to be a major mistake—a disaster for yours truly. Of course, that's life. *Que sera, sera.*

❖   ❖   ❖

Sometimes a scream is better than a thesis.

—Ralph Waldo Emerson (1803-1882)

There is no sincerer love than the love of food.

—George Bernard Shaw (1856-1950)

I don't even butter my bread; I consider that cooking.

—Katherine Cebrian

I have an existential map; it has 'you are here' written all over it.

—Steven Wright

Mr. Wagner has beautiful moments but bad quarters of an hour.

—Gioacchino Rossini (1792-1868)

Manuscript: something submitted in haste and returned at leisure.

—Oliver Herford (1863-1935)

I have read your book and much like it.

—Moses Hadas (1900-1966)

The covers of this book are too far apart.

—Ambrose Bierce (1842-1914)

Few things are harder to put up with than a good example.

—Mark Twain (1835-1910)

Hell is other people.

—Jean-Paul Sartre (1905-1980)

Now I am become death, the destroyer of worlds

—Robert J. Oppenheimer (1904-1967)
(citing from the Bhagavad Gita, after witnessing the world's first nuclear explosion)

Happiness is good health and a bad memory.

—Ingrid Bergman (1917-1982)

Friends may come and go, but enemies accumulate.

—Thomas Jones

The Inheritance

My father was a decent man, a first class guy. Although he only graduated from the eighth grade, most of his friends and business acquaintances assumed he was a college grad. He was kind, loving, wise, and extremely fair, especially with his family, the focus of his life. He was greatly admired and respected by friends and business associates, and loved by his family.

Of course, he was also one of those evil capitalists who made a profit by selling things for more than he had paid for them. He was one of those hated conservative Republicans, and a selfish merchant who hired people to work for him, so he could profit from their hard work. For this reason, he was an enemy of the poor and downtrodden. To make it worse, he was also a Christian. This made him a fanatical, right wing, fundamentalist Christian.

This tongue-in-cheek paragraph is how he would probably be described by today's far left Democrats, even those among his many progeny. They would completely ignore all his kindnesses and the help he gave to others, the employees who earned a decent living, his many charitable donations as well as the dozens of young people his business sent through college. Many of those colleges were ones that taught hatred for his kind of people, businessmen.

Early in 1954 I had been offered a dream job with American Potash and Chemical in Havasu City California on the Arizona border. It would have meant a substantial increase in salary, and a move to a brand new, rapidly expanding town on the Colorado river south of Hoover dam. It was the place where the famous London Bridge was moved. At the same time, I had the opportunity of joining my father in his dental supply company, Johnson Stipher, Inc. after the death of his partner, Henry Stipher. It did not take me long to make the decision to join my father. I turned down the job offer, and we moved to Cleveland in March of 1954. I never told my father about the job I gave up to join him, mainly because of the pay.

I enjoyed both the business and our close relationship during the eighteen years we were in business together. In 1958, I purchased his cottage on Tippecanoe Lake when he and my mother built a lovely year round home less than a quarter mile around a bend on the shore of the lake. The property he purchased was a full acre running between the lake on the west and the highway on the east. It was and still is a beautiful setting.

My father in his efforts to be fair to all three of his children, my two older sisters and myself, left the business to me in his will, mainly because I had put so much time and effort into the business. He left his lake property to my two sisters with my mother having ownership rights as long as she lived. He explained to me he thought this was fair even though the business was worth more than the lake property. It was also because I had become the largest shareholder after him, was President, and the driving force behind the business after he released the reins. Little did he realize what the future held in store for our business, for nearly all family businesses in our industry.

What my father did not realize or even understand in his later years was the major changes taking place in the dental industry. These changes, from family businesses to conglomerates, were gaining momentum in virtually all industries and businesses. Between 1965 and 1970, nearly two thirds of the privately owned dental manufacturing and supply businesses who were members of the American Dental Trade Association, were sold to major corporations or conglomerates. During this buying frenzy, a number of new supply house branches were opened, primarily in competition with successful family owned businesses. As a result of this assault, many family business failed and were picked up by conglomerates.

One example of those businesses, Fort Wayne Dental, was owned by the Link family for more than a hundred years. The Links were friends, active in the ADTA. Their little business went under quickly due to the onslaught of the "chains" and was picked up by Patterson Dental, one of the conglomerates. Dan Link confided in me that his family had personally lost in excess of $100,000 when they sold. That was the difference between the business liabilities and assets they had to make up with personal funds in order to complete the sale.

We managed to hold out until 1973 when we were expecting our second year of substantial loss. When we finished the sale of J&S to Codesco, we were in a negative equity position to the tune of about $15,000. This amount was to be taken out of my first three year's salary and bonuses. My retirement insurance cash value was to be held by Codesco until the $15,000 was paid.

I was made manager of the new branch. My new boss, Larry Valant, was the Codesco retail manager. He not only gave me a hard time, but even bragged to others in the company, saying, "I'm going to cut that Howard Johnson down to size." He didn't know I was told about his comment by friends who overheard him. He even denied saying it when I confronted him. His anger was all over an innocent personal incident where I bested him unintentionally in front of several others in the company. It occurred after I had given a speech at the ADTA national convention in Atlanta. He was angry over what I considered a minor incident over which I had no control. I didn't see a future with such a jack ass as a boss.

After I had quit as manager of their new branch in 1975, they sued me personally for $50,000 claiming breach of contract. It consisted of the $15,000 I owed them plus $35,000 in punitive damages. As an individual being sued by a huge corporation, I was nervous, but refused to knuckle under to their pressures. When I personally filed a counter suit charging them with everything, Don, my attorney could come up with, they decided we should meet and work out a compromise. During the meeting, they read all of the charges and then they made a monumental mistake. I couldn't believe those high-powered lawyers and accountants could make such a mistake. They presented their first offer of settlement and whoever wrote up the agreement either put the decimal point in the wrong place, or wrote in the wrong amount. My attorney noticed and asked to leave the room with me for a private consultation.

Outside in the hall Don smiled and stopped me. "Did you see what I saw? They put the amount of their settlement at $5,000 rather than the $50,000 in the original."

"You've got to be mistaken. They couldn't have made such a colossal blunder. What should I do?"

"I'd sign the agreement as quickly as possible and hope they don't notice. You'll also have to pay their Attorney fees which are listed at about $3,000. Still, that's $7,000 less than you owe them not even considering the punitive damages. I'll bet they told their attorneys to add $35,000 to the $15,000 you owe them and they simply misunderstood and entered $5,000. Maybe they put the decimal point in the wrong place. Who knows? I'll bet none of them even checked the amount."

"Are you certain we're not missing something?"

"I read through the document twice carefully. It covers everything. I particularly like the disclaimer about this as a final settlement of all debts."

"How does this work? I mean my insurance, what I agree to pay them and everything else."

"After the agreement is signed, you will issue them a sight draft for $8,035. They will then release your insurance policy documents, worth in excess of $15,000, and that's it."

"That's it? What about the sight draft? I won't have that kind of money until I cash my insurance policy."

"I have all the documentation we will need. The draft will not be cashed until you put the money in the bank. We can draw up the papers and sign the agreement right now. I checked. They have your policy in hand."

"What happens if I don't deposit the money from my Insurance?"

"You'll go to jail and the whole deal will be off. Only an idiot would do such a thing and you're no idiot."

"Let's do it"

"One caution, don't appear too anxious to sign. Show some reluctance. Act like you are getting screwed. We don't want them to notice anything until the papers are all signed and in hand. Once that's complete you can gloat if you want. The deal will be done."

The signing went off without a hitch. I acted as contrite and disappointed as I could. The draft was accepted. They turned over the Insurance policy, and we left. It took me several days to cash in my insurance and deposit the money in my bank to cover the draft. Don's bill was $500 for the hours spent, and now I had $6500 more in my bank account than I thought would be the best I could have done.

Of course, my inheritance from the sale of the business turned out to be my payment of $8,035 plus the $500 I paid my attorney. I was reminded of this whenever friends or family spoke about all the money I had received from the sale of the business. Other than my attorney, Don, no one ever knew I had inherited only a debt. At this late point in my life, it no longer matters, so I can include the story in my memoirs.

For the record, I did profit greatly from the new dental design business I started soon after I quit. I named my business, Howard Johnson and Associates. With the connections I had made and experience I had gained from twenty years in the dental business, It only took a couple of years to make a real success of my new dental design business. For 1976 I paid taxes on an income in excess of $100,000, the most I ever made before or since. It enabled us to buy the home on Blackfoot which, of course, I lost in the divorce. The year 1977 was a totally different story. The economic disaster brought on when Jimmy Carter was in the Whitehouse and Democrats controlled both houses of Congress effectively destroyed my business. I was wiped out because of that recession and Carter's double digit inflation with unemployment and interest rates above 20%,. There was almost no business to be had, and the loan I had made to expand my business was eating me up. I couldn't pay Don Wylie who came to work with me before the Carter disaster. He had to look for work elsewhere. Don was a hard worker and a real asset. I felt terrible.

I tried several ventures based on designing dental offices for the next few years with little success. In desperation I took over managing an engineering job in the Philippines in 1981-82, but that's another story. In 1985, I gave up the dental business entirely after trying to revive it with limited success for several more years. That business was the last vestige of a valuable inheritance. I know the small debt I received for J&S was far outweighed by the experience with my father, the friends I made, and the valuable lessons I learned that carried me through some trying times. Working with my father for those twenty years and seeing him pleased with what we accomplished together is a treasured memory. Given the chance, I would do it all again, albeit a bit differently. Hindsight is always 20-20.

Good old Larry Gets His Comeuppance

In August of 1975, shortly after Roberta and I were married, I was going away for a week to do one of the consulting jobs I had contracted before J&S was sold to Codesco, a wholly owned subsidiary of IU International. As much as my superiors disliked my doing these independent contracts, they had been included in the original employment contract. They earned me more in a week than I earned in a month in the dental business. This irked my managers, and especially the assistant branch manager, Dave, who had often been at odds with me many times for many years in the old J&S.

I may have made the mistake of mentioning to someone I would be making almost as much in the eight consulting jobs I had been permitted to complete the first year as my entire salary as manager. I may even have intimated I was considering leaving and concentrating on consulting. I don't remember doing so, and it might have all been conjecture on Dave's part, but he jumped on it and told the district manager. I received a call from John, president of the Codesco division and one of my long time friends in the industry. He asked me if this was true.

"Who told you such a thing?" I answered.

"Your assistant, Dave called Gene, your district manager who called me. Tell me, is it true?"

"The part about the income is quite true. I will realize almost as much from my consulting as my base salary. However, that is without considering the performance bonus and according to what we have done so far this first year, it should be a fairly good one."

"That doesn't concern me in the least. What we all want to know is are you thinking about leaving? Dave made that quite clear."

"If he did, he is lying. I know I never told him any such thing. Dave's had it in for me for a number of years. My guess is he's trying to get rid of me so he can have my job. He's resented and worked against me since I took over running the place."

"You know, you have made yourself quite unpopular with Gene and Larry, don't you?"

"Yes, but do you know how?"

"Something about your behavior at the meeting in Atlanta last fall."

521

I then told him the complete story of what happened, including Larry's night with a hooker. "That's all about some personal happenings. It had nothing at all to do with business. Besides, did you know Larry virtually ordered me to get rid of my Porsche Super Leggra? Apparently he considers it a better one than his and doesn't like a mere store manager to be driving such a car. I told him the car I chose to drive was none of his damned business. Is it?"

"Howard, you're in the corporate world now. Some things aren't generally considered appropriate."

"How about bragging about spending a night with a hooker? I suppose that's OK."

"No, that is not OK. Still, there is nothing I can do about it now."

"I wasn't trying to get anyone in trouble, I was simply trying to make a point. John, you've known me for almost twenty years. We've worked together on a number of committees. I've visited your lovely home in Phoenix with my wife. We've gone out socially a number of times. What do you think?"

"I see what you mean. You may be quite a maverick, but there's no denying the sales reports from your branch. You are doing a good job or at least some of your people are. Let me see what I can do."

Early Saturday morning, I parked my car inside the building for storage while I was doing consulting work at Mrs. Smith's Pies in Alabama. I arrived at my consulting job Saturday afternoon and immediately started working with my crew. They had driven there in our van from Chicago with all of our equipment, and had been there for several days. Working straight through, we finished up in time for me to catch a return flight Friday afternoon. It was nine o'clock Friday night when I arrived at the store to get my car.

When I tried to unlock the door, my key did not work. It didn't take me long to realize the lock had been changed. I had no way to get my car, so I walked about half a mile to an outside phone booth and called a cab. While I was waiting for the cab I called my friend John, the last Codesco person I had talked to. I blew my top when he answered. After I calmed down, he told me what had happened.

"Howard, I did what I could for you, but you have generated so much anger and animosity in those managers above you by a number of your actions nothing I said, even your excellent sales record, could sway anyone. You have been terminated for cause. That means . . ."

I interrupted, "Yes, I get the message. Codesco does not want aggressive, innovative managers who produce excellent results. They want compliant lackeys who suck up to their bosses and spit polish their shoes. All they should ever say is, "Yassuh, boss." Well John, I don't want to work for such an organization in any circumstances. I predict Codesco will not be in the dental business for more than five years." (They only lasted three)

"Howard, do you know what terminated for cause means? You will get no severance pay, no bonus, and your health care insurance will be terminated with no opportunity for you to take it over as an individual. You'll have to start from scratch."

"You said it was for cause. What specifically did I do except bruise the fragile egos of a couple of guys who are far more interested in bragging about themselves than the company, or the business. Of course, I also capitalized on a stupid mistake their entire negotiating team made when we closed the sale and wouldn't back down when they discovered their stupidity. Did they put forth any explanation?"

"They called your actions insubordination."

"How was I insubordinate, ever?"

"For one, you refused a direct order from your district manager."

"If you are talking about my changing my flight home from Atlanta from Friday evening to Sunday evening, I think this is still a free country and what I do on my private time that does not interfere in any way with my job, is none of the company's damned business, at least as long as it is legal."

"As one representing Codesco and being paid by us, we do have the right to determine your travel."

"John, my part in the ADTA program was put together long before your company became involved. It was specifically included in the sales agreement and did not cost Codesco a single penny. The only conceivable negative thing I did was to step accidentally on a couple of huge, fragile egos. I find it hard to believe grown men could be so petty. I lost all of the respect I once had for either of them."

"That was not my understanding. I'm sorry, I made some incorrect assumptions. I still doubt I can do anything to help you."

"John, I've always known you to be a sharp individual and a gentleman. I have great respect for you. I cannot say anything remotely similar about others I know in your company. You know who and what they are. Their actions define them. I am actually pleased they have decided to sack me for whatever reasons. They may have done me a big favor. However, not only will I take legal action, but I have other avenues available. I can use these to create problems for my ex employers. The local head of the federal government's labor department is a member of my church and a close friend. She is in charge of all suits involving unfair employer practices. I'm sure she will find my story interesting and will help me file a serious complaint. There were many friends of mine present during most of the events in Atlanta, each of whom I'm sure would be willing to testify or make depositions on my behalf. I suggest Codesco may want to think about the cost of their negative actions as balanced against the cost of a peaceful and amicable financial settlement. Incidentally, I will not speak to or negotiate with any of the three idiots who are the originators of all of these problems for you, and for me. I will accept you and or anyone you designate and say is fair."

"You said three, who is the other one?"

"The assistant manager, Dave. He's worked against me since soon after I joined the company more than twenty years ago. I'm certain he's the one who told Gene I was going to quit."

"OK, Howard. You have always been a straight shooter as far as I know and that's been for nearly twenty years. I will keep most of this under my hat while I stay abreast of what's happening. I can't say I can do anything to help, but you never know. Good luck."

"Thanks, John. I appreciate it."

I don't know for sure if John helped or not, but it certainly seemed as if he must have. I was not terminated for cause. A woman I did not know came in from the home office with papers spelling out our separation agreement including my financial settlement. It was not what I thought I should have received, but it was probably as close to it as I could hope to get and there were no legal meetings or fees. The same day as she arrived, the new manager, a suave little man named Jim arrived and took charge of the store. He was very self assured but exhibited a bit of what many call a *small man complex*. He was like a more controlled version of the assistant manager, Dave.

The battle was not over. When I asked him about removing the lock so I could get my property, he said he had turned the keys over to Dave and that I would have to see him. When I spoke to Dave, he refused to unlock the storage room saying, "Not until you pay back that $30,000."

"Your actions are in direct defiance of our agreement. I will speak to my attorney who will provide me a court order. Should you defy the court, you will be arrested."

Dave never flinched. He stood there with a look of defiance.

I soon had the court order directing them to open the storage room. The next Tuesday morning I took Roberta to the airport. She was going to Detroit for a two day sales training program put on for Codesco by the Kerr Company. As I dropped her off, I realized all the salesmen and managers would be away for two days. I soon hatched a plan.

I rented a large trailer, hooked it up to my van and drove to the store. I opened the door to the shipping dock and backed the trailer inside, leaving the van under the door, so it couldn't be closed. I took the court order in to our bookkeeper, who I hired about ten years earlier, and who would be in charge with everyone else gone. We had the following conversation:

"This court order seems genuine, but you are still out of luck. Dave has the only keys with him in Detroit."

"No problem," I replied as I turned and headed for my van to get the large set of bolt cutters I had borrowed. It took about a minute for me to cut the hasp and open the door. John, the service manager and a long time friend, helped load all my stuff on a mobile cart and wheel it to the dock where we loaded everything into the trailer. With John's help, it took me less than an hour to get all of my inventory, my cabinets, my furniture, my art and my sculptures out of the storage room and into the trailer. I even took three sets of steel shelving I had not planned on taking because there was more room than I thought there would be.

About fifteen minutes before we were finished loading, the bookkeeper rushed out and made a list of all of the things in the trailer and asked me to sign it. She must have initially thought I gave up when she told me about the keys because she was almost in a panic when she rushed out. I took the paper, read what she had written and then signed it, but held on to it. Incidentally, everything I took was my personal property, listed as such among the inventory documents in the final sales agreement.

I finished loading, checked the room to see if I missed anything, hollered thank you and goodbye to everyone, and got into my van. As I started to drive out, the bookkeeper came over to my door and demanded the signed list. I folded it carefully, three times and handed it to her out the window. I immediately drove off wondering when and if she would notice I had signed it, "Adolph Hitler."

I drove to my rented space, emptied and returned the trailer, all before five in the evening. The next day when I picked up Roberta at the airport I received a mixture of smiles and frowns from my ex employees as I cheerily greeted them when they walked by. Greasy Jim with his slicked back hair walked up to me and initiated the following conversation which was quite loud, so everyone heard.

"If you ever do anything like that again, you will be arrested and put in jail."

"If you had done so in defiance of the court order I have, you would be the one in jail. Furthermore, maybe **you** would repeat such a surprise, but **I** am smart enough not to do such a thing. You did not see this coming. It was a complete surprise. The next time it will be an even greater surprise which you will also not see coming. Have a nice day."

I walked away from a stunned and silent group to take Roberta home. During the next year, I earned nearly three times what I would have earned had I stayed with them. I also had the great pleasure to have taken four large equipment orders away from them. I started handling equipment specifically to compete with them and rub their noses in it. I found it to be a surprisingly profitable part of my business.

Oh yes, my old friend, Larry. I found out from one of my friends who still worked there, dear Larry was no longer general sales manager. The rumor was he was fired because of the sloppy way he handled the final price fiasco and his obsession with "cutting that smart ass Howard Johnson down to size." I didn't shed a single tear.

About three years later, IU sold Codesco to a New York take-the-money-and-run investor. This George Soros clone transferred every Codesco asset he could get his hands on into his own pockets. In the process, he stiffed suppliers big time, I was told millions of dollars. In two years, he stripped the inventory and declared bankruptcy depriving many employees of their last few paychecks and their jobs. A nice guy who pocketed millions legally stolen from many companies and people.

❖　　　❖　　　❖

You can get more with a kind word and a gun than you can with a kind word alone.

—*Al Capone (1899-1947)*

The gods too are fond of a joke.

—*Aristotle (384-322 B.C.)*

Distrust any enterprise that requires new clothes.

—*Henry David Thoreau (1817-1862)*

The difference between pornography and erotica is lighting.

—*Gloria Leonard*

It is time I stepped aside for a less experienced and less able man.

—*Professor Scott Elledge on his retirement from Cornell*

Every day I get up and look through the Forbes list of the richest people in America. If I'm not there, I go to work.

—*Robert Orben*

There are some experiences in life which should not be demanded twice from any man, and one of them is listening to the Brahms Requiem.

—*George Bernard Shaw (1856-1950)*

> If thou of fortune be bereft,
> And in thy store there be but left,
> Two loaves, sell one, and with the dole,
> Buy Hyacinths to feed thy soul

—*Muslih-uddin Sadi*

A Marital Train wreck

In 1975, I made one of the most disastrous mistakes of my entire life, one I would pay for dearly during most of the next five years. Hindsight, always 20-20, reminded me of all of the red flags I had ignored. Many were waved right in my face.

I met Roberta in an interview where I hired her as a sales person for my company. She was quite good at sales which she worked at diligently. She had a bubbly personality and was one of those people who made a positive and friendly first impression. People liked her right off the bat. It wasn't long before we were dating. It was one year after a long overdue divorce from my wife of 25 years. She was also the mother of five of my children. I was quite vulnerable. To make it worse, I should have divorced my first wife seven or eight years previously and married a delightful lady named Caroline. Unfortunately, I waited too long and so she married another. Big mistake number one. Honestly, it was probably number 200 to 2,000, but who's counting? There is more about Caroline in two other memoirs, *A Hidden Reality for My Family* and *A Life Changing, Wonderful Experience - Reunion with My Daughter, Kristen.*)

It wasn't long before I moved in with Roberta. The following summer, we were married. Things started downhill the afternoon after our wedding. My children, all of whom were there at our cottage on Lake Tippecanoe, wanted to look at some of the slides taken there at the lake. When I showed them, lo and behold there were a few with their mother in them. To satisfy Roberta, I guess I should have gone through them before showing them and burned all the ones with Dee in them. Anyway, Roberta went ballistic, or postal, or psychotic—whatever—and from then on, angry tension began to rise. It stayed there, beneath the surface, but able to rise up and bite me at any time, whenever we were together. I tried hard to make things work, but all of my efforts were useless. Our marriage was destined for failure.

For three years, I did what I could to save our marriage. The following is my description of one of the troubling situations I faced on a daily basis. In any argument, virtually every disagreement we had before was dragged out and entered into the argument, in chronological order starting with the slides on our wedding day. She never forgave or forgot.

I would tell her metaphorically, "You emblazon every real or imagined slight of mine on plaques which you then hang on the wall. When a new one comes along, usually unforeseen by me, you add another plaque and then proceed to read and enumerate each one of the previous plaques. It is a never-ending, cumulative process of punishment for each and every real or imagined conflict we have." I always thought that was a perfect description.

By the end of the third year, we were talking divorce, a simple parting with each taking what we brought into the marriage, and splitting evenly what we had acquired together. We even discussed using a single attorney to cover the legal requirements. This would conserve funds for both of us.

No more than a year after this my once successful dental office design business was devastated by Jimmy Carter's destructive administration. It went down the tubes along with many other businesses as inflation, interest rates and unemployment soared. With nothing to do and no money coming in, I jumped at the chance when I was offered a contract engineering job with the U. S. Navy that would take me to the Philippines for a year. I left in September of 1981. I returned home for a week at Christmas, but within a few days we were at each other's throats once more. We even battled all the way out to the airport when I left to go back.

In June of 1982 the Navy contract was finished, and I returned to the States. I immediately packed all of my belongings into my van and the thirty-five-foot cargo trailer I used to deliver the custom wood cabinets I sold in my dental design business. I left all of my business furniture and fixtures in a building I rented on Cleveland's west side. I would return to get them several months later.

I then moved my dental design company into a space I rented in the building next door to the engineering firm that had hired me for the Navy contract. I used most of my earnings from the Navy contact and bought a 25% stake in the company. I would work both companies and see which offered the best financial opportunity. Both were now in the same building in Franklin Park near O'Hare field.

I did nothing but work for nearly a year. I had been served divorce papers not too long after moving. From a friend I learned Roberta had hired one of the toughest female divorce attorneys in Cleveland and was out to get everything of mine she could grab. As soon as I returned from the Navy contract, I made an offer of settlement my attorney and I thought was most generous. Being a bit suspicious even while overseas, I had made arrangements to transfer my lake property to a trusted friend. One brilliant move as it turned out later. Before leaving for the contract in the Philippines, I removed a number of my most valuable possessions from our Euclid home and took them to my lake cottage in Indiana. The actual worth of my jewelry, paintings, sterling silver, coin collection, and other items was in the neighborhood of $20,000.

I had to travel to Cleveland several times for court appearances and then for a final deposition where, of course, her attorney would take me apart, or so I was told. My own attorney, Gene, was outstanding. Using the records from my first offer, he calmly outlined our position. As I looked across the table at Roberta and her attorney, I suddenly realized they looked so much alike they could have been twins. The thought made me laugh out loud.

This seemed to unnerve both of them. My attorney told me later my laugh set them up for what happened next.

Forcing a severe look, her attorney asked, "I see no listing among your assets of the house in Indiana or of several valuable items including sterling silver, jewelry, paintings, and a coin collection. Where are they?"

"I sold them."

"If you sold them, where is the money?"

"I spent it."

My attorney immediately grabbed my arm, excused us for a private consultation, and pulled me out of the room into the hallway. He closed the door and almost ran down the hall pulling me with him. By the time we reached a bench and sat down, he was laughing so hard tears were streaming down his face.

"What's this all about? What's so funny?" I asked.

"Let me catch my breath. I had to get out of the room before I burst into laughter. I almost didn't make it. I have run up against that woman several times before, and she is tough. I can tell you now she guaranteed your wife she would end up with both of your houses and the list of items she read off. You didn't sell the stuff to a family member, did you?"

"No, I sold it to a friend, a lady they don't know even exists. I sold it to her a full year before those papers were served."

"Damn, that was a smart move. I'm glad you didn't tell me about it. Now let's get back there and finish. I have some ammunition I'll bet you have forgotten about."

"What's that?"

"Remember telling me in our first meeting about her cheating on you with an old lover, and while you were still living together?"

"So?"

"Look at this," he said, handing me a paper.

I couldn't believe what he handed me. "You actually interviewed him and got him to sign this admitting to having sex with her while we were still together?"

"It wasn't hard to get him to sign. Apparently, she did something nasty to him as she did to you. This will headline our countersuit. Watch her attorney's face when she reads this."

I looked at him, grinned and said, "That's dirty pool."

"When she sees this, she will forget all about the lake house and other stuff. Let's go. I'll wager her part of the deposition is about finished, and I want to slip her this little paper as soon as I can, while they're still trying to recover from the shock about the lake house."

When we came into the room, they were in a corner talking. As soon as we sat down, her attorney asked, "Do you have proof you sold the place?"

"In the public records in Kosciusko County, Indiana.".

"How about the rest of those things/'

"Went with the house," I stated firmly.

After a few more back and forths, they said their part in the deposition was over. Then Gene handed over the paper he showed me and asked her to explain. Her attorney took one look, handed it to Roberta, and asked if it was true. That ended the deposition and the meeting. Later, Gene worked up the final settlement and sent me two copies to sign. Roberta had signed both copies already. Two weeks later, a final short court appearance and I was a free man. The final settlement gave Roberta at least $20,000 less than our first proposal. I almost felt sorry for her—no—not really.

Before I left for Chicago and after I had thanked him profusely, Gene told me that her attorney always charged at least $4,000 for a contested divorce. His bill for me was a paltry $750. As my father often told me, "If you fell into a sewer you would come out smelling like a rose." In this instance his words were dead on.

When it was all over Gene said, "You did most of the work. You know I could not have advised you to do what you did, don't you?"

"Yep."

"I'm also sure you know how happy I am you did it. I had no idea about your actions since you didn't tell me."

"It was done a full year before I was served, so I didn't think I needed to mention it."

"As it worked out I'm glad you didn't. I'm sure my genuine surprise took its toll on both of them. It was a perfect moment of retribution. I loved it."

"Wouldn't it have been much tougher without the paper you had?"

"Yep. Now aren't you glad you told me about her little escapade, so I could track the guy down and talk him into signing it?."

"Yep."

I headed back for Chicago happy as a clam. Never saw Gene again. I did run into Roberta a couple of times over the years—same old, nasty b----. Some folks never change.

A Cheatin' Dentist

Dr. Jeffrey Wooley came out of the blue one Wednesday when he walked into my design studio. As a dental office designer of some local note, I got most of my clients at dental meetings and recommendations of other clients. This was the first time a client had walked into my studio and said flat out he wanted me to design an office for him. There were no preliminaries, questions about what I did, how much I charged, how long it would take, or anything else I was usually asked to begin with.

After introducing himself, he handed me a plan of the space he had rented. "I recently signed the lease on this space. It's new, empty and out on Pearl Road in a small shopping center. I am a specialist and I want two operating rooms, a fair sized lab, a private office and the usual waiting room/office combination. I need seating for only four in the waiting room. What do you think?"

"Off the top of my head I can see it is a small space. I will have to play with it for a while first to see what I can come up with. This is not a simple problem seeing as you want so much in such a small space."

"A friend of mine said if anyone could do it, you could. That's why I came. He spoke highly of your work. Said you also designed and installed custom natural wood cabinets. I rather like the idea of wood rather than Formica."

"Oh, and who might that be? I'd like to thank him."

"He asked me not to say, said something about him being a private person. You know the type."

"Yeah, I suppose. Can you give me a set of these plans? I'd like to go over them and do some sketches first to see how things work."

"Sure, keep those. I have several more sets."

"My sketches are to scale. Once you have approved a layout we can finish out the plans, so a contractor can work with them. There is no fee until you approve a sketch. Then the fee to present final plans will be $2,000, half when the sketches are approved, the other half when you accept the finished plans. The fee includes some supervision of the construction and especially the plumbing. The plumbing is where most problems occur, particularly in spaces like this where the floor is to be poured after all utilities are properly located. It is important for me to check the locations of all plumbing and electrical before the floor is poured, extremely important."

"I'm sure there will be no problem with the inspection."

We had been standing, so I invited him to sit and discuss exactly what he wanted, and what kind of equipment he would be using. I offered to supply the equipment as well as the cabinets. He said for me to submit prices of the equipment, so he could see if I was competitive. We talked for at least an hour during which time I made up a considerable list of everything he wanted.

As he was leaving he asked, "How long before you will have the sketches ready? I'm rather in a hurry?"

"I'll call you as soon as I know, probably within the next two days."

"Great. I'll see you Friday afternoon if you are ready."

Friday he was pleased with one of the sketches. He asked me to design cabinets for the entire office. One request troubled me. "You want both upper and lower cabinets right across the back door in the lab? I don't think the fire Marshal would approve. It would certainly be against the code."

"We can hold off installing those cabinets and the top until after the final inspection. They won't know. We can install all the cabinets right up to the door and tell them the tops aren't ready yet. They'll give us an OK, I guarantee."

I didn't like it, but what the hell, it would be his funeral.

"It will take me about a week to complete the cabinet designs and prepare the price list for cabinets and equipment. If we get everything ordered within a few days of your approval, we should have no trouble completing everything on schedule. That's even leaving a week to spare in case of emergencies. Is next Friday at this same time good for you?"

"Sounds like a plan. You'll call me if there is a problem, won't you?"

"Absolutely. Now I will need the first check of $1,000 please."

He whipped out his check book and wrote the check. No problem. I thanked him, and he was on his way.

By Thursday noon I had everything ready. I did some neat things to provide everything he wanted in the tiny space. It came out better than I at first thought possible. I included part of a hall way in his private office. The particular part of the hall involved would never be used when he was in the private office, so I was able to provide a larger one than I at first though possible. I was satisfied the plans provided an excellent use of the space. I called him and said things were ready for him. He would be in early Friday. That way, I could start getting things ordered the same day.

He was extremely pleased with the final plans and was especially pleased with the private office which was designed to be closed off with a folding door across the open end. He brought out a list of equipment with prices from other suppliers. He compared my list with them and proceeded to order more than half of the equipment from me. I could hardly believe I was to supply most of the profitable pieces. The list of equipment came to about $12,000 and the cabinets to $14,800.

"I gave you the second check for your design services. I don't mind the $7,400 deposit on the cabinets, but I don't think I should have to pay for all of the equipment in advance."

"Dr. Wooley, I supply equipment mostly as a courtesy to my clients at competitive prices. Almost everything you ordered is special and I would have a hard time disposing of it if you didn't take it. I don't have the money to tie up in equipment for several months at such a narrow margin of profit. This has been my policy for several years, and besides, you'll be holding the $7,400 still due for the cabinets. It should be enough to guarantee the job gets done properly."

He grumbled a bit, but in the end gave me the amount I requested.

Everything went smoothly, mainly because the contractor was quite good. I picked the cabinets up with my cargo trailer from the cabinet maker in Nappanee, Indiana and then added the equipment. The trailer with everything in it was delivered to the job site a day before we were to begin installation. I estimated six days to install equipment and cabinets. We were finished by 3:00 on the sixth day. The only difficulty we experienced was installing the long cabinet in the lab. Because of the Doctor's demands, I had to wait several weeks until the inspection was completed before the lab cabinets could be installed. I couldn't instal the long top alone. It required at least two men.

There was a barber shop right around the corner about a hundred feet from the job. I had my hair cut there while on the job. The barber talked a lot about the neighbor hood where he had his shop for almost twenty years. He was very friendly.

I walked over to the shop and asked him, "Do you know of any reasonably strong healthy men who could give me a hand with one difficult cabinet top? It should go in easily with three pairs of hands. I can't manage it alone. There's twenty bucks apiece for those who will help. It shouldn't take more than ten or fifteen minutes."

The head barber smiled and said, "Would a pair of Barbers do? I'm about finished, Ray is sitting there and we have no more customers waiting. We could both use the money."

No more than fifteen minutes later we were done and back in the shop. I handed each one a twenty-dollar bill and thanked them. The head barber grinned and said, "If you need any more help like that, please call on us."

The job finished, I asked the doctor to check things out. After his thorough inspection and approval, I asked for a check.

"I'll send you one in the mail tomorrow. I don't have my check book with me."

After four weeks and many phone calls, it became obvious the doctor was going to stiff me. He never intended to pay. A little research and I discovered he had stiffed almost everyone that worked on his office, the contractor, the carpet man, the lady who decorated and furnished the drapes. Even the guy who delivered the water cooler. When he responded to one of my many calls with, "Sue me." I called my attorney.

His response? "That guy knows exactly what he is doing. It would likely cost you almost that much by the time he went through all the things you did wrong in a number of counter complaints. You would do well to file a complaint with the BBB and forget about it. You'll never see a penny and you could lose quite a bit if he manages to trot out enough plausible complaints about made up problems. We run into this kind of cheat many times."

I grumbled about that jerk for at least a month. Then one early Saturday morning around four, I was driving down Pearl Road past his office when an inspiration hit me. I turned into the parking lot, pulled up behind the Barber shop, walked over to the dentist's door. Then I squeezed a tube of super glue into both the doorknob key slot and the deadbolt key slot. No more than five minutes later I was back on Pearl Road headed for home. I was so inspired, I decided to call the Fire Marshal and tell him about the blocked door.

About a month after that I stopped in for a haircut at that barber shop. During my haircut the following conversation took place:

"You remember that Dentist office where we helped you with that big cabinet top?"

"Yeah. What about it?"

"He had quite a problem about a month ago."

"What happened?"

"He had scheduled an open house. Invited hundreds of people—had a big ad in the paper. It was even on the radio for a Saturday afternoon. Anyway, they arrived with their caterer, truck and all and couldn't get in the door. Both locks were frozen. They couldn't even get their keys into the locks. It took them an hour and a half to get a locksmith and even he couldn't get in. He figured some kids had filled the locks with glue. With that door being a security door, they couldn't even break it open. They hauled in a fork lift and used the fork to punch a hole through the door. By this time there was a crowd of people standing around.

Then it began to rain. The whole thing became a disaster. They used the fork lift to pull the door out, frame and all. By the time they cleared the trash out and could get in, it was past the time for the open house to end. That doctor was furious. He about lost it several times. We could hear him shouting, even inside our shop."

"Gee, that's too bad," I said with a grin.

"You one of those he stiffed? I hear he stiffed about everyone who worked on his office. He must have done it to you too."

"Yep. Therefor I find it hard to feel anything but good about his little mishap. I'd like to shake the hand of whichever kid did that. Shake his hand and hand him a fifty."

"That's not all the grief he had recently."

"No?"

"Two weeks after his open house and before they had even finished replacing the front door, the Fire Marshall calls and inspects his place. I heard he fined the doctor and ordered those cabinets removed and the back door made useable as an emergency exit. They had to cut apart that big top we helped you with."

"Such a shame. I warned him there could be big problems if he were to close off that door. Warned him several times."

"You don't seem too broke up about it."

"Would you if you'd been stiffed for several thousand bucks? Poetic justice I call it. True poetic justice. What goes around comes around. Not always, but when it does, like this one, it makes a body feel pretty good."

"He's not been very popular around here in the shopping center. Lots of the local merchants react much like you did when his name is mentioned. I still hear lots of negative comments. He's never come in here, at least not that I know of. I wouldn't even know him if I saw him."

And I looked, and behold a pale horse: and his name that sat on him was Death, and Hell followed with him.

—*Revelation 6:8*

Men are not disturbed by things, but by the view they take of things.

—*Epictetus (55-135 CE)*

What about things like bullets?
    —*Herb Kimmel, Behavioralist, Professor of Psychology, upon hearing the above quote (1981)*

Attention to health is life's greatest hindrance.

—*Plato (427-347 B.C.)*

Plato was a bore.

—*Friedrich Nietzsche (1844-1900)*

Nietzsche was stupid and abnormal.

—*Leo Tolstoy (1828-1910)*

I'm not going to get into the ring with Tolstoy.

—*Ernest Hemingway (1899-1961)*

Hemingway was a jerk.

—*Harold Robbins*

How can I lose to such an idiot?

—*A shout from chessmaster Aaron Nimzovich (1886-1935)*

Not only is there no God, but try finding a plumber on Sunday.

—*Woody Allen (1935-)*

I don't feel good.

—*The last words of Luther Burbank (1849-1926)*

Nothing is wrong with California that a rise in the ocean level wouldn't cure.

—*Ross MacDonald (1915-1983)*

Men have become the tools of their tools.

—*Henry David Thoreau (1817-1862)*

I have never let my schooling interfere with my education.

—*Mark Twain (1835-1910)*

It is now possible for a flight attendant to get a pilot pregnant.

—*Richard J. Ferris, president of United Airlines*

The cynics are right nine times out of ten.

—*Henry Louis Mencken (1880-1956)*

Mrs. Smith's Pies

As a consulting engineer of mainly commercial building roofing systems, I managed to get on and inside numerous manufacturing facilities as well as many Navy, Air Force, and NASA buildings. Observing these places inside and out I saw some interesting sights. These facilities included Kellogg' cereal plants, AT&T buildings of many types, NASA's gargantuan Vehicle Assembly Building (VAB), and the Plumbrook, Ohio vacuum sphere, and military facilities all around the Pacific. I went to Hawaii for several months every two or three years to do the buildings at Pearl Harbor and Hickam field. I spent nearly a year in the Philippines at Subic Bay and Clark Air Base. I even made it to Diego Garcia, a flat rock out in the middle of the Indian Ocean.

In all of these often unusual buildings in exotic places, one of the most fascinating observances I ever made was, by comparison, in an extremely mundane manufacturing facility. This smallish plant in Alabama produced, among other things, the product which gave the company it's name, Mrs. Smith's Pies. Imagine a medium sized building out in the middle of farm country in rural Alabama. Into this building, semi trailers and freight cars disgorged wheat, milk, sugar, various fruits and other savory items as well as cardboard and paper. From the same shipping dock emerged semi trailers and freight cars loaded with fresh boxed pies and other edibles.

Among these other edibles were thousands of boxes of Eggo Waffles. Now manufacturing Eggo Waffles does not bring much excitement to mind. One can easily imagine gigantic baking machines disgorging the waffles onto a conveyor belt which conveys them to a boxing machine where they are packaged and shipped. Actually, that is a rough description of the process, but the real baking machine was radically different and produced waffles exactly as they are produced in homes all over the country and even the world—with a few small differences.

In the clean room, we were permitted to visit there were two round, rotating tables about eighteen feet in diameter. Around each table closely spaced and with the lift handle for the top pointed out, were at least eighty of those round waffle irons. In size, shape and operation, they were identical to the family waffle irons in homes everywhere, but heated by gas not electricity. They were heated by hot air from burners beneath the table blown through tubes into the irons both top and bottom. There were two stations, one on each side of each table, where the batter flowed into the open iron from a dispenser tube that placed the right amount in each iron as it passed under the nozzle. An arm gently closed the lid, and the iron continued toward the next station as the table rotated slowly. At the opposing station, another arm lifted the lid and placed the finished waffle onto a conveyer belt that took it to be boxed.

As soon as the iron was emptied, a new batch of batter was poured, and the next waffle was made in that iron. During each batch of waffles, the table rotated 180 degrees. This meant that four waffles were produced about every ten seconds on the two tables. It also meant that at any one time, there were one hundred and fifty-two waffles being cooked in that room. The plant manager proudly informed us that every Eggo Waffle was produced in those ten-inch, round waffle irons like mom's, on those two tables in that room, in his plant. Well, there were a few minor differences. He also informed us that each table was shut down about once every eight weeks, so all of the waffle irons could be replaced. They were still usable, but regular replacement before they could cause problems was the most cost effective system. Apparently the costs associated with a single iron going bad were so high they were all replaced long before they could create any problems.

This waffle making process fascinated each of us. It was unbelievable that Eggo Waffles were made in the same waffle irons and by almost the same process as most people use in their homes, especially in this high-tech manufacturing world. Will wonders never cease?

I never miss a chance to have sex or appear on television.

—Gore Vidal

I don't want to achieve immortality through my work; I want to achieve immortality through not dying.

—Woody Allen (1935-)

Men and nations behave wisely once they have exhausted all the other alternatives.
—Abba Eban (1915-2002)

A consensus means that everyone agrees to say collectively what no one believes individually.
—Abba Eban (1915-2002)

To sit alone with my conscience will be judgment enough for me.
—Charles William Stubbs

Family Letter from the Philippines

Friday, January 22, 1982

Dear Family,

Once again, I send you greetings from the tropics. Presently, I am staying in San Miguel, the Communications Center for the Navy in the Philippines. It is a small base about thirty miles west of Subic Bay on the shores of the South China Sea. We are almost completed with our navy contract and should have everything wrapped up by Monday or Tuesday of next week. On Monday I will be going to Clark Air Base to start our contract there. If all goes well, we should be finished and headed home by March or April.

A rather chilling experience happened the night before last. I walked down to the beach a quarter mile from here to eat my supper and watch the sun set. It was a beautiful, sunny day in the eighties, and the long, sandy beach was deserted. Less than half a mile from where I sat on the beach was a small fishing village with twenty or so boats and a few shacks beyond the fence that marked the edge of the military reservation. I had been there nearly an hour, finished my sandwich, and was preparing to watch the sun set on the ocean when the unmistakable click of a shell being chambered in a rifle came from right behind me. I turned to see two Marines approaching from the base. They stopped and put their rifles at ease when they recognized I was an American. Somewhat unnerved, I asked what was going on. One asked if I had noticed any fast-moving boats on the horizon. When I said one went by, turned around, and headed back the other way, he suggested I get off the beach immediately. The boat was probably a pirate who may have spotted me on the beach and could be preparing to attack. A few months earlier, pirates had attacked the nearby fishing village, killed several men, and carried off everything of value, as well as two young women from the village. Their fast, long, narrow boats with outriggers and pivot engines on the back, could come from the horizon to the beach in less time than it would take me to run through the sand to safety. They would probably shoot me as soon as they hit the beach and then plunder my body, including knocking all the gold from my teeth and leave in their boats in about five minutes. It had happened before on this same beach. That's why they patrol the beach every morning and evening. The South China Sea is famous for pirates to this day. I thanked the Marines for the warning and viewed the beautiful beach in a different light.

I have learned a great deal about this part of the world and something of the people. I'm beginning to see why so many Americans marry Filipinos and even choose to live here as

539

civilians after they leave the navy. The area (Subic Bay) has many resorts and beaches, and the living costs are so low. I have made many new friends, both Filipino and American, and have been invited to their homes on many occasions. I've learned to enjoy Filipino and Chinese cooking, which is different from the Cantonese we call Chinese. I've taken many photos (hopefully the right ones) of all extremes of life from a temple in Manila to women doing their laundry in a river near men harvesting water plant roots for food. I've watched expensive cars racing down narrow roads past naked children, men plowing with carabao, and many people planting rice by hand in rice paddies. The scenery is sometimes beautiful—mountains, green rice fields, sunsets, slash-and-burn agriculture fires at night, seashores with warm blue waves breaking, sometimes ugly half-starved children, denuded hills, crowded dwellings in poverty areas with stinking and polluted waterways, and crude fleshpots.

It is now Wednesday, February 3, and I am in the Oasis Hotel in Clarkview, a small enclave of houses, shops, hotels, and restaurants right outside Friendship Gate of Clark Field. Today we went to Capas, Tarlac, leaving at dawn for this small village twelve miles north of Clark Air Base. The navy has a large transmitter station and listening receiver west of the village of Capas on land that was the WWII death camp of the Japanese. It is where American and Filipino prisoners of Bataan and Corregidor were taken after the infamous Bataan Death March. Of the 76,000 souls that started, only 68,000 made it to the camp where they were held under terrible conditions by the Japanese. The Americans were kept on one side of the compound, the Filipinos on the other. Only a few thousand survived to be freed at the war's end. There are two memorial mass graves there, one American and one Filipino. I remember so well reading about this in all the newspapers while in high school. Being there in person was a moving experience.

While working atop the administration building at Tarlac, we were accompanied by a Marine with an M16 rifle who instructed us to hit the deck immediately if he called a warning. He did not elaborate much but seemed wary of a real danger. There were about five buildings grouped on a small hilltop among small hills of gently rolling country denuded of all vegetation. There was not a tree or bush for at least a mile around the camp, and the rolling hills were covered with numerous, tall antennae. This was our radio listening spot for a huge area from New Zealand to China to Hawaii. We had been working for only half an hour or so when the Marine shouted, "Hit the deck!" At the same time, sharp, whistling sounds in the air above our heads announced we were being shot at. The crack-crack-crack of a semiautomatic rifle followed, answered by the Marine's M16. While I tried to bury myself in the gravel roof, the Marine assured us that we would be safe as long as we stayed low. Since they were shooting uphill at us, we were shielded from their fire by the parapet wall of the building roof. I turned my head so I could see the Marine. He was prone on the roof with his rifle resting on the low parapet wall, trying to see whoever was shooting at us. He was firing occasionally in return while talking rapidly to someone on his handheld radio, reporting the attack. After the first few rounds, we heard no more gunfire and lay in silence for what seemed like forever. Every few minutes, the Marine would fire a couple of rounds. He explained that was to keep them from getting set to fire at us. Then we heard the sound of a helicopter. The Marine said it would be over shortly. Within a few minutes, the unmistakable sound of a machine gun rent the air. After a short conversation on his radio, the Marine announced that our attackers had been eliminated, and we could resume our work safely as soon as the helicopter completed a sweep of the area.

I quite agree that Churchill was right when he said, "Nothing in life is so exhilarating as to be shot at without result."

By now we were sitting up on the roof, looking in the direction of the sound of the helicopter. We saw nothing but the gently rolling hills. The Marine explained that, occasionally, rebel insurgents would sneak in among the hills undercover of darkness, set up an observation post behind a hilltop, and make records of whatever they saw. Sometimes they would set up a rifle over the crest of a hill and take potshots at anyone in range. Over the last ten years or so, three people had been hit and one killed by these sporadic attacks. Frequently, but not every day, a helicopter or observation plane from Clark would cruise the area after dawn to discover the infiltrators before they could do any damage. They were easy to spot from the air as there is no cover of any kind within a mile and a half of the camp. He said they killed or captured fifteen or twenty infiltrators every year, mostly by helicopter, but sometimes using ground vehicles guided by aircraft. As he talked to us, the helicopter rose suddenly from behind the hill, obscuring our view, and made a slow circle of the area. A truck made its way toward the place the chopper had previously been on the ground. The Marine explained the truck would pick up the invaders.

I've enjoyed my time here working ten or twelve hours a day, six days a week. I only wish I had more time to get around in the countryside. The people are so friendly (for the most part) and like Americans, or is it their money? I've also learned more about military life, particularly the red tape and compounded bureaucracy. It's not all bad, but there are some real losers, male and female, in the lower enlisted ranks. I can see how they would have a difficult time living in the real world without the restraint and regulations of the military. I've also met several officers who, after they left the service, found out they couldn't make it in civilian life, so they rejoined. Seeing all these people with no family ties (many military and civilian employees seem to have no real family or home) makes me appreciate the family I have even more than before.

I even had the chance to visit with some Negritos, small dark people who were the original human residents of the islands. They were there long before the Malays, Orientals, and other light-skinned people came to the islands.

They are tiny people, few are more than four feet tall, who live in the bush with little contact with modern living conveniences. They shun the cities and live in tiny villages deep in the mountain jungles. I wonder how long they can survive the mounting population pressures that are destroying their jungle homes and nomadic way of life.

Now, in closing, I give you my poetic impressions of a local early morning and midnight in Japanese haiku style.

Morning	**Midnight**
Earth-lit moon,	White, black clouds
Pinking blue sky,	Black, black sky
Bird calls.	Gray-white moon
Volcano tips	Orion racing.
In gray cloud drips.	Gecko chirps,
Winding roads—misty	Waves on rocks,
Jeepneys, trikes,	splashing Invisible!
carabao carts, bikes.	Almost silence!
Children scampering,	Dreams.
Wet fields, shining	
Wind sound in Pines.	

Love, HJ

PS Wish I could bring you all here for a while.

❖   ❖   ❖

Sanity is a madness put to good uses.

—*George Santayana (1863-1952)*

Imitation is the sincerest form of television.

—*Fred Allen (1894-1956)*

Always do right- this will gratify some and astonish the rest.

—*Mark Twain (1835-1910)*

In America, anybody can be president. That's one of the risks you take.

—*Adlai Stevenson (1900-1965)*

I Died Last Saturday

Yep, that's what I told her. That's exactly what I said to the lady who answered my phone call to the doctor's office. The call went something like this:

"Dr. Phelp's office, Diane speaking. How can I help you?"

"Diane, my name is Howard Johnson, Howard E. Johnson on your records. I am calling about an appointment I have for next Tuesday. I won't be able to make it."

After a wait of several minutes—she was obviously looking up my records—she replied, "Oh? Would you like me to schedule another appointment for you? We'll need to see you soon."

"No. I don't think that would be possible."

"Let's see what were you scheduled for. . . .Oh, my goodness. You know, Mr. Johnson, your emphysema treatment is critical for your health. You must have one within no more than a week."

"I don't think so. You see, I died last Saturday."

Dead silence.

"Did you hear what I said? I died last Saturday. My obituary is in today's Warsaw Times Union paper. Look it up."

"If this is some sort of a joke, it is not funny. Who is this?"

"I told you, I'm Howard E. Johnson. I have an appointment scheduled for next Tuesday. I called to tell you that I won't be there because I died last Saturday. Look up my obituary in the paper if you don't believe me."

She hung up.

Let's go back a few months to a bitter cold, clear, dark February morning before dawn when I was driving my Dodge maxivan to a breakfast Kiwanis meeting in Warsaw. I was going up a low hill on clear, dry pavement. As soon as I crested the hill I saw the road on the other side was black with ice formed from water flowing over the road. I pumped the brakes, but even at only forty miles per hour I still realized I was not going to be able to make the curve right ahead. I was going to go off the curve and into a deep ditch. There was no guard rail. As I hit the grass before the ditch, I folded my arms against the steering wheel and forced my head against my arms as hard as possible. I learned that from accident training required to get my racing license. I was wearing my seat belt.

The next thing I knew I was in pain, wondering how badly I was hurt. Then I wondered how long it would be before anyone would see me and send help. I tried not to move at all, staying as still as possible. There was an EMS ambulance with me in a matter of a few minutes. I found out later that a gentleman in a nearby house had seen the accident and immediately called 911. The EMS ambulance had been on US 30 less than a mile away returning from the hospital. They came directly to my rescue. The medic who first arrived on the scene told me not to move—I already knew that. He examined the steering wheel—it had been pretzelled.

He looked at me and said, "How on earth did you keep your teeth? In all the accidents I've seen like this the driver always loses his teeth when his mouth hits the top of the steering wheel."

By this time the growing pains in my chest, back and legs overwhelmed me, and I couldn't answer. With the bright spotlight from the ambulance illuminating the inside of my van, I could see that the windshield was no more. I noticed my heavy, red tool box, that had been in the back, had flown forward and was now jammed into the grass outside of where the windshield had been. I was lucky that missile had missed my head as it flew past. The medics brought a back board in through the windshield frame, muscled it around and under me, strapped me in after securing my head, and then took me out through the same opening. The next day's Times Union paper would display a photo of my removal through that opening along with the medics doing their job.

I was in the emergency room of Kosciusko County Hospital no more than half an hour after the accident. The damage they found? I had compression fractures of three lower vertebra (that cost me the instantaneous loss of two inches in height), lacerations of both forearms (they took the blow that otherwise would have removed my teeth), a badly bruised chest, but no broken ribs, and a collapsed lower left lung. After several hours with several doctors for my various different injuries, and after being medicated and patched up, my friend, Al Hayes picked me up and drove me home. I never did make my meeting. I was wearing a plaid sport coat with two shredded sleeves.

The several specialists I was referred to for treatment of my various injuries included an orthopedist, a dentist (my teeth had sustained some damage from being jolted against each other when I hit), and a pulmonary specialist.

All except the pulmonary guy were excellent and took excellent care of my ills. The pulmonary doctor was a piece of work. My first and only visit to his office was a masterful demonstration of patronizing idiots going blithely about their business, totally oblivious to their patient. After filling out several pages of forms including a description of why I was there, a collapsed lung, I was interviewed by the doctor's nurse.

"Let me see?" she said as she examined the paper she held in her hands. " You're here for treatment of emphysema." That was a statement, not a question.

I corrected her. "No, I don't have emphysema. I'm here for treatment of a collapsed lung. I was injured in a car accident."

"How long have you had trouble breathing?"

"Since the accident, a week ago last Friday."

"Have you been using your inhaler like this one?" she asked, holding up a round, purple inhaler.

"No, I have never had one of those."

"You know, you need to be using your inhaler regularly."

By this time I realized we existed in different worlds, quite remote from each other. "I have never had, used or needed an inhaler," I said quite emphatically. I could swear she never heard a word I said. She had far more important things to do than listen to what an old man was saying.

"The doctor will be in to see you in a few minutes," she said as she rose and left the room. I was sure the doctor would soon straighten things out. WRONG!

"Good morning Mr. Johnson and how are we doing today?" the doctor said as he entered. I immediately knew things were about to go from bad to worse, so I steeled myself and prepared for verbal combat.

"Except for my **collapsed lung, we** are doing quite well, thank you." I hoped my strong emphasis on the significant words would make a point. They didn't.

"The nurse tells me that youhaven't been using your inhaler. We must use that regularly or your breathing problems will be getting much worse. We wouldn't want that to happen, would we?"

"Doctor Phelps, you must have made some mistake, maybe you have the wrong records. I have never been to this office before, and I have never used an inhaler of any kind."

"It's Howard E. Johnson, right?"

"Right."

"You're from Pierceton, right?"

"Wrong, I'm from Leesburg. That's about twelve miles southwest of Pierceton.

"You've moved recently? We'll have to change your address on our records."

"No, I have never lived in Pierceton, never. **You must have the wrong records**."

From then on things degenerated into massive noncommunication and confusion. It was as though nothing I said registered. I had the distinct feeling he decided I was a confused old man and then went blindly on, completely ignoring my words of protestation. He continued trying to treat me for emphysema. I thought possibly he had the records of a man with my same exact name who lived in Pierceton. Later, a little research on my part confirmed my supposition.

I walked out of the office with a bag containing two inhalers, instructions on using them, and a book on how to handle emphysema. I also had an appointment for two months later. I left with a cheery admonition from the nurse, "Use those inhalers regularly. Have a relative or friend read the instructions and check on you regularly. We'll see you in April."

Not in this lifetime. I thought as I walked out.

It was my same friend, Al, the one who picked me up from the hospital, who called me several weeks later. "I see you died last Saturday," he jokingly commented.

"What the hell are you talking about?"

"Look in the Times Union obits. There's one there for Howard E. Johnson. I thought maybe that was your ghost in church, Sunday."

"That explains a little problem I had with a doctor. I must have guessed right."

"What's that all about?" Al asked.

He laughed his deep, hearty laugh when I related to him, the story of my strange visit to Dr. Phelps. Then it struck me. I needed to make a phone call.

When I explained my plan to Al, he roared with laughter. "Howard, you'll be confusing those poor people even more. Let me know how it goes."

Funny, I never did receive the usual confirmation postcard for my appointment.

Copy from one, it's plagiarism; copy from two, it's research.

—*Wilson Mizner (1876-1933)*

When ideas fail, words come in very handy.

—*Goethe (1749-1832)*

The only truly stupid question is the one that is not asked.

—*Howard Johnson, 1960*

How to Find a Wife

I had been single for nearly fifteen years after a short, disastrous second marriage. It had been a serious mistake from the day we spoke our vows. Past sixty, I was happy living alone on a lake in Northern Indiana. A year before, I had broken up with a wonderful lady after a five-year relationship. We discussed marriage, but when she was ready, I wasn't, and when I was ready, she wasn't. We remained dear friends even after we broke off our relationship. Perfectly happy living alone, I rarely experienced feelings of loneliness. With many friends, I attended the nearby Leesburg United Methodist Church regularly, and sang in the choir. Living alone and being lonely are very different things. I certainly enjoy being among people, but have also experienced many pleasant and quite comfortable times being alone. A few times I had felt the terrible loneliness that sometimes strikes one among people, even in a crowd.

One morning, early in July, I sauntered into the offices of *The Paper* in the nearby town of Milford. I was there to place an ad for a salesman for the personal computer business I had started a few months before. A scruffy-looking man of about thirty stood at the counter talking to a lady I assumed to be an ad taker. He wrestled a crumpled scrap of lined yellow paper from his pocket, carefully flattened it on the counter, and spread a crumpled five-dollar bill atop the yellow paper. Paying no attention to their conversation, I waited until the lady behind the counter turned to me when the young man left.

She smiled as she spoke, "Can I help you?"

I unfolded the ad copy produced on my computer and handed it to her. "I'd like to place this ad for a computer salesman."

"If you'll wait a few moments, I'll have a want-ad taker speak to you."

I was a bit mystified. "Aren't you an ad taker? What was the bit with the guy who spoke to you?"

She laughed, "I run the singles column. He was giving me his requirements for a lady friend. Hang on a sec. I'll get someone for you."

"Wait a minute. Tell me about it," I sputtered as she turned to go. After all, I was single and unencumbered. I wanted to know what it was all about.

She turned and grabbed a paper from a stack behind the counter. Shuffling through until she found the right page, she spread it out on the counter in front of me. "Here's last week's column. You can get an idea by looking at these. Do you think you might want to place an ad?"

After skimming through the ads I commented, "I've never even considered anything like this. How does it work?"

"As you can see, there are no addresses or phone numbers in the ads. We provide a PO box number here at the bottom of the page. An interested party simply sends a letter to that PO box with the ID number in the ad. You come and pick up those letters, and then it's up to you. The usual first meeting place is a restaurant or other safe public place. We suggest a lunch as the first meeting. Then if you both want to pursue further contact, a date can be set."

"That seems simple enough."

"This month we're having a promotional special, two weekly ads for five dollars."

My curiosity was instantly piqued. "I think I'll take you up on your special." I then wrote out the following brief ad:

D/W/M, 62, 5'10", self-employed computer engineer/scientist/poet. I love the outdoors, sailing, fishing, waterskiing, biking, fall colors, fresh snow, the smell of spring, moonlight on the water, singing, and my large, loving family. I'd like to share an active life with a like-minded, independent woman. I'm not looking for a subordinate. Write Single No. 5653 in care of *The Paper*.

After paying my five dollars and also placing my ad for a salesman with the correct person, I headed for home wondering what this momentary lapse in my usual conservative nature would bring.

About two weeks later, I picked up more than thirty letters, to be followed several weeks later by another bundle of nearly thirty more. I couldn't believe there would be so many responses to an old coot. They had a wide range of both quality and content. One was written on a single page of lined paper, in pencil, with terrible grammar, and many misspelled words. The other extreme was one of two pages, impeccably composed with near-perfect grammar and spelling, and in beautiful handwriting. The engineer in me prompted the organization of the letters, rating them from one to ten based on my impressions of the writers. Naturally, I started at the top of the list with the only one that rated above an eight. Here is the text of that letter:

July 18, 1990

Hi,

I'm Barbara. I'm a forty-nine-year-young legal secretary who's still going to school. Just finished my first of two years at IUSB in the paralegal program.

My first love has to be music (big band) and ballroom dancing. I do enjoy light classical, and I sing in my church choir.

I'm a nature lover beyond a doubt and especially enjoy listening to and watching the birds at my feeder. Needless to say, I love flowers and have them everywhere, especially those that will attract God's critters! I look forward to being out of doors as much as is possible. And really enjoy camping and don't do enough traveling! Being near or on the water gives me my tranquility.

I enjoy gourmet cooking (and preparing and eating it). I like to try new dishes and make dining an experience rather than just something one must do.

To sum up, I'm an incurable romantic and a cockeyed optimist! I have blond hair (with some silver for highlight, of course) and blue eyes. I'm 4' 11 ½" tall and weigh 107 lbs., and I meet life head-on!

If you think we might have some things in common and could be friends, I can be reached at 848-5780.

When I called the writer of this letter, her mother answered and said that Barbara was out. I gave her my name and phone number and asked her to return my call. I continued down the list until I made several dates for lunch. As a result, I met several ladies resulting in a number of interesting experiences. The first was a quite timid, longtime secretary, several years younger than I. She spoke meekly when I could get her to speak. Tall and slender, she was pleasant but had never married and was definitely not my type.

The next lady was the opposite. By the time we finished lunch, she was planning our winters in Arizona. About sixty, this farmer's widow was small, muscular, and sported a blonde beehive hairdo that added at least four inches to her height and could probably have withstood a tornado. Her years on the farm showed in her local speech idioms and calloused hands. She was a formidable presence despite her diminutive size. That lunch was our first and last contact.

I then met a pleasant, plump, attractive lady almost my age. Her clothes, the car she drove, and her general demeanor showed at least the trappings of wealth. It soon was apparent to me that she was lonely and still seriously mourned her husband who died a few years earlier. Near the end of our lunch, I realized how vulnerable she was and became concerned she might fall prey to an unscrupulous man after her money. I asked if she was active in a church or other women's group. Her negative response prompted these words I recall exactly, "To me, you seem a pleasant but vulnerable lady. I don't think you are ready for the dating game. Why don't you join a church or other organization and become active with a group of women? When you feel secure in the fellowship of those women, then you might reconsider entering the dating game."

Nearly a year later, I received a phone call from her. She said my words had been some of the best advice she'd ever received and thanked me repeatedly. She had done what I suggested and was now active in a group of church women who had become dear friends. "I don't know if I'll ever look for a man again, but thanks to your advice, my life is now much fuller and more rewarding." That was a marvelous reward for speaking to a near stranger of a genuine concern.

During this time, Barbara, the lady with the letter I placed atop my list, had been playing phone tag with me. I discovered later that, during that time, she was in the hospital having sinus surgery. It was mid-August before we connected after exchanging many phone calls for at least three weeks. We set a Sunday afternoon dinner date at the Holiday Inn in Goshen. I arrived before the appointed time and waited in the lobby. After waiting and waiting and waiting, I was wondering if I had been stood up. About half an hour after I arrived, a young lady came over from the check-in counter.

"I hope this is not a prank, but would you happen to be Howard Johnson?"

When I replied in the affirmative, she said, "There's a lady on the phone who wants to speak to you." The ensuing phone conversation went something like this:

"Hello?"

"How's your patience quotient? I'm terribly sorry, but I got lost. I'm calling you from a filling station in New Paris."

New Paris is eight or ten miles from where I was waiting. "How'd you end up in New Paris? That's not even near US 33."

"I'm not familiar with Goshen and must have taken the wrong road."

I laughed. "You probably went straight where US 33 turns left. That's why you ended up on Indiana 15."

"How far away am I?"

"My guess is about ten or twelve miles."

"I'm so sorry. It will take at least twenty minutes to get there. Do you want to cancel?"

"No, I've plenty of time. Come on over." I then gave her directions to the Holiday Inn and sat down to wait.

Nearly a half hour later, a diminutive female walked swiftly and with determination across the parking lot and into the lobby. I knew it was Barbara as soon as she stepped out of her car.

"I'm so terribly sorry," she said as soon as we met. "You must be a patient and understanding man."

After our introductory conversation, we entered the nearly deserted dining room and sat down for dinner. I don't remember much of the conversation, but we hit it off right away and our dinner stretched out for several hours. We were definitely kindred souls, delighting in each other's company. We parted reluctantly after making arrangements to see each other again.

During the next few years, we were frequent companions and met each other's families and friends. Barbara soon started attending the Leesburg UM Church with me, joining me in the choir. We became deeply committed to each other. Friendship Sunday at church, in October of 1992, I stood up in the choir loft during the "Joys and Concerns" part of the service. After a short preamble, I bravely uttered the words, "I would like to ask the little lady seated in front of me if she will marry me."

No one including the minister, my friends, or Barb knew this was coming. Don Shanks, our pleasant but often tongue-twisted minister, blurted out, "I don't know if this is a joy or a concern." Turning to Barb he added, "Is there an answer?"

Caught a bit off guard, Barb replied, "What could I say In front of all these people but yes."

With her reply, the entire congregation applauded.

Wallace Huffman, a member of the congregation, was videotaping the service that Sunday and captured the entire thing, including the proposal and acceptance. The following May 29, 1993, we were married in an unusually emotional ceremony fashioned after the Quaker wedding of my grand niece, Deanna, that we experienced and loved a few months before. Another member of the church videotaped our wedding, and, in cooperation with Wally, presented us with a tape of both events as a wedding present. In the years since then, we have often viewed that video and relived those delight filled moments.

Barb and her boys march down the aisle, Michael on her right and Adam on her left

(Below) The happy couple with Ho's family outside the church. From left to right: Jessie, Diana, Jeff, David, Barb, Howard M., Ho, Danni, Robbie, Deb, Nicole, Bob, Stephanie, and Michael.

The Calling

I heard it in fifth grade when my teacher, Mrs. McManus, banished me to the boys' room after I turned in a poem mostly plagiarized. "Howard, you know that is not your poem," she scolded. "Take your notebook and pencil to the boys' room and don't come back until you have written your own poem."

After spending serious effort for two nights, rehashing someone else's work, I put it together as my own poem. Caught in this unforgivable act at such a tender age was a powerful lesson. Alone in the boy's room, I crafted my first real poem in less than half an hour and handed it to my teacher as instructed. It won the award for the fifth grade at Taylor School and was published in the school paper. My family was proud. I never told them the whole story. I first heard the call then but ignored it.

When I was a seventeen-year-old freshman at Purdue, I was placed in a special, advanced creative English course because of the results of my orientation tests. In mid semester, my professor took me aside one day after class to talk about a story I submitted about a boy and his dog.

"You have an unusual talent," he told me. "I knew it when I saw your orientation test results. Your work in this class confirms it. I strongly recommend you switch to journalism."

At that age, I knew that I wanted to be an engineer and ignored his suggestion. Again, I heard the call but ignored it.

While I was in college, I wrote several short stories for an English composition class. That got me going and soon I was writing stories for fun. I remember one story, "A Christmas Tree for Carol," about a blind girl who miraculously recovered her sight when her bed was struck by lightning during a Christmas snow storm. I particularly liked that story and was devastated when my sister told me that it was nonsense because lightning never struck during snow storms. Absolutely certain it did, and yet never having experienced it, I could call on no personal knowledge. A few years later, I was standing by the back door of my wife's parents' home in a blinding snow storm before Christmas when a brilliant flash lit up the whole sky. The loud thunder that followed confirmed it was lightning. As I stood there watching flash after flash, I remembered that story and my sister's words. I was vindicated. I have since learned thunder snow storms are not terribly

553

rare. Again the urge to write came over me, and I wrote several stories that Christmas season. Yet again, I ignored the call.

As the years went by, I collected quotes, witticisms, and poetry I liked. Occasionally I would write an essay, poem, or short story. Each time, I would wonder if I could write something of value. During the seventies, my daughter Deb sent me a poem she wrote entitled, "Enigma." I immediately answered her with my own poem, "Epilog to Enigma." Those two poems remain among my favorites. This experience triggered a short flurry of writing which included one wild poem titled, "Sound Rainbow." My poem won an award and was published in a collection of poetry in 2000. This time the call held me for a while, but once more I turned and ignored it.

In 1981, I spent nearly a year working in the Philippines. While there, I began writing descriptive letters and experimenting with haiku, a highly condensed Japanese form of poetry that uses few words to paint a picture. It was at this time the first ideas for a novel began going through my head when going off to sleep. I tried various ideas to start weaving into a novel. Though I heard the call quite loudly then, I still ignored it.

In 1984, I bought my first computer, an IBM XT clone made by Zenith. I used it with AutoCAD software as a design tool for the dental offices I was designing for a living. It was much quicker and easier than the paper and ink I had been using. In the software I received with the system was a word processor named Word Perfect. It wasn't long before I began using Word Perfect to write stories, essays, and poetry. Several times I even began the novel I dreamed about. Still, my main effort was not in that direction. I heard the call once more but still did not listen.

Soon I was all wrapped up in a struggling computer business I started with hardly a nickel to my name. Survival was the name of the game, and all my efforts were so aimed. In 1990, I met a lovely little lady who soon became the focus of my life. Both of us were soon singing in the choir of the little Methodist Church in Leesburg. On friendship Sunday in October of 1992, I stood up in the choir loft during our sharing of joys and concerns and proposed. Incredibly, one of our members was videotaping the service, so we have a complete record of the proposal and Barbara's acceptance on tape. The following May, we were married in the same church. I continued with the computer business while Barbara became a Methodist minister and was appointed to a nearby small country church. The call kept getting louder and more persistent.

In 1999, I started writing my long-considered novel in earnest. Working from five in the morning until about nine or ten, I wrote feverishly. Those four or five hours each morning flew by as the story miraculously appeared on the computer screen before my eyes. It became much more like reading a novel than writing. New characters and circumstances appeared out of nowhere as my main characters became real people to me experiencing real-life situations. Never in my life have I enjoyed doing anything more. Well into the story, I realized one book could not tell the entire story, so I decided to make it into a trilogy. This time I heard the call loud and clear and I answered. I was hooked. Writing soon became my main passion in life.

It took eight months to write the first book and fourteen more months to rewrite it. About the time I finished the first draft, I found and joined an international group of writers named Science Fiction Novelists who critique each other's work. This wonderful group of writers, from all walks of life, provided invaluable assistance in the rewriting of my book. Without their excellent (and sometimes brutal) critiques, my rewrite would not have gone nearly so well. I was so encouraged with my writing that I sold the PC business in June of 2000 to pursue a writing career full-time. We decided that with Barb's salary as a pastor and my social security, we could survive until my writing began to pay off.

Then misfortune struck as Barbara's health suddenly began to deteriorate. Soon after I sold the business and after nearly six years spent in the pulpit, she was incapacitated by the pain and weakness of post polio syndrome. On December 31, 2000, she stepped down from the pulpit. It was the saddest day of our lives together. Not too long after this emotionally draining and terribly devastating event, she began helping me by proofing my writing. Soon she was my invaluable aid in editing, proofing, and rewriting my work. Her expertise in English and totally different nontechnical view of the world from my own, became an essential factor in my work. Without it, my writing would not be nearly as readable or grammatically correct. In short, we became a writing team.

On January 8, 2002, I composed and sent the following letter:

To my treasured family and friends:

I do believe I have discovered who and what I am.

This Christmas, my daughter Debby gave me a book entitled *For Writers Only* by Sophy Burnham. I opened it immediately and read these words by the author, "I give this book then to all writers, to all creative people, to all of us poor troubled humans who are struggling with our doubts and love. I hope that it will live in your hands until it drops, stained and dog-eared, into dust, too yellowed and frayed even for the outdoor racks of second-hand bookstores. I hope you steal it from libraries and buy it in stores to give to your sons or wives or daughters or nephews or husbands or mothers, in order to encourage them to write the stories of their hearts.

"For we all have stories. And they must be told. In telling our stories, we affirm ourselves, our being, and thereby the purpose of our creator and our lives."

This book's cover says, "Inspiring thoughts on the exquisite pain and heady joy of the writing life, from its great practitioners."

A quick glance through the first few pages fulfilled the promise of marvelous thoughts from kindred souls. If ever I knew who and what I am, I do now. This book reached my soul and prompted me to say, "I am a writer." In one of the early pages, a John Gardner quote said, "True artists, whatever smiling faces they may show you, are obsessive, driven people." As I search for more quotes to share, I realize all these writers say what I so deeply feel. Each quoted paragraph is a footnote to the man I found myself to be. However, there is one repeated comment that misses me completely. Several writers describe the pain of the empty void that comes when a work is finished, the empty mind searching for a new verbal mission. I have never experienced that emptiness.

In fact, I know I will not live long enough to empty my thoughts of valid things to put to paper. Writing my first novel was an unbelievable joy. I hated even those interruptions for biological necessities such as food, sleep, and others. Typically, I wrote during the silence of the morning from five until nine or ten. It took eight months of every moment I could spare to complete the book. Then it took another fourteen months of equal dedication for me to rewrite it. Even that was a great joy as I carefully read and re crafted phrase after phrase and paragraph after paragraph. Robert Heinlein said in one of his books on writing that a novelist was a true storyteller who read more than wrote his story. I found that to be quite true. In fact, it was an unbelievable joy to write a story that created itself as it went along. New situations, new characters, and new actions constantly appeared spontaneously as the words seemed to place themselves on the screen. Although I certainly had a rough idea of the story, the details seemed to come from nowhere, like an actual happening. This never ceased to amaze me.

When a day starts as I sit in front of that magic screen, I am never sure what is going to happen, or even what I am about to write for that matter. After finishing *Blue Shift*, I started immediately on the second book of the trilogy. A third completed, that book now waits as I struggle to work other more pressing thoughts into organized words. During the last six months, I have completed a collection of quotes, poems, comments, and short stories into a book entitled *Words from the Lakeside* as well as two collections of essays, *Thoughts on the Cultures of Today* and *The Feudals*. The first collection contains a number of essays about things we don't often think about yet which profoundly affect our lives. My book, *The Feudals*, is a collection of essays and comments about the intellectual elite who control the media and much of the extreme political left. I call them Feudals after the political system of the Middle Ages to which they seem so bent on reviving. These books are not politically correct. I have tried to write with frankness. As a result, I have doubtless written something in each of them that will please or compliment as well as offend or insult each reader. That was done incidently, not intentionally. I call myself an equal opportunity pleaser/offender.

More recently I have been writing memoirs, mostly putting on record stories of my life which I have told and retold over the years, or which have particularly strong memories. I started out with a list of eight. By the time I had written four of those, the list had grown to twenty-five. The writing of a true story requires details which, when dredged up from deep in memory, bring to mind other memories that need to be told. My excitement continues to grow almost exponentially. To paraphrase—so many stories, so little time. Like the proverbial kid in the candy store, I hardly know where to start.

As a seventeen-year-old student in an advanced composition course in my freshman year of college, I was told I had the tools to be an effective writer. My professor urged me to transfer from engineering to journalism. My mind and heart were set on becoming an engineer, so I didn't listen then. Now, his words reverberate in my head. I have no regrets for my choice, for I have had an exciting, event-filled life of many joys and marvelous experiences with friends and an incredible family. Now, in my seventies, I am truly a writer. It's all I want to do. I think, or at least hope, I am good at it, and perhaps time and good fortune will now smile on me once more. Each moment stolen from writing hurts somehow. There is never enough time to say what must be said. Each word I read written by others demands a thousand in reply. I could not possibly live long enough to say what I must say.

Even if I never sell another book, article, story, or poem, I will still be pleased with what I have done. I have two closing quotes:

To believe your own thoughts, to believe that what is true for you in your private heart is true for all men—that is genius.

-—*Ralph Waldo Emerson*

The world has no room for cowards. We must all be ready somehow to toil, to suffer, to die. And yours is not less noble because no drum beats before you when you go out to your daily battlefields, and no crowds shout your coming when you return from your daily victory and defeat.

—*Robert Louis Stevenson*

Love to you all, Howard Johnson - January 8, 2002

February 4, 2002: At this point in time, my first novel, *Blue Shift*, is about to be released. I am well into the writing of the second book of the proposed trilogy which I am hopeful will be completed within the year. I am currently putting the finishing touches on several other books, mostly nonfiction, including; *Words from the Lakeside*, a collection of quotes, comments, poems, short stories, and other writings; *Images of Pain*, comments and responses, mostly e-mail, triggered by the September 11 attack; *Thoughts on the Cultures of Today*, essays about many of the problems now facing humanity; *The Feudals*, political commentary and opinions about the challenges facing Americans. As these words are written, Barbara and I are traveling westward from Texas through New Mexico on our way to Arizona and California. I set up my laptop at each motel and work on my short stories each morning until Barbara wakes up. Even as I write, the adventures continue.

July 8, 2002: Much has happened since the last chapter. Early in February, we arrived in Visalia, California, to visit my daughter Deborah and her husband, Michael. Barbara's hands and feet were becoming more paralyzed every day as her condition steadily worsened. Thanks to the suggestion of a doctor friend of my daughters' I took Barbara to the Sansum Clinic in Santa Barbara. After emergency spinal surgery, a heart attack, and twelve weeks in a halo device, Barb was much better than she was before we left home in January and was improving each week. While here, my book came out, and we received five hundred copies that now reside in Deb and Michael's garage. After calling on bookstores in Visalia and Fresno, I held my first signing at the Magic Dragon bookstore in Visalia. There, I sold and signed eleven books. It was an exciting learning experience, not terribly successful, but certainly not a failure.

July 21, 2002: Home at last. We arrived precisely six months to the day we left home. This is an ongoing tale. While I may add to it from time to time, someone else will have to write the final paragraphs.

January 26, 2007: Since my last entry, my life has been turned upside down and inside out. I lost my precious Barbara on October 16, 2005, after a long steady decline in her health and ability to get around. During this period, I shelved my writing except for a few essays and short stories. I have also written a lot more about Barbara in several stories and essays elsewhere in this book. After a period of recovery and adjustment to life alone, I have returned to completing the book, *Words from the Lakeside*, and have finished another, a nonfiction study of our energy crisis and what to do about it titled *A Convenient Solutions*. It is my hope that both books will be published by the end of next year.

As I again look toward restarting to work on the *Blue Shift* series and several other novels, I find myself with a new lady in my life, Daphne. We are both in a period of readjustment in our lives, having each lost a dearly beloved spouse. Who knows what the future holds for us, but for now, we are adjusting and enjoying being together, one day at a time. Each day a bonus is how we look at the future.

January 18, 2009: Finally, my book on energy, *A Convenient Solution*, is about to be released, almost two years after I thought it would be. If I had ever known how much work it was going to take to get this book into print, I might never have started. Believe me, fiction is far easier to write for many reasons. I had to rewrite virtually every page of this book several times—some dozens of times. Confirming research reports with multiple sources and then recognizing the differences between opinions and facts took a Great deal of effort and great gobs of time. Thanks to Daphne's efforts for the last two years, I have softened my expressions in the book of many of my positions on debatable subjects, primarily political. Unfortunately, even being neutral on some subjects can rile some people. For example, I am quite neutral about global warming. I do not think it is a serious or human caused problem, but only because the current science on the subject is quite flimsy, ambiguous, and inadequate. Show me some definitive and conclusive data confirming human caused global warming and I'll get on the bandwagon. Until such time, I stay neutral, being neither a subjective supporter nor denier.

Also, I'm putting the finishing touches on this book, *Words from the Lakeside*. I have four other novels in progress, Blue Shift II, Carol Hughes, The Crystal Feather, and an unnamed semi-historical novel involving native Americans. Writing these novels is a Great deal more fun than writing non fiction. 'Nuff said!

September 22, 2009: I find it hard to believe that I am still working on my book on energy, *A Convenient Solution*. I keep finding reasons to make minor changes both to correct small errors and to respond to suggestions about layout or content from people whose opinions I trust. Also, new, important information about energy is constantly coming to my attention. Information that simply must be added to the content. Numerous Internet links and references must be removed as they are no longer available. While I decry the continuing delays in publication, I am thankful these changes and corrections are being made **before** the book is released. As these words are being written, the final galleys of the book are being sent electronically to my publisher. The first copies of the book should now be available in October, probably while we are away on a major trip.

This book, *Words from the Lakeside*, is also in the final edit process and might even be available before the other, something I never thought possible. I have discovered how much work must go into finishing a book and making it ready for publication, especially when the author must rely on friends and his own effort to supply the necessary finishing touches.

Though the promotion of these two books will consume much of my time for the rest of the year and well into next year, I can hardly wait to get back to work on several writing projects that were set aside, so these two books could be finished. From now on I will stick to fiction and try to complete these five other projects. They are in various stages from beginning to about 60% completed. How much easier it is to conceive of a new project than to complete one already started. I have made a promise to myself not to start another one until at least two of the ones I have already started are either finished or abandoned as not being worthy of the effort.

July 27, 2010: Will my book on energy ever be released? After an unbelievable software glitch that has cost me quite a bit of money and a lot of valuable time, my final files are once more off to the printer. I still have 450 of the original books in boxes in the garage. They have such egregious errors they cannot even be used for promotion. Nearly a hundred errors had to be carefully found and corrected, errors caused by an errant word processor whose developer says what happened was impossible. When I sent them a copy of the file with the errors, they quit talking to me and wouldn't answer my calls. Thank you Corel.

With Word Perfect continuing to generate errors on its own, I had to develop my own system for creating errorless Adobe Acrobat pdf files. It was a lot of work and delayed the release for at least three months. To make certain there was no confusion, I renamed the book, *Energy, Convenient Solutions*. I also rewrote a number of sections because of new information. I have checked each of the 280 pages in the final pdf file for errors. Although I probably missed a few, it's going to press as it is. The new name is far more search-engine friendly than *A Convenient Solution*, a big deal in the age of rapidly growing Internet commerce.

December 10, 2010: *Energy, Convenient Solutions* is finished and available in hardcover, paperback, and ebook through most popular book outlets. This book, *Words from the Lakeside*, has been approved for publication and will be available in Mid January, 2011.

For my birthday my thoughtful Daphne gave me a canvas bag with the following appropriate quotes:

> "The only end to writing is to enable the readers better to enjoy life, or better to endure it."
> —*Samuel Johnson - 1709 - 1781*

> "When the itch of literature comes over a man, nothing can cure it but the scratching of a pen."
> —*Samuel Lover - 1797 - 1868*

November 26, 2012: (my 85th birthday) I have been quite busy since my last entry in this blog. Some two dozen articles I wrote about energy and my book, *Energy, Convenient Solutions*, have been published in trade and local magazines. The Tucson Citizen, the only sizeable newspaper to review my book, published a glowing review. I couldn't have done better if I wrote it myself.

This edition of my anthology/collection, *Words from the Lakeside*, has been published replacing the first edition which is no longer available. The explanation of the changes from the original is on page 1. The SciFi short stories were combined with an equal number of new short stories. The resulting collection is was released early in 2012. *Starring*, is the name I gave this collection of short stories, most of which are SciFi. I have D. Keith Howington, a member of my Science Fiction Novelists critique group to thank for that name which I like. Everything escept the short stories that were in the original ***Words from the Lakeside***, have been combined with many new memoirs and is titled, ***Memoirs from the Lakeside***. In this book, I share a number of interesting and often exciting life experiences from my long and active life. It too was released in 2012.

I also have two non fiction projects nearing completion. One is a collection of political essays titled, ***The Feudals***. (Those on the far left and far right would probably call it fiction) The other is a collection of essays about possible solutions to many of our painful problems titled simply, ***Solutions***. The first is nearing completion while the second is at least a year away from publication.

The Crystal Feather, a very different kind of SciFi novel was released in 2012. I am proud to announce that ***The Crystal Feather***, though unfinished at the time, won first place in the Florida Writers Association Lighthouse novel contest of nearly 400 entries. Required entry was the first two chapters and a synopsis of the rest. Work on my energy book delayed the completion of my other projects more than I liked. Once it was released I returned to my first love, fiction. Hopefully I can now get down to business finishing the projects I have started.

I currently have four novel projects underway, Three are SciFi including ***Blue Shift II and III*** which are each about half completed. ***Carol Hughes***, a distinctly different SciFi novel, is about a third completed. I am also working on a historical novel about native Americans (some of my forebears) from the late 1700's up to today. In this story, I plan to use some of my grandfather's tales about our Indian ancestors. I hope I live long enough to complete these projects and even others already coursing through my head.

In October 2011, I started what for me is an entirely different type of novel, a suspense thriller. The name of this novel is ***Days of the High Morning Moon***. It has become a welcome diversion from writing memoirs or Sci-Fi, my usual venues. I may never finish it, but it has already broadened my writing skills. Writing this has also inspired me to finish the four Sci-Fi novels I have already started. Whenever I get stuck (writer's block) on whichever of those I happen to be writing, I switch over to the thriller, and that clears my mind. When I go back to my Sci-Fi novels, the block is almost always gone.

Why don't you write books people can read?

—Nora Joyce to her husband James (1882-1941)

Some editors are failed writers, but so are most writers.

—*T. S. Eliot (1888-1965)*

Criticism is prejudice made plausible.

—*Henry Louis Mencken (1880-1956)*

It is better to be quotable than to be honest.

—*Tom Stoppard*

Being on the tightrope is living; everything else is waiting.

—*Karl Wallenda*

Opportunities multiply as they are seized.

—*Sun Tzu*

A scholar who cherishes the love of comfort is not fit to be deemed a scholar.

—*Lao-Tzu (570?-490? BC)*

The best way to predict the future is to invent it.

—*Alan Kay*

Never mistake motion for action.

—*Ernest Hemingway (1899-1961)*

I contend that we are both atheists. I just believe in one fewer god than you do. When you understand why you dismiss all the other possible gods, you will understand why I dismiss yours.

—*Sir Stephen Henry Roberts (1901-1971)*

Hell is paved with good samaritans.

—*William M. Holden*

The longer I live the more I see that I am never wrong about anything, and that all the pains that I have so humbly taken to verify my notions have only wasted my time.

—*George Bernard Shaw (1856-1950)*

Silence is argument carried out by other means.

—*Ernesto Che Guevara (1928-1967)*

Father's Day 2002

As I sit, editing and adding new thoughts to my book, *Words from the Lakeside*, it is Father's Day, June 17, 2002. Barbara and I are more than two thousand miles from home, visiting my daughter, Deborah, and her husband, Michael, in Visalia, California. We have been here and in Santa Barbara since early February for a planned six-week visit. Barbara's emergency spinal surgery and subsequent heart attack kept us here during her treatment and recovery. We will remain here until early July when we plan to head for home. We are most thankful to our gracious hosts for taking such loving care of us.

Two of Michael's daughters, Emily and Amelia, and Deborah's son, David are here to celebrate the day with us. After a delicious breakfast created by Michael and Deb, I decided to do the dishes to help. Both Deb and Michael admonished me, saying, "You shouldn't be doing dishes on Father's Day."

I replied, "Since I can choose to do what I wish, I choose to do the dishes."

That simple statement prompted thoughts of Father's Days long gone: family, children—now grown and no longer children—and the incredible family of which I am a member. My daughter, Diana's, birthday was a few days ago, a week after another daughter, Melinda's, birthday. For whatever reason, I didn't manage to get hold of either of them to wish them a happy day. Over the years, I have missed many family birthdays and anniversaries, even after having the best of intentions. I think about them, plan something special—a letter or card or such—and then find the day has passed with no action. So now, I'm going to ramble on about some related (and a few unrelated) things.

I know I can't make up for missed opportunities, but this Father's Day, I am going to honor my children as a group, and in order, from youngest to oldest. If it weren't for them, I wouldn't be a father. One valuable thing I learned about children is that they belong to themselves, not to their parents. As parents, we are charged with guiding the infant into childhood, the child into a young adult, and the young adult into a full person. It's an awesome responsibility for which so many are ill trained. Parents who try to *possess* their children make a serious mistake and frequently drive them away. God loans us children for a short while, charging us with guiding them to independent adulthood and equal status. They are for us to teach, enjoy, love, and set free.

Someone asked a wise man, "How does one hold on to love?"

He replied, "Lightly, but surely, like holding a bird in your hand. Hold it too tightly and it will smother and die. Hold it too loosely and it will fly away."

I now address my treasured brood directly:

I thoroughly enjoyed you all as children. You were a tremendous joy and source of pride. Now I can enjoy all of you, and your spouses, as truly independent adults and full equals. As they reach adulthood, I am beginning to know your children, my grandchildren, in the same way. I only hope my love for you all is evident. I certainly treasure you together and individually.

One of the joys most of you has yet to experience is that of knowing your children as grown people on an equal plane, one person to another. Don't be in a rush. Try to help your children enjoy childhood by not pushing them into adult activities as so many of today's selfish parents do. Such parents do this, not for their children, but to satisfy their own needs and failings. It is one thing to encourage excellence and quite another to push a child to fulfill a parent's needs vicariously. As far as I can see, you have all avoided that pitfall. A little suggestion: treat them as equals a bit early rather than late. They may stumble a bit, so be there for support when they need it.

Parental overprotection is another pitfall children face on the road to adulthood. A little trouble, pain, disappointment, and even illness will teach a child the world can be hurtful and dangerous. Try to prevent the most serious hurts and ills when you can. An unfortunate effect of overprotection for health purposes is emerging now according to health experts. Small children overprotected from colds and other diseases as infants and children do not develop the needed resistance to disease. Those little snot-nosed urchins playing in the dirt and catching all those childhood bugs are growing up far healthier than those protected from disease exposure.

I can cite myself as an example. As a child, I had many colds, bouts with the *croup* and other childhood diseases. I also played for most of the summer in a lake that was a veritable biological soup of infectious micro biota. I am quite certain that is why I have such strong resistance to diseases of all types. In 1981, I went to the Philippines with a crew of six men far younger than I. They all were sick for the entire first week and several other times. I never had even an upset stomach the entire year we were there. I now cringe seeing mothers keeping their children out of the same lake to avoid exposing them to the biota there, and *protecting them from illness.* Even in the area of health, overprotection can do real harm.

Sorry, gang. Even now I can't help doing a bit of *preaching.* Let's say I'm trying to pass the wisdom of the ages onto those I love. Whether you agree or disagree, you are certainly capable of accepting or rejecting my suggestions and will do so according to your own determination. As a group, you have all turned out quite well. You are each and all truly individuals, your own persons. No carbon copies in this crowd. With each of you individually and to all of you together, I feel an extraordinary bond. I love each of you in a special, different way that is truly unquantifiable. I take great pleasure in the particular relationship I have with each of you. I only regret our times together are so short, but maybe that makes them all the more precious. Hopefully, we will have many more joyous times together.

Certainly, I don't always agree with what you do. That's a father's prerogative, but you rightfully do not have to agree with all I do either. Sometimes you tell me about it in no uncertain terms. That's a right you have inherently which I grant you with boundless joy and satisfaction. To have six out of six children brought up in the last fifty years all turn out successful—with none destroyed by drugs, booze, or other evils—is short of miraculous. Hopefully, I had some part in that along with your mother and others in our family. It wasn't all blind luck, of that I am sure. I am so proud of whom you are and what you have done. Remember, most of what you have become is due to your own determination and hard work. We may have given you an early boost in the right direction, but you soon took the reins and made it on your own.

I want to say a special **thank you** for the Memorial Day weekend when three of you girls and your husbands came down and painted our house. That was such a delightful gift. Not the painting, but the time we all spent together. We certainly had a blast. That was one of the happiest times I can remember at Tippy. Mike, Stephanie, Mindy, and Joe—we all missed you, and you missed a truly enjoyable party. Now I will address you individually.

Melinda Jean, my baby, since I was the baby in my generation of kids, I know a bit about older siblings and being the 'baby.' Mindy, I think you outgrew that status long ago. Now you are a mother with a bright young son and a delightful new baby daughter. You never forget a birthday or holiday, sending loving remembrances to Barb and me by card or e-mail. Please know they are greatly appreciated. I see careful motherhood showing in your son, Joseph. Loving care always helps to build a small child into a caring adult. I know Chantel will be a joy to you and Joe and a loving sibling to Joseph in years to come.

I have always felt you were somehow shortchanged in your childhood by the problems that ultimately ended in your parent's divorce. Even though your mom and I were both careful not to use you children as agents in our problems, you had to have suffered some misuse and neglect. I know at times we were so engrossed in our own difficulties that we didn't spend the time we should have with you. In spite of this, you were a delightful and adorable child. To my diminished joy, I sorely missed so much of your growing up. This did not lessen my sense of responsibility toward you, or my caring.

Diana, you are my wild one, the *Auntie Mame* of our little family, full of energy, enthusiasm, and the joy of living. How worried I was of where you were headed as a teen.

How you soon turned that worry into pride when you started at Kent State and then chose a career. How kind, loving, and mature you have become. I remember so well those trips in the maroon van with so many of your friends as young teens. I so enjoyed those times. No chore was more joyfully undertaken. I'll never forget the buzz and happy squeals of a dozen or more young girls packed in that van as we tooled through the streets on the way to this dance or that party.

You grew to be the one always willing to speak out. There is never any doubt where you stand on most things. Your honesty and forthright opinions can be counted on. Even when we disagree you are never afraid to *let me have it*, something everyone needs on occasion. I may not always agree with you, but I treasure the fact that you are not afraid to speak your mind. I can always count on a frank opinion no matter how *off-the-wall* it may seem to be.

Remember when we met in Hawaii—the tour of the ship? Barb and I often reminisce about our dancing on the wharf to the music from the ship. We danced from when we waved goodbye until the ship's music could no longer be heard after you sailed off. No father could be more proud of his child. How Barb and I enjoyed staying at your little house on some of our visits to Michigan. How we treasure those rare visits to the lake, from the joy of arrival to the tears of goodbye. You must know how much you are loved.

Roberta, my quiet genius. I know you don't understand my saying that, but you do have a fabulous mind whether you realize it or not. How worried I was when you dropped out of high school. I never knew, nor do I ever need to know the real reason. All I knew was that you were hurting, and I didn't know how to help. You were so quiet about things that troubled you, at least to me. I worried about you for a long time, even as a young woman. Through all of that, you were so kind, loving, and affectionate.

When you were about ten, you were the cause of one of my most terrifying moments. I am quite certain you remember the time I accidentally started that old boat and thought I might have run over you. You had been playing in the water right beside the pier moments before. I remember standing up in the boat after it stopped, screaming for you at the top of my lungs. When you came running out of the cottage asking what's wrong, I collapsed on the seat in tears. Today I can still feel that terrible fear whenever I recall the incident.

Now you are a successful mother with three darling daughters and a caring husband. How we enjoy our holiday visits and those times when you all come to the lake. How quickly your three girls will grow up and be off into the world. Treasure the time you have with them. These will be the happiest moments of your lives. Hopefully, you will enjoy adult relationships with them as much as I do with each of you.

Howard—Mike—Noward, my only son and namesake. Oppressor of your younger sisters, victim of an older one. You had such a rough time growing up, ha! I only had one worry as you grew from child to teen—that was, were you going to survive. I have many memories: doing a cartwheel in the hall of the hospital upon hearing I had a son, a photo of a tiny boy in a knit blue and yellow hat, filling the DeSoto gas tank with water, climbing down the trellis—up the trellis after meeting my frat paddle, a totally devastated room suddenly clean and neat, unplugging the paper plugged toilet, fishing in the little boat many times, turtle hunts in grassy creek, and countless others.

Later years brought more joys and memories: two more grandsons—an Air Force pilot, a U2 pilot—and finally, a Delta captain and another grandson. Visits were few but joyous. One of the happiest times and fondest memories of my life was the year we worked together renovating the old cottage. That was a special time working with you. It was a time when we came to know each other as men—a treasured experience for us both. While this was a father and son working together, it was even more—two friends working with each other on a common project.

A wonderful son, so are you a great father with three sons of whom to be proud.

I can say about you that which I cannot say about any of your sisters, you are my favorite son. One thing I can say in all seriousness, you are certainly not pompous or a stuffed shirt. I see you wear your achievements humbly and without a trace of snobbery. It is a real treasure for me to see my son as a real person, a bit chauvinistic, but quite real. You certainly didn't expect to get nothing but compliments, did you? I may love you, but I'm not blind.

Deborah, my firstborn and senior member of the Johnson mob. I remember the day you were born: riding on the elevator in the hospital with your mother holding you. You were positively the most beautiful sight I had ever seen. You were beautiful. Your mother was also radiantly beautiful. I was totally unprepared for that experience. Other memories come to mind: crawling on the grass in Grandma Osborn's backyard, tears after bumping your forehead on the kitchen table on Nipomo, fingernail polish on a painted gray wall, a first day at school, dance lessons and recitals, waterskiing and joyous days with cousins and siblings at the lake, a young woman stepping down the stairs on first high heels (that one brought tears), a young you with your brother jumping the waves in your new outboard speedboat. These are but a few.

Then there was college, dental hygiene school, a professional at the Chicago Dental meeting, my first grandchild, another grandson, many visits to Bowling Green, special sharing visits at the lake, big life changes, bringing an old friend into the family, several moves ending in California. Now a teacher of dental hygiene and master's student—busy, busy, busy. Can you know how proud I am of you? A kind, caring daughter with two terrific sons, mostly grown, and of whom you can be justly proud.

I can see in you much of the women whose genes you share, your mom's and your grandmothers'. I see also the special person you have made of yourself. You have matured famously and with a great deal of class. I thank you especially for the care you have given Barb and me during our extended stay for her convalescence.

To the spouses of my children, I have many pleasant memories of all of you brought into our family. You were all clever and lucky to become part of our family. Maybe you think it is the other way around. No matter, we're all one friendly, loving family. You have shared with and loved those I love. You have contributed to my being a grandfather, welcomed me into your homes, shared your children, shared your time—in short, you also became one of my family. You all have a *mano a mano* relationship with me.

To all my grandchildren, from a few adults to a few infants, you also are a joy and source of pride. I pray that when you reach my age, you look on your life as joy-filled and wonderful. Rather than give advice, I am giving each of you my book ***Memoirs from the Lakeside*** and asking that you read it when old enough to understand. It contains a collection of sayings I consider wise and collected over most of my lifetime. Most are the words of others; a few are my own thoughts. It also contains a few short stories of experiences I had, most as a child or young man. Should you read this and not have the book, ask me for one. If I am no longer around, ask your parents. They'll find a copy.

For all of you, I remember birthdays, Christmases, weddings, games, parties, trips, boat rides, waterskiing, fishing, swimming, fun with friends and cousins, family meals around dinner and breakfast tables, times of quiet, times of great joy, and times of serious interaction. My bank of memories is overflowing with wonderfully joyous memories. No father ever more enjoyed his progeny than I. No father was ever more proud. You have made major contributions to my wonderfully joyful and rewarding life. So on this Father's Day, I honor each and every one of you. After all, if it weren't for you, I wouldn't be a father, or a grandfather. Thanks!

When my wife, Barbara, threw a huge surprise party for my seventy-fifth birthday, it overwhelmed me with joy. The party was held at the Lion's Meeting Hall in Leesburg where the ladies from church prepared the food. Many friends and family came along with my five children and six of my ten grandchildren. I can't believe how much I look like my grandfather Dickinson in the photo on the next page.

NOTE: When I wrote this letter, my daughter, Kristen, and I were estranged from each other. Because of this she was not mentioned in the above. In spite of this, she was sorely missed and loved in absentia. There is much about Kristen in later parts of this book . It was written with the boundless joy of a father reunited with a lost daughter at long last. See the chapter about Kristen starting on page 625.

My five children:

Deborah, Diana, Howard Michael, Melinda, and Roberta—standing behind me at my seventy-fifth birthday party.

Life, as inevitable as the universe itself, is defined as one thing only—that which can change its environment through the controlled use of energy. This seems a cold definition of life, but it could be that what we seek is a direction rather than a definition. There has only been one direction to life since its inception.

—Howard Johnson, 1961

Six of my ten grandchildren are in this photo. They are the ones that attended my seventy-fifth birthday party.

Back row: Howard Michael's boys, Daniel and Michael.

Front row: Roberta's twins Danielle and Jessica with sister, Nicole, and Melinda's Joseph.

Missing are Howard Russell (HoJo IV); Deb's boys, Jeffrey and David, and Melinda's new baby, Chantel.

A Halo for Barb: Barb's Miracle

Shortly after Christmas 2001, Barb and I were to leave to visit family and friends in Texas and Arizona on our way to a six-week visit with my daughter Deborah and her husband, Michael, in Visalia, California. A year earlier, my wife, Barbara, a Methodist pastor, had stepped down from the pulpit because of the effects of post polio syndrome. Barb was physically devastated by this strange disease which strikes many who had polio thirty to fifty years earlier. These polio survivors suddenly begin to experience severe muscle weakness and body-wracking pain requiring much rest and the use of a walker or wheelchair to get around. Her condition quickly worsened during the summer and fall of 2000.

As the year came to a close, she could no longer manage the demands of the profession she so loved. December 30 was her last Sunday in the pulpit. There was not a dry eye in the entire congregation of the little country church at the end of the service. It was a tear-filled, gut-wrenching, yet wonderfully loving experience neither of us will ever forget.

During the following year, we had to learn to deal with this devastating problem. It had changed a vivacious, independent, physically active woman into one dependent on a walker or wheelchair to get around. She had to depend on others to do much of what she previously enjoyed doing. It was a major adjustment for both of us. As the year ended, we looked forward to enjoying the trip we had planned for so long.

After our family Christmas visits, Barb was not feeling well, complaining of pain in her upper back and unusual weakness in her right leg and hand. We talked of cancelling our trip. After a visit to our family physician and a cortisone shot with little effect, we sat and talked about it. We delayed our departure about two weeks during which time she showed little improvement. Finally, she said to me, "Let's go ahead and trust in the Lord to help me through this." On January 21, 2002, we packed our big Mercury sedan, loaded Barb's wheelchair and walker into the trunk, hitched up our tiny cargo trailer packed with our other travel gear and clothing, and loaded Rags, our little black-and-white Shih Tzu dog into the pen I built in the backseat. Then we headed out from our home on Lake Tippecanoe in Northern Indiana. Our first stay would be with my son, Howard; his wife, Stephanie, and their son, Michael, in Colleyville, Texas.

Because of the limited time Barb could travel without severe weakness and pains the following day, we could manage no more than four hundred miles of travel daily. It took us

571

four days to drive to Texas. During a week of relaxation and wonderful visits, we managed to see Michael play a basketball game and watch Stephanie play racquetball. We drove to Waco to visit our dear friend, Bettye Baker, who had lost her husband, Ken, to cancer the preceding year. Ken and Bettye were members of the church Barb served and had become close friends. During this time, Barb became progressively weaker. She could no longer walk with her walker and was now limited to her wheelchair.

When we reached Arizona, we missed connections with family in both Tucson and Phoenix because of our delays in starting. After a visit to the Grand Canyon—Barb's first—we drove directly to Visalia. There, Deb and Mike gave up their first-floor master's bedroom to us since stairs were an impossible barrier for Barb. We settled in and even took a drive up the mountains to Sequoia National Park where I wheeled Barb through snow and up paths to see the Giant Sequoias. By this time, Barb was experiencing serious difficulties and could not even stand on her own. Her right hand was clenched tightly into a fist which she could not open. We knew something was seriously wrong and had to get her to a hospital or clinic soon. Deb's neighbor, an emergency room physician, was contacted and suggested the Sansum Clinic in Santa Barbara would be the place to take her. The people at the clinic were kind enough to arrange for an initial exam within a few days.

The following Tuesday, we drove the 250 miles to Santa Barbara only to find that the doctor who was to see Barb was ill and unavailable. The staff at the Sansum Clinic treated us with kindness and consideration. They, arranged for another physician to stay late and start the exam. After two days of tests, we headed back to Visalia with several more tests and an MRI scheduled for the next Tuesday. By this time, Barb had deteriorated to the point where she could not get out of her wheelchair without assistance. She could barely even feed herself as her left hand was beginning to close like her right. Through all her pain and paralysis, Barb remained calm and pleasant. "I know God will take care of me. I don't think he's through with me yet," she would say. Her indomitable spirit and inner strength certainly helped the rest of us handle her problems.

Early the following week, we returned to Santa Barbara and two more days of tests and MRIs. When all tests were completed, we went for a report to the office of the neurologist at the clinic. He displayed the MRI of her cervical spine, so we could see what it showed. He carefully pointed out her problem. Even our untrained eyes could see her spinal column was severely compressed from its normal round, nearly half-inch diameter to a thin crescent less than an eighth of an inch thick at the widest part. "You're going to see a neurosurgeon NOW," he announced as he got on the phone. Within a few minutes, I was wheeling her out of the clinic, down and across the street and into the office of the neurosurgeon, Dr. Thomas Jones.

After viewing the computer screen showing the MRI, Dr. Jones repeated the neurologist's explanation. "We're going to schedule you for emergency surgery as soon as possible," Dr. Jones said as he phoned the hospital to make arrangements. I then wheeled Barb out of his office, across the street, and into the Santa Barbara Cottage Hospital. Because she had eaten lunch, they couldn't do the surgery until the following day, so she was checked into the hospital where she would stay until her surgery. I was informed that her condition was quite precarious and dangerous. How much, if any, of the use of her hands and feet she would regain was unknown. It was even possible none of it would come back, but the doctor was hopeful she would recover fully. Time alone would tell after the two scheduled surgeries were completed.

I was taken into her room to see her soon after she came out of recovery from the first surgery. She looked so pale and weak, her neck in a plastic collar, her head softly cradled in the slightly elevated pillow. As soon as she saw me, she smiled, raised her right hand, and said, "Look!" Then she opened her hand for the first time in weeks. "God works miracles," she said softly. Tears flowed and prayers of thanks were given by both of us. After several moments of emotional struggling, we were able to talk. That little smile and soft voice spoke volumes. For the first time in many weeks, I knew my little lady was going to be all right.

During the week before her second surgery, she recovered well, steadily gaining more use of her hands. After the second surgery, we were told she would have to wear a halo device for several weeks to hold her neck in place while the plate, bone graft, and other changes healed. It was done while I had returned to Visalia for a few days to take care of some necessities there and return with fresh clothes and other items. The story of the installation of the halo device is best told in her own words.

"I was peacefully resting after breakfast when my doctor and two men suddenly entered the room. They sat me up, made several quick measurements, and began assembling this contraption on my chest, back, and around my neck in a flurry of activity. The four-pointed screws that were driven through my skin and into my skull were the only painful part. The slight crunching sound as the screws were tightened was almost worse than the pain which soon subsided. As suddenly as it began, the whirlwind of activity was over. After some instructions on care and behavior, they were gone, and I was left with my head locked rigidly in this black metal halo bound tightly around my chest. How on earth could I possibly sleep? My head couldn't even reach the pillow as it was suspended in this four-pointed vise."

The truly amazing thing was her high spirits. Driven by an indomitable faith, Barb fairly glowed to all who came to see her. With kind words, high hopes, prayers, and a bubbling sense of humor, she raised the spirits of all who entered her room. The limits imposed by the halo only slowed her down a little. She soon had a repertoire of funny quips about the device which she shared with all who entered her room. It wasn't long before hospital personnel were stopping to see her as they were going off duty or before coming on. She brightened the day for all who stopped by with her prayers, humor, and infectious pleasantness. I heard no words from her of gloom, despair, or complaint about her condition. Her only complaints were about poor service, tasteless food, or physical inconveniences, and those were few and far between. Mostly, the staff, services, and conditions in the hospital were outstanding.

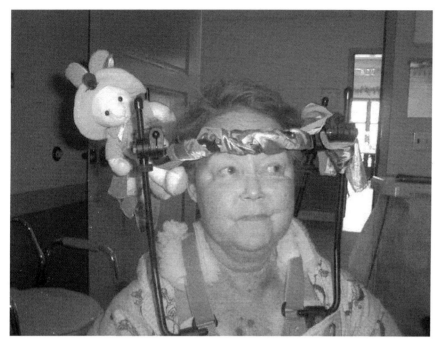

Barb decorates her halo device during her stay in the Santa Barbara Rehabilitation Institute.

After about two weeks, Barb was transferred to the rehabilitation section of the hospital where she was to remain until she was able to move to the Rehabilitation Institute. While there, she was able to have our little Shih Tzu, Rags, as a visitor. This raised both of their spirits as Rags had missed his "mom." One morning about ten days after Barb was moved to the rehabilitation section, I was awakened at four in the morning by a call from the hospital. They explained Barb had experienced a heart attack, was doing all right, and wanted to speak to me. I heard the following directions given in a soft, weak voice, "I'm okay, so don't rush down here. Get up at your usual time, take Rags for a walk, have your breakfast, and then come to the hospital." Yeah! Right! It took me about ten minutes to dress, pop a surprised Rags into the car, and get to the cardiac unit at the hospital.

There they informed me Barb was doing well, was in surgery, and that I should stay in the waiting room by the elevator. "Your wife will be brought from the elevator right past you when they're finished. You can't miss her," the nurse explained over the phone. For the next three hours, I sat unaccompanied, but certainly not alone in that cold, empty room across from the elevators. During those hours and with only an occasional interruption as someone came or went on the elevators, I had a continuous conversation with the man upstairs. It was not quite prayers, but a knockdown drag-out discussion about life in general and Barb in particular. I had much to say—no complaints mind you—just a whole lot of thoughts and feelings poured out as they overflowed my heart and mind. When the elevator door opened, and Barb emerged on a Gurney, I was extremely relieved to see she was still wearing the halo. That meant they had not done heart surgery.

I kissed her and heard her say, "See, I'm still here. I told you he is not done with me yet." As they wheeled her into the cardiac recovery section, the surgeon took me aside to explain what had happened.

"She took a major hit," he explained. "We placed a stent in the main artery feeding the entire lower part of her heart. Her arteries are so tiny even the smallest one was difficult to place. We looked at the rest of her heart and found all her arteries clear and without any plaque."

At four foot ten, with tiny feet and hands, most of Barb's physical proportions are on a small scale. Apparently a blood clot, possibly because of her transfusions, had lodged in and blocked the main coronary artery. Fortunately, she was treated quickly and though the attack was a major one we were told she would probably have a complete recovery. I continued to be amazed at how calm, pleasant, and hopeful she remained even after this serious setback.

After a brief stay in the cardiac unit, she was moved to the cardiac rehabilitation section where therapy for her surgery would continue and new therapy for her heart would be started. It was the same story wherever she went in the hospital. Barb positively radiated hope, peace, and high spirits with her enthusiastic Christian faith.

After two weeks in cardiac rehab, Barb was transferred to the Santa Barbara Rehabilitation Institute, a few blocks from the hospital. There she was to undergo physical therapy, occupational therapy, and periodic exams for several more weeks until her halo could be removed and she could go home. While there, she used her walker more and more, going on walks around the building and attending meals in the dining room. She was soon raising the spirits of other patients with her pleasant smile and hopeful words. We played gin rummy in the dining room and outside on a patio when it was warm enough and not too windy.

Six weeks after her initial surgery, Barb and I went to her neurosurgeon to see if the halo device could be removed. We went to his office with the latest X-ray of her neck in hand. After studying the film, he explained her spine had not grown enough bone around the graft, and she would have to wear the halo for at least six more weeks. Barb was deeply disappointed as she was being discharged the next day and would have to wear the halo back to Visalia. This meant at least a six-week delay in our returning home.

On my fifteenth return trip to Visalia from Santa Barbara, my little lady was with me for the multi stop, leisurely drive. Not knowing how she would handle the drive with the halo, we planned frequent rest stops and a lunch en route. The trip went better than expected, and we pulled into Deb and Mike's driveway late in the afternoon. The Sierras were a beautiful backdrop as we drove the last quarter mile to their house.

In the weeks that followed, Barb walked more and more with her walker, finally covering nearly a quarter mile at a time. By Mothers' Day, she was able to join us for a drive into the foothills for a celebration meal at a lovely little restaurant beside a swiftly flowing river. By this time the weather was quite warm and pleasant, and we spent a great deal of time out on the patio playing gin and talking. Six weeks after returning to Visalia, we again headed for Santa Barbara for final checkups at the clinic and hopefully removal of the halo.

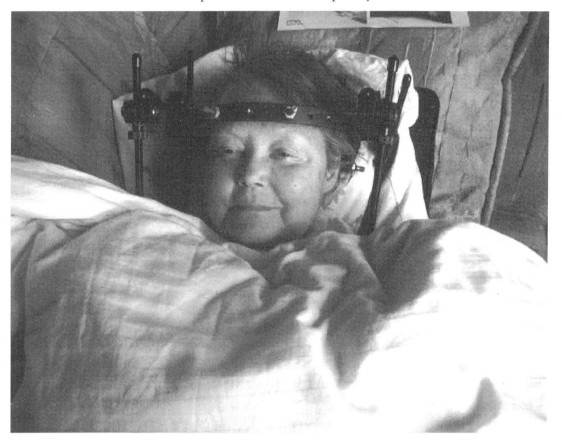

Barb trying to keep warm in Deb's cool house in Visalia. She would get out of the halo device in about two weeks.

Removal of the halo was quite painful for the few seconds it took to loosen the screws. I had to steady Barb's head as they were loosened and the device was carefully disassembled. Finally! Freedom at last. We stayed an extra day to see her cardiologist and have blood work completed.

Medical necessities finished, we took a leisurely sightseeing tour up the California coast on Highway 1 and visited San Francisco before returning to Visalia. I purchased one of those fold-up steel and plastic cots which Barb would use to lie down and rest periodically during our trip. We planned to stop for an hour for her to rest horizontally about every hundred miles. Being in the car for more than three hours without horizontal rest caused Barb considerable discomfort for the entire day that followed. These planned stops would enable us to travel farther each day. This became our guide for the entire motor trip home in July.

Recalling Christmas Memories, 2004

It's four in the morning, Monday, December 27, and I am sitting in Buford (our motor home) looking out the windshield at our silent, darkened campground. Our wide horizontal dashboard is draped with cotton "snow" while several scenes Barb's careful, loving hands have placed display the season. There is a manger scene Barb has had forever with a huge new star hanging directly above it. There are trees and carolers, angels and candles, a musical Christmas lighthouse (thanks Fees), a musical Christmas clock (thanks Bob and Bobbie), a metal angel festooned with flowers (thanks Claudia) and holly leaves, a large fabric Santa with ho-ho-ho facing outside (thanks Mom True), and two soft angel bears (thanks Fee girls), one on each side of the windshield. We miss our Dickens village that previously brightened our living room along with our brightly lit and decorated tree, but mobility has its limits. Lifestyle changes can leave many treasured experiences behind. Fortunately we have photos and memories.

Buford's dash with the decorations described above.

Notice the driveway of our winter home through the windshield.

Over the last month or so, I oftentimes thought about the wonderful treasure trove of Christmas memories I have collected throughout the years. I'm sure it's a melding of several early Christmases, but my first fuzzy memory is of coming down the stairs Christmas morning and seeing the brightly lit blue spruce in our living room. After Daddy had gone down to "check to see if all is well" and turn on the lights, I went first as the smallest with Bobbie and Sister following close behind. I called my sister Lois "Sister" during most of my childhood, hardly knowing her as Lois for years. Wide-eyed with wonder, we took down our stockings and spilled out the wonders the strange round and bumpy lumps promised. There were oranges, tangerines, Brazil nuts, walnuts, almonds, hazelnuts, and pecans all crying to be peeled or broken open and eaten. Each stocking also held a few small presents among the edibles. While we were permitted to open the small presents in our stockings and to play with those presents open and not wrapped in the brightly bundled packages under the tree, the major package opening would wait until after breakfast.

Oh! That agonizing wait as we contemplated the wonders hidden in those brightly colored packages stacked beneath the tree. It was almost too much for a small boy. Somehow we all endured through our traditional family Christmas breakfast of fried oysters, eggs, and toast as anxious excitement steadily grew. Finally, the magical moment would come as we gathered 'round the tree and Daddy passed out the presents one by one, carefully reading each tag. "This one's from Mother. This one's from Sister. This is from Bobbie. This one's from Santa." There were always a number of presents from Santa who also had filled our stockings. I always wondered how he got those presents down our chimney and out through our tiny fireplace. There was no wild frenzy of everyone opening at once as Mother directed us each in turn to open one present and give a proper thank you to the giver. Ribbons were gathered up, and papers were carefully folded to be saved for next year. Depression Christmases taught us to save and recycle all we could. There were no black plastic bags filled with hastily ripped-off paper back then in the early thirties.

The first toy I remember was a shiny, red painted, metal race car with headlights that lit when a lever was moved to connect the lights to the battery inside the car. The headlights were two flashlight-type screw-in bulbs that were shaped like real headlights. I clearly recall carefully placing the car on the floor in the darkest spot I could find under the piano. I then moved back and admired my wonderful race car, imagining it racing around the Indianapolis speedway track. There were many other toys that thrilled a small boy: a game with a set of bright green marbles, a sand derrick that lifted sand in tiny metal buckets and dumped it into a bin that opened to fill a dump truck, a tiny bright red die-cast metal Duesenberg Roadster, my "Duzy," and other metal and rubber cars. Then there was the dream of every boy, coming down to find a Lionel electric train running on a figure 8 track beneath the Christmas tree. Years later I would hear stories of Daddy and his brother, my uncle Chase, waiting for me to go to bed in the days before that Christmas, so they could set up my train and play with it, "to make sure it worked right." I still have that train all packed away in its original box.

The real joys of the season were the wonderful family gatherings. There was always the visit to the Ramseys; my aunt Nell, Daddy's sister; Uncle Alfred, and cousins Carol, David, and Laura Lee. Several years older, David always showed some wondrous electric toys to my rapt attention. Christmas Eve was the Dickinson family party at Grandma and Grandpa

Dick's. How different was their open-branched pine tree from our blue spruce. I only remember one present from those parties, a beautiful metal speedboat with a windup engine that drove a propeller that moved the boat in the water. A gift from my uncle Harry, I would remember it several summers later, the spring motor rusted and useless, the propeller gone, the paint chipped and streaked. To a small boy in the water, it was still a great toy boat to be run through the water by hand accompanied by the motor sound *rrrrrrrr* from my lips.

After the gift exchange and food, we all would go out caroling in the neighborhood. We had our own instrumental accompaniment with Harry's horn, cousin Dale's trombone, cousin Donna's accordion, and Uncle Oscar's trumpet. At least that's how I remember it. Those family occasions were missed after we moved away to Cleveland in 1936. I believe we may have gotten back for one Christmas several years later, but I'm not sure. I did have one more Christmas visit to Grandma and Grandpa Dick's during my first year at Purdue when there wasn't time for a trip home.

Another memorable Christmas was in 1940 when we were suddenly awakened by snowballs hitting our windows at three in the morning. Harold, Lois, and two tiny babies bundled up in baskets had driven from Indiana through the snowy night to surprise us. We were all overwhelmed with joy, thankful they had made the trip safely in their Plymouth sedan. That was a truly joyous Christmas at our house on Blanch Avenue, the only Christmas we would spend there. I received two memorable gifts that year, a target air rifle that shot safe, marble-sized cork balls for indoor target practice (a BB gun was out of the question), and a battery-powered telegraph set with two terminals connected by wires with green and white waxed string insulation. Bobbie and I soon learned Morse code and spent hours telegraphing each other between her room on the third floor and my seat at the bottom of the stairs. We still try to reach each other with messages, sometimes successfully.

Another special Christmas memory was the Heights a Capella Choir Christmas concert. Bobbie sang in the choir with her friend Dottie Donner, and I enjoyed it so much and was so proud that my sister was in the choir. Even though I was six years younger, I had a terrible crush on a little blonde girl who stood in the front row. I don't even remember her name, but I do remember thinking she was about the most beautiful girl in the world.

I remember one Christmas when we went to Indiana. We drove at night through swirling snow for many hours. We used gasoline ration stamps saved up for the occasion to obtain enough fuel. The thirty-five miles per hour war-time speed limit was about as fast as Pop wanted to go on the slippery, snow-covered roads, so it took a long time to get there. I don't remember if Bobbie was with us on that trip or not.

Then there were Christmases on Idlewood when I was delivering the Plain Dealer through the snow and during which time Bobbie married Bob Grimm and moved away. When I started at Heights, one of my dreams was to sing in the choir. Even though I played an instrument and was in band and orchestra at Roxboro, I gave that up and joined the chorus, a prerequisite to being in "the choir." When I auditioned and earned a spot in the second tenor section, I was thrilled, knowing I would be in the next Christmas concert. Another choir member was Dolores Osborn, a beautiful girl I had a terrible crush on since the eighth grade. In fact, I had decided in junior high that I was going to marry her. Though I had taken her to the movies a few times, she was not interested in me. I had given up even asking her for dates and was reduced to merely admiring her from afar.

December 17, 1943—everything changed. The choir was to sing downtown at Higbee's department store, an annual event. Given permission to drive the family car, I asked several members of the choir including one girl who lived near Dolores. When she asked if I had room for one more and told me Dolores needed a ride, I, of course, agreed. After the performance, the choir was treated to a short party with refreshments in a room at Higbee's before we headed home. I sat with Dolores during the party where she did not seem so standoffish as before. As I took everyone home, I happened to end with Dolores as my last passenger. We spent the next two hours in the car in front of her house. I told her how I had felt about her since we met in the eighth grade. Before long we shared our first kiss. Much later, when I walked her to her door, my feet wouldn't touch the ground, my mind was whirling, uncontrolled, and I was deliriously in love.

At the choir Christmas party after the concert, I gave her a card with the words "A penny for your thoughts" and enclosed a bright new penny. Sometime later, she returned the favor and the penny. I still have that penny. We were to share many Christmases together. We would also walk onto that stage together at Heights as choir alumni many times, sing "Emitte Spiritum Tuum," and share memories with choir friends. We managed to do that one last time in 1996 at our fiftieth class reunion with about forty choir alums.

Then there was my first year at Purdue. Because of the wartime schedule of classes, we had only Christmas Day off, so there wasn't time for me to go home. I wondered how Mom and Pop felt with all their children gone at Christmas for the first time. It was a lonely one for me, but I did manage to get to Indianapolis to spend part of Christmas with Grandma and Granddad Dick. The other years at Purdue, the trip home for Christmas vacation was a major event of warmly happy visits with family and friends. Christmas 1946 was special. After hitching a ride home with some friends, Mom, Pop, and I took a train trip to Rochester, New York, to spend Christmas with Bobbie and Bob.

I remember the steam swirling alongside the train in the bitter cold and then a long delay in Batavia, New York—something about a switch being frozen. I remember walks through the snow-covered campus of Colgate, playing cards in the apartment and the warmth of a loving family. Bobbie and Bob taught me to play bridge that Christmas. It started in me a passion for the game which I played every chance I had all through college, including many all-night sessions.

I remember other Christmases at the farm with the Swensons and their two adorable little identical twin girls. No single Christmas stands out, but I remember many happy times. There was one moonlit walk through deep, powdery snow down the lane and up a tiny frozen creek to the woods. That mental picture of Lois and Harold playing with me in the snow is vivid.

Christmas 1949 was another memorable one. Our first Christmas together as man and wife, Dee and I had a bedroom in the home of a Lutheran minister and his family in Oak Park, Illinois, the Reverend Lorand V. Johnson, his wife, and three children. They were wonderful people and immediately became like family. Our kitchen/dining room was in their basement. We cooked on one of those old cast-iron laundry double burners most of you have never seen, used a card table set for meals and boxes for our dishes and utensils. I hung a green burlap curtain between our dining room and the rest of the basement to make things a bit more pleasant. I was going to paint the concrete floor, but the Johnsons gave us an old carpet to use to warm things up. We had no car, so we walked everywhere and brought groceries home in a small two-wheeled cart. The El and bus took us to work and out for entertainment. We hung a few Christmas tree lights around the basement for decoration.

The next Christmas I remember was in 1951 when we were living with Dee's folks on Saybrook Road in University Heights. I had taken a job with Lubrizol in Wickliffe, and we had moved back home with our tiny baby, Debbie. There were presents under a beautiful tree in their living room, and my folks joined us for Christmas dinner at the Osborns'.

Christmas 1952, we were in a small rented house on Nipomo Street in Long Beach, California. We had a tiny tree and took our little Debbie up to Mt. Baldy to see the snow. I remember her asking, "Who put all that white stuff out there?" I don't remember Christmas 1953, but Dee was pregnant, and I think we drove to Tucson to visit the Swensons.

Christmas 1954, our Christmas card was a photo of Deb and Mike in front of the fireplace in our rented duplex at 1111 Eddy Road in East Cleveland. It was our first Christmas after moving back to Cleveland to join my Dad at Johnson Stipher for the next twenty years. Once more we had a joyous family Christmas dinner at the Osborns' with my folks.

Christmas memories during those years were not so vivid, blurring together into a montage. I remember several Christmases: a Christmas dinner on Curry Drive in Lyndhurst with two little ones bundled up in snow suits playing outside the front window; an aluminum Christmas tree with pink ornaments in our home on Hawthorne Drive in Mayfield Heights, the first home of our own; a tiny round-faced Robbie in a canvas and wrought-iron wing chair; Deb and Mike helping hang lights in the shrubbery in our front yard; a tiny Dee Dee reaching for ornaments from the silver tree; and several Heights Choir Christmas concerts where we sang "Emitte" with a dwindling number of high school friends. During those years, I remember a happy, joyful family Christmas with the Grimms in Erie. Obstreperous cousins joined parents and Johnson grandparents in a marvelous Christmas dinner and celebration. I believe that was in 1957.

Christmas '63 found us moved into our big house on Lynn Park in Cleveland Heights. Once more, there are memories that are blurred in time and sequence. I still see tall Christmas trees in the front window; a stack of soup cans; Dee sleeping on the couch wrapped up in her robe after decorating and wrapping all night; Our loveable little beagle, Gussie, tearing into her wrapped toys; friends and family visiting; boisterous dinners with four, then five happy children, friends, and family; our fifth child, Mindy, taking her place in the family; a wide-eyed Mindy hearing a recording of Santa saying "ho-ho-ho" as he took his cookies and milk.

Then there were some not so happy Christmases as our marriage fell apart, and our lives took different paths. The next merry Christmases I remember were in Lagrange, Illinois, with Iola and her daughters, Kim and Kara. We tried, but in spite of many happy times, we never quite put our lives together. Christmases 1986, '87, and '88 I spent alone in the folks home on Tippy. Those years I enjoyed Christmas Day with friends, Norma and Al Hayes, who invited me to join them for family Christmas dinners. By this time, I was attending Leesburg United Methodist Church (thanks for inviting me, Norma and Al) and singing in the choir. The next few Christmases would have been bleak, were it not for friends from the church. I remember singing Christmas carols in the town center, candlelight services, warm fellowship, and a tiny porcelain Christmas tree from my sister Lois in the front window. That tree was my only decoration for several years.

Christmas 1989 dawned with a new lady in my life. Barbara and I decorated our little artificial tree, the same one we placed outside Buford's door this year. On that tree, we carefully placed a special lighted ornament with the words "Our first Christmas together." This was also our first time singing together in the church choir Christmas Eve service. Christmas 1993, we celebrated our first Christmas together as man and wife. There were many happy Christmas celebrations with each of our families over the ensuing years. There were Christmas services and several memorable Christmas pageants at Morris Chapel when Barb became a Methodist pastor, serving Morris Chapel for six years. The warm fellowship with the congregation at Morris Chapel brought heartfelt experiences of costumed Christmas celebrations and delicious dinners.

Many happy Christmas celebrations were enjoyed with our children and grandchildren in Michigan and Dayton as we shared time with our growing family. There were joyous gatherings with the Johnson girls and their families, usually at the Fees's among mounds of toys and Christmas decorations. There were happy dinners in Dayton with Barb's family. Our decorations at home grew as Barb's Dickens village was expanded each year with gifts from her mom and her sons, Michael and Adam. We decorated bigger trees, and a growing light tree display graced our front yard.

Christmas 2000, the last Christmas of the second millennium, was a saddening trial as Barb served her last Christmas and Sunday as pastor. Her worsening health forced her to step down from the pulpit. It was a tearful congregation that walked past us and said their goodbyes after that last Sunday service of 2000. These delightful people had become close friends and truly our local family. Many of them remain as family to this day.

Christmas 2001, I built a four-by-eight-foot table to display our complete Dickens village, and we decorated a large tree with lights built in. That year saw us off on a new adventure as we headed for a six-week tour of the southwest. We enjoyed Christmas with the Howard Johnson III family in Texas. The paralysis in Barb's hands was increasing. By the time we reached Deb's in California in early January, it was obvious something serious was wrong. Two surgeries to her spine, six weeks in the hospital, and twelve weeks with her head in a halo device and our six-week trip had extended into a six-month adventure. It all started before Christmas. When we arrived home in July, all of our Christmas decorations were still in place—certainly the longest-lasting Christmas display ever. It was also the last time we used all those decorations.

This year saw our first Christmas Day apart from all of our families, including our church families. After a Thanksgiving day/Christmas at the Fees's with our three Michigan girls, their husbands, and five grandchildren, we worked preparing to head to Dayton and then south. We spent a happy pre-Christmas celebration at Adam's with Barb's entire family. Her mom, her brother Bob and his wife Rose, son Michael, son Adam and his wife Shelly, and granddaughters—Nina (5), Bailey (4), and Diana (1)—all shared in the festivities. After the celebration, we headed home to complete preparation of the annex for our delayed trip south.

Christmas Eve, we watched several old movies on TV including *Holiday Inn* (renamed *White Christmas* for today's audience, *Sound of Music*, and *It's a wonderful Life*. We noted that Hollywood's offerings have certainly deteriorated since those warm, love-expounding movies were made.

Far from families and church friends, we were fortunate to share Christmas Day with Donna and Sam Parks, our campground friends from last year. After two days of forty- and fifty-degree cloudy, misty weather, the skies opened up on the day before Christmas, and it poured off and on, mostly on, until late on Christmas Day. Despite the foul weather, we had a memorable Christmas, spending a day of good food, friendly conversation, and playing cards with the Parks in their RV. The day after Christmas, we returned the favor and spent the day in Buford with them, a near repeat of Christmas Day as we ate Christmas leftovers and again played cards. The skies had cleared, but it was still cold and windy.

So this is my collection of Christmas memories—at least those that are still fresh in this old mind. I know there are many more buried below the surface. These may or may not surface at another time. The most important of these memories are those of the love shared with family and friends—of Christmas meals, music, celebration, and relaxation and many times singing, "Happy Birthday, Dear Jesus" each Christmas morning. After all the gifts, glitz, and glitter, it is still the celebration and the meaning of his birth that is most important.

So I wish you all a very merry Christmas and a happy, healthy, and prosperous new year.

❖　　　❖　　　❖

Well done is better than well said.

> —*Benjamin Franklin (1706-1790)*

The average person thinks he isn't.

> —*Father Larry Lorenzoni*

Heav'n hath no rage like love to hatred turn'd, Nor Hell a fury, like a woman scorn'd.
> —*William Congreve (1670-1729)*

A husband is what is left of the lover after the nerve has been extracted.
> -*Helen Rowland (1876-1950)*

Learning is what most adults will do for a living in the 21st century.

> —*Lewis Perelman*

Dogma is the sacrifice of wisdom to consistency.

> —*Lewis Perelman*

Sometimes it is not enough that we do our best; we must do what is required.
> —*Sir Winston Churchill (1874-1965)*

The man who goes alone can start today; but he who travels with another must wait till that other is ready.

> —*Henry David Thoreau (1817-1862)*

There is a country in Europe where multiple-choice tests are illegal.

> —*Sigfried Hulzer*

Ask her to wait a moment - I am almost done.
> —*Carl Friedrich Gauss (1777-1855), while working, when informed that his wife is dying*

A pessimist sees the difficulty in every opportunity; an optimist sees the opportunity in every difficulty.

> —*Sir Winston Churchill (1874-1965)*

I think there is a world market for maybe five computers.
> —*Thomas Watson (1874-1956), Chairman of IBM, 1943*

There are three unforgetable moments, good ones, bad ones, and embarrassing ones.

> —*Howard Johnson, 1967*

Rags: A Love Story

It's July 4, 2005. Barb and I are mourning the loss of our little boy, Rags. Yesterday morning, I buried him beneath one of his favorite spots in our backyard between the maple tree at the end of the hedge and the red bud tree about six feet from the maple. He was a loving companion and a constant joy to both of us for the thirteen years he was a part of our family. The tears will dry slowly, and we will miss his love, companionship, cute littl face, and proudly raised tail for a long time.

Picture of Rags taken in the Goshen Hospital parking lot June 30, 2005. The last photo of our baby.

It was spring of 1992 when Barb called me from the bank in North Webster where she was working. She told me a customer had a small dog that needed a new home and asked if I could come there right away. The young man who owned him said that his little Shih Tzu was a people dog, and with him working two jobs, he couldn't give him enough time. He said we could take him home if we liked and see if Bud Light might fit into our household. I loaded this little bundle of long black-and-white fur into my car and headed home while Barb stayed at work. In a few minutes, he walked into the house and promptly plopped down in the middle of our living room as if he belonged. It was love at first sight. From that moment on, he was our little boy—curious, friendly, happy, and fiercely independent. I promptly called Barb who soon made arrangements with his previous owner for his papers, toys, leashes, and food dish.

By the time Barb arrived home from work, the little guy and I had become friends. He tirelessly retrieved the ball I threw for him both in the house and outside in the yard. As she walked in the door, Barb was greeted with a friendly woof, a happy smile (yes, he could smile), and a wildly wagging tail with long hair flying. Soon after the greeting, he once more plopped down on the floor prompting Barb remarked, "He looks like a pile of dirty rags." Indeed, with his long and tangled hair, it was hard to tell his head from his rear unless he barked or wagged his tail. He was desperately in need of a bath and haircut. Barb then added, "If you stuck a mop handle in him, you could mop up the floor with a rag mop." This prompted us to start calling him Rags, and it stuck. We liked Rags a whole lot more than his previous name, Bud Light. It didn't take him long to become accustomed to his new moniker.

We had him groomed as soon as we could get an appointment. With the matted hair gone, a complete bath, a "puppy" cut, and bows in his hair, he was about the cutest thing we had ever seen. With his new "doo," he seemed to prance about proudly as if to say, "Look at me. Aren't I cute?" And he certainly was. It didn't take us long to recognize his distinctly independent nature, his happy-go-lucky personality, and often impish behavior. He did all the cute things most small dogs do—the head cocked to a new sound; the rear up, head down attitude of play; the friendly bark when he wanted your attention, and one other unique action. He could grin. It wasn't a grin like a person would make, but he would look at you intently, his lips would part slightly, and his teeth would show through his slightly open mouth. It certainly looked like he was grinning.

Before long I placed a long line between the front corner of the house and a tree in the front yard about a hundred feet away. I fashioned a long leash, about twenty feet of rope and five feet of bungee cord. A small snap for his collar was attached to the end of the bungee cord and a large brass snap to the loose end of the rope. With the large snap hooked on the long line, it slid easily, and he could romp through most of our big front yard. A similar arrangement in the backyard gave him lots of running and playing space which he used and seemed to enjoy greatly. He soon had several favorite grassy spots where he would lay and survey his domain. He always wanted to be with us and particularly liked the long walks we would take with him on a leash. We soon discovered he needed to be leashed at all times. Otherwise, he would disappear even when we were with him. He would wonder around the neighborhood for hours before returning home. Several times he managed to sneak away and worried us to death until he announced his return home with a sharp bark at the back door. We discovered that a little dog can disappear and become impossible to find in our neighborhood. The busy highway at the rear of our place was a constant worry. During one of his forays, I discovered him jogging happily down the side of that highway on his way home, oblivious to the danger of death passing a few feet away. Unfortunately, Rags seemed utterly fearless of both vehicles and other even much larger dogs. The only things that seemed to frighten him were loud noises, mostly thunder and fireworks.

We have a multitude of chipmunks living under our concrete slab porches, and Rags spent a lot of time chasing them back under the porches. He never managed to catch one, but I'm sure he scared many of them out of their wits. Rags would chase any small animal he saw—rabbits, squirrels, and an occasional raccoon and possum. One morning as I was taking Rags out to his run in the front yard, a rabbit sauntered across the yard about twenty feet away. On seeing the rabbit, Rags jerked loose from my grip before I had attached his leash and took off like a shot after the rabbit. The bunny also took off and headed for our neighbor's, the Steurys, front yard with Rags right on its heels. Around the house it ran and, the first time by Rags, was less than a yard behind. The second time, the rabbit had gained at least twenty feet on Rags and was pulling away. By the next time around, Rags lost sight of the rabbit which was now more than a house-length ahead. This time the rabbit didn't complete the circuit, running across their backyard and slipping under the tool shed some seventy feet behind the house. The next time I saw Rags, he was panting rapidly with his nose to the ground, tracking the rabbit by smell as he was regaining his breath. After tracking the rabbit directly to the shed, he stuck his head under the shed and began barking. It was obvious the space under the shed held the rabbit but was too small for Rags to get under. He began digging fiercely. At this point I walked over and gently pulled him out and tucked him under my arm. He was not happy with me, but I was not about to let this chase go any farther. Over the years, there would be many such chases, all unsuccessful as far as I know.

A few memorable ones were quite humorous. I had a PC business in a tiny building in nearby North Webster and always had Rags with me when I was there. Using one of his long leashes from home, I usually tied him to one of the posts by the front door. The neighbors, my landlords, had a yellow cat that took great joy in taunting Rags and staying barely out of his reach. No matter how much Rags stretched the bungee cord on the leash, that cat would sit inches away from his farthest possible advance. While Rags exhausted himself stretching the bungee as far as he could while barking furiously, the cat would sit there, nonchalant and clean herself inches away. She was seemingly oblivious to the ruckus a few inches away. It must have infuriated Rags. I would pick up my angry pup and carry him inside as he grumbled and occasionally barked at his nemesis.

One day I was putting Rags out on his leash when I noticed the cat sitting in her usual taunting spot. Not thinking about the situation, I now attached Rags' leash to an eye on the side of the house at least six feet closer to the cat than ever before. A soon as Rags saw the cat, he headed for her at a high rate of speed. Both he and the cat expected the bungee to stop him a few inches away with no harm done. Instead, Rags plowed into the cat at full speed, knocking her flying head over heels and scaring the daylights out of both of them. As soon as she hit the ground, the cat sprang straight up at least five feet in the air, flailing the air in a vain attempt to get away. Rags hit the house after overrunning the cat and struggled to regain his footing. About the same time as the cat landed, Rags regained his feet and was after the cat in an instant. The cat headed for a tree about five feet away and soon disappeared far up in the branches with Rags barking fiercely at its base. I never again saw that cat.

Our little guy surveys the road from his usual perch on Buford's dash as copilot

Another time, I was letting Rags out on the same leash when Rags spotted a squirrel in the yard a few feet away. The squirrel saw Rags at the exact same moment and immediately ran about six feet up a nearby tree with Rags in hot pursuit. At this point, the squirrel turned with his head downward and taunted Rags with a long, loud series of squirrel chatter. The two glared at each other as the squirrel chattered and Rags barked. Once more I had to pick up my naughty little boy and carry him inside over serious objections.

Another memorable cat chase happened one cold winter day with about ten inches of fresh snow on the ground. Once more, I was letting Rags out on his leash at the front of my PC store when a gray cat sauntered by. Before I could snap on his leash, Rags pulled free and took off after the cat. Both were bounding high to make headway in the deep snow. By the time I caught up with them, Rags was barking at an opening under a nearby outbuilding where the cat had taken refuge.

After picking him up and settling him down, I headed back toward my store half a block away. I was walking along a frozen rut in a drive that extended across the slope behind my store. The entire rut was glare ice, but easier to walk in than the deep snow. With Rags beginning to relax, I put him down to walk on the rut, knowing he would walk back to the store if I stayed behind him. As he walked, he kept looking over his right shoulder toward where the cat had last been seen. While he was so distracted, another cat came out of some bushes ahead and started toward us in the other rut. At about the same instant, Rags and the cat saw each other. Rags immediately started running down the rut toward the cat.

He happened to be on a grassy pause in the glare ice, so he was soon at top speed as he hit glare ice once more. The cat took off toward us rather than try to turn on the slippery ice. Shredded ice flying, the cat dug its claws into the ice and soon gained speed. Rags, seeing the cat now passing by, reversed field with disastrous results. For an instant, Rags was running north but continuing south at a great rate of speed on the ice-filled rut. Stability on the ice under this maneuver was impossible, and immediately Rags was on his back, sliding tail first in the direction opposite to the way he wanted to go. Feet flailing rapidly, but unable to gain any purchase while sliding on his back, Rags slid about ten feet on the rut into some snow that had collapsed ahead of him as he slid into it. Rags then tried to scramble to his feet and promptly fell down again before he could renew the chase. After repeating this fruitless activity at least six times, Rags was rescued by my gloved hands as I scooped him up and headed toward the store. Immediately he glared at me, his dignity damaged, possibly grumbling, "I could have done it myself." Rags always grumbled when I picked him up after any similar activity.

Another frightening incident took place one morning as I was about to hook him up to his backyard leash. Sitting in our drive was a German shepherd. The instant Rags saw him, and before I had fastened his leash, he headed straight for the large dog. I quickly envisioned Rags getting chewed up as I tried to get him away from the shepherd, so I rushed out to his aid. There was no need. The last I saw of the German shepherd, Rags was chasing after him and down the drive to Walkers Park, barking fiercely. They disappeared around a corner long before I got there. As I started to turn to head for home and our car to hunt him down, Rags pranced around the corner like he owned the place and seemed to say, "See! I can take care of my place," as we headed home. I was still afraid one day he would try that with the wrong larger dog and pay dearly for it. Fortunately, that never happened

Rags had several places to sleep in our house. The corner of the foot of our bed was one, another was a small fabric bed he barely fit in. We kept this one on the floor of our bedroom. Another was a large fabric pillow in the living room by the TV. His big bed, about twice the size of the little one, was also kept in our bedroom under Barb's sewing desk. Rags did not like to be disturbed when he was in any of his beds, particularly the sheepskin pad on the foot of our bed. Any slight movement that disturbed him while there would be greeted with a grumble or two. Too much disruption would be greeted after several grumbles with a loud "kerplop" as Rags jumped down from the bed and sought a quieter place to sleep. We laughed many times as that kerplop seemed to signal the disgust emanating from our little guy. As the years passed and our aging little boy became unable to make the jump from the floor onto the bed, I built a carpet-covered box, so he could easily jump onto the box and then the bed. We used the same box in our trailer and then in our motor home, Buford.

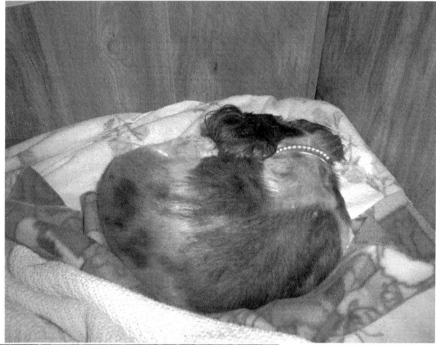

Rags snoozes at the foot of our travel trailer bed

Our boy braves the cold wind on the front porch in January 2002. Notice his "flag" blowing in the wind.

Rags was versatile, once earning the nickname Electrician Dog. I was running new electric cable under the Grimm cottage next door to ours. A supply cable for a new main circuit breaker box needed to be run under the cottage from one side to the other. I looked under the cottage and prepared to crawl through the narrow clearance between the ground and floor joists, dragging the cable with me. Little Rags, looked up at me and gave me an idea. I attached the end of the cable to the end of his leash and told him, "Sit." Then I walked around to the other side of the house and called him. Dutifully he walked under the house, dragging the end of the cable to me, tail wagging. Now I had my cable under the house, and I didn't have to crawl under it. I immediately praised him and called him my "electrician dog."

Rags handled all kinds of responsibilities. In one of Barb's early sermons as a lay speaker, she used him as an object of her sermon. He pranced down the aisle of the church, sat down next to the pulpit, and dutifully filled his role as the object of her sermon. He never moved until her sermon was finished, proudly walking out with her to stand in the greeting line as everyone left the service. Rags could be well behaved on certain occasions and a rascal during others. No matter what, he was always loved.

Rags loved to travel, and we took him everywhere we went. Whether by car, boat, or RV, he went with us joyfully, quickly finding the best seat where he could see out the windows. When we went out on our pontoon boat, he would sit way out front outside the railing, so he had an unobstructed view. When we caught a fish, he would bark to let that fish know who was boss. Normally not much of a barker, he barked furiously at any critter he saw from beetles to fish to squirrels, deer, and other dogs. Rags got along quite well with other dogs, even big ones. He approached any dog absolutely fearlessly, tail wagging and barking a friendly greeting. One time we were in Virginia when he got out of the car and took off running before I could hook him up to his leash. Soon he was amidst four local dogs about five or six times his size. My immediate worries were allayed when, by the time I got to where they were, the five of them were running and jumping together in play, Rags right in the thick of it. It took me some time to round him up and get him back in the car, so we could continue our trip.

We had to watch him carefully because our little guy loved to explore and would walk slowly and nonchalantly off, sniffing everything he came across for markings left by other dogs. He didn't move very fast, but could disappear in an instant and was hard to find, even around the house. Many times Barb took him out when she was out in the front yard or working in the garden. He would lie down in the grass nearby, often for hours at a time. Seeing his chance, Rags would sometimes sneak away and go exploring. Discovering him gone, we sometimes looked for him, mostly unsuccessfully. When we found him, it was usually because we spotted his distinctive black-and-white coat down by the lake or around a neighbor's house. When we didn't find him, he would eventually show up at the back door looking sheepish and sometimes dirty. More than a few times, it took a stint in the shower to get the dirt and smell out of his fur. He knew we didn't like him to disappear like that, and he hated the indignity of the bath that usually followed his foray.

When Barb was the pastor at Morris Chapel, she frequently had meetings and services at night. When this happened, Rags often curled up on the floor next to my chair in the opening to the hallway and rear door, his chin resting on his front paws facing the back door. Many times I saw him raise his ears suddenly and then sit up facing the door. I knew Barb would soon be home. It was apparent he recognized the sound of her car from as much as half a mile away. When his tail began wagging furiously, Rags and I knew Barb would soon come through the door. Until his hearing failed, he never missed.

Rags greeted everyone who came to our door with a bark or two the first time they came inside. Once we welcomed them into our home, Rags remembered them and greeted them with tail wagging, but no more woofs. This included service and delivery people as well as new friends. Summer of 1999, we took a vacation in Branson, Missouri, staying in a motel that accepted people with pets. We paid a small deposit in case of dog damage. The deposit was to be returned when we left after someone from the hotel examined the room. As we were preparing to leave, a lady came, inspected our room and gave Rags a friendly pat. Soon we received the following phone message, "You tell Mr. Rags he can bring his family down for a stay with us whenever he wants."

Rags and Mom at a stop on our way home from Branson.

Fiercely independent, he was definitely not a lap dog. He hated being picked up or held by anyone for any reason. Except for rare occasions, he would growl and nip at anyone other than me trying to pick him up. While he grudgingly allowed me to pick him up, he usually even grumbled at me although he never nipped at me. One warm summer day, Rags was at his usual spot in my PC store when a woman with two small girls came in looking for a new PC. The little girls immediately headed for Rags to pet him. I warned them not to rush up, but wait for him to come to them. Then I turned to answer the woman's questions. Suddenly and without warning, the five-year-old grabbed Rags and tried to pick him up. In an instant, Rags barked and clamped down on her tiny bare arm. I had immediate visions of lawsuits as the little girl screamed and grabbed for her mother. Expecting to see blood and puncture wounds, her mother and I examined her arm. We were amazed at what we saw. Her tender skin was not only unbroken but completely unmarked. Rags had merely mouthed her arm as a warning and deliberately not bitten down. I praised him for being a good boy while the little girl calmed down. Fearful the next time could turn out differently, I closed Rags in my office whenever small children came in after that.

The end of 2000, Barb had to step down from the pulpit because of deteriorating health. Now at home full-time, Rags became her constant companion. Wherever she went, Rags followed. When she went to bed, he was right there at the foot of her bed. When she went anywhere in her power wheelchair, he was not far behind. He seemed to understand she was in pain and needed companionship. The only time he wasn't with her was when we let him out on his leash line or when a thunderstorm was in the area. At the first clap of thunder, Rags would run and jump in my lap if I was seated or stand right at my heels if standing. He would not go far from me until there was no more thunder.

When Rags was about twelve, he seemed to have developed arthritis and even had trouble climbing steps. He could hardly get up into the bed even with the step I had built. I started giving him the same glocusamine-chondroitin I was taking, but in appropriate amounts. I was amazed when after a few weeks his problem was gone. From then on, he ran and jumped up steps like he had before his arthritis hit. About the same time, we noticed his hearing was failing and he began developing cataracts in both eyes. Fortunately it didn't seem to have slowed him down. He ran as fast after toys thrown. The only change we noticed was he quickly tired and wouldn't retrieve toys after three or four tosses. While almost completely deaf for the last two years, his vision didn't seem to be getting any worse. He could still spot dogs and other animals several hundred feet away.

April 2005, Rags turned fifteen. About that time we took him to a vet in Wauchula, Florida, to be groomed. When I picked him up, the groomer asked me to feel the left side of his skull. It felt bumpy while the right side was quite smooth. She had called it to the attention of the vet who then examined him. The vet said he seemed to have a slow-growing brain tumor in his left brain. Evidence was the lumpy left side of his skull and his sudden deafness a few years before. She asked if he had any personality change in the last three years. When I said he had suddenly begun to snap at us instead of merely grumbling when he was disturbed, she said the vet believed the tumor had been growing slowly for three or four years. He also felt it had not involved the skull until quite recently. I was terribly saddened when she told me the vet felt Rags had only three to six months left. There was no chance of stopping the tumor as it had probably been inoperable for at least two years. I decided there was no need to share this with Barb until it became obvious and we would then decide when to put him down. We had already discussed his age and the probability of our putting him down, but that could wait for the time being.

Rags seemed as loving, happy, and playful as ever, staying next to Barb as her pain and health problems increased the last few weeks we stayed in Florida. During Barb's emergency stay in the Tifton hospital, Rags and I stayed on Buford in the hospital parking lot the entire four weeks Barb was in the hospital. Once in the middle of her stay, she was brought out in a wheelchair and had a chance to visit with Rags. I know he lifted her spirits with that short visit. We left for home on Friday, June 10. Once we arrived home on June 14, Rags was happy to be in his own yard, exploring his old haunts and remarking his territory. After living in Buford for five days while I aired out the house, I moved Barb into our bedroom Saturday, June 18. Rags immediately took up residence on the foot of her bed.

The morning of June 19, one day after she moved into the house, Barb called me at about four in the morning, saying she didn't feel good. Checking her blood pressure, I couldn't get a reading, so I called EMS who took her to Goshen General Hospital. Telling Rags I was leaving, I jumped in our pickup and headed for the hospital. As soon as they stabilized her, they moved her into intensive care. Once her vitals, and especially her blood pressure, were back near normal, I headed home with a friend to pick up Buford and Rags. Once more we were living in Buford in a hospital parking lot. We stayed there in the parking lot for Barb's entire stay until she was moved to Lakeland Rehabilitation Center on Friday, July 1. I then moved Buford back home. My friend Al Hayes drove me to Goshen to pick up our truck and bring it home. That afternoon, I brought Rags in to visit her, carrying him under my arm as I walked in. They were delighted to see each other. That night, Rags and I stayed in our house.

The next day, I spent the morning replacing the steps in Buford. As I was tightening the last bolt, my cell phone rang. It was Barb telling me her blood pressure was once more low and that EMS was going to transport her back to Goshen Hospital. I pulled myself out from under Buford, went inside to clean up, and left for the hospital. I left Rags inside the back door, telling him to be good while I was gone. As I turned to shut the door, that little imp must have quickly run out the door behind me unseen onto the porch. He had done this several times during the last few days, but I always caught his deception and put him back in the house. Preoccupied with Barb back in the hospital, I failed to check the porch for Rags. That little slip would seal his fate a few hours later.

I stayed with Barb until about ten-thirty when I headed for home. I saw the last shots of several fireworks displays during the half hour it took me to get home. As soon as I was in the house, I whistled for Rags several times. When he didn't appear, I searched high and low to no avail. Stepping outside, I whistled for him several more times. Then I got into the pickup and drove around Walkers Park and up and down the highway behind our house, still no Rags.

I reasoned he must have sneaked out as I described earlier and, because of the fireworks, might be huddled in fear under a nearby house or garage. I slept fitfully during the night, worrying about my little buddy. Early the next morning, I went out in Buford and, using our PC, printed a handful of "lost dog" notes with his picture. I started distributing these all down Walkers Park and onto Black's Landing. Suddenly I saw our neighbor to the old cottage running toward me, one of my notes in her hand. She bore the terrible news that Rags may be the dog they saw killed on the road behind their cottage during the rush of traffic away from the fireworks display. She and two of her friends walked with me back to the place where Rags lay. It was right beside the Walkers Park sign in the spot where I often waited for my father's car seventy years previous as a small boy. It was obvious he had died instantly and never knew what hit him. It happened about ten-thirty as I was leaving the hospital to come home. I walked home to get the pickup, tears flowing freely.

I picked him up and tenderly laid him in the bed of my truck. As soon as I was home, I got shovel, ax, and fork to dig a grave for our little boy exactly where Barb and I had decided he should be buried some day. I marked off a spot about two feet by four feet and began digging. It was hard digging through the roots and rocks, but I was on a mission. The digging gave me something to do to help stem the tears. It took me about an hour and a half to dig down about four feet into the fine beach sand beneath fill dirt and muck.

Lifting him gently from the pickup bed, I carried him to his grave, and placed him carefully with his head facing the house and the lake. He would rest below one of his favorite spots. While I filled in his grave, the tears began once more to flow. Now I was dreading the heart-wrenching job of sharing the hurtful news with Barb. After cleaning up, I headed for Goshen Hospital with a terribly heavy heart. I hoped to find an appropriate time to tell Barb, but what time is appropriate for such hurtful news? I then called ICU to inquire how she was doing. When they reported she was doing quite well, I asked them to explain I had been delayed with car trouble and would be there in an hour or so. When I walked into ICU, I saw that Barb looked bright and much better than when I last saw her. I sat and talked with her, hiding my heavy heart and waiting for the right moment to share my painful revelation. No appropriate moment came for some time, and Barb drifted off to sleep.

It was three hours later when Barb awoke. Now it was getting late, and I prayed she would not ask about Rags. I didn't want to share the bad new so late in the day. I could save it until tomorrow and give her most of the day to adjust. About ten-thirty I headed for home, glad that Barb hadn't learned the terrible truth. It was tearing me up inside, but thankfully I was able to hide the turmoil and didn't show it enough that Barb noticed. I spent a painful night of fitful, interrupted sleep, getting up about five in the morning. Every way I turned there were reminders of our little guy. His toys, food dish, leashes, rope runs, even the towel Barb decorated for him that we used to wipe him off when he went out in wet weather, each of them brought a twinge of heartbreak.

Our little boy looks out Buford's door at the Goshen Hospital, June 30, 2005.

I secured Buford for the trip to Goshen, leaving at about eight. At eight-thirty, I pulled Buford into the parking lot and threaded it into an appropriate parking space. I headed for the elevator, dreading the chore I had to perform. Barb was sitting up finishing breakfast when I arrived. We talked for a while as I waited for the right moment. My heart sank when she asked about our little Raggy. At this point I had to tell. It was both a hurtful exchange and a great relief. We shared a lot of tears before we began recounting how much joy and companionship our little guy brought into our lives. After thirteen wonderful years with this little bundle of energy and love, we agreed he had been the best critter either of us ever had and would be impossible to replace. We even loved his independent nature. He will be sorely missed and definitely be a tough act to follow.

I'm sure we will have and come to love another canine baby, probably another Shih Tzu, but that will have to wait 'til Barb is able to come home. I doubt the place he made in our hearts will ever be filled.

❖   ❖   ❖

Outside of a dog, a book is man's best friend. Inside of a dog it's too dark to read.

—Groucho Marx

If you pick up a starving dog and make him prosperous he will not bite you. This is the principal difference between a dog and man.

—Mark Twain

**One of the best pictures of our little guy in our S10 pickup
in Florida in May 2005.**

❖   ❖   ❖

Happiness is a warm puppy.

—Charles M. Schulz

I think it would be a good idea.
—Mahatma Gandhi (1869-1948), when asked what he thought of Western civilization

The only thing necessary for the triumph of evil is for good men to do nothing.
—Edmund Burke (1729-1797)

I'm not a member of any organized political party, I'm a Democrat!
—Will Rogers (1879-1935)

If a Person Lives With Criticism

    He becomes one who condemns.

If a person lives with hostility

    He becomes one who fights.

If a person lives with ridicule

    He becomes shy.

If a person lives with shame

    He becomes one who feels guilty.

If a person lives with tolerance

    He becomes more patient.

If a person lives with praise

    He becomes one who appreciates

If a person lives with fairness

    He learns more of justice.

If a person lives with security

    He becomes one who has faith.

If a person lives with approval

    He becomes one who likes himself.

If a person lives with acceptance and friendship

    He becomes one who finds love in the world.

—Paraphrased words of Dorothy Law Nolte by Howard Johnson, 2001

They who can give up essential liberty to obtain a little temporary safety deserve neither liberty nor safety.

—Benjamin Franklin

The U. S. Constitution doesn't guarantee happiness, only the pursuit of it. You have to catch up with it yourself.

—Benjamin Franklin

Barbara Johnson's Last Days

The last days of Barbara Johnson, my wife, lover, companion, spiritual mentor, best friend, sharpest critic, and strongest supporter, for the far too-short period of fifteen years we shared.

My heart has been wrenched from within me and slammed against the wall. *God help me to bear the pain.* Sunday morning, October 16, 2005, at eight-forty, my dear, sweet Barbara passed from this life into the next. May God receive and hold her precious soul in his gentle hands. She waited until Sunday, her day, to leave us. Born on Sunday as well, I can't help thinking she wanted to leave us on a Sunday. Both of her boys, Michael and Adam, as well as my daughters—Robby, Diana, and Mindy and her husband, Joe—were with me at her bedside as she passed. My daughter Deb was with us until three, Sunday morning, when she had to head back home to California. Barb lost her struggle with countless health problems and great pain she suffered for most of the last seven years.

Her final journey started a week ago on Friday morning when she was taken to the ER of Goshen Hospital. Her nurse at Lakeland Rehabilitation and Healthcare Center called to tell me that Barb had been rushed to the ER because her blood pressure became so low. This had happened several times before, and though I was concerned, I was also sure she would rally as before. I arrived at her bedside at about the same time as the doctor, and though she was weak, she managed a smile and the words, "Well, here we are again." As usual, she was joking with the ER personnel. Somehow, Barb always had a jovial or uplifting thing to say no matter how bleak things seemed. The previous week, she had been having intestinal problems and had become weak and dehydrated. Also, as usual, they had tremendous difficulty getting an IV into her tiny veins. When they succeeded, the IV soon began to bring her blood pressure back to acceptable levels.

Fully aware, she asked for the phlebotomist to draw her blood after the nurse tried twice unsuccessfully. Even good ones had trouble with her tiny veins and fragile skin, so she was pleased when the gal who took her blood successfully several times before came in. A tiny woman four foot ten in height, Barb had tiny hands, feet, and most other proportions of her body. She always joked about her tiny "nothing" nose. Having gone through this many times in many places, I was neither surprised nor much afraid as she had always come through them all. I felt certain the hospital staff would soon have her back in Lakeland where she could continue to improve and soon come home. At Lakeland, she had been slowly regaining her strength and was walking more and more with her walker.

The last photo taken of Barb with me in her room at Lakeland.
Deb took the picture about August 13, 2005.

When the ER staff had given her enough fluids to bring her blood pressure into the normal range, she was transferred to ICU where she had been in June. Tests indicated she had an infection, probably in her lungs. There was some fluid around her lungs, and her liver enzymes were somewhat elevated. She was given antibiotics and medication to help her liver. By Saturday noon, her vitals were stabilized, but her breathing was still labored, a bit less than before. She was given Lassix to help eliminate excess fluids and her kidney output was soon doing well. We were all encouraged by her progress.

As always, Barb charmed those around her with her warmth and Christian affection. Few who spent time with her avoided falling under the spell of her persistent, loving charm. Her marvelous knack of touching the hearts of those around her worked its magic no matter how weak or sickly she became. Several of the nurses remembered her fondly from her previous stay in ICU in June. They stopped by to cheer her up even though they were not assigned to her care. This was the essence of Barbara, touching all who came near her with love and kindness based in a strong and generous Christian faith. A pastor, but not a "preacher," she spoke more with people than at them, even in her sermons. This remarkale talent was a rare gift.

Saturday through Monday, they ran many tests including blood, an ultrasound of her heart and arteries, a special QV test similar to a CT scan, but less stressing. They expressed concern about her liver, and did a complete ultrasound examination of that organ. The results would not be read until Tuesday, but a preliminary report was encouraging. During this period, many of our friends came to visit and offer encouragement.

Tuesday morning: Her doctor told me the terrible news that her liver was failing, and she would be with us only a few more days at most. There is no painless way to hear those words, and no words can describe the flood of hurt and pain coursing through my body. Devastated doesn't begin to express what I felt at that moment. The doctor explained that all they could do was make her comfortable and relieve her pain for the time to come. The remainder of that day is a jumble of activities in my memory. I hardly knew what to do next. I soon went in to be with her, and we shared the dreadful news. She looked into my tear-filled eyes and said, "I know it is time. I've made my peace with the Lord, and I'm ready to go home." I have no idea how long I held her. It could have been hours, but was probably only a few minutes.

Somehow I managed to call all our family members; drive home and get Buford, our motor home; take Charlie, our little dog, to stay with friends, and head back to the hospital. Fortunately, our dear friend Helen Smoker was there to be with her while I went home to do what I had to do. With Buford in the parking lot, I could stay there as I had in June, less than a minute from her bedside. I hurried back to her room to be with her as much as possible. Helen stayed with us for a while, which was a great comfort. Pastor Mike Heath came and spent time with Barb several times, but when is a casualty of my disrupted memory. I prayed she would last long enough for her boys and my girls to get here. I had no concept of what the next few days held in store for us. Totally crushed physically, emotionally, and mentally, I stayed and talked with Barb until the medication mercifully put her to sleep.

At the nurses urging, I trudged out to Buford and collapsed on the couch in my clothes. They assured me that I would be called on my cell phone should there be any change or should she awaken. I slept fitfully until about six in the morning.

Wednesday: When I woke up, I shaved, washed my face, and put on my red shirt, one that Barb liked. I wanted to look my best for my sweetie. She was still asleep when I arrived in ICU. Once more the day was a muddle in my memory. She awoke soon after I arrived at her bedside, and we talked. I showed her the red shirt and said I wore it especially for her. Mostly I told her how much I loved her. I know several friends came, but as before, the day is such a muddle of memories, I'm not sure even what happened on what day. Mostly I prayed for our family to be here before she passed. I told her that her boys, her mother, her brother, and my girls were on their way to see her. It was obvious she was failing, and I prayed that she would last until they arrived.

During the day they removed her IV and stopped giving her meds. She would only be given Atavan to relieve her anxiety and morphine for her pain as long and as much as was necessary. Fluids would be withheld as her kidneys were failing, and fluid buildup in her body would certainly make her uncomfortable with no real purpose. It was quite clear her organs were shutting down, and there was nothing more they could do medically but try to make her comfortable. To free a bed in ICU, they moved her to the fourth floor next to the room she had been in during her stay in June. They explained that everything being done in ICU could be done on the fourth floor and that they needed the ICU bed for another patient. It hurt, but I realized they were right.

By nighttime she had declined considerably. Her breathing was more labored, and she spoke only with difficulty. Before she went to sleep, she looked directly at me and kept repeating, "Go home! Go home!" That about ripped the heart right out of me. Some time back, she had said she wanted to look out her window at the lake once more. I would have gladly taken her home to die, but of course, it was not practical. Those words in her weak, plaintive voice, "Go home! Go home!" will repeat in my head for a long time. I shall try to continue sharing this experience as I remember it.

Thursday: Dear Barbara, it is morning and I am praying your sons and my girls arrive before you depart. I've said a thousand times, "I love you!" and shed a thousand tears. I feel like falling apart, but must remain strong and supportive for you. I can see you look at me, but your voice is silent. Please hear me as I say "I love you!" over and over. I know no other words to say. I'm quite sure that you hear me as I can see a response in your eyes.

Thank God! Adam, Mike, and your mother have arrived. There were tears and hugs, kisses and more words, "I love you."

"Mindy! How glad to have you here."

My youngest daughter has arrived with more hugs, kisses, and tears. Thank you, God, for the wonderful families both Barb and I have and treasure. Sharing doesn't lessen the tears, but does make them more bearable. I can't imagine going through this without someone for support. I kiss my dear Barbara on the forehead, cheeks, and lips, hoping to relieve her pain and anxiety. I hold and kiss her tiny hand, realizing full well it may be for the last time. My God, how I am going to miss her. Her passing will leave a large, empty hole in my life.

We borrow a CD player from ICU and play soft music in the background. Barb loved music, and I pray she hears the gentle music and is comforted. Friend Holly goes to her car and brings up several hymn books. We sing many of her favorite hymns, hoping they are of comfort. She opens her eyes and moves on occasion. We have no way of knowing whether she sees or hears us, but we speak to her as if she does. How I hope she hears our songs and words of love. Or should we be praying she is asleep and unaware? It is so painful to have these thoughts as we know so little about what she understands and she can't communicate with us. I now pray for her to go to sleep. I spend the night in her room with Adam. She seems to sleep through the night with little stirring.

Friday: As I write these words, my dear, sweet Barbara lies here—eyes closed, breathing fitfully—in her hospital bed, her tiny hand clutched in mine. I hold and kiss that small soft hand slightly more than half the size of mine that will soon pass from me forever. I watch as her body struggles to maintain life. Her eyes have not opened for a full day, but she occasionally stirs and utters a low moan. I pray that she is not aware and suffering, but fear she may be. The nurses are giving her morphine through her IV on a regular schedule so that she does not feel pain. Her breathing continues in gurgling gasps, sometimes as long as thirty seconds apart. My daughter Deb arrives, and we share a teary greeting. The growing support in the room from family and friends is comforting. I continue praying to God, asking that he take her NOW! The pain of seeing her struggling and suffering is unbelievable. The nurse assures me that she feels no pain, but still, I worry.

Saturday: Another day much like Friday, but with her breathing becoming slightly more labored. Words of affection and love—words of compassionate prayer, no words of hope. I look at her frail little arms, purplish brown bruises from the slightest touch, skin like crinkled tissue paper (all from her long use of prednisone), muscles shrunken from the ravages of PPS and months of little exercise. I kiss her tenderly on the forehead and notice the *dimples*, as she called them, from the penetrating screws of the halo device she wore for the twelve weeks of recovery from major cervical spine surgery in California in the spring of 2002. That time also she walked a narrow line between life and death.

I hold her hand more now, reluctant to let go, fearing and knowing soon I will never be able to do so again. As she sleeps peacefully, all in the room are talking more now, sometimes pleasant, sometimes joking. Then, when she stirs and tries to open her eyes, the pain in our hearts bursts forth in tears and sobs as we all are attentive to the beloved little lady as her body struggles and clings to life. Now we are all praying she will go to sleep and suffer no more. Grief and pain come over us in waves. *God, please end her suffering and take her now!* I pray silently as I gaze at her beloved face.

Sunday: Daughters Dee Dee and Deb stay with me for the night. At three, Deb leaves to drive to Detroit to catch a plane back to California. She kisses my sleeping lady, gives me a long hug and a kiss, and I walk her down to her car. Dee Dee stays with Barb, so she won't be alone. About seven o'clock, the rest of the family returns to her room. She is still sleeping peacefully, her head propped against a rolled-up towel to keep her in a comfortable position. I am reminded of the times when she was wearing the halo device during her recovery from neck surgery when I rolled towels and placed them under her head to make her more comfortable. This is quite different.

Between eight and eight thirty, she stirs several times and seems to be trying to open her eyes. We all express our love in word and deed as best we can. Tears flow freely as it is apparent the end is near. After a few struggling gasps and a low moan, her breathing stops. A few moments later, the nurse checks her and pronounces she is gone. It is eight-forty as she passes into the hands of our waiting Lord. We all say our tearful goodbyes.

A message from Barb? I am driving Buford home with Adam beside me. The words "Take your meds! Take your meds!" come into my head clearly. I had not taken them as yet. Those words had never popped into my head before. My memory doesn't work that way. About fifteen minutes later, the same words are ringing in my head once more, "Take your meds! Take your meds!" Suddenly I realize what is happening. Those are the same words Barb used so often to remind me to take them. I told Adam, and we agreed it must be Barb. Surely, my new guardian angel is on my case, reminding me.

Monday morning, daughter Diana (Dee Dee) and I are meeting with the funeral director and going over Barb's carefully thought-out plans for her funeral. I had two handwritten pages she had gone over with me several times, and the funeral director had a sheaf of papers in his folder. My dear little lady didn't want anyone to have to deal with planning. She told me over and over, "I want this to be a celebration of my life—a happy recollection—a loving gathering." Everything down to the smallest detail was written on those papers she started writing about the time we were married. Over several years, she updated it frequently with changes in black ink over the original blue. The changes were on our copy and on the one the director held.

When we were finished, the funeral director said, "There is only one thing she didn't mention, what she is to wear."

What she was to wear was missing from the paper I held. At first I decided that, in the absence of instructions, I would have her wear her favorite red dress.

Then a sudden inspiration popped into my head—her wedding dress. I immediately said, "Wedding dress."

Diana and I conferred quickly and decided that would be it, the beautiful dress she wore in our wedding. After the director had written this information in his records, he gathered the instructions up and noticed a single sheet with her earliest instructions written in 1993. He turned the sheet over, looked, and showed us a note on the back with the words *wear wedding dress* written in her clear hand. I cannot remember our discussing this though we may have. I like to think my dear, sweet, guardian angel was once more prompting me.

Family reunion at Tippy, August 13, 2005. Barb's last visit to our home.

Left to right: Ho, Barb, Chantel, Jeff, Joseph, Deb, David, Dave, Diana, Danni, Bob, Mindy, Robbie, Joe, Nicole, and Jessie

❖   ❖   ❖

If stupidity got us into this mess, then why can't it get us out?

—Will Rogers (1879-1935)

I never met a man I couldn't like.

—Will Rogers (1879-1935)

The backbone of surprise is fusing speed with secrecy.

—Von Clausewitz (1780-1831)

Democracy does not guarantee equality of conditions - it only guarantees equality of opportunity.

—*Irving Kristol*

There is no reason anyone would want a computer in their home.

—*Ken Olson, president, chairman and founder of Digital Equipment Corp., 1977*

The concept is interesting and well-formed, but in order to earn better than a 'C', the idea must be feasible.

—*A Yale University management professor in response to student Fred Smith's paper proposing reliable overnight delivery service (Smith went on to found Federal Express Corp.)*

Who the hell wants to hear actors talk?

—*H. M. Warner (1881-1958), founder of Warner Brothers, in 1927*

We don't like their sound, and guitar music is on the way out.

—*Decca Recording Co. rejecting the Beatles, 1962*

Everything that can be invented has been invented.

—*Charles H. Duell, Commissioner, U.S. Office of Patents, 1899*

Denial ain't just a river in Egypt.

—*Mark Twain (1835-1910)*

A pint of sweat saves a gallon of blood.

—*General George S. Patton (1885-1945)*

After I'm dead I'd rather have people ask why I have no monument than why I have one.

—*Cato the Elder (234-149 BC, AKA Marcus Porcius Cato)*

He can compress the most words into the smallest idea of any man I know.

—*Abraham Lincoln (1809-1865)*

Don't let it end like this. Tell them I said something.

—*last words of Pancho Villa (1877-1923)*

The right to swing my fist ends where the other man's nose begins.

—*Oliver Wendell Holmes (1841-1935)*

Postscript to a Beautiful Life

More than a year before we married, Barbara asked me to help her with some important decisions she needed to make. At the time we had not yet seriously discussed marriage, but I feel certain we both knew it was in our future. We sat at my dining room table—she with several pens, a lined notepad, and a few other papers, and me with a curious, open mind.

"What's this all about?" I asked my beautiful sweetheart.

"I want you to help me write a new will and plan my funeral," she replied in a strong, steady version of her sweet voice.

A planner and organizer, Barbara liked to have essential things planned and those detailed plans in place. Being with Barb, I realized early on that when she set her mind to something, there were only two viable options for me. I either helped or got out of the way. Though our two strong wills sometimes crossed swords, this would not be one of those times. There is no recollection of my answering words, but they certainly contained encouragement and an offering to help.

She had mentioned wanting to do this several times over the preceding months and now was crunch time. She planned a meeting for us with Gary Eastland, a local funeral director, in the near future, and we would now bring things together in preparation for that meeting. This time we discussed a number of details in general terms. She asked about my feelings toward burial, internment, and cremation. I don't believe we planned many details as she wanted to be prepared for our meeting with Gary. That would be the time to iron out the details. She did urge me to follow her example with a new will and funeral plans. One thing I remember we heartily agreed was that we wanted a celebration—a celebration of life and of good times remembered and shared.

Soon after we met with Gary and worked out a more detailed plan, Barb bought an insurance policy to cover her final expenses, payable to the funeral home. She didn't want anyone to have to worry about what to do or how to pay for it. That was the quintessential Barb, concern for others, even after she would no longer be with us. I struggled with my own personal decisions about those final days and would wait many years before following her example.

Several times during the years that followed, she dug out those papers and updated them. She always prompted me to develop my own plans, and I always put off that uncomfortable task. With her help, I did write a new will, but planning my funeral was always put off. Thirteen years older than Barb, I was certain I would go first, and she would be there to help with my funeral. I did state and write down that, like her, I wanted to be cremated and have my ashes sprinkled in Lake Tippecanoe. Gradually she won me over, and I too sat with the funeral director, made plans, and bought and assigned an insurance policy to relieve survivors of that solemn responsibility at a time of distress.

After she became a minister, she expanded her instructions with a detailed memorial service including decorations, scriptures to be read, music, and the names of colleagues she hoped would speak at the service. She couldn't have known how touchingly beautiful that service, crafted by her tender loving heart, was to be. Her aura would permeate the Leesburg United Methodist Church on that day to come.

Several times after she became ill, she had us sit down and go through her instructions. In a joking, loving manner, she always said she would be watching me, and that I had better follow her instructions. As her illnesses brought on more pain and weakness, she often reminded me where her important papers were, *just in case*. She always insisted on *the folder* being with her at the hospital and nursing home. In late summer of 2004, we once more met with a representative of the funeral home to bring her records up-to-date with her latest instructions. We discussed each detail thoroughly. She frequently asked if I approved. I would carefully consider my answer and then agree. We had no disagreements about any of the instructions. She even persuaded me to work on my own service, plan the music and a few scriptures. My overall instructions for the funeral home were put in place.

She instructed me verbally in the placement and disposition of a number of small items, mostly jewelry and photos. She made me pledge to see to it that her service was truly a celebration. I agree most heartily with that premise and ask for the same at my funeral. I told her many times that I have had a full and rewarding life with few regrets. If I should go tomorrow, I want all to know it's been a blast! We both have vast treasures in our children, our grandchildren, and our many dear friends.

The following was written by Barb in September 2003 as she once more edited her instructions:

"I would like it [the celebration service] to be a real celebration—I have had a great ride—with some side trips that only served to strengthen me."

See the next page for the bulletin prepared by our dear friend, Norma Hayes, who is also the church organist:

A service of celebration for the life of Barbara True Galbraith Johnson - October 21, 2005

Prelude	Norma Hayes	
Gathering	Pastor Michael Heath	
The Word of Grace	Pastor Michael Heath	
Prayer	Pastor Michael Heath	
*Hymn	"How Great Thou Art"	Overhead
Scripture	Psalm 139:1–10, 13–16	Rev. Ted Blosser
*Hymn	"Be Still My Soul"	Overhead
Scripture	★Psalm 103:1–5; 46: 10–11	Rev. Paul Newman-Jacobs
		Rev. Chris Newman-Jacobs

★*"The Lord gave me this verse when I first started going downhill in 1993."* Barbara Johnson

*Hymn	"Blessed Assurance"	Overhead
Scripture	1 Peter 5:7–10	Rev. Milton Gould
*Hymn	"Because He Lives"	Overhead
Scripture	1 John 5:11	Rev. Don Shanks
*Hymn	"It Is Well with My Soul"	Overhead
Scripture	2 Timothy 4:6–8	Rev. Michael Heath
Scripture	1 Thessalonians 4:16–18	Pastor Barb Lloyd
Special Music	"I'm Working on a Building"	The Willows
Scripture	Psalm 139:1–10, 13–16	D. S. Bob Dexter
Witness		Pastor Michael Heath
Dance of Celebration		"I Hope You Dance"
	Roberta Grimm	
Special Music	"Look for Me Around the Throne"	The Willows
Benediction		Pastor Michael Heath
Postlude		Norma Hayes

Rev. Barbara True Galbraith Johnson

Born: Sunday, February 26, 1941, in Dayton, Ohio

Married Howard Johnson on May 29, 1993

Passed from this life: Sunday, October 16, 2005

The service was indeed a beautiful celebration of Barb's life. I'm certain there was a brand-new angel watching and enjoying the service with an approving gaze. I could almost feel her seated beside me, watching, smiling, crying, and loving with all of us. I thank everyone who was there, those who weren't there except in their hearts, and everyone who supported her in so many ways during her continuing battle for health and life. The many cards, calls, letters, e-mails, visits, and prayers were each treasured. They brought joy and encouragement to Barb at times when things didn't look good, when she was weak, in pain, and so often near death.

Some information about the participants:

Norma Hayes: Longtime friend, church organist, accompanist, and guide for the Willows, church secretary, Norma is a friend you can always count on. She prepared the bulletin and helped in many ways with the service.

Pastor Michael Heath: The current minister of Leesburg United Methodist Church, Mike visited Barb many times in both Goshen Hospital and Lakeland Rehabilitation and Health Center. He and Barb met for the first time when he visited her in Goshen Hospital. He was with us when we shared the terrible news with Barb.

Rev. Ted Blosser: A colleague and then district superintendent when Barb was an active pastor, Ted was also a dear friend. They attended many meetings and conferences during her period of service.

Rev. Paul Newman-Jacobs: He was the pastor of LUMC during the period Barb was secretary, when she attended the Lay Speaker Academy and when she answered the call to the ministry. His support and encouragement were instrumental in helping her during this period.

Rev. Chris Newman-Jacobs: Paul's wife and the pastor who preceded Barb at Morris Chapel United Methodist Church, her guidance often helped Barb at her new charge.

Rev. Milton Gould: A colleague and dear friend, Milt and his wife accompanied Barb on her trip to the Holy Land in 1999. Barb shared many stories of their exploits on that wonderful journey.

Rev. Don Shanks: Don was the pastor at Leesburg UMC when I proposed to Barbara in the choir loft, and he officiated at our wedding. I'm sure both were memorable occasions where Don was caught surprised and delighted. His pleasant face and words grace the videos we have of both occasions.

Pastor Barbara Lloyd: She was the pastor at Leesburg UMC, and when we returned after Barb's health problems forced her to step down from her ministry, it became Pastor Barb ministering to Pastor Barb. Her words of support and comfort were a great help to Barb and me.

The Willows: Our quartet was formed when Barb and I joined friends Al and Norma Hayes and Carol West (then Shively) to sing at early informal services at LUMC. Active to this day, the Willows sing gospel and other religious music wherever and whenever we have the opportunity. Barb's sweet, beautiful voice can be heard clearly in both of these recordings. While *I'm Working on a Building* was one of her most enthusiastically performed numbers, she told us many times she wanted *Look for Me Around the Throne*" to be performed at her final celebration. It was with utter regret that she relinquished her role with the Willows because of the demands of her new assignment as a Methodist pastor. Helen Smoker soon stepped in with her strong voice, so the Willows continue performing to this day.

Robert Dexter: District superintendent of the neighboring Michiana District of the Northern Indiana Conference of the United Methodist Church, Bob visited Barb several times during her last stay at Goshen Hospital while our local DS was traveling and was not available. He was a comfort to me during her last days.

Roberta Grimm: My sister Bobbie immediately close to Barb after they met and became *the sister I never had*, as Barb told me many times. She and her husband, Robert, a retired minister and executive of the Council of Churches, have been dear bulwarks of support to both Barb and me in many times of dire need. Among her many talents, she is a liturgical dancer and has published a book on liturgical dance. When she asked me if she could perform this dance for the service, I knew it would be something Barbara would support enthusiastically. Her dance was a beautiful and emotional experience for me as for the entire audience.

A few loving comments about my dear Barbara:

Always concerned for others, she had an almost magical knack of reaching and relieving pain in others. It was a talent few possess that will now be sorely missed by those whose lives she can no longer touch directly. There are several anecdotes that spoke volumes of whom Barbara was and how she touched others. Later, I may add others, but these that follow are the ones that first come to mind.

During one of her first hospital stays after we were married, I remember Barb talking to a nurse as I came into the room. Barb's operation had been over but a single day, and I heard these words spoken clearly if a bit weakly, "Dear God, help this kind woman in her time of need. Comfort her pain and relieve her suffering. I ask this in Jesus's name. Amen."

It didn't surprise me for this was Barbara, through and through. I noticed the nurse wiping a tear from her eyes as she stood and thanked Barb before leaving the room.

"What was that all about?" I asked.

"That poor girl has serious marital difficulty, and I asked for God's help for her," Barb replied, as if it was an ordinary occurrence. "It's the least I could do."

"You're lying there in pain and starting to recover from serious surgery, and you are comforting your nurse? How'd you come to be discussing her personal problems anyway?" I was curious, not at all upset.

Barb smiled sweetly. "During our conversation I sensed she was in torment and asked her what was wrong. She soon blurted out that her husband had walked out on her. I had to try and help." That is what Barb was so often about, helping others. She seemed to sense when people were in distress and always tried to relieve it.

During a long-planned trip to visit my daughter Deb in Visalia, California, Barb started losing the function in her right hand. In a few days, the numbness progressed to include her arm and then her legs. She could not stand up. A consultation with Deb's neighbor, an ER doctor, soon led us to the Sansum Clinic in Santa Barbara. Two hundred and twenty miles from Visalia, Sansum has a national reputation for excellence. My story, *A Halo for Barb: Barb's Miracle*, tells that story in detail.

During the eight weeks she spent in the hospital in Santa Barbara, California, I traveled the route over the mountains from Visalia to Santa Barbara many times. The route took me north on 99, then over *the Grapevine* on I-5, down the Ventura freeway to the coast, and up the coast highway to Santa Barbara so many times I could probably have driven it blindfolded. During our visits, I began noticing members of the hospital staff stopping by to visit with Barb. Many of them had nothing to do with her care or support; in fact, their only reason for visiting was to talk to "the little lady with the big heart" as one young nurse's aide explained to me. At least a dozen of these people visited her regularly, and she loved it. Her ministry of Christian mission and service continued even from her hospital bed.

This held true in every institution in which she spent time. People in both rehabilitation institutions in California felt her Christian touch. I saw the same love and affection expressed in nearly all the ER and hospital stays across the country during our travels.

When Dolores, my ex-wife and mother of my five children, passed away early in 2004, my children asked Barbara to help them with the service and officiate as the pastor. My gang and even my ex had become close to Barbara over the years. She and Dolores had become friends, even shopping together for Christmas presents for the family on several occasions. She treated my children and grandchildren as her own, becoming Grandma Barb to the next generation. My entire family adored her and will all miss her terribly.

Such were the essence, the spirit, and the love of the woman I was privileged and honored to call my wife for the twelve short years we shared. She taught me much of the value of seeking the best in everyone and helped me to like and respect myself. I never had any idea how I would react under times of intense, personal crisis where constant loving care was essential. During those last years, I was with her 24/7 and caring for her every need. There was never a time when I had the least resentment for having to care for her. It was a great joy doing all those things for her. Of course I wished she was well and didn't need all that care, but I never minded doing for her the things she couldn't do for herself.

On numerous occasions, she would say, "How can you look at me and kiss me with my swollen face and bruise-blotched arms? I look so terrible."

My truthful reply was always, "When I look at you, I see that beautiful woman who sang *Til There Was You* to me during our wedding vows," or maybe, "I see that happy dancer in the red dress jitterbugging at the bishop's retreat." No matter what, I always saw the real beauty in that kind, loving face. Her warm smile and sweet face even showed in the last photo Deb took of the two of us in her room at Lakeland a few weeks before her passing. No, Barbara, you have always been beautiful in my eyes as you will always be.

Barb had an unusually strong faith and rarely complained. Throughout the years of pain and suffering, her Christian faith never wavered in the slightest. Even when she learned she only had a few days to live, she thanked God for what she had been given. With Pastor Mike Heath and I at her bedside, she said, "I'm ready to go home." That was a few days before her passing. Her strong faith was matched by her indomitable spirit that remained uplifting to the end.

Pastor Mike told me that on one of his visits during her last days, he found himself being ministered to by Barb as she lay in her bed. "I can hardly believe how she touched my heart, turned the tables, and comforted me when I was there to comfort her." That indeed was who Barb was, and how so many will remember and honor her memory.

Some months or maybe even years back, she told me, "If and when I go first, I will ask to be your guardian angel. Remember that," she said with a grin. Then, wagging her finger at me, she announced, "You had better behave!"

I promise to try, Barb, but it will be difficult. Such was the lady I had the phenomenal good fortune to call my wife for the last twelve years

Some months before her final attack, Barb had a few days of feeling unusually low. Several times over the years, she had expressed some concern about my feelings for her because of her health problems and how her appearance had changed. In a roundabout way, she was asking how I could still love her, and I knew it. I never had a good answer until that day when an inspiration opened my heart.

I kneeled down at her bedside, put my arms around her, looked directly into her beautiful big blue eyes, and said the following, "My dearest Barbara, I can honestly say that I have loved you without reservation since that August day in 1990 when we first met and spent those carefree hours getting to know each other. My love has grown constantly over time and never once has there been even a hint of doubt of that in my heart. I love you even more today than I did yesterday. I will love you more tomorrow than today, and it will continue so until the day I die."

After some tears of joy, we allowed that our feelings were mutual. I then explained how I had always felt about love. "I know only one way to love, and that is totally and without reservation. I know of no person I have ever really loved that I ever stopped loving. I think that quite impossible for me. Those both living and dead I still love include not only my family, but the several women who were at one time central in my life. I may have parted company with them, may never even see them again, but I will always love them." I had spoken to Barbara about this on numerous occasions over the years, but I knew in that moment she truly understood.

"Your words are beautiful," she replied through her tears.

We continued talking for several hours, sharing our deepest feelings with warmth, love, and candor. It was truly a magical moment, like many close and often unexpected moments we shared in our love over the years. I don't believe I ever felt closer to her than during those intimate hours. She certainly experienced the same wonderful feelings. It comforted me when I recalled those moments during the last few weeks, thankful we had that loving experience and revelation before she passed.

There was another incident several years ago that is rather a strong testimonial to her faith. Somehow, while we were watching a show on TV, we had a conversation about death and what it meant. Barb knew I was not as comfortable about that inevitability as she was.

She tried to help me feel less apprehensive. As we were talking, there was a death in the show followed by the following words: "This is not the end. This is not the beginning of the end. This is the end of the beginning."

Barb looked at me, repeated those words softly and then said, "That's what I believe it is." I've thought about that many times during the last few days.

I have one last incident to include in this little essay. On the day we learned her time was so short, we had a lengthy, sometimes teary, sometimes joyful, always loving talk. During those hours, Barb reminded me of her promise to be my guardian angel.

She said, "The first thing I'm going to do when I get up there is to apply for a job as your personal guardian angel."

Then she said something that totally blew my mind, "I'm going to find a wonderful lady for you to share the rest of your life."

That was so like Barb, always thinking of the welfare of others—she was as completely selfless as any person I have known.

After the gathering: It is now Sunday evening, a week after Barb passed. The last of my supportive family left a few hours ago, and the house is now incredibly quiet. It is the first time I have been alone for at least ten days. During that time, all of my children—including my three sons-in-law—both of Barb's sons, my sister and brother-in-law, six of my grandchildren were in and out of the house. It was a busy place with lots of love and support at a time when it was desperately needed. They have gone away, yet I know their comforting hearts are still here with me. Our precious friends are still close by for comfort should I need it. They all made their concerns clear and, asked that I contact them should any need arise. I know they will contact me frequently to ask if there is anything they can do.

The writer in me cried for expression, so here I am writing, creating, recording; putting words about love and comfort down to share with others. I wonder about what the coming days hold in store. I plan not to make any changes in my life until things settle and the mind-numbing fog that now envelops me clears. I do have some immediate plans. My grandson, Howard Russell Johnson—HoJo IV, Russ to most—is getting married in two weeks, and I will go to Washington DC to attend. I also plan to spend Thanksgiving with my three Michigan daughters and their families. After that, I may make the trip west that Barb and I planned to make when she was well enough to travel. My California daughter, Deb, has invited me to visit her. I plan to spend as much time writing as I can as I have several unfinished books, and of course, the completion of Barb's book is high on my list. I can't conceive of living long enough to put down all the words, sentences, paragraphs, and chapters that lay waiting in my writer's mind. My writing will desperately and sorely miss my fantastic editor, proofreader, critic, and promoter. I shall try to write as if she were looking over my shoulder. Sadly, her magic touch and peerless red pencil edits will be missing from my work.

I say a sad and humble farewell—until we meet again—to my dear Barbara whose faith never wavered, hope never dimmed, and good spirit never failed. I am hopeful of her words mentioned earlier, "This is not the end. This is not the beginning of the end. This is the end of the beginning." So life goes on.

A Fast Trip to California

After I lost my dear Barbara in October, 2005, Charlie and I puttered around the lake house, trying to get used to Barb not being there. There was a large hole left in my life, and probably in Charlie's too. That sweet little lady we spent so much precious time with was no longer around.

Early in November, I sent Charlie *to camp* at a kennel in Warsaw and flew to Washington for the happy occasion of my grandson, Russ's wedding to Jeanelle James. After the excitement of the wedding, of traveling around sight seeing in our nation's capital, and all the parties in celebration, it was back to our lonely home at the lake. We drove to Michigan for Thanksgiving and the celebration of my seventy-eighth birthday with the whole gang living around Detroit. It was a quiet, pensive trip back to Tippy and our empty house. To my great delight, my daughter, Debby, invited us to come spend the winter with her, but I had some things to do before I could leave.

We stayed there on the lake until before Christmas. Charlie took the opportunity to frolic in several heavy snows that fell in December. We had one humongous snowstorm that was Charlie's delight.

Charlie loved snow. Here he is bounding through snow deeper than he is tall. On this day, he never wanted to come in, staying outside on his own nearly all day.

615

Four days before Christmas, Charlie and I headed out for California in Buford, our motor home. Our plan was to drive nearly straight through, on I-80, stopping only for biological necessities and a little sleep each night, courtesy of several Wal-Marts. We had all the food and supplies we would need for the entire journey. It was a cold, gray day when I finished loading all of our stuff into Buford and headed northwest to I-80. The weather looked promising, as there didn't seem to be any snow forecast for our entire route.

We drove uneventfully south of Chicago, across Illinois, across the Mississippi River into Iowa, speeding over those long, rolling hills of brown fields, stopping only once for fuel. After sunset, we crossed the Missouri River into Nebraska and about midnight I found a Wal-Mart parking lot, our stopping place for the night.

Charlie sits in Buford's navigator seat as we rumble west on I-80 through Nebraska, December 22, 2005. Notice all the plants we took with us.

Early the next morning we filled up the gas tank and headed out, all before sunrise. We were still in Nebraska when daylight caught up with us. By noon we were chugging up the hill out of Cheyenne, Wyoming, a road I remember well from when we battled a blizzard on an earlier trip to San Francisco. This time the road was clear and dry, but we were battling high cross winds all the way across the prairies and over the mountains. Buford was being tossed around making for some tense driving. When we came out of the Mountains and into Salt Lake City, our short daylight hours, and the wind, were gone. I realized then that most of the trip was made in the dark. Before reaching, Nevada I stopped to catch a few winks before the third and last leg of our trip. This time we stopped at an all night eatery in a small town.

Charlie and I make an early morning *pit stop* in Nevada after our harrowing, blind ride through a foggy night on the tail of a semi.

About three in the morning I woke up and decided to head out—into a thick fog. On an unknown road in fog and darkness, my heart sank thinking how this would delay my trip. I was inching along at about thirty when a semi blasted past doing twice my speed. I slammed down the accelerator and managed to catch up with him enough to see his red tail lights. I followed those lights far enough back to be able to panic stop, but close enough so I didn't lose sight of them in the fog. We rolled through Nevada at between sixty and eighty, constantly adjusting Buford's speed to stay with the semi, but not dangerously close. About three hundred miles later, daylight had dissolved the fog, and I relaxed. The air was now clear of fog and the sky was sunny.

We continued across Nevada and into California and over Donner Pass. There was lots of snow up there, but I-80 was clear and dry. We continued down the mountain to the junction of I-80 with I-5 in Sacramento, and from there, south to Stockton.

We arrived in Stockton at Deb's around noon on the 24th. As we drove down the street, there was Deb, out in the middle of the street, waving enthusiastically and showing us where her house was.

After parking Budord in the street beside her house, I discovered that the fierce wind we encountered in Wyoming had blown the plastic roof out the rear half of the driver side. A trip to an RV dealer and we learned it would cost about $5,000 to fix the roof. My insurance would cover that less the $1,000 deductible. I got on the Internet and soon learned how to fix the roof myself. I bought about $250 worth of parts including the following: a roll of 16-inch-wide, heavy, plastic roofing; an aluminum extrusion to hold one side of the plastic; and about a dozen tubes of two types of caulking, I spent about a day and a half making Buford's roof better than new. I figured the insurance check of almost $4,000 was adequate pay for my effort. I also knew the repair was done correctly. It's held up through all kinds of weather ever since.

Deb fixed me up with a computer desk right next to hers in a sun room converted to an office. We had a fabulous time together and did a whole lot of California style running around. It was with a bit of regret that I headed back to Indiana and that lonely house in late April of 2006. Deb and I enjoyed our time together. Oh yes, Deb's dog, Jake, and Charlie became close buddies. They often took me for walks around the neighborhood.

Here are the buddies, Jake and Charlie, pulling me along on one of our daily walks at Deb's house.

An experience on the Anhinga Trail in the Everglades
- A photo essay

In March, 2007, Daphne and I visited Flamingo in Everglades National Park in our RV for a week. During our visit, we walked the Anhinga Trail. While on the trail, we had a most unusual and interesting experience. As we walked on the wooden section of the trail, we photographed a cormorant drying its wings on a branch above the water.

A Cormorant dries its wings while a couple of gators look on hungrily

Suddenly the bird dived into the water and disappeared for nearly a minute. It then struggled to the surface with a large catfish in its bill, flew out of the water and landed on the boardwalk a few feet away from us.

The catfish struggled mightily, but the cormorant held on, its beak clamped onto the head of the catfish by its gills. As we watched fascinated, the bird maneuvered the catfish against the wooden planks until its beak was deep into the gills. Then it repeatedly stabbed directly into the heart of the fish with its pointed lower bill. Having many times cleaned catfish, I knew exactly where the heart was as obviously did the bird. Soon the catfish lay motionless and quite dead.

Cormorant maneuvering the catfish and stabbing it repeatedly directly in its heart to kill it so as not to get stabbed. (Above)

It finally lay still and was now obviously dead. (Below)

At this point, the cormorant relaxed its grip on the fish, turned it and grabbed it by the head, lifted its head and the ill fated catfish straight up toward the sky and proceeded to swallow it, headfirst. The bulge in its throat that was the catfish, moved slowly and steadily down the bird's throat to disappear into its body. The bird walked around for a few minutes, probably adjusting the position of the rather large meal, then flew nonchalantly right past a dozen amazed spectators onto the handrail where it paused for a moment before flying away to a perch to digest its prey. That prey probably weighed nearly a quarter of what the hungry bird did before its meal.

What was so amazing to us was how skillfully and swiftly the bird dispatched the catfish that otherwise could have done it great harm if it had been alive and struggling when swallowed. Quite obviously the cormorant had handled this kind of dangerous prey before.

❖   ❖   ❖

The difference between fiction and reality? Fiction has to make sense.

—*Tom Clancy*

It's not the size of the dog in the fight, it's the size of the fight in the dog.

—*Mark Twain (1835-1910)*

It is better to be feared than loved, if you cannot be both.

—*Niccolo Machiavelli (1469-1527), The Prince*

Whatever is begun in anger ends in shame.

—*Benjamin Franklin (1706-1790)*

The President has kept all of the promises he intended to keep.

—*Clinton aide George Stephanopolous speaking on Larry King Live*

We're going to turn this team around 360 degrees.

—*Jason Kidd, upon his drafting to the Dallas Mavericks*

Half this game is ninety percent mental.

—*Yogi Berra*

There is only one nature - the division into science and engineering is a human imposition, not a natural one. Indeed, the division is a human failure; it reflects our limited capacity to comprehend the whole.

—*Bill Wulf*

There's many a bestseller that could have been prevented by a good teacher.

—*Flannery O'Connor (1925-1964)*

He has all the virtues I dislike and none of the vices I admire.

—*Sir Winston Churchill (1874-1965)*

Write drunk; edit sober.

—*Ernest Hemingway (1899-1961)*

I criticize by creation - not by finding fault.

—*Cicero (106-43 B.C.)*

My New Home town, Saint Augustine, Florida.
Looking north on Saint George Street in downtown Saint Augustine

I am a new resident of St. Augustine from northern Indiana where I lived on Tippecanoe Lake. 2011 was is my sixth winter here, and I love it. St Augustine has to be one of the best kept secrets of America. There is an unbelievable variety of things to do and see here from centuries old structures and forts to the charming, narrow streets of the old town to specialty sights like the lighthouse, Ripley's museum, and the Alligator farm. Tour trains and horse drawn carriage rides that wind through the narrow streets are a delight while cultural attractions and wide, sandy beaches appeal to a wide range of tastes. All these attractions are readily accessible in a smallish town with a friendly atmosphere and plenty of good restaurants. Other interesting places and sights in Florida, Georgia and South Carolina are within a day's drive. Of course, the main attraction for me here is my lady, Daphne Fox.

**Daphne's lovely home in Vilano Beach with Buford
and my beemer in her driveway**

❖   ❖   ❖

All the gold which is under or upon the earth is not enough to give in exchange for virtue.
—*Plato*

If you ignore the faith-based content of the Bible, it becomes a valuable history, a study of human nature, and an excellent guide to human behavior.
—*Howard Johnson, 1967*

A Life Changing, Wonderful Experience
Reunion with My Daughter, Kristen

On Saturday, July 11, 2009, I read the following email, sent on Friday, July 10, 2009:

> Hello after all these years....
>
> I hope this email from the family blog site still works to reach you.

It seems like a strange convergence of events that have brought me to writing this email.

A few weeks ago I received an update email from Classmates. It listed a small amount of information about the individuals that had searched my name. It listed an 81-year-old from Leesburg, FL. Florida? 81? I quickly came to the conclusion that you had searched on my name.

Last Thursday on our drive down to the Great Sand Dunes, my eldest daughter (Cadence, 6) asked *"Do you have a daddy?"* *"Where is your daddy?"* To which I answered *"Yes, I do. I am not sure if he is still alive, however, I think he is and he might be living in Florida."* She has come to realize that she has *Opa* on her father's side of the family, but no grandfather on her mother's side of the family. She then asked why we didn't see you when we were in Florida, and I told her that was because we didn't know exactly where you were.

Then on Wednesday evening, my mom and I took the girls (yes, two girls – Cadence Emily 12/21/02 & Madeleine Claire 11/5/04) out to dinner. While sitting on the patio eating dinner, Mom says to me, *Right now Cady's eyes look just like your fathers.*

So today at work I decided to Google *Howard Everett Johnson FL* and I finally found a link into your life. I have looked before, however, it is a little difficult for Google to sort out the seemingly millions of references to Howard Johnson hotels in the world. Had to get the information just right (Howard Everett Johnson Indiana didn't get me this far) to get the search gods to smile down upon me.

So, if you would like to converse, talk, meet...just reply and let me know.

It's been a long time. People make choices. I'm not here to worry about the past. Can't say I don't have questions, but if you would like to get to know me, your son-in-law, your granddaughters....this is where you can start.

kristen

After reading that email, I sat and bawled like a baby for at least ten minutes. It was something I had prayed for and dreamed about for many years. As soon as I gathered myself together, I responded:

- - - - - - - - - -

Saturday, July 11, 2009

My dear Kristen:

I hope you have some concept of the joy your email brought into my life. It is overwhelming! I have dreamed about having contact with you countless times. Several years ago I even wrote a short story of a fanciful meeting. Of course I would love to talk to you and your family, see you---anything. I am at your disposal as to what, when, where, etc.

I will be here at the lake in Leesburg, Indiana (not Florida) until early October when we head south to St Augustine, Florida.

Right now my heart is full, and my mind is struggling with what to say—overflowing. In spite of my nature to blurt out everything that comes to mind, I'll leave all those things for another time. Should you want to know a bit about my views and who I am, goto **http://tdbookshj.blogspot.com/**. It has links to many of my writings in blogs. Of course, it may have much more than you have the time or care to dig into. It's a path to a lot of information.

Kristen, you opened a doorway I have prayed for many years to be opened. My hope is it will lead to many delightful things.

Love, Dad

- - - - - - - - - -

Monday, July 13, 2009, I received the following:

Hi Dad,

It just seems so surreal to be writing those words after all of these years. Powerful emotions overwhelm me...good ones, not bad. So many words, stories, anecdotes come to mind that I have not been able to type this until now. I am very excited to be starting on this journey of discovery.

I think I'll start with current info and we can get to all the past stuff a little later.

I live just west of Denver, Colorado with my husband of ten years (Vince) and our two daughters, Cadence & Mady. We are just finishing a two-year long endeavor of remodeling our house. We came home from our Florida vacation on May 23, and our basement flooded the next day. It was only rain water, but we are now *re-remodeling* about 1/3 of our basement. Vince mountain bike races for Rocky Mountain Race Team (known as RMR) and works for Level3 Communications in

Broomfield. I work for Avant Datacomm Solutions in Golden, CO. We are a specialty contract firm that installs voice/data communication cabling and security. I also try to get my exercise by competing in sprint triathlons. I will be competing in my 5th this August. Cadence will be starting 1st grade this fall. Mady is twenty-two months younger, but you would never know it!

I am attaching some pictures, some more recent than others. So there is a quick recap and I hope all of the pictures will make it through. There will be more later, but right now I literally have to run!

love, Kristen

- - - - - - - -

After a rather long phone call with many ideas and plans for meetings discussed I received the following:

Monday, July 20, 2009

Hi Dad,

It was great talking to you Friday night. I have cleared everything on this side, and it looks like we are a go for August as long as the following works for you.

I will fly out on Saturday, August 15 to Dayton. From there I will go to Columbus for the night to visit my friend that has been in the hospital almost three months. I'll drive from Columbus to Leesburg on Sunday (August 16) after lunch, arriving late afternoon. Then stay with you Sunday and Monday night. I will need to leave for the Dayton airport around 11am on Tuesday (August 18).

Just let me know, and I will make it official.

love, k

- - - - - - - - -

Monday, July 20, 2009

Hi Kristen!

F Boom and Yippee! I now have a smile on my face that no one can erase. That itinerary sure works for me.

Daphne leaves for Florida tomorrow morning (Doctor appointments), so Charlie and I will be batching it for a couple of weeks. I'll try to finish her list of things I'm to do before she gets back. <grin>

You have already changed things around here. Looking at me from the wall at the back of my cubby hole PC desk, I have that lovely photo (8 x 10) of your family with the waves breaking behind you. That is one gorgeous picture! I also now have two new granddaughters gracing my rogue's gallery in the living room. They fit in right along with the other ten grandkids—all beautiful.

I'm putting together a package of stuff you might find interesting. In it are a few things that need explanation. That evil-looking guy on the cover of the Passion Play program is none other than yours truly. I played Tiras, a chief priest of the San Hedron for exactly 100 performances over ten years. It was an extremely rewarding experience. For years I've been performing in amateur theatrical productions or singing groups. For the last fifteen years, I've sung with *The Willows*, a local gospel quartet. We used to sing at churches and civic events. We decided to disband this year for several reasons when one of our members began having health problems.

I sing in two choruses in FL: *Singers by the Sea* in Jacksonville Beach, and the *St. Augustine Community Chorus*. I guess I'll keep singing 'til my voice goes away.

The book, *Words from the Lakeside* is self explanatory. It's been written and collected over at least the last fifty years—most in the last ten. I write both fiction and non and find the freedom in calling my work fiction lets me tell stories, even true ones. If labeled non fiction, these stories would prompt many to remark, *You expect me to believe that?* I expect it to be published late this fall, but the copy I'm sending is quite close to what the finished product will be. It was essentially completed when Barbara passed away, but it took me a long time to have the spirit to hone it into a finished product while I was putting my life back in order.

Nearly five years of increasing care of a treasured wife changed me a great deal. I am gratified to learn I have that quality of love and dedication. I don't think anyone truly knows how one will react in a situation like that until facing it. To watch a once vital and extremely active lady with a bubbly personality slowly deteriorate to the point of being bed and wheelchair bound is a personally devastating experience, yet it was immensely gratifying. I found that I was so intent on helping and caring for her it was a rewarding experience. The last few months we were closer than we had been for many years. We were together 24/7 for most of the last three years. In spite of the terrible pain and growing weakness, Barb's spirits were high, and she always had a quick quip or two to brighten the moment. There is much about Barb in the last section of the book.

Here I go rambling on. I guess that's the writer in me. Once I get started, the words seem to pour out in a flood. At least that's how it seems. The other day I snapped awake at the keyboard, looked up at the clock, and saw it was after eleven. A look at the window startled me because it was bright out. I looked at the computer screen, and it was appropriately filled with Zs. The last thing I remembered was noticing four o'clock a.m. on my clock. I had started writing about seven in the evening, worked 'til some time after four in the morning and then fallen asleep on the keyboard. With Daphne away overnight I had no one coming in tapping me on the shoulder and asking, *Aren't you ever coming to bed?* I don't have a clear sense of time, especially when I'm writing or reading.

I'm loving looking forward to your visit. It's a bit unusual meeting someone I have loved in absentia for so many years.

Love, Dad

- - - - - - - -

The August visit was fantastic. It was the first contact I had with Kristen since she was fourteen, 27 years ago. A bit of explanation by way of some personal history may be appropriate at this point.

- - - - - - - -

In the 60s, my wife, Dolores, and I had some serious marital difficulties which led to several separations and ended in divorce years later after the last long separation. We worked desperately to hold our marriage together for the benefit of our children whose welfare certainly gained from our efforts. I was personally devastated. I did not handle our problems well and soon embarked on a dangerous path driven by a mixture of anger and self pity. I will not reveal the details in this piece. I became a not so nice person. Well into this personal decline I met Caroline, a kind and delightful lady who soon captured my heart. We fell in love. Without her influence, I would probably have continued on my downward path, and gone even further into the depths. She certainly helped turn my life around for the better—much better.

We had a beautiful daughter, Kristen, in February 1968. For some valid and some foolish reasons which I will not enumerate I did not do what I should certainly have done. As a result, Caroline and I ended our relationship. From that date forward, my self esteem deteriorated to the point where for several reasons including Caroline's request, I stepped completely out of Kristen's and her lives.

I was a mess, and with a second, sudden and foolish marriage on the rocks I moved to Chicago. Shortly after that, I spent a year, working in the Philippines. It was an opportune time for me to be far away as I began a steady recovery. By the time I returned from the Philippines and received a second divorce, I was regaining control of my life and my self esteem was beginning to come back.

In 1982 during the early part of my recovery period and during my move to Chicago, I stopped to visit Caroline and Kristen at their apartment in Rocky River. I had fleeting thoughts of trying to get back together, but I was broke and had terribly low self esteem. After seeing them, I decided they would be far better off with me out of their lives. This was the second crucial mistake in judgement I made involving two people I dearly loved. Since one can take but a single path in life, there is no way of knowing how the other path might have turned out. It is entirely possible they were better off with me out of their lives.

Now gun shy of women I kept to myself, immersing my efforts in my work. After a year or two I found myself once more wanting companionship and joined PWP, Parents Without Partners, a group with many social activities and opportunities to meet singles. It was in this group that I met a terrific lady who helped me regain my self respect, and put my life back together. Iola and I were together for several years. I adored her two teenage daughters who in time became almost like my own. For several nebulous reasons we never married and finally decided to part company. We have remained faithful friends to this day.

After we went our separate ways, I met and then married Barbara. There are several articles about Barbara near the end of this book, including how we met, our marriage, and her tragic death. Barbara helped me regain my self esteem, personal confidence, and to respect myself once more. She also helped me to become a writer, a creative endeavor I have come to love that is most rewarding. It was a life-changing experience. After Barb had passed away, I met Daphne Fox, a wonderful, classy lady from St Augustine, Florida. We have been together enjoying the warmth of love and companionship ever since.

Kristen and her dad during her visit to the lake, 2009

When Kristen arrived Sunday, August 16, we shared a long, teary hug. It was a breathtaking experience. The two days she spent with Daphne and me were among the happiest days of my life. It was almost the Biblical story of the prodigal son all over again, factually quite different, but each father obviously had similar feelings.

Now I have a fabulous daughter back in my life, two beautiful new granddaughters, and a new son-in-law. In a few weeks I will be in Denver to meet my new family in person for the first time. My excitement knows no bounds. In most respects, life has been kind to me as well as rewarding. The wonderful happenings far outweigh the hard knocks. Now that my youngest daughter is back in my life bringing with her two beautiful granddaughters and a son-in-law I have yet to meet, life is indeed sweet.

Saturday, November 21, 2009, 9:00am - I am in Denver, or rather Lakewood Colorado on Green Mountain several miles west of Denver in Kristen and Vince's lovely home. My daughter is up, and my two beautiful little granddaughters, Cadence and Madeleine, are eating their breakfast right beside me.

This has been a spectacular week with more beautiful surprises than I could imagine. It started last Saturday when I rode down the escalator in the Denver airport to be greeted by Kristen and my two youngest granddaughters. I had been extremely nervous hoping and praying I would make a good impression. It's not often one meets one's granddaughters, ages seven and five, for the first time. My fears were unfounded as two beautiful little girls greeted me with big smiles and several hugs as I reached the bottom of the escalator. I was in emotional overload.

This unbelievable week continued as I got acquainted with Vince and those two little sprites. When Vince called me *Dad* it touched me. He is the first one of my son-in-laws to call me Dad. I loved it. Five year old Mady came up with her own original name for me. A name that was accurate, original, and that deeply touched me. All on her own she started calling me *Mommy's Daddy*. We played games, read, talked, went out to dinner, fixed breakfast together. It was a warm and loving family experience made all the more precious since it was our first time together—ever.

I was a bit nervous about meeting Kristen's mother, Caroline, because of our history, and because I knew she was coming over to pick up the girls. I expected a polite, cooly friendly greeting when she came. I was totally unprepared for the warm, friendly hug I received, even before we spoke. In an instant, thirty eight years of assumed hostility evaporated into nothing. It was two old friends, estranged for many years because of misunderstanding on both parts suddenly discovering each still cared for the other and so instantly renewed that close friendship. Once more I went into emotional overload because of the spiritual beauty of the realization of what had happened. It was a real joy not to be excluded from this wider family in Denver because of the past. It was truly a beautiful and highly emotional happening. It was an incredible relief from those years of tension experienced by each of us—that tension, gone in an instant of realization. Life would now be different for us forever, all because of that simple email message from Kristen.

The Allens - Mady, Kristen, Vince, and Cady - Christmas 2010

❖   ❖   ❖

For centuries, theologians have been explaining the unknowable in terms of the-not-worth-knowing.

—*Henry Louis Mencken (1880-1956)*

Pray, *verb.*: To ask that the laws of the universe be annulled on behalf of a single petitioner confessedly unworthy.

—*Ambrose Bierce (1842-1914)*

Charlie

About a month after our little Rags' untimely death, we decided to look for another companion dog. I called the local humane society in Warsaw as a first step.

"I'm looking for a small dog, a Shih Tzu or Lhasa Apso, preferably."

The answering voice replied in a friendly, female tone, "We almost never have any of those breeds. If we do get one, it's gone almost immediately."

"Well, I'm not in a hurry. Would you call me if you ever get one?"

With a smile in her voice, she answered, "I can do that. Give me your phone number."

After I gave her my number and said goodby, I put it out of my mind. With my wife, Barbara, in deteriorating health in a local nursing home I had more important things to do than look for a dog.

No more than a week later, I received a phone call from the lady at the humane society.

"Are you still looking for a dog?"

After the few moments it took for her question to register, I said, "Why yes. I am."

"Well, we had a little caramel colored male Lhasa Apso turned in. I can hold him for a few hours, but that's about it."

"Say no more. I'll be there in less than half an hour," I said before dashing out the door and heading for the Humane Society.

When I arrived and gave the lady my name, she said, "Go through that door and walk to the second aisle. He'll be in the last cage on the right."

As soon as I opened the door, a cacophony of barking from dozens of dogs greeted me. As I walked down the row of cages, each dog greeted me with excited barking. It sounded like a madhouse for dogs. When I glanced into the last cage, there against the back wall sat a quiet, seemingly nonchalant little bundle of caramel colored fur, watching me intently. As soon as I called to him, he came over to greet me, his ample tail flag wagging happily. He licked my outstretched hand, and looked up at me with big brown eyes that seemed to say, *Love me! Love me!* I was hooked.

Charlie, the charmer. "What's not to love?"

After signing a number of papers, I paid to remove him from incarceration, hooked up his new leash and headed out the door. I immediately discovered he was a strong, friendly, exuberant little dog. When I picked him up to place him in my car, I realized how compact and solidly muscled he was. My first action after picking him up was to head to Milford to introduce him to Barb at the nursing home. During the twenty minutes it took me to drive there, he was jumping all over the car, front seat, back seat, on the floor, even in my lap. When we arrived at the nursing home, he headed for the front door with me in tow. He seemed to know where he was going. He pranced through the front door, down the hall, greeting each person he passed with full body wags.

Barb was asleep when we got to her bedside, so I sat down in the chair and kept him on a short leash. The name on his papers was *Tank*, appropriate for his burly physique, but not for his personality. Out of the blue I started calling him, *Charlie*.

Barb woke up and saw me in the chair. "Why didn't you wake me when you got here?"

"I wanted to let you sleep. I'll be here all afternoon." She couldn't see Charlie on the floor. "I brought you a surprise," I said as I reached down to pick up the bundle of fur at my feet.

"Where'd he come from? He sure is cute," Barb said with a smile as I placed Charlie on her bed.

"This is *Charlie*," I explained as he immediately snuggled up to her and gave her a bunch of doggie kisses. "I picked him up at the Humane Society. His name was *Tank*, but I started calling him *Charlie*. He looked like a *Charlie* to me. What do you think?"

"He certainly is friendly," Barb said as he snuggled up to her with even more kisses. I think you're right. He does look like a *Charlie*. *Tank* is definitely not him."

"OK, mutt. Barb says you are *Charlie*, so *Charlie* it is."

From then on, Barb and I took Charlie for a walk around the grounds of the nursing home every afternoon. He took to his new name quickly and even when we let him run loose. He usually came when called. Of course, when there was another dog or animal around he paid no attention to our calls, so we soon kept him on a long leash, even when we played cards in the Gazebo in front of the home.

After we lost Barb in October, I took Charlie with me to pick up her things at the rest home. Charlie went eagerly back to Barb's room with me and seemed confused when she wasn't there. He looked all around and even led me outside to the Gazebo. I sat down in the Gazebo and cried for about ten minutes. How do you explain someone's gone to a furry family member?

We stayed in Indiana until just before Christmas. My daughter, Debby, invited us to come spend the winter with her, but I had some things to do before I could leave. We also had a humongous snowstorm that was Charlie's delight. Four days before Christmas, Charlie and I headed for California in Buford, our motor home. Our plan was to drive nearly straight through, stopping only for biological necessities and a little sleep each night, courtesy of several Wal-Marts. The details of our trip are described in *A Fast Trip to California* on page 615. We arrived in Stockton at Deb's around noon on the 24th. As we drove down the street, there was Deb, out in the middle of the street, waving enthusiastically and showing us where her house was.

After a few territorial arguments and a few more over the possession of toys, Deb's dog, Jake, and Charlie became the best of friends wrestling and play-fighting constantly. They soon became protective of each other when other dogs came near. They were great buddies.

Big changes then came into our lives. We left Deb's hospitality in April and headed back to our empty home in Indiana. I then had the good fortune to meet a lovely lady named Daphne Fox who lived in St Augustine, Florida. We soon established a wonderful relationship and put our lives together. Charlie now had a new lady to love, and a new place to live when we weren't at the lake in Indiana. Since this story is about Charlie, I will provide the details in my story about Daphne and how we came to be together.

Charlie is an extremely exuberant character. He loves everyone he meets and accepts much abuse at the hands of children without complaint. I'm sure his tolerance has limits, but so far they haven't been reached. He gets excited around other dogs, but seems to get along quite well, when not on a leash. Something about being on a leash makes him more aggressive toward other dogs. He aggressively chases birds and small animals including cats. This has led to several altercations where he came out second best. He is now a bit more cautious about how close he gets to those knife equipped critters. Charlie is the best communicator for a dog I have ever known. He almost speaks in actions and sounds. He has a well developed set of communication activities and repertoire of different barks and other noises. He also has an amazing understanding of what Daphne and I say. We have spelled out several activities like, b-i-k-e-r-i-d-e for some time. He has even learned what b-i-k-e means and acts accordingly when we spell it out..

Here are four photos from one excursion with Charlie:

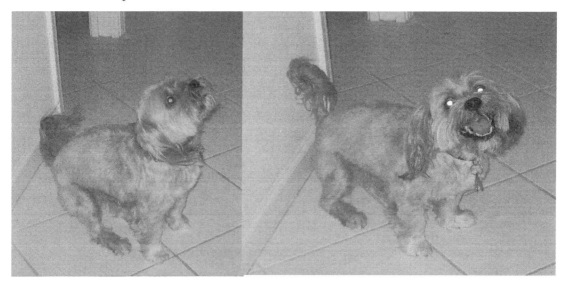

Waiting impatiently by the garage door.   Notice werewolf Charlie in this photo.

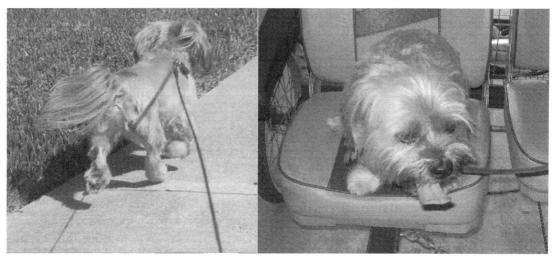

Charlie runs at the end of his leash.   The heat wore him out, so he rode home.

We have a four-wheeled two-person pedal vehicle called a Rhoades car. Charlie has decided it belongs to him. He goes wherever it goes with obvious enthusiasm. Hint at a bike ride and he immediately leads you to the door to the garage where he literally and repeatedly bounces almost five feet into the air until the door is opened. Open that door and he heads quickly for the Rhoades, prancing around in eager anticipation as I attach his leash and move the Rhoades toward the driveway. As we start down the driveway, he pulls with all his considerable strength, running down the street as fast as those powerful little legs can carry him. At his vet's recommendation, I reign him down to a dog trot after he runs full tilt for about 100 yards. We follow a figure eight, mile and a half, route on the streets of the development. In hot weather, Charlie slows down to a walk shortly after we are half way through the figure eight. When it's cold, he maintains his dog trot almost for the entire route.

When on this route he is on a mission, not slowing or turning except to mark his territory or answer nature's call. Even when we pass other dogs, he seems more intent on his route than paying attention to them. Of course, he has a few buddies he stops to greet, and a few others he avoids. He becomes insistent on going for his ride when it gets late in the day. He does the same when we are at the lake where we have a little bit shorter route over gravel roads.

Charlie loves and expects treats after dinner. He behaves well when we are eating, not begging or bothering us in any way. As soon as dinner is over he heads for the kitchen, bouncing and prancing with anticipation, especially when Daphne heads away from the table. She usually saves him some meat from the dinner, but if not, I give him some chicken strips. We have trained him to wait until we say OK. I hold a small piece between my fingers down at his level. He waits patiently until I say OK then he gently takes the piece from my fingers. Sometimes I test him by making him wait for as long as a minute. He waits, maybe not patiently, but he does wait.

Charlie is death on toys. It took us a long time to find toys he couldn't dismantle in a few minutes. He destroys most regular tennis balls in short order. The ones packaged as dog toys seldom last even one minute. We found one plastic ball covered with knobby protrusions that he loves and seems unable to even faze. He still has the first one of these we got him. Some stuffed toys with squeakers last a day or two at most before the squeaker is silenced. Sometimes we find pieces of white fuzzy stuffing spread all over the floor after he has found a vulnerable spot.

When we travel in Buford, our motor home, Charlie either rides on the wide dash by the front window or sleeps on the couch right behind Daphne's passenger side seat. Charlie is an enthusiastic traveler and loves riding in virtually any vehicle.

Charlie standing on the dashboard in Buford's front window while we camped in Key West over Christmas in 2008. Note the bay out front.

Charlie is not shy about anything. I'll be working at the computer and suddenly he forces his head up between my legs and into my lap to let me know he wants to go out or for a bike ride. He lets us know emphatically when anyone is outside or at the door, or when the pool pump has started up. Of course, when he is outside on his run he only barks at passing dogs or other animals. Should a stranger come to the house, he walks up to them, happy as can be, wagging his tail eagerly without making a sound. Some watch dog.

He chases his ball endlessly as long as someone will throw it for him and often throws his toys all by himself when he is not chewing on them. He always wants to go with us, but at the lake when we are going out (he always knows) he will trot out on the back porch and hop up on the couch until we return. When we are in St Augustine, he goes out into the garage and climbs into his bed under the same circumstances. He is infinitely better behaved than he was a few years ago, but still, he will run if left outside on his own.

He is so happy, so cute, and so loveable you can't get angry with him, even when he misbehaves. His facial expressions clearly express how he feels from sad to eager to happy. I have never known a dog who could so clearly communicate both what he wants and how he feels. All in all, Charlie is a thoroughly delightful companion who loves everyone.

Daphne Fox, The Lady from St Augustine,
A Marvelous, Romantic, Late-life Relationship

I have had a number of truly amazing and intensely rewarding relationships with wonderful women in my life. These include a mother, grandmother, sisters, aunts, children, grandchildren, and many dear friends. I have also had six committed love relationships and one major disaster. These included three marriages and four intense and significant relationships outside the formality of marriage. My second marriage was a total disaster from the start. I'll say nothing more about it except it was a lesson in human frailty and foolishness that kept me from ever making a similar mistake. All the rest were and are beautiful realities making for pleasant memories, many related elsewhere in this book. While my first marriage ended in divorce, we remained on relatively friendly terms. Neither of us ever said a bad thing about or against the other to our children, a blessing that made all of our lives much more pleasant.

I know it runs counter to what most people believe about love and relationships, but I never will stop caring for any of these six women. This story is about my current love relationship, so I'll leave the others to their proper place elsewhere in the book. In any event, those people who are important to me, know most all of the significant details, so they need not be repeated here.

After Barb died and as she specifically requested, I decided to search for a new mate. I was not in a hurry, but at 78 it's not realistic for one to take a whole lot of years waiting for the right time. Besides, I had almost four years of grief as my Barbara slowly deteriorated before my eyes. I signed up on match.com, an Internet matchmaking service. I was surprised at the number of women interested in a man like me, especially at my age. Over the next few months I met a number of decent ladies, and a few duds. Typically, we would meet for coffee or lunch at a neutral spot after deciding from our successful communication on the Internet. I even formed a few friendships, pleasant, non romantic friendships. I found myself at first interested in one lady who turned out to have so many negatives and red flags that I took off and never looked back. One disastrous relationship per lifetime is enough.

Daphne, with whom I had a fascinating email exchange, was so far away in Florida, I almost didn't continue in spite of being attracted by our email communication. Then it dawned on me, I fly free on Delta so why not? What a fortunate decision that was. We exchanged many emails over the next few months and that led to my trip to Saint Augustine in March. Another long visit a few weeks later and I was definitely hooked on this classy, beautiful person. I had a wonderful time during our visits. Daphne was even more pleasant and gracious in person then via email. After, or maybe even during my second trip to St Augustine, I decided, I would give up my other female interests and concentrate on this lovely lady from St Augustine. We hit it off extremely well, and once more I knew quite early that she was **the** one. After recovering from one little rough spot, we were off on our way to a satisfying relationship. I was soon very much in love as was Daphne. Of course, she was much more reserved and cautious than I with the *jump-in-with-all-four feet—here I am—what you see is what you get—let it all hang out* approach to life of yours truly. Daphne is a private, socially rather conservative person. It took some time and effort for me to tone down my open, incautious lifestyle and mellow things down between us. Realistically, we hit a sort of compromise, and before long were both in a committed relationship.

To this day I strive to accommodate this kind and gracious woman, even to the way I am writing this section. She will read and approve my writing about us before it becomes a part of my book. I do this because of my love and my great respect for the lady she is. My desire to please and never disappoint her determines that. Suffice to say we have come a long way since those first few months more than four years ago. I have never wavered in my love, or my commitment to Daphne.

One incident, among the many that put an exclamation point on our relationship, happened at the Johnson family reunion at the lake in 2007. My son, Howard, and I were sitting at the dining room table visiting when the subject of my new lady came up. It was the first time he had met Daphne.

While looking at me intensely, he shook his head and said firmly, "Dad, you are the luckiest SOB I know of."

I allowed that his statement was quite accurate and added, "You can say that again. She certainly is a jewel. I consider myself one lucky old goat."

Another incident that warmed my heart happened during one of our visits to Gainesville where three of her daughters live. Daphne and I were staying with Missy, twin sister of Abbie for some occasion. I was seated alone in her dinning room when Missy came up to me.

"Howard, I want to apologize for treating you so cooly up until now." Then she put her arms around me and gave me a big hug.

Understanding where she was coming from, I replied. " Oh for God's sake, Missy. She's your mother. That was nothing personal against me. You were being protective. That's a natural and loving thing."

✻　　　✻　　　✻

Fast forward to the present, November of 2010. Daphne and I have been together for more than four years. My constant hope now is that she is half as happy to be with me as I am to be with her. I could not possibly have found anyone with whom I could be more pleased. Yes, we have many differences, amplified by long years of being who we are. One of our initial concerns was the difference in our religious beliefs. I now go to mass with her each Sunday, and she goes to church with me when we are in Indiana. For all practical purposes it became a nonissue. She keeps a clean and tidy home, so I am trying to mend my sloppy ways with some success, maybe not fast enough for Daphne, but at least I am trying. This even holds true at the lake house in Indiana. She's a positive influence on me.

Daphne loves to travel, and we have taken many tours and Elderhostels together. Let's see, we've been to Greece, sailed the Mediterranean in a clipper ship, and even spent a Saturday night in Dublin. (She's Irish.) We traveled to San Francisco, Seattle, Calgary, and the Grand Canyon for Elderhostels. We took a cruise up the inner passage to Alaska and Glacier Bay. Also, on that cruise we took the train out of Skagway up to the famous pass to the Klondike. April 20, 2009 we started off on a 9 thousand mile, 9 week RV tour clear across the south, the southwest, and up the California coast. On that trip we visited my son, Howard, and two nieces in Texas and Arizona before going to San Diego to visit Daphne's son Scott, and his family. We then drove up the coast to Monterey, San Francisco, and then Stockton where we visited my daughter, Deb. We then drove across America's waistline on US 50, called *The loneliest road in America*. In fact, we only passed four vehicles on one stretch of more than a hundred miles. We arrived at the lake house in mid July. Traveling with her by whatever means we choose is a marvelous experience we both enjoy immensely.

On our RV trip we visited the Grand Canyon, Yosemite, and Arches National parks, Monument Valley, Sante Fe and Taos in New Mexico, along with numerous other spectacular sights. We even visited the Wizard of Oz museum in Kansas and spent a day in Hannibal Missouri of Tom Sawyer and Mark Twain fame.

Since then we traveled to Tahiti and boarded the Star Flyer for a sailing tour of the islands of French Polynesia including Bora Bora and the Tuomatu Atolls. Our last trip was an informative guided tour of Germany. The jewel of the trip was the Passion Play at Oberamergau. This was followed by a week on our own in and around the Bavarian Alps. We have reservations for a trip to England for a tour in April of 2011. While there, we will visit with my Grandson, Russ, his wife, Jeanelle, and their baby girl, Jameson, whom I have never seen.

Daphne created a beautiful scrapbook of our initial exchanges made on match.com that initiated our relationship. That scrapbook was the first in a series of beautiful scrapbook records she made of our many travels and family experiences.

Since Daphne brought me to St Augustine, I have become active in several groups. I sing in two local choruses, Singers by the Sea in Jacksonville Beach, and the St Augustine Community Chorus. I am also an active member of the Florida Writers Association, the Socrates discussion group, the River House Players, a senior drama group, and the River House Writers group who write and critique memoirs. I'm also an occasional lecturer at the Council on Aging.

All in all, I have been unbelievably fortunate, first, to have met Daphne, second, to have the unbelievable good fortune to have hit it off so well with her and her family. They are a terrific group of people about the same size and temperament as my own.

Life can truly be spectacular and fulfilling at any age, if you let it.

❖   ❖   ❖

Woman was God's second mistake.

—Friedrich Nietzsche (1844-1900)

This isn't right, this isn't even wrong.

—Wolfgang Pauli (1900-1958), upon reading a young physicist's paper

We are not retreating - we are advancing in another Direction.

—General Douglas MacArthur (1880-1964)

There are only two tragedies in life: one is not getting what one wants, and the other is getting it.

—Oscar Wilde (1854-1900)

There are only two ways to live your life. One is as though nothing is a miracle. The other is as though everything is a miracle.

—Albert Einstein (1879-1955)

I think he had one more vote than any other, and that placed him at the head of the committee. I had the next highest number, and that placed me the second. The committee met, discussed the subject, and then appointed Mr. Jefferson and me to make the draft, I suppose because we were the two first on the list.

The subcommittee met. Jefferson proposed to me to make the draft. I said, 'I will not,' 'You should do it.' 'Oh! no.' 'Why will you not? You ought to do it.' 'I will not.' 'Why?' 'Reasons enough.' 'What can be your reasons?' 'Reason first, you are a Virginian, and a Virginian ought to appear at the head of this business. Reason second, I am obnoxious, suspected, and unpopular. You are very much otherwise. Reason third, you can write ten times better than I can.' 'Well,' said Jefferson, 'if you are decided, I will do as well as I can.' 'Very well. When you have drawn it up, we will have a meeting.'

—John Adams, 1822 letter to Timothy Pickering, published 1850.

The Clever Ladies of COA River House Writers

There are times when simple happenings have a tremendous impact, even a life changing impact. Other times such happenings turn on new realizations, thoughts or feelings—experiences that turn on a light in a mind or soul—an *aha* moment.

Such an occurrence happened to me on May 5, 2010 in a writing class I was attending at the time. There were about 15 ladies in the class. There was one other gentleman who sometimes attended. Our teacher, leader, guide, was Peter, a reporter for a local paper.

Peter was absent, so several of the ladies read their work. Their stories had tragedy and comedy, sadness and joy, with bits of humor, pain, laughter and tears, all steeped in human emotions. These women, these fellow wordsmiths, these new friends, shared parts of their lives with each of us. They shared their pain, their innermost feelings, so much more than the words they read. For a tiny portion of our lives, we all were in each other's lives, sharing and experiencing the joy and the hurt. Lives, that until that moment when we met, and spoke, and read, and shared, were no part of our beings.

We were Samuel Johnson and Elizabeth Barrett, Edgar Allen Poe and Harriet Beecher Stowe and Barbara Getty. Shaw was among us, and Emerson, and Bronte, and Frost, and Clemmons, and scores of other accomplished painters of word pictures. There were also among us a thousand other voices no one has ever heard, names we will never hear or know. They were there with us—sharing.

I read for them a piece named, *The Calling,* I had written some ten years ago about finding myself as a writer. It starts on page 553 in this book. I have been pleased to be the writer of that piece. It is a confession of slow growth of understanding of one part of whom I am. I think only writers or those with the spirit of writers, or maybe story tellers, would truly understand my words—or maybe not. Those ladies, my friends, each took their own understanding from my words. The beautiful thing was that they listened. That gave power to my words, depth to my feelings.

One of the ladies asked why I was in the group. Maybe because I come across as so experienced, so sure of myself. The implication was that I didn't need to be there, that I was too advanced. I tried a short, simple answer, but the truth is far too complex, too emotional, too thought provoking, to be explained in a few words. The truth to me is a lesson I continue to learn and appreciate. I have found that the more I know, the more I learn, the more I experience, the more I hear—all of these *mores*—make me realize how much more there is that I do not know, how much I have yet to learn, to experience, to hear. I also realized painfully how much, how inconceivably much there is I will never know, learn, or experience. Each of these ladies, my friends, had something of value to add to who I am. Hopefully, I have and will reciprocate. Is this not the essence of living? Is this not what makes us the human creatures that we are?

Life is an extremely complex set of interrelated actions and reactions. It is far too complicated to describe in a few sound bytes, or even in a thousand volumes. I have tried to explain so much of life to my children, my grandchildren, even my friends, particularly those in difficulty or grief. I am certain that most people have tried to do the same. Sometimes it's difficult. Other times it is impossible. Yet still we try.

I have truly had a wonderfully enjoyable life, mostly through no effort of my own. At my advanced age, I am in a romantic relationship with a spectacular lady. We met on the Internet four years ago. She brought me to St Augustine which I have grown to love and enjoy immensely. But then, I have always been a hopeless and optimistic romantic, so why not.

Several years back, after my wife, Barbara, passed away, I went to see my personal physician for a checkup. Dr Andy had then been my physician for nearly forty years. We were friends, and he knew at least as much about my life as anyone.

After the exam was over we were talking and said to me, "Howard, I wish I had your life."

I replied, "Why do you say that, Doc? Right now I'm broke, living alone, and trying to put my life back together after taking care of Barb full time for nearly five years. In the process, I lost everything I owned."

"This is not the first time, you know. You'll bounce back like before," he remarked with a grin.

Then he rattled off a number of the things I had experienced and gone through during those forty years. Included were three marriages, two divorces, three serious relationships, six incredible children and being widowed.

I grinned back at him and said, "When you put it that way, I guess I have had quite a ride."

He also reminded me how I had been broke twice before, and bounced back.

To that I pleaded, "Yeah, but I was twenty years younger the last time I was broke."

He laughed at me. "Somehow, I think you'll manage to come out OK."

When I was asked that question about why I was in the writing class, I remembered the story I then told him. Here's that story, reconstructed from old memories.

My maternal grandfather was a wonderful friend for a young boy. He was a master story teller who could keep me occupied listening for a long time, an eternity to a small boy. Many a time, as he would be telling me a particularly juicy tale, I would hear my grandmother call from the kitchen or other part of the house, "George! You quit filling that boy's head with your nonsense." Granddad would lower his voice and keep right on telling.

One evening, when I was probably seven, he and my grandma came to our house for dinner. I managed to sit right beside him as always. Among other of my mother's offerings, she had stewed turnips, a dish I did not particularly care for.

When it came my turn at the turnips I declined, probably saying, "I hate turnips," or something of that nature.

Granddad's face turned to a scowl as he looked at me, bent down, and whispered privately, "Howard, You should never say things like that. Never!"

"Why not? That's what I think."

He then gave me a lesson in human nature I shall never forget. "Well, you think wrong, young man. You think wrong. Unless you change, you will deny yourself many enjoyable things."

I held my grandfather in awe and adoration, so I was extremely attentive as he continued.

"What you should say is, I love stewed turnips. . . . Say it now, I love stewed turnips."

It made no sense to me, but I said it anyway, "I love stewed turnips."

"Repeat it, several times. Announce it to everyone."

"I love stewed turnips. I love stewed turnips. I love stewed turnips," I dutifully remarked for all to hear.

"OK. Now, ask your mom for some turnips—right now. And say you love her stewed turnips when you ask. Go ahead. Do it."

Though I was skeptical, I did what Granddad told me to do. "Mom. May I have some turnips? I love your stewed turnips."

Mom had a quizzical look on her face as she placed the serving of turnips on my plate, but said nothing,

Granddad leaned down and whispered, "Now keep saying, I love stewed turnips, as you eat them. You'll be amazed."

Well, I was amazed. Not only were they edible, but I even took a second helping. They were delicious. To this day, I love stewed turnips.

While he and I sat on our porch after dinner, Granddad tried to explain why this was so. "Howard, you ought to apply what you learned to everything. Consider all the things you say you don't like. Think about treating them the same way. Know why this works?"

I was in wide eyed attendance to his words. "No, I don't, so why did that work. I really did like those stewed turnips. Still do. That's hard for me to believe."

Granddad smiled as he continued. "It's quite simple. It's the power of commitment. Once you announce something to your friends, to anyone, you then must back up your statement with actions. Let me ask you, do your friends like school?"

"Nah. They all hate school. Drather be doin' almost anything else."

"That's what they say, and once they've said that, it's a commitment. They then have to prove it to their friends. They must act as if what they said were true. Each time they say they hate school, they must act as if it were true, and that reinforces their feelings. That's human nature, and it is so with nearly everything in our lives."

"I don't know that I can do that with everything. Some things, like bad fish, taste terrible. They even smell bad."

"That's quite true, but you'll be surprised how many things, even people, are not nearly as bad as you say. Take school for instance. Do you like school?"

"I guess I do like learning about new things—even have fun solving word problems in math. School's not really so bad."

"Don't your pals tell you they hate school? How about you. Do you ever say you hate school?"

"They always say they hate school. I do too, sometimes."

"Every time any of you say that, it's a commitment. Then you have to prove it—continually. The next time you talk with your friends, tell them that you love school and keep on telling them that. See what happens."

"I can't do that. They'll all laugh and make fun of me."

"You'll knuckle under and do what they want you to do. Is that it? You can't do anything they won't let you do?"

I bristled a bit at his words. "No, It's not that way, not at all. They don't tell me what to do."

"Sure sounds to me like they do."

Fortunately, I was already fiercely independent. I had to prove I was not under the control of my friends, or anyone for that matter. "OK, Granddad. I'll do it. I'll tell all my friends I love school."

"Now you're on the right path. You'll understand better as you grow up."

Much later I would realize Granddad was teaching me two important life lessons, the power of commitment and the force of peer pressure. It would take years for the full effect of these lessons to work into who I am. From that day on I always said I loved school—and it worked. I also learned to ignore taunts of others and make my own decisions about what I like or dislike. As a result, I have never been a joiner or member of a cause. Well, maybe some independent causes for what I believe to be the better good. Peer pressure? I can take it or leave it as it suits me. There is no doubt in my mind that those simple lessons my Granddad taught me have had a major positive effect on my entire life. To this day, I eat with relish virtually any food that is well prepared. I love everything edible.

How does all this tie into my first comments and the question from the ladies in my writing course? It's all about how I see people as individuals. The question posed implies I might set my self somehow above those who have had less experience, those who may know less than I about any subject or idea.

In my lifetime I have never met another human being from whom I could not learn something I didn't know, something of inestimable value. I have never reread anything, even my own words, from which I have not gained at least something. Those who do not listen, or value, the words of others, regardless of how less knowledgeable they may seem, are missing out. No matter how large yawns the supposed chasm between intellects or experience, nonlisteners (Been there. Done that.) shut themselves off from a vast store of information and emotions they may never feel, know, or learn of. I have a consuming desire to find, know, and feel these things.

The Front view of St. Augustine COA Riverhouse from the river.
The Riverhouse Writers meet here Wednesday mornings.

When wondering about someone they don't know, many tend to ask for statistical information. They want to know about their job, wealth, education, age, looks, religion, politics etc. I certainly see and recognize these things. However, what I would really like to know are things like the following: What does her voice sound like? What music does he love best? Does she collect butterflies? Does he talk with his children? Does she dream in technicolor? Does he cry at beautiful experiences? What does she hate and what does he love? What tragedy has bruised his life? What inconceivable joy has brightened her outlook?

So, ladies, and in particular the questioner, do you now understand why I am in this class?

❖　　❖　　❖

Get the facts first. You can distort them later.

—Mark Twain

I am for those means which will give the greatest good to the greatest number.

—Abraham Lincoln

It's not good enough that we do our best; sometimes we have to do what's required.

—Winston Churchill

Opportunity is missed by most people because it is dressed in overalls and looks like work.

—Thomas Edison

The greater the difficulty, the more glory in surmounting it. Skillful pilots gain their reputation from storms and tempests.

—Epictetus (55-135 CE)

The important thing is not to stop questioning.

—Albert Einstein

The most serious mistakes are not being made as a result of wrong answers. The truly dangerous thing is asking the wrong question.

—Peter F. Drucker

Dogs never bite me. Just humans.

—Marilyn Monroe

The greatest deterrent to Communism is an affluent, intelligent, and well-informed middle class.

—Karl Marx

Charlie and Friends

For the past four years I have lived in Villages of Vilano, a gated community on the barrier island between the Inter Coastal Waterway and the Atlantic Ocean north of St. Augustine, Florida. It is a safe, comfortable community of mostly older and retired people. I live there because of the lovely and gracious lady who owns the house. We have put our lives together after having both been widowed from precious loved ones. During much of the summer, we live in my home on a lake in Northern Indiana. I am one extremely lucky man to have found Daphne and to have such a phenomenal, loving relationship at my age.

There is another contributing member of our family. Charlie is a delightful, exuberant, little bundle of energy who loves everyone. His ancestors were caravan dogs in Tibet where they were expected to hunt small animals to feed the drivers and provide alarms so the protector dogs, Tibetan mastiffs, could do their thing. Lhasa Apsos are muscular little dogs with powerful hunting instincts and the ability to jump amazingly high for their size. They are notoriously friendly to people and are excellent with children. Charlie has been mauled by my small grandchildren with nary a complaint, or hint of a growl. He loves the attention.

Charlie has a thick, soft coat which grows long without shedding, so he must be brushed regularly, and groomed when his coat gets too long. We let his coat grow long in the winter, but crop him close for the heat of summer. Each of his substantial front paws is white. Otherwise, he is totally light caramel in color. He has large, expressive brown eyes, a black, inquisitive nose, and floppy ears with long hair. Charlie is the happiest dog I have ever known. He is an amazing communicator and understands so many of our words we have to spell things like *ride* or *bike*, and the mention of *go* brings about a dash for the door where he bounces like a rubber ball. He has no trouble telling us what he wants, be it food, water, to go out, or if someone's at the door.

Charlie loves to travel and will quickly jump into any car with the door open. Sometimes we have difficulty getting him to come out once he's entered a vehicle. When we travel in our motor home, he usually sits or roams about on the large, flat area between the dash and the windshield where he can see everything. He even sleeps there sometimes while we're moving.

There is a figure eight loop of streets in the Villages of Vilano plus a few cul-de-sacs providing access to the homes. One complete passage through the figure eight rolls up a mile and a quarter. For the last three years, I have tried to pedal our four-wheeled bike-like conveyance called, a Rhoades car at least once around that figure eight each day. The Rhoades

649

is a steel-framed, six-speed cycle type of conveyance with two seats side-by-side, and a large basket in back of the seats. It is a hefty vehicle weighing in at about 170 pounds. Each seat has access to a set of pedals that drive the rear wheels through a six-speed derailleur system, one unit for each seat.

On the first time we took the Rhoades for a spin around the figure eight, we took Charlie with us on his retractable leash. Maybe I should say, Charlie took us. From that moment on Charlie decided that the Rhoades belonged to him. No one would be able to get on that bike without Charlie going with them. He loves to run and can pull me on the Rhoades up to 8 miles per hour all on his own. This usually takes place at the start of each of our outings. I quickly throttle him back, as instructed by our vet, so he doesn't over exert.

We are now settled into a routine. He runs as fast as he can with me pedaling furiously to keep up for the first fifty yards after which he settles back to a comfortable dogtrot. This makes both Charlie and his vet happy.

Except during our frequent travels, when Charlie goes to *camp*, we have done this regularly while in St. Augustine for the last three years. During rides around those loops, we have greeted many people and their dogs. Usually the other dogs are being walked slowly so we breeze by with Charlie's little legs churning. By the time we have completed most of the loops and are on the last block or so toward home, Charlie has slowed to a walk and is making regular stops to mark his territory and tell other dogs, *I was here.*

Sometimes things happen that put one's perspective in sync with reality. Such a thing happened to me this morning. It certainly was not a boost for my ego. Charlie and I started out early for our bike jaunt, so as to miss the heat of the day. The sky was clear, the air still, and the streets of our figure eight were quite empty. As we headed east on the third leg of our journey, a man on a regular bicycle rode past going the opposite direction. I did not recognize him, but as he rode past he smiled and gave a cheery, "Hi Charlie."

I asked, "Charlie, who was that?" Charlie, of course, did not answer.

Later, on the part of our circuit where he runs on the sidewalk, Charlie pranced up to a woman walking the other way. She stopped, bent down, gave Charlie a pat on his head, and said, "Good morning Charlie. How are you?"

Charlie greeted this attention with a kiss and frantic wagging before sauntering off as if nothing happened. As we passed, we exchanged good mornings and went on. I did not recognize the woman, but she certainly knew Charlie. I have no idea whether or not either of Charlie's friends knew me other than as the man who rides that odd bicycle with Charlie every day. There was no doubt they knew Charlie.

As we headed down the last street toward home, I wondered more and more why and how they knew Charlie, but I didn't know them. Maybe I'm the guy Charlie leads around every day, I don't know. It certainly is humbling when people know your dog by name, no matter how loveable he is, but don't know you.

But then, Charlie is no ordinary dog. You knew I was going to say that, didn't you? As I write this, Charlie is curled around my right foot, his chin resting on my left. That's where he is much of the time I'm writing. Sooner or later, he will thrust his head up between my legs and the keyboard, look me in the eyes and say. *OK, dad. Time for me to go outside, or for a bike ride, or something else.* Of course, he can't actually vocalize it, but make no mistake. Charlie is not subtle. He lets me know clearly what he wants.

Charlie just back from the groomer

❖   ❖   ❖

The nearest way to glory is to strive to be what you wish to be thought to be.

—Socrates

The only man who makes no mistakes is the man who never does anything.

—Theodore Roosevelt

The pessimist sees difficulty in every opportunity. The optimist sees opportunity in every difficulty.

—Winston Churchill

Find something nice to say to everyone, each time you meet. Then say something nice about someone else to that person. If you do these two things consistently, your life will be pleasant, you will have many friends, and no one will ever find fault with you.

—Howard Johnson, 1980

Accuracy is the twin brother of honesty; inaccuracy, of dishonesty.

—Charles Simmons

People often say that motivation doesn't last. Well, neither does bathing-that's why we recommend it daily.

—Zig Ziglar

In England I would rather be a man, a horse, a dog, or a woman, in that order. In America I think the order would be reversed.

—Bruce Gould

Good advice is something a man gives when he is too old to set a bad example.

—Francois de La Rochefoucauld

It is difficult to know what counts in the world. Most of us count credits, honor, dollars. But at the bulging center of mid-life, I am beginning to see that the things that really matter take place not in the boardrooms, but in the kitchens of the world.

—Gary Allen Sledge

If the only prayer you ever say in your whole life is "thank you," that would suffice

—Meister Eckhart

To accuse others for one's own misfortunes is a sign of want of education. To accuse oneself shows that one's education has begun. To accuse neither oneself nor others shows that one's education is complete.

—Epictetus (55-135 CE)

The New Cell Phone

Ihave a number of serious problems with my AT&T wireless service. They all started when I lost my old phone. First of all, I notified AT&T my phone was lost. Immediately I could not access my account on the Internet. It would not have been a problem had it not been for subsequent events.

I then called about changing my service and seeing about getting a new phone. Had I done what I wanted to do and had my account added to my daughter's account, things could have been much more satisfactory. When I asked, the AT&T rep said I would have to sign a new contract and get a new phone before transferring my account. I have since found out that is untrue. I argued with the rep commenting the information made no sense, but he or she insisted I must sign a new contract first. My guess is the sales person wanted the commission for selling a new two year contract and didn't give a damn how much it cost me or what difficulties it might create. I don't know, maybe she was one of those outside independent sales people AT&T uses so they can say, "Well, it's not our fault since the salesman you used is not an AT&T employee."

Regardless, a new two year contract was shoved down my throat. To save money, I was trying to get my account transferred before my next billing date. Now, because of the duplicitous efforts of AT&T I not only had to pay another month, but today, May 10, I will have to pay still another.

Here's a record I made of my experiences. It is written in play format as I may polish it and use it as a script for a comedy vignette. I'm sure many folks will be able to identify with my frustration.

At about noon on Friday, April1 (April fools day, so I should have known better) I went on the internet and reported my lost cell phone to AT&T. They immediately shut down my account including Internet access. When I tried to see about getting another phone via the AT&T website—of course—my account was shut down, and I couldn't access it. No matter. I can call them on the phone—yeah! All of the initial conversations are between me and an AT&T machine answering system.

HO: *Dialed the AT&T customer support number*

MACH: Thank you for calling AT&T about your wireless service (repeated en Español)

Please enter your wireless number starting with the area code. For other options press 1.

Entered number

MACH: You have entered 574-453-3307. If that is correct, press 2. If it is not correct, press 9 and reenter your number.

Pressed 2

MACH: Our records show that you recently ordered a new phone or device that needs to be activated. If you have the device and are ready to activate it from the device, press 2. If you are ready to activate it from another device, press 3. If you haven't received your device yet, please call back when you have, and we will help you to activate it. To return to the main menu, press 5, otherwise, hang up.

Pressed 5

MACH: To pay your bill press 1

To ask for help with voice mail, press 2

To find out how many useable minutes you have left, Press 3

To order additional services or a new phone, Press 4

For other options, including talking to a service representative, Press 5

To return to the main menu, Press 6.

Pressed 5

MACH: Thank you for calling AT&T customer service. We are here to help you and answer your questions. Please state your question clearly so we can direct your call.

HO: I lost my cell phone and want to order another and . . .

MACH: (interrupting) To report a lost or stolen phone, Press 2.

For sales, Press 3

For other questions, Press 4.

To return to the main menu, Press 5

HO: *Pressed 4*

MACH: Thank you for calling AT&T customer service. We are here to help you and answer your questions. Please state your question clearly so we can direct your call.

HO: I want to change my service and get a new phone.

MACH: For sales, Press 3

           For other questions, Press 4.

           To return to the main menu, Press 5

HO: *Pressed 4*

MACH: Thank you for calling AT&T customer service. We are here to help you and answer your questions. Please state your question clearly so we can direct your call.

HO: I want to speak to a real person, NOW!

MACH: I don't understand your question. Please restate it carefully.

HO: I WANT TO SPEAK TO AN AT&T REPRESENTATIVE, NOW!

MACH: You wish to speak to one of our customer service professionals, right?

HO: YES! RIGHT!

MACH: All right. I'll transfer your call to our next available representatives. . . All of our representatives are currently busy helping other customers. Stay on the line and someone will be with you shortly.

HO: *(TO HIMSELF OUT LOUD)* Yeah! I'll bet.

ATT REP   Hello. How may I help you?

HO: I certainly hope you **can** help me. And it's how **can** I help you. Not how **may** I help you, unless you are speaking a language other than English. That little Wal Mart mis-worded slogan grates on my ears whenever I see or hear it.

ATT REP   What can I do for you today.

HO: Much better. I lost my phone, and I need to order a new one.

ATT REP   You can easily report a lost or stolen wireless device on line, but I can help you, What is your wireless number?

HO: 574-453-3307. I have already reported my phone lost. I now want to make some changes in my account and order a new phone.

ATT REP I can help you. I'm Liz. What is your name?

HO: Johnson, Howard Johnson.

LIZ: And the last four digits of your social security number?

HO: 5555.

LIZ: Ah! I have your account. I see you are eligible for an upgrade. You can get an upgrade to one of the new phones, absolutely free.

HO: That's not true. AT&T doesn't give anything away free. I'll bet I have to sign a new, 2 year contract for more than a thousand dollars to get a *free* phone, right?

LIZ: It's just a continuation of your old contract. There are just a few small charges to cover the shipping and process of your new phone. Your old contract charges remain the same.

HO: Eau contraire, Miss. By signing a new contract I am committing to pay AT&T at least a thousand bucks. Right now I could cancel and not pay another cent. But I still need a new phone, and I want to make changes in my account, so let's get on with it.

LIZ: What changes do you want to make in your account?

HO: My daughter wants to put me on her account as a family member. We figure that would save me about $25 a month. How do I do that?

LIZ: That's easy, but first you must sign a new contract and order a new device. Then, once your new device is activated we can add you to her account.

HO: Are you sure? That makes no sense to me. Why can't I transfer my account to hers and then order the phone?

LIZ: That would take much longer, maybe several weeks. You would be without your phone all that time. If you sign the contract first, we could have a new phone in your hands in at most, three days.

HO: Are you sure it wouldn't cost me more money? It's already April 1, and my next account payment is due on the 12th. I want this all completed before my next bill is due. Also, what would be the charge on my daughter's account if I sign a two year contract?

LIZ: There's plenty of time, Once I sign you for a new contract and order your new phone it would be about three days, four at the most before you received your phone. Then your daughter would be able to start the transfer of your phone to her account. That should be completed in two days at most The monthly charges would depend on what services she has on her account. Usually it costs no more than $12 a month to add a family member to an account. That's a lot less than you are paying now for individual service.

HO: OK. You talked me into it. Let's get going.

I was soon on the AT&T website on my PC while she walked my through the process of signing a new contract and selecting and ordering a phone. Liz was helpful and stayed with me on the phone during the entire process which took about twenty minutes until I placed the order for the phone.

I have ordered many things on the Internet for years and thought I was aware of all the pitfalls. WRONG! I selected my phone, clicked on the check out button and carefully checked my order. The shipping address they showed was my Indiana address, so I dutifully changed it (as per instructions on the ordering page) to my Florida address. I checked everything carefully before clicking on the *submit* button. The instant I pressed the submit button, the shipping address reverted back to my Indiana address, and the page changed to one with the BIG message, SHIPPED. There was no acknowledgment, no confirmation, no option to check to make sure everything was correct and as I wanted, nothing.

I immediately told Liz what had happened and asked her to stop the shipment and correct the address. After several moments, the following exchange ensued:

LIZ: I'm sorry. I am unable to stop the shipment or correct the address.

HO: Then cancel the order, and I'll reorder with the correct address.

LIZ: I'm afraid I can't do that. It's already in the system. It's on its way to the Post Office.

HO: What do you mean, you can't stop it? The order was placed less than a minute ago.

LIZ: Once an order has been submitted, there's nothing we can do to stop it. You'll have to wait 'til you receive the phone.

HO: Even when the monument to ignorance you call a website screwed up my order?

LIZ: There's nothing I can do. You'll have to wait until you receive the phone and then ship it back to us.

HO: Your company's system screwed up the order and now I am expected to pay for their mistake? You can't be serious.

LIZ: It's just the way the system works.

HO: Liz, you sound like a reasonably intelligent human. Do you think this is the way a system operated by a huge, high-tech company with lots of computer expertise should work? I mean make it impossible to correct their own mistake?

LIZ: My opinion doesn't really matter, It's just the way the system works. I must deal with that reality.

HO: Well, I don't like it one damn bit. Is there anyone there who can bypass the system and correct this ridiculous mistake?

LIZ: Honestly, Mr. Johnson, I could transfer your call to a supervisor, but they will operate the same way and with the same system I must deal with.

HO: So I will have to wait until I receive the phone even if it takes a year?

LIZ: It shouldn't take long, only a few more days.

HO: No? Your system shouldn't have changed my shipping address either, but it did.

LIZ: I'm sorry that happened. Let me contact the Postal Service and see if they can help you.

HP: [sarcastically] OK. I have great confidence in the Postal Service of our unbelievably efficient government. See what you can do.

LIZ: Let me have a phone number where I can reach you after I've contacted them. I promise to call you right back as soon as I have any information.

I gave her my home phone number and hoped she would call back. I have to say, Liz was courteous and seemed genuinely concerned in trying to help. I hate to be cynical, but she was probably trained to sound that way. About fifteen minutes after we hung up, she called.

LIZ: Mr. Johnson?

HO: Speaking.

LIZ: It took a while, but I have discovered what you must do. I have a tracking number for your phone. If you call the postal service and give them this tracking number, they can tell you all about the shipment of your phone. Let me know when you are ready as it is a long number.

HO: OK, shoot.

LIZ: The number is 9-1-0-1-9-0-0-0-4-2-3-1-1-4-0-8-0-9-9-6. Got it?

HO: Yep! Let me read it back to you. 9-1-0-1-9-0-0-0-4-2-3-1-1-4-0-8-0-9-9-6. Correct?

LIZ: Yes. When you contact them, ask for Recall Mail and give them that number. They should be able to help expedite the priority mail shipment. You should have the phone in no more than three more days.

HO: You're sure.

LIZ: That's what the post office told me.

HO: You mean you are not sure, and in fact, haven't a clue what is going to happen. I personally wouldn't believe anything any agency of our federal government told me under any circumstances. If I receive the phone before the end of April, I will be amazed. I would not be surprised if I never receive the phone.

LIZ: I must take the word of the Post Office. Phone them and see what happens. I'll call back on Monday to see where things are. Good Luck.

HO: My experience tells me that I will need a lot more than good luck. Thanks for your help.

I called the postal service using the number Liz gave me. After making selections from five nested menus, all machine driven, I reached a level where the machine said:

MACH: Please enter your tracking number.

HO: *keyed in 91019000423114080996*

MACH: That is not a valid tracking number. Please enter a valid tracking number.

HO: *keyed in 91019000423114080996*

MACH: That is not a valid tracking number. Please enter a valid tracking number.

There were no options, no other choices, no reasons why the tracking number was not valid. I called AT&T customer service with the following familiar exchange:

MACH: Thank you for calling AT&T about your wireless service (repeated en Español)

        Please enter your wireless number starting with the area code. For other options press 1.

        Entered number

MACH: You have entered 574-453-3307. If that is correct, press 2. If it is not correct, press 9 and reenter your number.

        Pressed 2

MACH: Our records show you recently ordered a new phone or device that needs to be activated. If you have the device and are ready to activate it from the device, press 2. If you are ready to activate it from another device, press 3. If you haven't received your device yet, please call back when you have, and we will help you to activate it. To return to the main menu, press 5, otherwise, hang up.

Pressed 5

MACH: To pay your bill press 1

To ask for help with voice mail, press 2

To find out how many useable minutes you have left, Press 3

To order additional services or a new phone, Press 4

For other options, including talking to a service representative, Press 5

To return to the main menu, Press 6.

Pressed 5

MACH: Thank you for calling AT&T customer service. We are here to help you and answer your questions. Please state your question clearly, so we can direct your call.

HO: I need a tracking number for an order I placed earlier today.

MACH: One moment, please, while you are transferred to one of our Customer service representatives. This may take a few moments, so please be patient.

ATT Rep: (after a wait of several minutes) Can I help you?

HO: I need to confirm a tracking number for an order I placed. I used the one I was given, and USPS says it is not a valid number.

ATT Rep: What is the wireless number you are calling about?

HO: 574-453-3307

ATT Rep: And what are the last four digits of your social?

HO: 5555.

ATT Rep: OK, I have your account information. Just to check and make sure, what is the tracking number you have?

HO: It is 9-1-0-1-9-0-0-0-4-2-3-1-1-4-0-8-0-9-9-6

ATT Rep: [After a short wait] I can see what the problem is. You have missed a single digit. The number should start 9-1-0-1-9-0-1-0-0. All of the remaining numbers should be the same.

HO: Somehow I must have missed a digit even when Liz read it back to me.

ATT Rep: It happens. Try the new number. It should work.

HO: Thank you.

I went to the USPS website and entered the new number in the proper box. I read the following from the resulting dialog box:

Label/Receipt Number: 9101 9010 0042 3114 0809 96

Expected Delivery Date: April 6, 2012

Class: Priority Mail®

Service(s): Delivery Confirmation™

Status: Forwarded

Your item was forwarded to a different address at 9:26 am on April 06, 2012 in NORTH WEBSTER, IN 46555. This was because of forwarding instructions or because the address or ZIP Code on the label was incorrect. Information, if available, is updated periodically throughout the day. Please check again later.

- **Detailed Results:**
- Forwarded, April 06, 2012, 9:26 am, NORTH WEBSTER, IN
- Electronic Shipping Info Received, April 06, 2012
- Out for Delivery, April 06, 2012, 8:06 am, NORTH WEBSTER, IN 46555
- Sorting Complete, April 06, 2012, 7:56 am, NORTH WEBSTER, IN 46555
- Arrival at Post Office, April 06, 2012, 6:56 am, NORTH WEBSTER, IN 46555
- Processed through Sort Facility, April 05, 2012, 9:47 pm, INDIANAPOLIS, IN 46241
- Processed through Sort Facility, April 04, 2012, 5:29 pm, FORT WORTH, TX 76161
- Acceptance, April 03, 2012, 1:08 pm, FORT WORTH, TX 76161

It is now Wednesday, April 13, seven days after the phone was supposed to have been forwarded from North Webster, and the phone has still not arrived. I tried to cancel my scheduled payment to AT&T on April 11, one day before the payment was supposed to be taken from my bank account. Guess what? For the first time ever, AT&T debited my bank account a day early, on the eleventh. I called AT&T with the following result:

HO: *Dialed the AT&T customer support number*

MACH: Thank you for calling AT&T about your wireless service (repeated en Español)

Please enter your wireless number starting with the area code. For other options press 1.

HO: *Entered number*

MACH: You have entered 574-453-3307. If that is correct, press 2. If it is not correct, press 9 and reenter your number.

Pressed 2

MACH: Our records show you recently ordered a new phone or device that needs to be activated. If you have the device and are ready to activate it from the device, press 2. If you are ready to activate it from another device, press 3. If you haven't received your device yet, please call back when you have, and we will help you to activate it. To return to the main menu, press 5, otherwise, hang up.

Pressed 5

MACH: To pay your bill press 1

To ask for help with voice mail, press 2

To find out how many useable minutes you have left, Press 3

To order additional services or a new phone, Press 4

For other options, including talking to a service representative, Press 5

To return to the main menu, Press 6.

Pressed 5

MACH: Thank you for calling AT&T customer service. We are here to help you and answer your questions. Please state your question clearly, so we can direct your call.

HO: I lost my cell phone and want to order another and . . .

MACH: (interrupting) To report a lost or stolen phone, Press 2.

For sales, Press 3

For other questions, Press 4.

To return to the main menu, Press 5

HO: *Pressed 4*

etc., etc., etc . . .

✳   ✳   ✳

On or about April 17 I received my phone. As soon as I opened it, I charged it as instructed. After it was charged, I called and had it activated. Then I set about trying to get used to this new phone. My wife has an LG phone which I found easy to operate. My new Sharp FX is much harder for me to use, and far less intuitive. I have had tremendous difficulty dialing the phone, entering phone numbers, in short, it was difficult for me to operate, so of course, I took it down to the friendly AT&T store here in Saint Augustine. I showed the young lady who came to help me, how and why I was having such difficulty. Mostly I could hardly get the keys to register to my touch. I had and still have to try several times before I can enter a number to make a call. I had no such problem with my wife's phone. The young lady spoke to me for several minutes, and I demonstrated how difficult it was to get the keys to register and how often I received two numbers when pressing a number on the keypad. I also showed her how difficult it was to use or make entries in the directory. Sometimes, no matter how I tried, it wouldn't register. She had the same difficulty, but did manage to get it to work a few times. As I remember, she said something about using it for a while would probably solve the problem. I tried for a while and it didn't.

Assuming the exchange policy was 30 days after activating the phone, I called to try to get my problems straightened out by selecting another phone. (Incidentally don't ever assume anything reasonable when dealing with AT&T) A precise repeat of the first phone call described earlier put me in touch with someone from customer care. (What a misnomer that is! It should be called the AT&T bull department.) I was then informed I had to take the phone to an AT&T store to have it examined. I was told to remove the battery and read the number on one of the bar codes. I was told there was a colored sticker inside the back of the phone. This sticker determined if the phone had been dunked in water. I found no such sticker. I probably went through explaining to the rep at least eight or ten times, that there was no sticker other than the white one with all the bar codes. She seemed to pay no heed to my words and kept telling me, "There is a colored sticker at the lower right corner between two screws."

I kept repeating to her, "There is no colored sticker." She, of course, ignored my protestations and continued asking me for the color of the non existent sticker.

I gave up in frustration and asked what I could do to get another, different kind of phone. That's when she informed me the time when I could exchange the phone had expired. When

I protested that because of the stupid errors in shipment (AT&T's fault described earlier), I only had the phone for a bit more than a week. She then told me the clock on those thirty days started when the phone was shipped. I don't recall ever being so informed. When I complained bitterly she told me, *"We don't give a damn what you think, those are our rules, TS."*

No, those are not her actual words, but the meaning was quite clear. The following quotes are in the same pattern. The actual meaning is in quotes even though the words spoken were quite different.

I was informed they would only replace the phone with the same lousy model, if they found it defective. In other words, *"We don't care about your problems, take it or leave it. Those are our rules. They apply even if we screw up so badly you never even get your phone."*

So then I was transferred to the department that arranges for replacement of a defective phone. The clearly oriental voice that spoke to me was almost unintelligible. I had to ask her to repeat everything several times. Obviously she was in Bankok, Darien, Manilla, or some other oriental (cheap) location and couldn't speak American. I was on the phone with this female for quite some time. She was probably as frustrated as I at her inability to communicate. The only thing she was able to make clear to me was my thirty days had expired, and there was no way I would get anything but another one of those junk phones, like the one I had. I gave up in disgust and hung up. AT&T was by now at the top of my s— list.

*   *   *

Email received May 9, 2012:

Re: Wireless from AT&T Customer Email - North Central - [CUST] (KMM32916072V56335L0KM

Dear Mr. Johnson,

Thank you for taking the time to e-mail AT&T regarding issues with your device and replacement options. I apologize for all the frustration regarding this matter. My name is Cynthia White, and I am happy to help you with your inquiry.

AT&T values our relationship with you and offers a return policy for equipment and services purchased directly from AT&T. It may not reflect the policies of authorized retailers of AT&T. Please visit http://www.wireless.att.com/returns for complete details of our Return policy.

If the equipment you purchased does not meet your expectations, simply return it either to a store or by mail within 30 days from the date the equipment was purchased or shipped. I have reviewed your account and see the device was shipped on 4/3/2012. Unfortunately, your device is no longer under the 30 day return/exchange period.

Please note wireless devices carry a one-year warranty from the date of purchase. To make a warranty claim after the first 30 days from purchase, call 1-800-801-1011. Apple branded equipment is covered by Apple's one-year Limited Warranty; for details refer to **http://www.apple.com/support/iphone**.

Mr. Johnson, if you need to contact us again regarding a new issue, please send us another email via the contact link through your online account. Again, my name is Cynthia White, and I thank you for being a valued AT&T customer for several years. We will do our best to ensure your wireless experience is a success.

I encourage you to visit our web site (www.att.com/wireless) often to view current and previous monthly statements, make payments and to shop for new product and service offerings.

Sincerely,

Cynthia White

AT&T

Online Customer Care Professional

❊   ❊   ❊

5-09-2012 email message response sent to AT&T:

I am extremely disappointed and angry at AT&T. I received an email from Cynthia White. (Above) The first two and the fifth paragraphs are mostly bull. Are AT&T personnel allowed to care about anything but AT&T mantras and money? I have been misled, misdirected, sold a junk phone, and promised things that were ignored for the last six weeks since I lost my phone, and tried to change my account. I would like to speak to someone in authority with a brain, not some automaton communicating from a script.

❊   ❊   ❊

My own comments below are in [brackets]. On May 9 I received the following email:

NOTE:. Please do not reply to this automated email.

Thank you for contacting AT&T. [*bull*]

We appreciate your business [*more bull*] and know your time is valuable. [*still more bull*] An AT&T Online Specialist has been assigned to your case and will respond to your concern within 2 business days; however, our response may be sooner. We will do our best to exceed your expectations. [*I am waiting with bated breath. At this point I expect nothing good from AT&T*]

Do you know about the free, fast and easy way to check your minutes or balance, or to make a payment from your wireless phone? It's available 24 hours a day, 7 days a week. If not, go to:

http://www.wireless.att.com/learn/basics/choosing-features-services/starservice s.jsp [*Asking for payment is obviously at the top of your concerns list while customer care must be somewhere down at the bottom between cleaning out the urinals and removing dog poop from the lawn. You even ship customer complaints overseas, don't you? Perhaps there is an advantage to using phone people who are so hard to understand. Most of those seeking the misnomered customer service, or having complaints, will give up their pursuit of assistance in frustration.*]

****17658490**** End of email message.

Of course, you took my money, $59.17 at 4:25 AM this morning (May 10) even though I have been without service since April 1. Of course, that's just the way it is, right?

One more little problem: I began wondering why I was receiving no phone calls, none at all. Then I noticed I had several voicemail messages. I listened to them and realized my phone had not rung—ever. In checking this out I discovered the phone had a tiny switch on the side with a + key to increase the volume and a - key to reduce it. The volume was set at its lowest. The phone did not ring. I cranked it up to near maximum volume, turned the vibrator on, and dropped it in my pocket. Problem solved—or was it. A day later, same problem. The phone would not ring, and again I had several voicemails. Once more the volume was turned all the way down. Another problem with the lousy phone? Again I was directed to take it to the AT&T store to see what was wrong. Apparently, soon after dropping the phone in my pocket, the ringer volume returns to zero all on its own. I have yet to catch it in the act. Incidentally, the phone is the only object in my pocket, and I am quite certain I am not accidentally turning it down by pressing the button, but the button may still be the culprit. I tried setting the phone on my desk, and the ringer volume stayed put, at least for one full day. Go figure.

To this point, nothing has been solved or resolved. I am about to make another trip to the AT&T phone store.

I was on the phone with Sandra who must be some kind of account troubleshooter for AT&T. I'll not repeat the long conversation we had, but suffice to say, they backed off and will let me exchange the phone. Unfortunately, they have decided to reverse the entire transaction and have me start over from scratch. She informed me that is the only way it can be done as I now want an iphone. I must return the phone I have, wait for confirmation it has been received at their shipping warehouse, and then, after the records are changed and credit issued, I will be able to order an iphone as I should have done in April. I will be without a phone until I receive my new phone, probably in two or three weeks. In the mean time I will, of course, continue to pay for service I haven't had since late March.

I did what I should have done in the first place, let my daughter, Deb, handle this for me there in California. She was able to steal some time, (she is a busy lady) take the information I gave her, and go to her local AT&T store there in Stockton.

As a result, I received my brand new iphone4 on Tuesday, June 28 after being phoneless for more than three full months while still paying AT&T. They insist on taking my money even though I was without service because of their lousy system. On the plus side, the iphone is an amazing piece of technology and I like it a lot. It's not an AT&T product, thank God. Of course it may take me a year to get used to all the things you can do with it. So far, I've barely scratched the surface.

Don't we all love dealing with giant organizations and their red tape machines? Bad as they can be, the people who deal with the public in these corporations, are incredibly efficient and considerate compared with those on any government department, agency, or organized unit of any kind. Can you even imagine how bad and inhumane dealing with government healthcare will be? With all those anal retentive paper pushers deciding who gets treated, and how and when, our once fantastic health care system will soon make treatment extremely hard to get, especially for seniors. As in all countries that have government run healthcare, the death rate for many kinds of serious diseases has gone up drastically, especially among the elderly.

Under the control of Obamacare, doctors, long a special person particularly close and familiar with our lives, will become impersonal, bureaucratic body mechanics who know patients only as clinical things to be dealt with as quickly as possible. That's one way to help relieve the population problem. The new Government Health Care System will probably kill off more people than virtually any kind of war, primarily the elderly. You see, there's a rosy side to almost anything if you look far enough.

We cannot continue to rely on our military in order to achieve the national security objectives that we've set. We've got to have a civilian national security force that's just as powerful, just as strong, just as well-funded.

—Barack Hussein Obama - In a speech in Colorado on July 2, 2008

It is interesting to note that among the numerous well-hidden provisions of the Obamacare law, there is one establishing and funding just such a force. Unlike the military, this force will be able to enforce federal law as police, and not forbidden to act as law enforcement as is the military. I wonder if they will be wearing brown shirts? I also wonder why the media has neglected to inform the public of this? Hmmmm?

—Howard Johnson, ol;p

When only the police have guns, it's called a police state.

—Robert Heinlein

Your children's children will live under Communism. You Americans are so gullible. No, you won't accept Communism outright, but we'll keep feeding you small doses of socialism until you will finally wake up and find that you already have Communism. We won't have to fight you; we'll so weaken your economy, until you fall like ripe fruit into our hands. Your cities, factories, farms, and workforce will belong to us, undamaged or diminished by warfare.

—Nikita Krushchev, 1959

The European Commission has just announced an agreement whereby English will be the official language of the European Union rather than German, which was the other possibility.

As part of the negotiations, the British Government conceded that English spelling had some room for improvement and has accepted a 5- year phase-in plan that would become known as 'Euro-English'.

In the first year, 's' will replace the soft 'c'. Sertainly, this will make the sivil servants jump with joy. The hard 'c' will be dropped in favour of 'k'. This should klear up konfusion, and keyboards kan have one less letter. There will be growing publik enthusiasm in the sekond year when the troublesome 'ph' will be replaced with 'f'. This will make words like fotograf 20% shorter..

In the 3rd year, publik akseptanse of the new spelling kan be expekted to reach the stage where more komplikated changes are possible.

Governments will enkourage the removal of double letters which have always ben a deterent to akurate speling.

Also, al wil agre that the horibl mes of the silent 'e' in the languag is disgrasful and it should go away.

By the 4th yer people wil be reseptiv to steps such as replasing 'th' with 'z' and 'w' with 'v'.

During ze fifz yer, ze unesesary 'o' kan be dropd from vords kontaining 'ou' and after ziz fifz yer, ve vil hav a reil sensibl riten styl.

Zer vil be no mor trubl or difikultis and evrivun vil find it ezi tu understand ech oza. Ze drem of a united urop vil finali kum tru.

Und efter ze fifz yer, ve vil al be speking German like zey vunted in ze forst plas.

—unknown Internet source

A few recent photos of our family and travels.

Our first photo together, taken in Daphne's back yard during my first visit in April of 2006. Daphne's daughter Abbie took this photo.

My 82nd birthday celebrated at Daphne's daughter, Missy's home in Gainesville. Daphne's daughter, Abbie, Missy's twin, is the one right behind me. Abbie's daughter, Emily, is standing beside Daphne.

A family portrait taken during our RV trip across America. It was taken while we were on an Elderhostel at the Grand Canyon May 2009.

White water rafting in Costa Rica - February 2009

Daphne and Ho at Bloody Mary's on Bora Bora in French Polynesia

Bobbie and Bob during their visit with us in January 2009. We visited Green Cove Springs where Bob ended up after the war where many American war ships were stored.

We celebrate the New Year 2010 by taking Charlie for a run with us on the Rhoades car, our quadricycle. Charlie has decided the Rhoades is his personal property and constantly pesters us to take him out for a *bike ride*. (he gets incredibly excited)

**Daphne and Ho at the clock shop in the Black Forest,
May 31, 2010 during our tour of Germany.**

The Johnson family reunion at Lake Tippecanoe, August 15, 2010

Standing, left to right:

Kristen, Vince, Mike, Debby, Mindy, Joseph, Audrey, Joe, Dan, Moira, Jeff

Second row: Diana, Ho, Daphne, David (Archer)

Front row: Dave (Heller) Cadence, Madelaine, Kelan, Chantel, Charlie

View the next page to see what they are really like.

I am so proud of all of the wonderful members of my family—even when they are acting a little crazy. That of course includes those not in the photo, those a bit more distantly related, and those no longer with us. They are a treasure far beyond measure. I consider myself to be one of the luckiest souls on this earth to be part of such a group.

Our *tent city* in the front yard at the lake during our last ever family reunion there in August of 2012.

Most of us miss out on life's big prizes. The Pulitzer. The Nobel. Oscars. Tonys. Emmys. But we're all eligible for life's small pleasures. A pat on the back. A kiss behind the ear. A four-pound bass. A full moon. An empty parking space. A smile from a small child. A crackling fire. A great meal. A glorious sunset. Hot soup. Cold beer. Then there is the top prize, wonderful to experience. That prize is to do something nice for someone in complete and absolute secrecy and then have it discovered by accident.

—unknown

"Do Unto Others as You Would Have Others Do Unto You."

The golden rule (Matthew 7:12 and Luke 6:31) is a direction for a way of life. It is simple and direct yet difficult for most to comprehend. It is a way for a person to act toward others, not something to be expected of them. One who applies the golden rule is a giver in the finest sense of the word and takes pleasure in the *giving*, with no thought of receiving anything whatsoever in return. Acceptance of the gift alone is the greatest kindness that a giver can receive.

Too often, we forget that others do not see the world as we do and are troubled that they do not seem to reciprocate our own imagined goodness. The golden rule has no component that says we should expect *anything* from another person. That is truly a function of *their* world, not ours. To expect any act from another is to burden ourselves with a potential disappointment and create tension in our interpersonal relationship with that individual. It is, in fact, the single most unkind action one can take toward another person.

Courtesy, consideration, friendship, affection, love—these are all giving things as *it is only in giving that we receive*. What we do receive is the pleasure of knowing what we have done and lies entirely within our own psyche with no external component. To expect or ask anything in return is to debase oneself and bring negative, displeasing specters into one's being.

"Judge not, that ye be not judged" (Matthew 7:1).

—Howard Johnson, 1978

The golden rule is of no use to you whatever unless you realize that it is your move!

—Dr. Frank Crane

A designer is an emerging synthesis of artist, inventor, mechanic, objective economist and evolutionary strategist.

—R. Buckminster Fuller

This I beheld, or dreamed in a dream: - - There spread a cloud of dust along a plain; and undereath the cloud, or in it, raged a furious battle, and men yelled, and swords shocked upon swords and shields. A prince's banner wavered, then staggered backward, hemmed by foes. A craven hung along the battles edge, and thought, "Had I a sword of keener steel—that blue blade that the king's son bears,—but this blunt thing!" he snapt and flung it far away and lowering, crept away and left the fray.

Then came the king's son, wounded, weak, and weaponless, and saw the broken sword hilt-burried in the dry and trodden sand, and ran and snatched it, and with banner lifted afresh he hewed his enemy down and saved a great cause that heroic day.

—Edward Rowland Sill

PANDORA'S SONG

Of wounds and sore defeat I made my battle stay;
Winged sandals for my feet I wove of my delay;
Of weariness and fear I made a shouting spear;
Of loss and doubt and dread and swift oncoming doom
I made a helmet for my head and a floating plume.
From the shutting mist of death and failure of the breath,
I made a battle horn to blow across the vales of overthrow.
O hearken where the echoes bring down the gray disasterous morn
Laughing and rallying!

—Willliam Vaughn Moody

Four things a man must learn to do
If he would make is record true;
To think without confusion clearly,
To love his fellow-men sincerely;
To act from honest motives purely;
To trust in God and Heaven securely.

—Henry Van Dyke

When the frost is on the punkin and the fodder's in the shock,
And you hear the kyouck and gobble of the struttin' turkey-cock,
And the clackin' of the guineys, and the cluckin' of the hens,
And the rooster's hallylooyer as he tiptoes on the fence;
O, it's then's the times a feller is a-feelin' at his best,
With the risin' sun to greet him from a night of peaceful rest,
As he leaves the house, bareheaded, and goes to feed the stock,
When the frost is on the punkin and the fodder's in the shock.

They's something kindo' harty-like about the atmusfere
When the heat of summer's over and the coolin' fall is here --
Of course we miss the flowers, and the blossums on the trees,
And the mumble of the hummin'-birds and buzzin' of the bees;
But the air's so appetizin'; and the landscape through the haze
Of a crisp and sunny morning of the airly autumn days
Is a pictur' that no painter has the colorin' to mock --
When the frost is on the punkin and the fodder's in the shock.

The husky, rusty russel of the tossels of the corn,
And the raspin' of the tangled leaves, as golden as the morn;
The stubble in the furries -- kindo' lonesome-like, but still
A-preachin' sermons to us of the barns they growed to fill;
The strawsack in the medder, and the reaper in the shed;
The hosses in theyr stalls below -- the clover overhead! --
O, it sets my hart a-clickin' like the tickin' of a clock,
When the frost is on the punkin, and the fodder's in the shock!

Then your apples all is gethered, and the ones a feller keeps
Is poured around the celler-floor in red and yeller heaps;
And your cider-makin's over, and your wimmern-folks is through
With their mince and apple-butter, and theyr souse and saussage, too!
I don't know how to tell it -- but ef sich a thing could be
As the Angles wantin' boardin', and they'd call around on me --
I'd want to 'commodate 'em -- all the whole-indurin' flock --
When the frost is on the punkin and the fodder's in the shock!

—*"The Hoosier Poet" James Whitcomb Riley (1849-1916)*

A Closing Jumble of Messages

This is a family email message, November 26, 2010. (My birthday) It is a last, late entry of mixed information to close this book, my last word.

Dear ones and all, my treasured family and friends:

Two interesting things coincided early this morning as I checked my email. I received the poem, *Crabby Old Man*, from Don Wylie, (see below) and I read a big promotional ad about Grandparents Day this Sunday.

Someone took the poem, *Too Soon Old*, renamed it, *Crabby Old Man*, created the following false story about it, and started it circulating on the Internet.

Here is the Internet story:

When an old man died in the geriatric ward of a small hospital near Tampa, Florida, it was believed he had nothing left of any value.

Later, when the nurses were going through his meager possessions, They found this poem. Its quality and content so impressed the staff, copies were made and distributed to every nurse in the hospital.

One nurse took her copy to Missouri. The old man's sole bequest to posterity has since appeared in the Christmas edition of the News Magazine of the St. Louis Association for Mental Health. A slide presentation has also been made based on his simple, but eloquent, poem.

And this little old man, with nothing left to give to the world, is now the author of this *anonymous* poem winging across the Internet.

The story about the old man (in some versions described as 100 years old) is a fabrication.

NOTE: There is a similar tale, also false, described under the title, *The Most Beautiful Will ever Written*. It was originally written by Williston Fish, and starts on page 14.

This poem, titled *Too Soon Old*, was written by Dave Griffith of Fort Worth, Texas. Griffith told **TruthOrFiction.com** he wrote the poem more than 20 years ago and meant for it to be simple, and to the point, from youth through old age in his own personal life, high school football, Marines, marriage, the ravages of his own disabilities.

Someone took the poem from his site, created a false story about it, and started it circulating on the Internet. None of this took away any of the value of the poem.

Griffith is the author of more than 500 poems, now posted on his personal website.

http://www.palletmastersworkshop.com/

I contacted him and received permission to use his poem in this book.

Too Soon Old - (Crabby Old Man)

What do you see nurses? . . . . . What do you see?
What are you thinking . . . . . when you're looking at me?
A crabby old man . . . . . not very wise,
Uncertain of habit . . . . . with faraway eyes?

Who dribbles his food . . . . . and makes no reply
When you say in a loud voice . . . . . "I do wish you'd try!"
Who seems not to notice . . . . . the things that you do,
And forever is losing . . . . . a sock or shoe.

Who, resisting or not . . . . . lets you do as you will,
With bathing and feeding . . . . . the long day to fill.
Is that what you're thinking? . . . . . Is that what you see?
Then open your eyes, nurse . . . . . you're not looking at me.

I'll tell you who I am. . . . . as I sit here so still,
As I do at your bidding, ... as I eat at your will.
I'm a small child of ten . . . . . with a father and mother,
Brothers and sisters . . . . . who love one another.

A young boy of sixteen . . . . . with wings on his feet.
Dreaming that soon now . . . . . a lover he'll meet.
A groom soon at twenty . . . . . my heart gives a leap.
Remembering, the vows . . . . . that I promised to keep.

At twenty-five, now . . . . . I have young of my own
Who need me to guide . . . . . and a secure happy home.
A man of thirty . . . . . my young now grown fast,
Bound to each other . . . . . with ties that should last.

At forty, my young sons . . . . . have grown and are gone,
But my woman's beside me . . . . . to see I don't mourn.
At fifty, once more . . . . . babies play 'round my knee,
Again, we know children . . . . . my loved one and me.

Dark days are upon me . . . . . my wife is now dead.
I look at the future . . . . . shudder with dread.
For my young are all rearing . . . . . young of their own,
And I think of the years . . . . . and the love that I've known.

I'm now an old man . . . . . and nature is cruel.
Tis jest to make old age . . . . . look like a fool.
The body, it crumbles . . . . . grace and vigor, depart.
There is now a stone . . . . . where I once had a heart.

But inside this old carcass . . . . . a young guy still dwells,
And now and again . . . . . my battered heart swells.
I remember the joys . . . . . I remember the pain,
And I'm loving and living . . . . . life over again.

I think of the years, all too few . . . . . gone too fast,
And accept the stark fact . . . . . that nothing can last.
So open your eyes, people . . . . . open and see.
Not a crabby old man . . . Look closer . . . see ME!!

—© *David L. Griffith - 1990*

Boy, could I identify with that poem. I'm not quite there in the nursing home crowd yet (I'm active and can still feed myself), but I am getting perilously near. Grandparents Day? That's a construct designed to benefit the flower, gift, and card industries. So please, this is not a request or reminder to send cards or flowers, so save your time and money. I would be pleased if you hit the reply button on this email and wrote Hi!

I am in the process of getting rid of not only boxes of greeting cards of all type of my own, but also several boxes of Barb's, and Mom's cards. Being a saver of personal treasures must run in the genes, or go with the territory.

This poem is filled with sadness. Were I to write such a poem, it would be much more filled with joy and happiness as you can probably tell from reading this book. I have had a life filled with incredible people who thrill my heart. (OK, so there were a few duds and sore heads, but that's life.) My heart is now filled with as much excitement and love as it ever was.

The difference is now, nearly all of those in my heart are younger. Those from my younger years are now but treasured memories. That's life as well.

I remember my dad, when in his seventies, telling me how he felt and thought about himself as he had when he was a boy and then a young man. I now fully understand what he meant for I feel much the same. I have the same heart and mind as when I was twenty with a lifetime of experience and memories added. Here are a few of my reflections about ages:

At twenty you dream marvelous dreams of the future, and talk about what is to come and what you want to accomplish.

At forty those dreams have changed, some are gone forever; some new ones, a bit more modest, have appeared. You talk about the past a bit, but the future still dominates your words.

By age sixty, memories have overcome dreams, dreams are fewer and less optimistic. You speak more of the past and those who are no longer with us, yet there is still active hope for the future.

At eighty and beyond, dreams have all but vanished and involve making the remaining years pleasant. Friends and family from one's youth are all gone. (with a few exceptions) Visits with friends and family have become important and buoy the spirits when they happen. One speaks almost exclusively of the past where almost all of one's life lies. One also realizes how short and precious life is.

A personal message: I still look with great hope for the future, short as it may be, to visits with family and friends, to heart-warming experiences and little successes, and, hopefully, to writing and successful publications. I still hold fast to this big dream. In addition to my marvelous, treasured family and friends, the dream still fills this old heart with action and hope. To me, my writing is already a huge success, even if I never sell a book or story. To have been accomplished and shared makes me pleased and thankful. It has become one of my major passions and a refuge.

My novel, ***Blue Shift***, sold few copies in spite of rave reviews in several newspapers. The primary reason is, my part in the marketing program put together by my publisher never took place. Barb's illness and surgeries took over my life when we were in California. I was, of course, quite willing to give my time to care for her when she was in such need. It was my highest priority. Because I was so completely occupied with loving care, I couldn't make the schedule of talks, public appearances, interviews, and book signings my publisher had set up for me.

My book on energy, ***Energy, Convenient Solutions***, has consumed unbelievable amounts of time and hard work. Now, after ten years of off and off effort, it is finished and has been released to the public. It is available in both print and ebook form through most outlets for publications and is now available on my own website, **www.senesisword.com**. Take a gander at this website when you get a chance. It will keep you up to date with my writing and with my new micro publishing company. My book, this book, **Words from the Lakeside**, is also published and available. Both publishing dates are years behind what I had hoped, but they are now accomplished and I can move on to new things.

I have several other book projects in the works including, **The Crystal Feather**, a hard SciFi novel that won first place in the Florida Writers Association Lighthouse novel competition in 2008. It is about 80% completed. **Blue Shift II**, is the second novel of the Blue Shift trilogy and it is about 60% done. **Days of the High Morning Moon** is a thriller involving murders and theft on Lake Tippecanoe. It should be released early in 2003. **Carol Hughes** is another hard SciFi novel of a quite different type. It is about half complete. Several others are at least started including **Blue Shift III** and an unnamed semi historical novel about Algonkian Indians and their forced movements from the East to mostly Ohio and Indiana. It also involves some experiences from my own life along with stories my grandfather told me.

The lake house is now free of most of my collection of stuff with that remaining confined in a locked closet. I planned on emptying the garage after returning in late September, but failed to get the job done. I have realized there are many things attached to my heart that I have no one left with which to share. I am taking photos of many of these before sending them to Goodwill. Some are now hanging and decorating the walls of my little den.

I have moved my personal office in its entirety into the little house in the back. It is crammed with stuff with which I could not bring myself to part, as well as the active requirements of a writer's den. It has again become my place to write, as it was first built to be before Barb became ill and I moved back into the house. It and the office in Daphne's home, are places where I can dream, create, and put my thoughts and stories on record.

I am a fortunate old man. My body may be deteriorating, my memory may be faltering, my thoughts may be slower, but the fire within still burns fiercely bright. It would not be so without the love and support of my treasured loved ones. I am so fortunate to be sharing my life with such a beautiful soul as Daphne. Please don't begrudge me my time spent writing. It's almost all I have left where I can create. Also, come into my life with visits, emails, phone calls, Skype, Facebook, or any other of the rapidly expanding means of communicating now called social interaction. I live for these contacts, as well as writing, and sharing with others.

Love to you all, Ho

> Writers love to write.
> Writers live to write.
> Writers love to live.
> Writers live to love.
> - - - - - Sometimes we cry.

—Howard Johnson - 2008

ACKNOWLEDGEMENTS

The quotes of other writers are acknowledged right below their quote. There may be a few erroneous credits even though I took great pains to properly identify the originator. I found a number of quotes attributed incorrectly by my sources. Mark Twain seemed to be the one whose words were most often attributed to later writers. Plato was another with about half the number of Mark Twain.

About a third of the short quotes came from a single source on the Internet. That source was a collection made by Dr. Gabriel Robins, Professor of Computer Science, Department of Computer Science, School of Engineering and Applied Science, University of Virginia. The web site is **http://www.cs.virginia.edu/~robins/**

Made in the USA
Charleston, SC
09 April 2013